Machine Learning and Artificial Intelligence in Fluid Mechanics

Machine Learning and Artificial Intelligence in Fluid Mechanics

Guest Editor

Filippos Sofos

MDPI

Basel • Beijing • Wuhan • Barcelona • Belgrade • Novi Sad • Cluj • Manchester

Guest Editor
Filippos Sofos
Department of Physics
University of Thessaly
Lamia, Greece

Editorial Office
MDPI AG
Grosspeteranlage 5
4052 Basel, Switzerland

This is a reprint of the Special Issue, published open access by the journal *Fluids* (ISSN 2311-5521), freely accessible at: https://www.mdpi.com/journal/fluids/special_issues/978S3F3MO5.

For citation purposes, cite each article independently as indicated on the article page online and as indicated below:

Lastname, A.A.; Lastname, B.B. Article Title. *Journal Name* **Year**, *Volume Number*, Page Range.

ISBN 978-3-7258-6458-4 (Hbk)
ISBN 978-3-7258-6459-1 (PDF)
https://doi.org/10.3390/books978-3-7258-6459-1

Contents

About the Editor

Filippos D. Sofos

Filippos D. Sofos is an Associate Professor at the Department of Physics, University of Thessaly, Greece. He studied Electrical and Computer Engineering at the Democritus University of Thrace, BSc and integrated Master's (DUTH, 1999), where he also obtained a second MSc in Informatics and Digital Systems (DUTH, 2002). He received his PhD from the Civil Engineering Department, University of Thessaly (U.Th., 2009). The topic of his PhD Thesis was related to the investigation of flows at the nano/micro-scale with the molecular dynamics method. His current research interests include the application of deep learning, machine learning, and data science techniques for fluid dynamics property calculation and computational integration, both with data-driven and physics-based approaches. He has been involved in numerous EU and national research projects, with actions covering advanced data analysis techniques, simulations at the nano/meso-scale, multiscale simulations, solid/fluid interactions, nanofluidics, and computational Hydraulics. To date, he has authored more than 100 scientific publications in international journals and conference proceedings.

Preface

Nowadays, fluid mechanics research has been reshaped as traditional methodologies, such as theory, experiment, and high-fidelity simulation, are integrated with computational intelligence. What was considered a novel trend only a few years ago has matured into a state-of-the-art component of research. This Special Issue rides on this evolution, suggesting the framework where advanced data science, artificial intelligence (AI), and machine learning (ML) are integrated with the fundamental physical laws governing fluid dynamics.

The main focus is to move beyond traditional "black-box" modelling, presenting the new, most powerful applications of AI/ML which are informed by the underlying physics. From foundational reviews that map the future of AI-aided Computational Fluid Dynamics (CFD) to the implementation of Physics-Informed Neural Networks (PINNs), this Special Issue explores ways to resolve complex flows, such as turbulence and separated flows, both in big-data and data-scarce environments.

The scope of this research spans an impressive range of scales and applications. New ML models and Symbolic Regression methods are providing the first analytical equations for the electrical conductivity of ionic liquids and the rheological properties of nanofluids. Advanced architectures, including Generative Adversarial Networks (GANs) and Convolutional Neural Networks (CNNs), are achieving super-resolution in 3D turbulence cases and high-accuracy flow field predictions. The practical utility of these tools is presented through innovative applications in hydraulic structure design, coastal breakwater optimization, and automotive engine calibration. Moreover, new research explores the power of Digital Twins, i.e., integrated AI/CFD frameworks that enable autonomous control and real-time optimization of complex systems. As we move toward a future of increasingly massive datasets, the methodologies presented here offer a roadmap for scaling analysis methods without sacrificing physical accuracy.

We hope this Special Issue provides researchers and practitioners with both the theoretical foundation and the practical inspiration to continue pushing the boundaries of what is possible in fluid mechanics.

Filippos Sofos
Guest Editor

fluids

MDPI

Editorial

Advancing Fluid Mechanics with Artificial Intelligence and Machine Learning

Filippos Sofos

Condensed Matter Physics Laboratory, Department of Physics, University of Thessaly, 35100 Lamia, Greece; fsofos@uth.gr

Fluid mechanics research is currently undergoing a significant transformation, driven by the integration of advanced computational intelligence. While theory, experiment, and high-fidelity simulation remain essential, artificial intelligence (AI) and machine learning (ML) now provide a powerful set of tools for extracting patterns and building models from large datasets. What was, only a few years ago, a novel trend has rapidly matured into an essential, and nearly standardized, component of the modern research toolkit [1]. These techniques have opened the way to new approaches to solve the most persistent and complex problems in fluid dynamics, such as turbulence modeling, flow control, drag reduction, combustion instability, and aerodynamic optimization [2–5]. Nonetheless, the current focus on data presents a significant challenge: scaling analysis methods to effectively process the massive volumes of information from high-fidelity simulations and advanced experimental techniques. The central theme of this Special Issue is, therefore, to achieve a new synthesis: to join advanced data science, AI, and ML techniques with core fluid mechanics problems. In all papers included, AI/ML methods are not just applied, but they are integrated in a way that explains and incorporates the underlying physical laws.

Four review articles set the basis of AI/ML connection to fluid dynamics. It is a fact that AI/ML methods need to be adjusted to fit specific problems, and it's essential to test their accuracy and understand their limitations. The key is to go beyond simple predictions and help us understand why things happen. In the review of Drikakis and Sofos [6], the regions where AI can aid fluid dynamics have been pointed out. A key research focus should be exploring how ML can speed up numerical simulations while also investigating its accuracy limits in both experiments and simulations. It's also vital to figure out how to apply ML methods effectively in both massive datasets scenarios and data-scarce situations. Beyond just using these tools, a better theoretical understanding of how ML works needs to be established. In Aly [7], an in-depth analysis of recent advancements in integrating AI/ML with computational fluid dynamics (CFD) is provided. One of the most impactful contributions of ML to CFD is the drastic reduction in computational time. Increasing accuracy in turbulence modeling, which remains a critical challenge in CFD, particularly for RANS simulations, is another point. The development of data-driven and physics-driven surrogate models are also presented, aiming at solving complex fluid dynamics or uncovering simplified representations of physical processes.

In cases where obtaining complete spatiotemporal datasets is difficult, particularly when modeling turbulent flows, physics-informed neural networks (PINNs) are employed to address this challenge. In El Hassan et al. [8], various fluid mechanics applications with sparse data are examined, and PINNs have been found to be accurate in resolving the implied flow fields. The authors conclude that the future of fluid mechanics partly depends on how data and physics-driven approaches will be embedded on current theoretical and computational frameworks. On the other hand, the choice of an AI/ML framework is not always the best choice. There are complex cases like separated turbulent flows where current joint ML/Reynolds-Averaged Navier-Stokes (RANS) methods are not yet mature enough to outperform established simulation techniques, especially for extrapolation. In Heinz [9] the alternative of minimal error continuous eddy simulation (CES) methods is suggested, to properly account for causal relationships inherent to separated turbulent flows.

Starting from applications close to the atomic scale, a ML model capable of reproducing and predicting the properties of ionic liquids is presented [10]. Along with interpretable, explainable and generalizable symbolic expressions derived with Symbolic Regression (SR) methods [11,12], the electrical conductivity of ionic liquids is given as an analytical equation for the first time, overcoming the high cost and computational demands of experiments and molecular simulations. In another application the viscosity and surface tension, which characterize the rheological properties of liquids, are employed to calculate the Ohnesorge number from dispensed droplet images [13]. Seven different convolutional neural network (CNNs) architectures are constructed and employed to extract information from spatiotemporal data, reaching an accuracy of nearly 95%. A further application in rheology involved predicting the viscosity of a BN-diamond/thermal oil hybrid nanofluid with four traditional "black-box" ML algorithms (Random Forest, Gradient Boost, Gaussian Regression, and Artificial Neural Network), with the ensemble tree-based Gradient Boost achieving the best performance [14].

As we move towards the macroscale, PINNs are widely used in fluid dynamics research during the past years to predict flow fields, especially in high-Reynolds number flows where turbulence dominates. Even when trained with low-resolution RANS equations, sufficient models emerge and the results are comparable to high-resolution DNS data [15]. By employing a generative adversarial network (GAN) and combining it with physics-informed constraints inside the loss function, a high-performing super-resolution framework is suggested in the paper by Ward [16], resulting in superior performance in both accuracy and computational efficiency for 3D turbulence cases. From an algorithmic point of view, it has been also shown that fully connected neural networks (FCNNs), whose loss function consists of both data and physics constraints, perform better compared to data- or physics-only models [17]. Moreover, the PINNs' potential for real-world applications is highlighted through an ocean engineering context [18]. The hydrodynamic nonlinear Schrödinger equation is embedded inside the loss function and works accurately even with limited data availability.

For application-specific research in hydraulic engineering, the work of Salemnia et al. [19] applies various ML algorithms to predict the dynamic pressure coefficients of vertical water jets, using data from controlled hydraulic laboratory experiments. With ML methods, it overcomes the limitations of traditional numerical models in complex turbulent flows and suggests more resilient hydraulic structure designs. In coastal engineering, ML is employed to guide breakwater design, predicting the value of the reflection coefficient and complementing other traditional methods. It is concluded that the reflective power of the underwater breakwater is increased analogous to its length, while wave attenuation reaches the maximum at shallow depths [20]. Machine learning is also used to replace the manual fine-tuning calibration required on pre-production motorcycles [21]. Trained on rolling bench data, the ML model (ANN or Random Forests) learns to accurately estimate air mass flow, a sensitive parameter crucial for torque and catalyst efficiency. This method successfully saves many days of calibration work without sacrificing accuracy, even when starting from a completely random (worst-case) calibration. Beyond that, to account for high emissions from gasoline direct injection engines during cold starts, another application employs ANNs to predict fuel spray behavior. The ANN successfully predicted detailed spray dynamics, including liquid penetration and width, with excellent accuracy [22]. Finally, an integrated AI/CFD framework with Deep NNs is constructed to forecast temperature and humidity values inside a ventilated room, while master control is carried out by a digital twin [23]. CFD data are used for model training and control. The findings validate the effectiveness of AI-based digital twins for control applications, highlighting their potential to improve system autonomy, optimize performance, and increase operational efficiency, making them key technologies for modern control engineering.

Acknowledgments: The Guest Editor would like to thank all contributing Authors, Reviewers and Journal Editors for their invaluable efforts in composing and enhancing the quality of this Special Issue.

Conflicts of Interest: The author declares no conflicts of interest.

References

1. Sofos, F.; Drikakis, D. A review of deep learning for super-resolution in fluid flows. *Phys. Fluids* **2025**, *37*, 041303. [CrossRef]
2. Zou, Z.; Xu, P.; Chen, Y.; Yao, L.; Fu, C. Application of artificial intelligence in turbomachinery aerodynamics: Progresses and challenges. *Artif. Intell. Rev.* **2024**, *57*, 222. [CrossRef]
3. Kumar, J.K.A.; Prabhu, K.S.; Veerendra, A.S.; Shastry, Y.B.S.; Srinivas, G. A comprehensive review of aerodynamic performance evaluation for aircraft wings and engines: Insights into experimental, numerical, and theoretical approaches. *Aerosp. Syst.* **2025**. [CrossRef]
4. Vinuesa, R. Perspectives on predicting and controlling turbulent flows through deep learning. *Phys. Fluids* **2024**, *36*, 031401. [CrossRef]
5. Bhattacharya, A.; Majumdar, P. Artificial Intelligence–Machine Learning Algorithms for the Simulation of Combustion Thermal Analysis. *Heat Transf. Eng.* **2024**, *45*, 176–193. [CrossRef]
6. Drikakis, D.; Sofos, F. Can Artificial Intelligence Accelerate Fluid Mechanics Research? *Fluids* **2023**, *8*, 212. [CrossRef]
7. Aly, A.M. Machine Learning Reshaping Computational Fluid Dynamics: A Paradigm Shift in Accuracy and Speed. *Fluids* **2025**, *10*, 275. [CrossRef]
8. El Hassan, M.; Mjalled, A.; Miron, P.; Mönnigmann, M.; Bukharin, N. Machine Learning in Fluid Dynamics—Physics-Informed Neural Networks (PINNs) Using Sparse Data: A Review. *Fluids* **2025**, *10*,226. [CrossRef]
9. Heinz, S. The Potential of Machine Learning Methods for Separated Turbulent Flow Simulations: Classical Versus Dynamic Methods. *Fluids* **2024**, *9*, 278. [CrossRef]
10. Karakasidis, T.E.; Sofos, F.; Tsonos, C. The Electrical Conductivity of Ionic Liquids: Numerical and Analytical Machine Learning Approaches. *Fluids* **2022**, *7*, 321. [CrossRef]
11. Dong, J.; Zhong, J. Recent Advances in Symbolic Regression. *ACM Comput. Surv.* **2025**, *57*, Article 290. [CrossRef]
12. Angelis, D.; Sofos, F.; Karakasidis, T.E. Reassessing the transport properties of fluids: A symbolic regression approach. *Phys. Rev. E* **2024**, *109*, 015105. [CrossRef]
13. Sardana, P.; Zolfaghari, M.; Miotto, G.; Zengerle, R.; Brox, T.; Koltay, P.; Kartmann, S. DropletAI: Deep Learning-Based Classification of Fluids with Different Ohnesorge Numbers during Non-Contact Dispensing. *Fluids* **2023**, *8*, 183. [CrossRef]
14. Ali, A.; Noshad, N.; Kumar, A.; Ilyas, S.U.; Phelan, P.E.; Alsaady, M.; Nasir, R.; Yan, Y. Application of Machine Learning Algorithms in Predicting Rheological Behavior of BN-diamond/Thermal Oil Hybrid Nanofluids. *Fluids* **2024**, *9*, 20. [CrossRef]
15. Pioch, F.; Harmening, J.H.; Müller, A.M.; Peitzmann, F.J.; Schramm, D.; el Moctar, O. Turbulence Modeling for Physics-Informed Neural Networks: Comparison of Different RANS Models for the Backward-Facing Step Flow. *Fluids* **2023**, *8*, 43. [CrossRef]
16. Ward, N.J. Physics-Informed Super-Resolution of Turbulent Channel Flows via Three-Dimensional Generative Adversarial Networks. *Fluids* **2023**, *8*, 195. [CrossRef]
17. Gustafsson, E.; Andersson, M. Investigating the Effects of Labeled Data on Parameterized Physics-Informed Neural Networks for Surrogate Modeling: Design Optimization for Drag Reduction over a Forward-Facing Step. *Fluids* **2024**, *9*, 296. [CrossRef]
18. Ehlers, S.; Wagner, N.A.; Scherzl, A.; Klein, M.; Hoffmann, N.; Stender, M. Data Assimilation and Parameter Identification for Water Waves Using the Nonlinear Schrödinger Equation and Physics-Informed Neural Networks. *Fluids* **2024**, *9*, 231. [CrossRef]
19. Salemnia, A.; Boldaji, S.H.; Atashi, V.; Fathi-Moghadam, M. Machine Learning for Dynamic Pressure Coefficient Prediction in Vertical Water Jets. *Fluids* **2024**, *9*, 205. [CrossRef]
20. Loukili, M.; El Moumni, S.; Kotrasova, K. AI-Driven Optimization of Breakwater Design: Predicting Wave Reflection and Structural Dimensions. *Fluids* **2025**, *10*, 34. [CrossRef]

21. Savioli, T.; Pampanini, M.; Visani, G.; Esposito, L.; Rinaldini, C.A. Engine Mass Flow Estimation through Neural Network Modeling in Semi-Transient Conditions: A New Calibration Approach. *Fluids* **2024**, *9*, 239. [CrossRef]
22. El Marnissi, Y.; Lee, K.; Hwang, J. Acceleration of Modeling Capability for GDI Spray by Machine-Learning Algorithms. *Fluids* **2024**, *9*, 267. [CrossRef]
23. Luis-Gómez, J.; Martínez, F.; González-Barberá, A.; Mascarós, J.; Monrós-Andreu, G.; Chiva, S.; Borrás, E.; Martínez-Cuenca, R. AI-Driven Surrogate Model for Room Ventilation. *Fluids* **2025**, *10*, 163. [CrossRef]

fluids

Review

Can Artificial Intelligence Accelerate Fluid Mechanics Research?

Dimitris Drikakis [1,*] and Filippos Sofos [2,*]

1 Institute for Advanced Modelling and Simulation, University of Nicosia, CY-2417 Nicosia, Cyprus
2 Condensed Matter Physics Laboratory, Department of Physics, University of Thessaly, 35100 Lamia, Greece
* Correspondence: drikakis.d@unic.ac.cy (D.D.); fsofos@uth.gr (F.S.)

Abstract: The significant growth of artificial intelligence (AI) methods in machine learning (ML) and deep learning (DL) has opened opportunities for fluid dynamics and its applications in science, engineering and medicine. Developing AI methods for fluid dynamics encompass different challenges than applications with massive data, such as the Internet of Things. For many scientific, engineering and biomedical problems, the data are not massive, which poses limitations and algorithmic challenges. This paper reviews ML and DL research for fluid dynamics, presents algorithmic challenges and discusses potential future directions.

Keywords: artificial intelligence; machine learning; deep learning; neural networks; fluids

1. Introduction

In the past few years, machine learning (ML) and deep learning (DL) have been pursued in most social, scientific, engineering, and industry branches. Initially established at the point where computer and statistical science meet, ML evolution has been driven by advancements in Artificial Intelligence (AI) and its related algorithms, fortified by the "big data" and access to cost-effective computational architectures. ML-based approaches span fields such as construction engineering, aerospace, biomedicine, materials science, education, financial modelling, and marketing [1,2].

Machine learning techniques fall into three groups: unsupervised, semi-supervised, and supervised [2,3]. This paper considers DL as a subset of the broader ML field. The supervised approaches rely on labelled data to train the ML algorithm effectively. This is the most common category, with wide applicability in science and technology [4–7]. Unsupervised ML involves extracting features from high-dimensional data sets without the need for pre-labelled training data. Well-established clustering methods, such as K-means, hierarchical agglomerative clustering, and DBSCAN [8,9], apply here. Furthermore, dimensionality reduction techniques such as proper orthogonal decomposition (POD) [10], principal component analysis (PCA) [11,12] and dynamic mode decomposition (DMD) [13] emerged from fundamental statistical analysis, and Reduced Order Model techniques have also been bound to unsupervised ML. Meanwhile, semi-supervised learning combines elements from supervised and unsupervised ML and has been successfully incorporated in applications concerning time-dependence and image data [14,15].

At the heart of ML theories, the algorithms proposed range from linear feed-forward models to complex DL implementations. To distinguish the complexity of various models, the algorithms may fall into shallow or deep ML, with possible combinations. Shallow learning (SL) algorithms, both in classification and regression tasks, involve linear and multiple linear algorithms, their linear variants such as LASSO and Ridge, support vector machines (SVM), stochastic methods based on Gaussian processes and Bayes' theorem, and decision trees [16]. Ensemble and super-learner methods, combining multiple shallow algorithms in more complex instances, such as the random forest, gradient boosting, and extremely randomized trees, have also increased in popularity in the past years [17].

Notwithstanding its wide acceptance, the applicability of SL is limited to small-to-medium-sized data sets, which might be a problem in demanding fluid mechanics tasks. Thus, the need for more robust methods has emerged, and DL has come to the fore [18]. The idea of DL is based on the multi-layer perceptron, the artificial neural network (ANN) architecture that can be seen as the corresponding artificial mechanism that mimics the biological functions of the neural networks of the human brain [6].

DL models have embedded complex mathematical concepts and advanced computer programming techniques in a great number of interconnected nodes and layers [19], making them an appropriate choice for dealing with applications that involve image and video processing [20–23], speech recognition [24], biological data manipulation [25], and flow reconstruction [26]. Nevertheless, a physical interpretation of the outcome is still lacking, which limits their application in physical sciences, where interpretability has a central role. In such cases, it would be beneficial to adopt models that produce meaningful results, i.e., mathematical expressions, which can spot correlations with existing empirical relations and propose an analytical approach bound to physical laws.

Conventional neural networks operate solely on data and are often considered "black box" models. On the other hand, symbolic regression (SR) is a class of ML algorithms in which mathematical equations bound exclusively to data are derived. In contrast to usual regression procedures, SR searches within a wide mathematical operator set and constants to connect input features to produce meaningful outputs [27] based on evolutionary algorithms. This process of extracting results that can be interpreted and exploited simultaneously has led to the development of numerous genetic programming-based tools that implement SR for disciplines from materials science to construction engineering to medical science [28].

ML methods have been employed in fluid dynamics research but have not been in engineering practice. Computational fluid dynamics (CFD) discretizes space using meshes: as the grid density increases, so does the accuracy, but at the expense of computational cost. Such techniques have benefited the aerospace, automotive and biomedical industries. The challenges are numerical and physical, and several reviews have been written on specific industrial applications or areas of physics [29–32]. However, despite significant advances in the field,there is still a need to improve efficiency and accuracy.

The most critical example is turbulence, a flow regime characterised by velocity and pressure fluctuations, instabilities, and rotating fluid structures (eddies) at vast, macroscopic and molecular scales. Accurately describing transient flow physics using CFD would require many grid points. Such an approach where the Navier–Stokes equations are directly applied is called direct numerical simulation (DNS) and is accompanied by tremendous computational overhead. As such, it is limited to specific problems and requires large high-performance computing (HPC) facilities and long run times. Consequently, reduced turbulent models are often used at the expense of accuracy.

Furthermore, while many such techniques exist, no universal model works consistently across all turbulence flows, so while many computational methods exist for different problems, the computational barrier is often unavoidable. Recent efforts in numerical modelling frequently focus on how to speed up computations, a task that would benefit many industries. A growing school of thought is to couple these numerical methods with data-driven algorithms to improve accuracy and performance. This is where ML becomes relevant.

The topic of ML and its applications in fluid mechanics is broad, and a single review article does not suffice to cover everything. Our primary aim is to touch on some applications as indicative examples, summarise the ML methods used, and identify challenges and future perspectives.

2. A Brief Overview and Methods Classification

Many articles and reviews are produced each year on subjects related to ML. This review covers the past five years when a vast increase in reviews on these subjects was observed. Figure 1 shows the number of publications referring to ML, DL and the number of relevant reviews, with data taken from Scopus. Further analysis of the results is presented in the pie charts in Figure 2. The key terms used in the Scopus search were: *Machine Learning and Review,* from *2017 to 2023*.

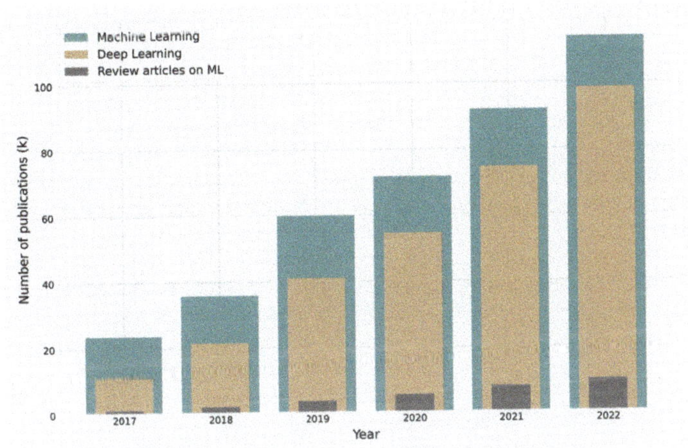

Figure 1. Publications and reviews on machine and deep learning from 2017 to 2022.

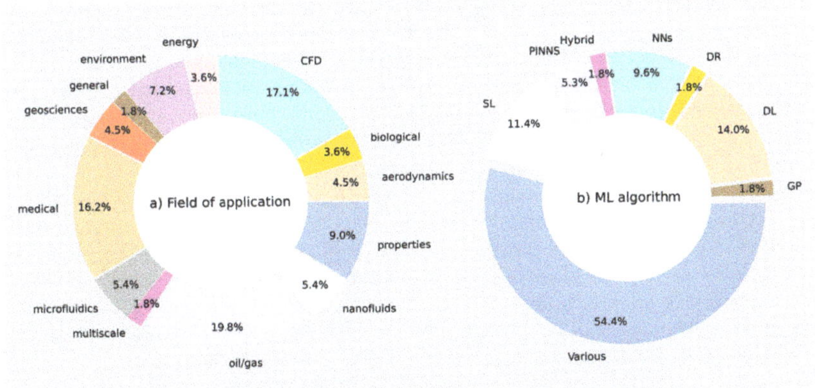

Figure 2. (**a**) Field of application and (**b**) algorithms employed in review articles on ML and fluid mechanics, from 2017 to 2023.

The findings from this survey revealed some interesting points. A great number of reviews meant that an even greater number of original research articles had been published. If we limited the results of the Scopus search to reviews related to fluids, three major fields of application emerge that attracted research interest in applying ML methods to (a) theoretical and industrial oil and gas [33–52], (b) CFD simulations [53–68], and (c) health and medical applications [69–86] (Figure 2a).

Another popular subject is fluid property estimation with ML algorithms, a promising alternative to prohibitively expensive experimental measurements.Most reviews have exploited historical data from experiments or simulations to predict fluid behaviour and calculate the properties of interest [87–96]. Furthermore, heat transfer applications, concerning the addition of nanoparticles in simple fluids [97–102] or the investigation of flows in microfluidic channels [103–108] play a central role. Of equal significance, though with fewer instances, several recent reviews refer to aerodynamics [109–113], geosciences, geoengineering and environmental subjects [114–126], energy applications [127–130], biological processes [131–134], control processes [135–137], multiscale simulations [138–141], and general ML [2,142].

Common to all these applications is the difficulty of obtaining accurate data; industrial processes usually obtain them from field sensors. In addition to data acquisition and resolution, compression might be needed to handle quantity effectively, especially in real-time applications [143]. Data availability is a prerequisite because, for example, a lack of available data in industrial multiphase oil and gas flows makes it hard to experiment with ML techniques, and the need for interdisciplinary collaboration between industry and academia is highlighted [42]. Using only high-quality data is important to avoid incorporating erroneous, missing, or redundant data points. Before data are applied on a specific ML algorithmic platform, an extensive statistical analysis is conducted on the implied dataset, and feature engineering is employed to curate the data. Feature engineering involves raw data manipulation and extracting characteristic features before transforming them into an acceptable ML format. This process encompasses a set of statistical tools (e.g., feature importance selection, correlation tests, and variable transformation and normalization) that can be used to predict the desired outcome and enhance the performance of the model.

All the studies above have incorporated several supervised ML algorithms, while only a few unsupervised algorithms have been exploited for dimensionality reduction (DR) tasks. In Figure 2b, most of them went through various algorithm reviews that argued for their applicability to a specific subject. Property prediction with historical data can be realized with SL algorithms such as decision trees, linear regression [144] and simple NNs like the multi-layer perceptron (MLP) [11]. In general, the most widely incorporated SL algorithms in fluid research can be categorized into five areas [96]: (i) linear models based mainly on linear regression techniques and their variants such as LASSO and Ridge, (ii) algorithms that employ kernel functions to transform input data into a higher-dimensional space (e.g., support vector machines (SVMs)), (iii) decision-tree methods, which exploit the non-linear tree-structure to achieve accurate predictions, (iv) neural networks built on the perceptron architecture, and (v) instance-based, which exploits the proximity of instances inside the training set such as K-nearest neighbours.

A new category, the ensemble technique, has emerged to boost prediction accuracy further. This approach incorporates more than one base ML algorithm to construct a combined estimator, ensuring accuracy, robustness and lower variance [145]. Individual predictions are combined in classification or averaging in regression tasks. In this way, each base learner in an ensemble model can achieve better results as it is trained in a specific region of the dataset, thereby improving prediction accuracy and generalizability. Some widely applied ensemble techniques involve random forest (RF), gradient boosting machine (GBM), and extreme gradient boosting (XGBoost) [146].

On the other hand, a DL architecture with convolutional NNs (CNNs) or Recurrent NNs (RNNs) is incorporated to address flow reconstruction problems [26]. Moreover, as imaging applications in most branches of science, engineering and medical studies become more demanding, DL techniques are further combined with ensemble techniques [147]. Physics-informed neural networks (PINNs), another class of algorithm, are gaining ground since they add physical intuition in common NNs, and can bind experimental data with Navier–Stokes equations for fluid mechanics [140,141,148–150]. Algorithms based on genetic programming (GP) are best suited for flow control in various applications [151] and can be an excellent choice when the output needs to be interpreted as a mathematical equation [8].

Another research direction that has attracted attention is Physics-informed machine learning (PIML), which integrates ML algorithms with high-fidelity numerical simulations [58]. PIML could potentially reduce computing time in complex turbulence simulations.

3. Applications of ML in Fluid Mechanics

As mentioned in the Introduction, a long-standing problem in fluid dynamics is turbulence, a flow regime characterised by unpredictable, time-dependent velocity and pressure fluctuations [152–155]. The use of data-driven methods has increased significantly over the last few years, providing new modelling tools for complex turbulent flows [156]. ML has been used to model turbulence, such as the use of ANNs to predict the instantaneous velocity vector field in turbulent wake flows [157] and turbulent channel flows [158,159]. The popularity of many ML methods, especially deep learning, as a promising method for solving complex fluid flow problems is also supported by high-quality open-source data sets published by researchers [160–162].

Much effort is currently put into using ML algorithms to improve the accuracy of Reynolds-averaged Navier–Stokes (RANS) models for which GPs have been used to quantify and reduce uncertainties [163]. Tracey et al. [164] used an ANN that inputs flow properties and reconstructs the SA closure equations. The authors studied how different components of ML algorithms, such as the choice of the cost function and feature scaling, affected their performance. Other examples include using DL to model the Reynolds stress anisotropy tensor [165,166]. Bayesian inference has also been used to optimize RANS model coefficients [167,168], as well as to derive functions that quantify and reduce the gap between model predictions and high-fidelity data, such as DNS data [169–171]. Furthermore, a Bayesian neural network (BNN) has been used to improve the predictions of RANS models and to specify the uncertainty associated with the forecast [172].

In the field of large eddy simulations (LES), an ANN was used to replace these subgrid models [173,174]. Convolutional neural networks (CNNs) have also been used to calculate subgrid closure terms [175]. Moreover, DL modelling using input vorticity and stream function stencils was implemented to predict a dynamic, time- and space-dependent closure term for the subgrid models [176]. More specifically, starting from the detailed DNS simulation output, the main focus was to employ low-resolution LES data to train a NN model and reconstruct fluid dynamics to replace DNS where possible. To this end, many relevant applications have emerged. Bao et al. [177] have suggested combining physics-guided, two-component, PDE-based techniques with recurrent temporal data models and super-resolution spatial data methods. An extension of this approach includes a multi-scale temporal path UNet (MST-UNet) model that can reconstruct both temporal and spatial flow information from low-resolution data [177].

Fukami et al. [178] have applied super-resolution techniques, common in image reconstruction applications, on 2D course-flow fields and managed to reconstruct laminar and turbulent flows. A more recent version of this method employed autoencoders to achieve order reduction for fluid flows [179]. A complete super-resolution case study is presented in another work, covering all the details of a CNN-based, super-resolution architecture for turbulent fluid flows with DNS training data for various Reynolds numbers [180]. Nevertheless, generative models are often incorporated because DNS data are hard to obtain. In the review of Buzzicotti [53], three prevalent models are discussed: variational autoencoders (VAE), GANS, and diffusion models. These models are based on CNNs, and their implementation is focused on reproducing the multiscale and multifrequency nature of fluid dynamics, which is more complex than classical image recognition applications.

Another exciting application of ML in fluid mechanics is related to food processing [181]. The ML models aim to optimize process parameters and kinetics for reduced energy consumption, speed up the processing time, and continuously improve product quality. Several food processing operations, including drying, frying, baking, and extrusion, each require the development of ML algorithms and their training and validation against accurate data.

Furthermore, ML has found application in many chemical engineering flows. These applications include transport in porous media, catalytic reactors, batteries, and carbon dioxide storage [182]. Many of these problems are linked to energy transition efforts such as batteries, where long-term safety, energy density, and cycle life are particularly interesting. Diverse ML applications in fluid mechanics include nanofluids and ternary hybrid nanofluids [183] as well as metallurgy processes [184]. The domain of applications is broad and goes beyond the scope of a single review article.

4. Challenges

The data from fluid simulations or flow sensors are vast and have prohibitive degrees of freedom to cope with. Bearing in mind also that 2D flow fields in the pictorial form are currently gaining ground as research items for fluid flow characterization, with added complexity when they evolve in time (e.g., multiple images), it becomes clear that great effort is needed to harness and derive meaning from processing and analysis. On the contrary, there are cases where information from data is sparse, and the implied ML algorithm must be accurate. Methods and techniques from reinforced learning could be exploited here, such as GANs [185,186], which can train an ML model with fictitious data without affecting accuracy.

We summarized the challenges of ML in fluid mechanics in Figure 3 and detail them below.

- ML algorithms are well-defined mathematically and widely supported by software. An open question on their applicability in high-precision fluid mechanics, such as in biological or aerodynamic flows, is the availability of high-fidelity data to train ML models effectively [187].
- Open databases should become available to the scientific community and play a crucial role in machine learning, enabling knowledge sharing and advanced development.
- Modern computers and experimental devices have higher speed, power, and flexibility, making access to data easy. However, concerns about storage, retrieval, post-processing, and other challenges may arise, especially when dealing with massive turbulence data.
- ML in fluids extends beyond image processing as fluid imaging is not static and includes dynamic information [3]. Furthermore, when flow images are provided in time sequence, the computational load increases dramatically, and algorithmic selection is essential.
- Algorithms perform differently depending on the problem they are called on to solve. For example, there is no need to dive into a complex DL architecture when the task is the property prediction of a specific fluid based on several variable-type input parameters [188].
- Differential equation solutions with PINNs may fail in complex physical phenomena, such as turbulence, compared to traditional numerical methods [189]. There seems to be much to do to consider PINNs as the dominant PDE solver for fluid dynamics.
- At the nanoscale, problems originate from time and length scales. Complex aqueous environments and fluid/solid interfaces can only come to light through atomistic simulation techniques with first-principles accuracy. Machine learning potentials have been successfully introduced over the past years and are now close to standardization as ab-initio simulation alternatives [190].
- There are many instances in the physical sciences where decision-making is based on empirical relations. The challenge for future ML methods is to provide accurate data-derived physics-based equations rather than empirical ones. Symbolic regression can help towards this goal. [191].

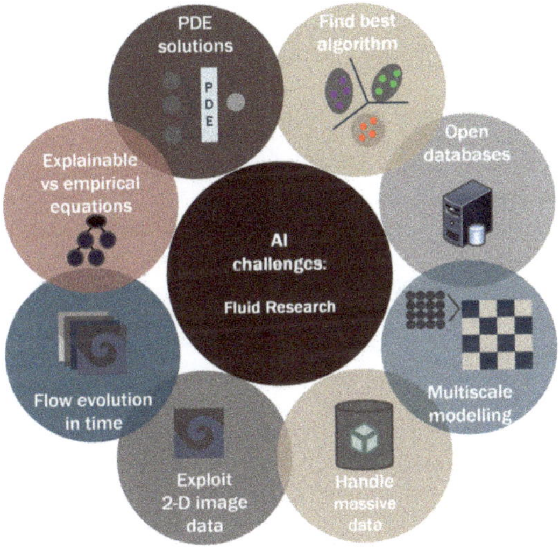

Figure 3. Machine learning challenges in fluid mechanics.

5. Perspectives

Following the discussion on state-of-the-art ML in fluid dynamics and considering the emerging challenges, it is imperative to summarize where scientists and engineers should focus next for ML fluid dynamics applications.

As was made clear in the literature review, AI and ML methods have been widely incorporated in the industry - from oil and gas to energy production and aerospace engineering -particularly in research. There is ample historical and newly generated data to be analyzed in these fields, and ML has much to offer to promote domain knowledge. The statistical nature of ML, either in interpolation or short extrapolation instances, will continue to evolve.

Nevertheless, these applications exploit traditional ML methods. Although widely incorporated, they were developed in a different era and cannot keep pace with current research needs. Moreover, issues originating from big data, such as size, heterogeneity, real-time processing speed or various complex forms (e.g., 2D and 3D), will continue to pose serious obstacles [192].

Novel algorithmic implementations are expected to continue to adapt to these profound demands. This is unavoidable from a computer scientist's and engineer's point of view since new versions of the traditional algorithms should be proposed to handle big data. But here comes the impact of domain knowledge. Data and hidden patterns inside it may be enlightening, but it is doubtful if it can replace physical understanding. Therefore, fluid dynamics researchers should work in parallel to embed fluid concepts and physical laws onto the backbone of the proposed algorithms.

The increasing interest in SR algorithmic implementations can significantly boost computational science by suggesting practical and physically meaningful analytical equations that outperform established empirical or approximate equations [8]. This direction unveils the "black box" nature of ML, enabling these approaches to play a prominent role in predicting properties and guiding fluid dynamics research. This field of investigation still has much to offer, and extensive research on real fluids and mixtures, although computationally demanding and challenging to interpret, is to come.

In another direction, Vinuesa and Brunton [193] have highlighted areas of future research on DNS acceleration, turbulence closure modelling improvement, and new reduced-order model (ROM) establishment, i.e., the areas where numerical computations are highly demanding and in many cases impossible to perform. Care has to be taken here since physics-informed procedures only sometimes adhere to physical consistency. When complex NNs are constructed for an intensive application, it is not guaranteed that they will lead to generalization. Although NNs may produce valuable predictions, they remain a "black-box" model.

Another engineering area of increasing importance concerns the effort and cost of performing expensive numerical simulations and laboratory experiments. This cost could be reduced by using ML models that use coarse data to produce refined predictions. This is particularly important when data are limited or exhibit noise or both. The above factors significantly influence the accuracy of interpolation methods and reduced-order models, including ML.

Poulinakis et al. [194] presented a study on these issues and concluded that ML and DL can be significantly more promising than interpolation under certain data sparsity and noise conditions. Otherwise, ML and DL can be more inaccurate than simple interpolation. For example, DL is significantly more robust to noise than cubic Splines. Therefore, DL can lead to a 'true' function hidden under the noise, thus making it a valuable tool in engineering applications. Furthermore, we know that increasing the data will increase the accuracy of DL to a certain level. What this level is will depend on the application problem. Therefore, future emphasis should be placed on establishing the limitations and best practices for DL methods in different applications. The explainability of DLs remains problematic, and research should be carried out in this direction.

6. Conclusions

As AI methods and algorithms evolve, simulation and experimental data will play increasingly important roles in practical designs, covering demanding applications in science and technology. Fluid dynamics research should take advantage of AI development but carefully explore the possibilities without hyping the potential benefits. ML and DL methods should adapt to specific problems, and their accuracy and limitations should be investigated. The key is that AI and ML approaches should fly above simple prediction tasks and dive deep into interpretability and causation. Their potential impact will be high only if the output abides by physical laws.

Although many of the issues discussed here apply generally to ML and DL methods, there are some specific areas where these methods can benefit fluid dynamics and areas where caution should be exercised:

1. Exploring how to accelerate numerical simulations by introducing ML.
2. Investigating the accuracy limitations of ML methods in experiments and numerical simulations.
3. Exploring ML methods when using big fluid dynamics data and when data are scarce.
4. Developing a better theoretical understanding of ML methods that will allow better explainability of the results.
5. Avoiding AI hyping such as "new equations will be discovered through AI", as first principles will drive fluid dynamics (and other physics) research.

Research in AI evolves fast, and fluid dynamics will benefit from it. Still, its applicability should follow rigorous verification and testing and allow researchers to publish not only their successes in using AI but also their failed attempts.

Author Contributions: Conceptualization, D.D. and F.S.; methodology, D.D. and F.S.; formal analysis, D.D. and F.S.; investigation, D.D. and F.S.; resources, D.D. and F.S.; writing, D.D. and F.S.; supervision, D.D. and F.S.; project administration, D.D. and F.S.; funding acquisition, D.D. and F.S.; contribution to the discussion D.D. and F.S. All authors have read and agreed to the published version of the manuscript.

Funding: This research received no external funding.

Data Availability Statement: Not applicable.

Conflicts of Interest: The authors declare no conflict of interest.

Abbreviations

The following abbreviations are used in this manuscript:

AI	Artificial Intelligence
BF	Basis Function
CFD	Computational Fluid Dynamics
CNN	Convolutional Neural Network
DL	Deep Learning
DNN	Deep Neural Network
DMD	Dynamic Mode Decomposition
ELU	Exponential Linear Unit
FNN	Feedforward Neural Network
FFC	Feedforward Fully Connected
GANs	Generative Adversarial Networks
GCV	Generalized Cross Validation
GELU	Gaussian Linear Unit
GP	Genetic Programming
NN	Neural Network(s)
MARS	Multivariate Adaptive Regression Splines
ML	Machine Learning
MLP	Multi-Layer Perceptron
MSE	Mean Squared Error
PINNs	Physics Informed Neural Networks
POD	Proper Orthogonal Decomposition
ReLU	Rectified Linear Unit
RNN	Recurrent Neural Network
SL	Shallow Learning
SNR	Signal to Noise Ratio
SR	Symbolic Regression
SVM	Support Vector Machine

References

1. Pugliese, R.; Regondi, S.; Marini, R. Machine learning-based approach: Global trends, research directions, and regulatory standpoints. *Data Sci. Manag.* **2021**, *4*, 19–29. [CrossRef]
2. Frank, M.; Drikakis, D.; Charissis, V. Machine-Learning Methods for Computational Science and Engineering. *Computation* **2020**, *8*, 15. [CrossRef]
3. Brunton, S.L.; Noack, B.R.; Koumoutsakos, P. Machine Learning for Fluid Mechanics. *Annu. Rev. Fluid Mech.* **2020**, *52*, 477–508. [CrossRef]
4. Singh, M.P.; Alatyar, A.M.; Berrouk, A.S.; Saeed, M. Numerical modelling of rotating packed beds used for CO_2 capture processes: A review. *Can. J. Chem. Eng.* **2023**. [CrossRef]
5. Guo, S.; Agarwal, M.; Cooper, C.; Tian, Q.; Gao, R.X.; Guo, W.; Guo, Y. Machine learning for metal additive manufacturing: Towards a physics-informed data-driven paradigm. *J. Manuf. Syst.* **2022**, *62*, 145–163. [CrossRef]
6. Nazemi, E.; Dinca, M.; Movafeghi, A.; Rokrok, B.; Choopan Dastjerdi, M. Estimation of volumetric water content during imbibition in porous building material using real time neutron radiography and artificial neural network. *Nucl. Instrum. Methods Phys. Res. Sect. Accel. Spectrometers Detect. Assoc. Equip.* **2019**, *940*, 344–350. [CrossRef]
7. Leverant, C.J.; Greathouse, J.A.; Harvey, J.A.; Alam, T.M. Machine Learning Predictions of Simulated Self-Diffusion Coefficients for Bulk and Confined Pure Liquids. *J. Chem. Theory Comput.* **2023**, *19*, 3054–3062. . [CrossRef]
8. Sofos, F.; Charakopoulos, A.; Papastamatiou, K.; Karakasidis, T.E. A combined clustering/symbolic regression framework for fluid property prediction. *Phys. Fluids* **2022**, *34*, 062004. [CrossRef]
9. Papastamatiou, K.; Sofos, F.; Karakasidis, T.E. Calculating material properties with purely data-driven methods: From clusters to symbolic expressions. In Proceedings of the 12th Hellenic Conference on Artificial Intelligence (SETN '22), Corfu, Greece, 7–9 September 2022; Association for Computing Machinery: New York, NY, USA, 2022; pp. 1–9. . [CrossRef]
10. Amsallem, D.; Farhat, C. Interpolation Method for Adapting Reduced-Order Models and Application to Aeroelasticity. *AIAA J.* **2008**, *46*, 1803–1813. [CrossRef]
11. Sofos, F.; Karakasidis, T. Nanoscale slip length prediction with machine learning tools. *Sci. Rep.* **2021**, *11*, 12520. [CrossRef]
12. Scherl, I.; Strom, B.; Shang, J.K.; Williams, O.; Polagye, B.L.; Brunton, S.L. Robust principal component analysis for modal decomposition of corrupt fluid flows. *Phys. Rev. Fluids* **2020**, *5*, 054401. [CrossRef]
13. Schmid, P.J. Dynamic Mode Decomposition and Its Variants. *Annu. Rev. Fluid Mech.* **2022**, *54*, 225–254. [CrossRef]

14. Deo, I.K.; Jaiman, R. Predicting waves in fluids with deep neural network. *Phys. Fluids* **2022**, *34*, 067108. [CrossRef]
15. M S, V.M.; Menon, V. Measuring Viscosity of Fluids: A Deep Learning Approach Using a CNN-RNN Architecture. In Proceedings of the First International Conference on AI-ML Systems, AIML Systems '21, Bangalore, India, 21–23 October 2021; Association for Computing Machinery: New York, NY, USA, 2021; pp. 1–5. [CrossRef]
16. Stergiou, K.; Ntakolia, C.; Varytis, P.; Koumoulos, E.; Karlsson, P.; Moustakidis, S. Enhancing property prediction and process optimization in building materials through machine learning: A review. *Comput. Mater. Sci.* **2023**, *220*, 112031. . [CrossRef]
17. Hadavimoghaddam, F.; Ostadhassan, M.; Sadri, M.A.; Bondarenko, T.; Chebyshev, I.; Semnani, A. Prediction of Water Saturation from Well Log Data by Machine Learning Algorithms: Boosting and Super Learner. *J. Mar. Sci. Eng.* **2021**, *9*, 666. [CrossRef]
18. Guo, K.; Yang, Z.; Yu, C.H.; Buehler, M.J. Artificial intelligence and machine learning in design of mechanical materials. *Mater. Horiz.* **2021**, *8*, 1153–1172. [CrossRef]
19. Xu, H.; Zhang, D.; Zeng, J. Deep-learning of parametric partial differential equations from sparse and noisy data. *Phys. Fluids* **2021**, *33*, 037132. [CrossRef]
20. Krizhevsky, A.; Sutskever, I.; Hinton, G.E. ImageNet Classification with Deep Convolutional Neural Networks. In *Advances in Neural Information Processing Systems*; Pereira, F., Burges, C., Bottou, L., Weinberger, K., Eds.; Curran Associates, Inc.: Red Hook, NY, USA, 2012; Volume 25.
21. Farabet, C.; Couprie, C.; Najman, L.; LeCun, Y. Learning Hierarchical Features for Scene Labeling. *IEEE Trans. Pattern Anal. Mach. Intell.* **2013**, *35*, 1915–1929. [CrossRef]
22. Szegedy, C.; Liu, W.; Jia, Y.; Sermanet, P.; Reed, S.; Anguelov, D.; Erhan, D.; Vanhoucke, V.; Rabinovich, A. Going Deeper with Convolutions. In Proceedings of the IEEE Conference on Computer Vision and Pattern Recognition (CVPR), Boston, MA, USA, 7–12 June 2015.
23. Jiang, Y.G.; Wu, Z.; Wang, J.; Xue, X.; Chang, S.F. Exploiting Feature and Class Relationships in Video Categorization with Regularized Deep Neural Networks. *IEEE Trans. Pattern Anal. Mach. Intell.* **2018**, *40*, 352–364. [CrossRef]
24. Hinton, G.; Deng, L.; Yu, D.; Dahl, G.E.; Mohamed, A.r.; Jaitly, N.; Senior, A.; Vanhoucke, V.; Nguyen, P.; Sainath, T.N.; et al. Deep Neural Networks for Acoustic Modeling in Speech Recognition: The Shared Views of Four Research Groups. *IEEE Signal Process. Mag.* **2012**, *29*, 82–97. [CrossRef]
25. Leung, M.K.K.; Xiong, H.Y.; Lee, L.J.; Frey, B.J. Deep learning of the tissue-regulated splicing code. *Bioinformatics* **2014**, *30*, i121–i129. [CrossRef] [PubMed]
26. Callaham, J.L.; Maeda, K.; Brunton, S.L. Robust flow reconstruction from limited measurements via sparse representation. *Phys. Rev. Fluids* **2019**, *4*, 103907. [CrossRef]
27. Tohme, T.; Liu, D.; Youcef-Toumi, K. GSR: A Generalized Symbolic Regression Approach. *arXiv* **2022**, arXiv:2205.15569. https://doi.org/10.48550/ARXIV.2205.15569.
28. Angelis, D.; Sofos, F.; Karakasidis, T.E. Artificial Intelligence in Physical Sciences: Symbolic Regression Trends and Perspectives. *Arch. Comput. Methods Eng.* **2023**, *30*, 3845–3865. [CrossRef] [PubMed]
29. Rider, W.; Kamm, J.; Weirs, V. Verification, Validation, and Uncertainty Quantification for Coarse Grained Simulation. In *Coarse Grained Simulation and Turbulent Mixing*; Cambridge University Press: Cambridge, UK, 2016; pp. 168–189.
30. Drikakis, D.; Kwak, D.; Kiris, C. Computational Aerodynamics: Advances and Challenges. *Aeronaut. J.* **2016**, *120*, 13–36. [CrossRef]
31. Norton, T.; Sun, D.W. Computational fluid dynamics (CFD) ,Äì an effective and efficient design and analysis tool for the food industry: A review. *Trends Food Sci. Technol.* **2006**, *17*, 600–620. [CrossRef]
32. Kobayashi, T.; Tsubokura, M. CFD Application in Automotive Industry. In *Notes on Numerical Fluid Mechanics and Multidisciplinary Design*; Hirschel, E.H., Krause, E. , Eds.; Springer: Berlin/Heidelberg, Germany, 2009; Volume 100.
33. Bai, K.; Fan, H.; Zhang, H.; Zhou, F.; Tao, X. Real Time Torque and Drag Analysis by Combining of Physical Model and Machine Learning Method. In Proceedings of the 10th Unconventional Resources Technology Conference, Houston, TX, USA, 20–22 June 2022. [CrossRef]
34. Bikmukhametov, T.; Jäschke, J. First Principles and Machine Learning Virtual Flow Metering: A Literature Review. *J. Pet. Sci. Eng.* **2020**, *184*, 106487. [CrossRef]
35. Gul, S. Machine learning applications in drilling fluid engineering: A review. In Proceedings of the ASME 2021 40th International Conference on Ocean, Offshore and Arctic Engineering, Online, 21–30 June 2021; Volume 10. [CrossRef]
36. Belazreg, L.; Mahmood, S.; Aulia, A. Random forest algorithm for CO_2 water alternating gas incremental recovery factor prediction. *Int. J. Adv. Sci. Technol.* **2020**, *29*, 168–188.
37. Khan, A.; BinZiad, A.; Al Subaii, A. Boosting algorithm choice in predictive machine learning models for fracturing applications. In Proceedings of the SPE/IATMI Asia Pacific Oil & Gas Conference and Exhibition, Virtual, 12–14 October 2021. [CrossRef]
38. Olukoga, T.; Feng, Y. Practical Machine-Learning Applications in Well-Drilling Operations. *SPE Drill. Complet.* **2021**, *36*, 849–867. [CrossRef]
39. Zhong, R.; Salehi, C.; Johnson, R. Machine learning for drilling applications: A review. *J. Nat. Gas Sci. Eng.* **2022**, *108*, 104807. [CrossRef]
40. Agwu, O.; Akpabio, J.; Ekpenyong, M.; Inyang, U.; Asuquo, D.; Eyoh, I.; Adeoye, O. A critical review of drilling mud rheological models. *J. Pet. Sci. Eng.* **2021**, *203*, 108659. [CrossRef]

41. Anifowose, F.; Mezghani, M.; Badawood, S.; Ismail, J. Contributions of machine learning to quantitative and real-time mud gas data analysis: A critical review. *Appl. Comput. Geosci.* **2022**, *16*, 100095. [CrossRef]
42. Arief, H.A.; Wiktorski, T.; Thomas, P.J. A survey on distributed fibre optic sensor data modelling techniques and machine learning algorithms for multiphase fluid flow estimation. *Sensors* **2021**, *21*, 2801. [CrossRef] [PubMed]
43. Cao, W.; Zhang, W. Data-driven and physical-based identification of partial differential equations for multivariable system. *Theor. Appl. Mech. Lett.* **2022**, *12*, 100334. [CrossRef]
44. Chen, G.; Zhu, G.; Zhu, Y. Advances in safety assessment and risk management for deepwater oil and gas exploitation. *J. China Univ. Pet. (Ed. Nat. Sci.)* **2019**, *43*, 136–145. [CrossRef]
45. Guo, J.; Lu, Q.; He, Y. Key issues and explorations in shale gas fracturing. *Nat. Gas Ind. B* **2023**, *10*, 183–197. [CrossRef]
46. Mata, C.; Saputelli, L.; Badmaev, D.; Zhao, W.; Mohan, R.; Gönczi, D.; Schweiger, A.; Manasipov, R.; Schweiger, G.; Krenn, L.; et al. Automated Reservoir Management Workflows to Identify Candidates and Rank Opportunities for Production Enhancement and Cost Optimization in a Giant Field in Offshore Abu Dhabi. In Proceedings of the Offshore Technology Conference, Virtual and Houston, TX, USA, 16–19 August 2021. [CrossRef]
47. Mawlod, A.; Memon, A.; Varotsis, N.; Gaganis, V.; Anastasiadou, V.; Nighswander, J.; Al Shuaibi, M. Reducing Composition Characterization Uncertainties Through Advanced Machine Learning (ML) Techniques—Data Clustering. In Proceedings of the ADIPEC, Abu Dhabi, United Arab Emirates, 31 October–3 November 2022. [CrossRef]
48. Mehmani, Y.; Anderson, T.; Wang, Y.; Aryana, S.; Battiato, I.; Tchelepi, H.; Kovscek, A. Striving to translate shale physics across ten orders of magnitude: What have we learned? *Earth-Sci. Rev.* **2021**, *223*, 103848. [CrossRef]
49. Ojeda, L.; Olubode, M.; Karami, H.; Podio, T. Application of Machine Learning to Evaluate the Performances of Various Downhole Centrifugal Separator Types in Oil and Gas Production Systems. In Proceedings of the SPE Oklahoma City Oil and Gas Symposium, Oklahoma City, OK, USA, 17–19 April 2023. [CrossRef]
50. Osarogiagbon, A.; Khan, F.; Venkatesan, R.; Gillard, P. Review and analysis of supervised machine learning algorithms for hazardous events in drilling operations. *Process. Saf. Environ. Prot.* **2021**, *147*, 367–384. [CrossRef]
51. Syed, F.; AlShamsi, A.; Dahaghi, A.; Neghabhan, S. Application of ML & AI to model petrophysical and geomechanical properties of shale reservoirs – A systematic literature review. *Petroleum* **2022**, *8*, 158–166. [CrossRef]
52. Dindoruk, B.; Ratnakar, R.; He, J. Review of recent advances in petroleum fluid properties and their representation. *J. Nat. Gas Sci. Eng.* **2020**, *83*, 103541. [CrossRef]
53. Buzzicotti, M. Data reconstruction for complex flows using AI: Recent progress, obstacles, and perspectives. *EPL* **2023**, *142*, 23001. [CrossRef]
54. Duraisamy, K. Perspectives on machine learning-augmented Reynolds-averaged and large eddy simulation models of turbulence. *Phys. Rev. Fluids* **2021**, *6*, 050504. [CrossRef]
55. Rabault, J.; Ren, F.; Zhang, W.; Tang, H.; Xu, H. Deep reinforcement learning in fluid mechanics: A promising method for both active flow control and shape optimization. *J. Hydrodyn.* **2020**, *32*, 234–246. [CrossRef]
56. Zhao, Y.; Xu, X. Data-driven turbulence modelling based on gene-expression programming. *Lixue Xuebao/Chin. J. Theor. Appl. Mech.* **2021**, *53*, 2640–2655. [CrossRef]
57. Hammond, J.; Pepper, N.; Montomoli, F.; Michelassi, V. Machine Learning Methods in CFD for Turbomachinery: A Review. *Int. J. Turbomach. Propuls. Power* **2022**, *7*, 16. [CrossRef]
58. Sharma, P.; Chung, W.T.; Akoush, B.; Ihme, M. A Review of Physics-Informed Machine Learning in Fluid Mechanics . *Energies* **2023**, *16*, 2343. [CrossRef]
59. Vadyala, S.; Betgeri, S.; Matthews, J.; Matthews, E. A review of physics-based machine learning in civil engineering. *Results Eng.* **2022**, *13*, 100316. [CrossRef]
60. Andrés-Pérez, E.; Paulete-Perianezez, C. On the application of surrogate regression models for aerodynamic coefficient prediction. *Complex Intell. Syst.* **2021**, *7*, 1991–2021. [CrossRef]
61. Ma, L.; Guo, Q.; Li, X.; Xu, S.; Zhou, J.; Ye, M.; Liu, Z. Drag correlations for flow past monodisperse arrays of spheres and porous spheres based on symbolic regression: Effects of permeability. *Chem. Eng. J.* **2022**, *445*, 136653. [CrossRef]
62. Panchigar, D.; Kar, K.; Shukla, S.; Mathew, R.; Chadha, U.; Selvaraj, S. Machine learning-based CFD simulations: A review, models, open threats, and future tactics. *Neural Comput. Appl.* **2022**, *34*, 21677–21700. [CrossRef]
63. Panda, J. A review of pressure strain correlation modeling for Reynolds stress models. *Proc. Inst. Mech. Eng. Part C J. Mech. Eng. Sci.* **2020**, *234*, 1528–1544. [CrossRef]
64. Razdan, S.; Shah, S. Optimization of Fluid Modeling and Flow Control Processes Using Machine Learning: A Brief Review. In *Advances in Mechanical Engineering and Material Science*; Lecture Notes in Mechanical Engineering; Springer: Singapore, 2022; pp. 63–85. [CrossRef]
65. Singh, S.; Podder, P.; Russo, M.; Henry, C.; Cinti, S. Tailored point-of-care biosensors for liquid biopsy in the field of oncology. *Lab Chip* **2022**, *71*, 44–61. [CrossRef]
66. Yu, J.; Yan, C.; Guo, M. Non-intrusive reduced-order modeling for fluid problems: A brief review. *Proc. Inst. Mech. Eng. Part G J. Aerosp. Eng.* **2019**, *233*, 5896–5912. [CrossRef]
67. Chen, Q.; Wang, Y.; Wang, H.; Yang, X. Data-driven simulation in fluids animation: A survey. *Virtual Real. Intell. Hardw.* **2021**, *3*, 87–104. [CrossRef]

68. Gibou, F.; Hyde, D.; Fedkiw, R. Sharp interface approaches and deep learning techniques for multiphase flows. *J. Comput. Phys.* **2019**, *380*, 442–463. [CrossRef]
69. Alhaddad, A.; Aly, H.; Gad, H.; Al-Ali, A.; Sadasivuni, K.; Cabibihan, J.J.; Malik, R. Sense and Learn: Recent Advances in Wearable Sensing and Machine Learning for Blood Glucose Monitoring and Trend-Detection. *Front. Bioeng. Biotechnol.* **2022**, *10*, 876672. [CrossRef]
70. Gerraty, R.; Provost, A.; Li, L.; Wagner, E.; Haas, M.; Lancashire, L. Machine learning within the Parkinson's progression markers initiative: Review of the current state of affairs. *Front. Aging Neurosci.* **2023**, *15*, 1076657. [CrossRef]
71. He, B.; Lu, Q.; Yang, Q.; Luo, J.; Wang, Z. Taylor Genetic Programming for Symbolic Regression. *arXiv* **2022**, arXiv:2205.09751. https://doi.org/10.48550/ARXIV.2205.09751.
72. Mir, A.; Sarwar, A. Artificial intelligence-based techniques for analysis of body cavity fluids: A review. *Artif. Intell. Rev.* **2021**, *54*, 4019–4061. [CrossRef]
73. Moradi, H.; Al-Hourani, A.; Concilia, G.; Khoshmanesh, F.; Nezami, F.; Needham, S.; Baratchi, S.; Khoshmanesh, K. Recent developments in modeling, imaging, and monitoring of cardiovascular diseases using machine learning. *Biophys. Rev.* **2023**, *15*, 19–33. [CrossRef]
74. Arzani, A.; Dawson, S. Data-driven cardiovascular flow modelling: Examples and opportunities. *J. R. Soc. Interface* **2021**, *18*, 20200802. [CrossRef] [PubMed]
75. del Real Mata, C.; Jeanne, O.; Jalali, M.; Lu, Y.; Mahshid, S. Nanostructured-Based Optical Readouts Interfaced with Machine Learning for Identification of Extracellular Vesicles. *Adv. Healthc. Mater.* **2023**, *12*, 2202123. [CrossRef]
76. Balana, C.; Castañer, S.; Carrato, C.; Moran, T.; Lopez-Paradís, A.; Domenech, M.; Hernandez, A.; Puig, J. Preoperative Diagnosis and Molecular Characterization of Gliomas with Liquid Biopsy and Radiogenomics. *Front. Neurol.* **2022**, *13*, 865171. [CrossRef] [PubMed]
77. Benton, S.; Tesche, C.; De Cecco, C.; Duguay, T.; Schoepf, U.; Bayer, R. Noninvasive Derivation of Fractional Flow Reserve From Coronary Computed Tomographic Angiography: A Review. *J. Thorac. Imaging* **2018**, *33*, 88–96. [CrossRef]
78. Bhattacharjee, G.; Gohil, N.; Shukla, M.; Sharma, S.; Mani, I.; Pandya, A.; Chu, D.T.; Bui, N.; Thi, Y.V.; Khambhati, K.; et al. Exploring the potential of microfluidics for next-generation drug delivery systems. *OpenNano* **2023**, *12*, 100150. [CrossRef]
79. Bratulic, S.; Gatto, F.; Nielsen, J. The Translational Status of Cancer Liquid Biopsies. *Regen. Eng. Transl. Med.* **2021**, *7*, 312–352. [CrossRef]
80. Gottlieb, E.; Samuel, M.; Bonventre, J.; Celi, L.; Mattie, H. Machine Learning for Acute Kidney Injury Prediction in the Intensive Care Unit. *Adv. Chronic Kidney Dis.* **2022**, *29*, 431–438. [CrossRef] [PubMed]
81. Khan, S.; Manialawy, Y.; Wheeler, M.; Cox, B. Unbiased data analytic strategies to improve biomarker discovery in precision medicine. *Drug Discov. Today* **2019**, *24*, 1735–1748. [CrossRef]
82. Ko, J.; Baldassano, S.; Loh, P.L.; Kording, K.; Litt, B.; Issadore, D. Machine learning to detect signatures of disease in liquid biopsies-a user's guide. *Lab Chip* **2018**, *18*, 395–405. [CrossRef]
83. Sandys, V.; Sexton, D.; O'Seaghdha, C. Artificial intelligence and digital health for volume maintenance in hemodialysis patients. *Hemodial. Int.* **2022**, *26*, 480–495. [CrossRef]
84. Serafim, M.; dos Santos Júnior, V.; Gertrudes, J.; Maltarollo, V.; Honorio, K. Machine learning techniques applied to the drug design and discovery of new antivirals: A brief look over the past decade. *Expert Opin. Drug Discov.* **2021**, *16*, 961–975. [CrossRef] [PubMed]
85. Sreedharan, S.; Zekry, S.; Leipsic, J.; Brown, R. Updates on Fractional Flow Reserve Derived by CT (FFRCT). *Curr. Treat. Options Cardiovasc. Med.* **2020**, *22*, 17. [CrossRef]
86. Tasoglu, S. Toilet-based continuous health monitoring using urine. *Nat. Rev. Urol.* **2022**, *19*, 219–230. [CrossRef]
87. Cassola, S.; Duhovic, M.; Schmidt, T.; May, D. Machine learning for polymer composites process simulation—A review. *Compos. Part B Eng.* **2022**, *246*, 110208. [CrossRef]
88. Yu, C.; Bi, X.; Fan, Y. Deep learning for fluid velocity field estimation: A review. *Ocean. Eng.* **2023**, *271*, 113693. [CrossRef]
89. Jirasek, F.; Hasse, H. Machine Learning of Thermophysical Properties. *Fluid Phase Equilibria* **2021**, *549*, 113206. [CrossRef]
90. Desgranges, C.; Delhommelle, J. Towards a machine learned thermodynamics: Exploration of free energy landscapes in molecular fluids, biological systems and for gas storage and separation in metal-organic frameworks. *Mol. Syst. Des. Eng.* **2021**, *6*, 52–65. [CrossRef]
91. Ahmadi, M.; Kumar, R.; Assad, M.; Ngo, P. Applications of machine learning methods in modeling various types of heat pipes: A review. *J. Therm. Anal. Calorim.* **2021**, *146*, 2333–2341. [CrossRef]
92. Souayeh, B.; Bhattacharyya, S.; Hdhiri, N.; Alam, M. Heat and fluid flow analysis and ann-based prediction of a novel spring corrugated tape. *Sustainability* **2021**, *13*, 3023. [CrossRef]
93. Yang, B.; Zhu, X.; Wei, B.; Liu, M.; Li, Y.; Lv, Z.; Wang, F. Computer Vision and Machine Learning Methods for Heat Transfer and Fluid Flow in Complex Structural Microchannels: A Review. *Energies* **2023**, *16*, 1500. [CrossRef]
94. Koutsoukos, S.; Philippi, F.; Malaret, F.; Welton, T. A review on machine learning algorithms for the ionic liquid chemical space. *Chem. Sci.* **2021**, *12*, 6820–6843. [CrossRef] [PubMed]
95. Upot, N.; Fazle Rabbi, K.; Khodakarami, S.; Ho, J.; Kohler Mendizabal, J.; Miljkovic, N. Advances in micro and nanoengineered surfaces for enhancing boiling and condensation heat transfer: A review. *Nanoscale Adv.* **2022**, *5*, 1232–1270. [CrossRef]

96. Sofos, F.; Stavrogiannis, C.; Exarchou-Kouveli, K.K.; Akabua, D.; Charilas, G.; Karakasidis, T.E. Current Trends in Fluid Research in the Era of Artificial Intelligence: A Review. *Fluids* **2022**, *7*, 116. [CrossRef]
97. Adun, H.; Wole-Osho, I.; Okonkwo, E.; Kavaz, D.; Dagbasi, M. A critical review of specific heat capacity of hybrid nanofluids for thermal energy applications. *J. Mol. Liq.* **2021**, *340*, 116890. [CrossRef]
98. Hemmati-Sarapardeh, A.; Varamesh, A.; Nait Amar, M.; Husein, M.; Dong, M. On the evaluation of thermal conductivity of nanofluids using advanced intelligent models. *Int. Commun. Heat Mass Transf.* **2020**, *118*, 104825. [CrossRef]
99. Gonçalves, I.; Souza, R.; Coutinho, G.; Miranda, J.; Moita, A.; Pereira, J.; Moreira, A.; Lima, R. Thermal conductivity of nanofluids: A review on prediction models, controversies and challenges. *Appl. Sci.* **2021**, *11*, 2525. [CrossRef]
100. Ma, T.; Guo, Z.; Lin, M.; Wang, Q. Recent trends on nanofluid heat transfer machine learning research applied to renewable energy. *Renew. Sustain. Energy Rev.* **2021**, *138*, 110494. [CrossRef]
101. Sharma, P.; Said, Z.; Kumar, A.; Nižetić, S.; Pandey, A.; Hoang, A.; Huang, Z.; Afzal, A.; Li, C.; Le, A.; et al. Recent Advances in Machine Learning Research for Nanofluid-Based Heat Transfer in Renewable Energy System. *Energy Fuels* **2022**, *36*, 6626–6658. [CrossRef]
102. Wang, H.; Chen, X. A Comprehensive Review of Predicting the Thermophysical Properties of Nanofluids Using Machine Learning Methods. *Ind. Eng. Chem. Res.* **2022**, *61*, 14711–14730. [CrossRef]
103. Fani, M.; Pourafshary, P.; Mostaghimi, P.; Mosavat, N. Application of microfluidics in chemical enhanced oil recovery: A review. *Fuel* **2022**, *315*, 123225. [CrossRef]
104. Galan, E.; Zhao, H.; Wang, X.; Dai, Q.; Huck, W.; Ma, S. Intelligent Microfluidics: The Convergence of Machine Learning and Microfluidics in Materials Science and Biomedicine. *Matter* **2020**, *3*, 1893–1922. [CrossRef]
105. Tsai, H.F.; Podder, S.; Chen, P.Y. Microsystem Advances through Integration with Artificial Intelligence. *Micromachines* **2023**, *14*, 826. [CrossRef]
106. Gao, J.; Hu, Z.; Yang, Q.; Liang, X.; Wu, H. Fluid flow and heat transfer in microchannel heat sinks: Modelling review and recent progress. *Therm. Sci. Eng. Prog.* **2022**, *29*, 113096. [CrossRef]
107. Srikanth, S.; Dubey, S.; Javed, A.; Goel, S. Droplet based microfluidics integrated with machine learning. *Sens. Actuators A Phys.* **2021**, *332*, 113096. [CrossRef]
108. Zhong, J.; Riordon, J.; Wu, T.; Edwards, H.; Wheeler, A.; Pardee, K.; Aspuru-Guzik, A.; Sinton, D. When robotics met fluidics. *Lab Chip* **2020**, *20*, 709–716. [CrossRef] [PubMed]
109. Zerouaoui, J.; Alaoui, A.; Ettaki, B.; Chakir, E. Assessing the Improvements Brought by Artificial Intelligence on the Prediction of Aerodynamic Coefficients. In *Artificial Intelligence and Smart Environment*; Lecture Notes in Networks and Systems; Springer: Cham, Switzerland, 2023; Volume 635, pp. 254–263. [CrossRef]
110. Kou, J.; Zhang, W. Data-driven modeling for unsteady aerodynamics and aeroelasticity. *Prog. Aerosp. Sci.* **2021**, *125*, 100725. [CrossRef]
111. Le Clainche, S.; Ferrer, E.; Gibson, S.; Cross, E.; Parente, A.; Vinuesa, R. Improving aircraft performance using machine learning: A review. *Aerosp. Sci. Technol.* **2023**, *138*, 108354. [CrossRef]
112. Li, J.; Du, X.; Martins, J. Machine learning in aerodynamic shape optimization. *Prog. Aerosp. Sci.* **2022**, *134*, 100849. [CrossRef]
113. Zhu, H.; Yang, B.; Zhang, Q.; Pan, L.; Sun, S. Wind engineering for high-rise buildings: A review. *Wind. Struct. Int. J.* **2021**, *32*, 249–265. [CrossRef]
114. Tahmasebi, P.; Kamrava, S.; Bai, T.; Sahimi, M. Machine learning in geo- and environmental sciences: From small to large scale. *Adv. Water Resour.* **2020**, *142*, 103619. [CrossRef]
115. Nallakukkala, S.; Lal, B. Modeling of seawater desalination by gas hydrate method. In *Gas Hydrate in Water Treatment: Technological, Economic, and Industrial Aspects*; Wiley: Hoboken, NJ, USA, 2022; pp. 77–111. [CrossRef]
116. Irschick, D.; Christiansen, F.; Hammerschlag, N.; Martin, J.; Madsen, P.; Wyneken, J.; Brooks, A.; Gleiss, A.; Fossette, S.; Siler, C.; et al. 3D visualization processes for recreating and studying organismal form. *iScience* **2022**, *25*, 104867. [CrossRef] [PubMed]
117. Brookfield, A.; Hansen, A.; Sullivan, P.; Czuba, J.; Kirk, M.; Li, L.; Newcomer, M.; Wilkinson, G. Predicting algal blooms: Are we overlooking groundwater? *Sci. Total Environ.* **2021**, *769*, 144442. [CrossRef] [PubMed]
118. Li, J.; Yuan, Y.; Shen, H.B. Symbolic Expression Transformer: A Computer Vision Approach for Symbolic Regression. *arXiv* **2022**, arXiv:2205.11798. https://doi.org/10.48550/ARXIV.2205.11798.
119. Singh, N.; Yadav, M.; Singh, V.; Padhiyar, H.; Kumar, V.; Bhatia, S.; Show, P.L. Artificial intelligence and machine learning-based monitoring and design of biological wastewater treatment systems. *Bioresour. Technol.* **2023**, *369*, 128486. [CrossRef]
120. Umenweke, G.; Afolabi, I.; Epelle, E.; Okolie, J. Machine learning methods for modeling conventional and hydrothermal gasification of waste biomass: A review. *Bioresour. Technol. Rep.* **2022**, *17*, 100976. [CrossRef]
121. Zaghloul, M.; Achari, G. A review of mechanistic and data-driven models of aerobic granular sludge. *J. Environ. Chem. Eng.* **2022**, *10*, 107500. [CrossRef]
122. Chandra, D.; Vishal, V. A critical review on pore to continuum scale imaging techniques for enhanced shale gas recovery. *Earth-Sci. Rev.* **2021**, *217*, 103638. [CrossRef]
123. Zhang, J.; Li, J.; Chen, X.; Li, Y.; Tang, W. A Spatially Coupled Data-Driven Approach for Lithology/Fluid Prediction. *IEEE Trans. Geosci. Remote Sens.* **2021**, *59*, 5526–5534. [CrossRef]

124. Viswanathan, H.; Ajo-Franklin, J.; Birkholzer, J.; Carey, J.; Guglielmi, Y.; Hyman, J.; Karra, S.; Pyrak-Nolte, L.; Rajaram, H.; Srinivasan, G.; et al. From Fluid Flow to Coupled Processes in Fractured Rock: Recent Advances and New Frontiers. *Rev. Geophys.* **2022**, *60*, e2021RG000744. [CrossRef]
125. Bhattacharya, S. Summarized Applications of Machine Learning in Subsurface Geosciences. In *Primer on Machine Learning in Subsurface Geosciences*; SpringerBriefs in Petroleum Geoscience & Engineering; Springer: Cham, Switzerland, 2021; pp. 123–165. [CrossRef]
126. Khan, H.; Srivastav, A.; Kumar Mishra, A.; Anh Tran, T. Machine learning methods for estimating permeability of a reservoir. *Int. J. Syst. Assur. Eng. Manag.* **2022**, *13*, 2118–2131. [CrossRef]
127. Wu, Q.; Guo, Y.; Liu, Y.; Wang, G. Review on the cavitating flow-induced vibrations. *Kongqi Donglixue Xuebao/Acta Aerodyn. Sin.* **2020**, *38*, 746–760. [CrossRef]
128. Lopes, N.; Chao, Y.; Dasarla, V.; Sullivan, N.; Ricklick, M.; Boetcher, S. Comprehensive Review of Heat Transfer Correlations of Supercritical CO2in Straight Tubes Near the Critical Point: A Historical Perspective. *J. Heat Transf.* **2022**, *144*, 120801. [CrossRef]
129. Mostafa, K.; Zisis, I.; Moustafa, M. Machine Learning Techniques in Structural Wind Engineering: A State-of-the-Art Review. *Appl. Sci.* **2022**, *12*, 5232. [CrossRef]
130. Drikakis, D.; Dbouk, T. The Role of Computational Science in Wind and Solar Energy: A Critical Review. *Energies* **2022**, *15*, 9609. [CrossRef]
131. Maksymov, I.; Huy Nguyen, B.; Suslov, S. Biomechanical Sensing Using Gas Bubbles Oscillations in Liquids and Adjacent Technologies: Theory and Practical Applications. *Biosensors* **2022**, *12*, 624. [CrossRef]
132. Du, Y.H.; Wang, M.Y.; Yang, L.H.; Tong, L.L.; Guo, D.S.; Ji, X.J. Optimization and Scale-Up of Fermentation Processes Driven by Models. *Bioengineering* **2022**, *9*, 473. [CrossRef]
133. Zottl, A.; Stark, H. Modeling Active Colloids: From Active Brownian Particles to Hydrodynamic and Chemical Fields. *Annu. Rev. Condens. Matter Phys.* **2023**, *14*, 109–127. [CrossRef]
134. Zou, Y.; Chu, Z.; Guo, J.; Liu, S.; Ma, X.; Guo, J. Minimally invasive electrochemical continuous glucose monitoring sensors: Recent progress and perspective. *Biosens. Bioelectron.* **2023**, *225*, 115103. [CrossRef]
135. Lv, H.; Zhang, S.; Sun, Q.; Chen, R.; Zhang, W. The Dynamic Models, Control Strategies and Applications for Magnetorheological Damping Systems: A Systematic Review. *J. Vib. Eng. Technol.* **2021**, *9*, 131–147. [CrossRef]
136. Anush, C.; Yashwanth, K.; Shashank, S.; Venkat Reddy, M.; Kumar, A. Bottle Line Detection using Digital Image Processing with Machine Learning. *J. Phys. Conf. Ser.* **2021**, *1998*, 012033. [CrossRef]
137. Akavalappil, V.; Radhakrishnan, T. Comparison of current state of control valve stiction detection and quantification techniques. *Trans. Inst. Meas. Control* **2022**, *44*, 562–579. [CrossRef]
138. Vowinckel, B. Incorporating grain-scale processes in macroscopic sediment transport models: A review and perspectives for environmental and geophysical applications. *Acta Mech.* **2021**, *232*, 2023–2050. [CrossRef]
139. Zhang, Y.; Xu, J.; Chang, Q.; Zhao, P.; Wang, J.; Ge, W. Numerical simulation of fluidization: Driven by challenges. *Powder Technol.* **2023**, *414*, 118092. [CrossRef]
140. Asproulis, N.; Drikakis, D. Nanoscale materials modelling using neural networks. *J. Comput. Theor. Nanosci.* **2009**, *6*, 514–518. [CrossRef]
141. Asproulis, N.; Drikakis, D. An artificial neural network-based multiscale method for hybrid atomistic-continuum simulations. *Microfluid. Nanofluid.* **2013**, *15*, 559–574. [CrossRef]
142. Alzubaidi, L.; Bai, J.; Al-Sabaawi, A.; Santamaría, J.; Albahri, A.; Al-dabbagh, B.; Fadhel, M.; Manoufali, M.; Zhang, J.; Al-Timemy, A.; et al. A survey on deep learning tools dealing with data scarcity: definitions, challenges, solutions, tips, and applications. *J. Big Data* **2023**, *10*, 46. [CrossRef]
143. Goh, G.D.; Sing, S.L.; Yeong, W.Y. A review on machine learning in 3D printing: Applications, potential, and challenges. *Artif. Intell. Rev.* **2021**, *54*, 63–94. [CrossRef]
144. Bhattacharya, S.; Verma, M.K.; Bhattacharya, A. Predictions of Reynolds and Nusselt numbers in turbulent convection using machine-learning models. *Phys. Fluids* **2022**, *34*, 025102. [CrossRef]
145. Md, A.Q.; Kulkarni, S.; Joshua, C.J.; Vaichole, T.; Mohan, S.; Iwendi, C. Enhanced Preprocessing Approach Using Ensemble Machine Learning Algorithms for Detecting Liver Disease. *Biomedicines* **2023**, *11*, 581. [CrossRef]
146. Mahdaviara, M.; Sharifi, M.; Bakhshian, S.; Shokri, N. Prediction of spontaneous imbibition in porous media using deep and ensemble learning techniques. *Fuel* **2022**, *329*, 125349. [CrossRef]
147. Rahil, M.; Anoop, B.N.; Girish, G.N.; Kothari, A.R.; Koolagudi, S.G.; Rajan, J. A Deep Ensemble Learning-Based CNN Architecture for Multiclass Retinal Fluid Segmentation in OCT Images. *IEEE Access* **2023**, *11*, 17241–17251. [CrossRef]
148. Cai, S.; Mao, Z.; Wang, Z.; Yin, M.; Karniadakis, G.E. Physics-informed neural networks (PINNs) for fluid mechanics: A review. *Acta Mech. Sin.* **2021**, *37*, 1727–1738. [CrossRef]
149. Raissi, M.; Karniadakis, G.E. Hidden physics models: Machine learning of nonlinear partial differential equations. *J. Comput. Phys.* **2018**, *357*, 125–141. [CrossRef]
150. Raissi, M.; Wang, Z.; Triantafyllou, M.S.; Karniadakis, G.E. Deep learning of vortex-induced vibrations. *J. Fluid Mech.* **2019**, *861*, 119–137. [CrossRef]
151. Agostini, L. Exploration and prediction of fluid dynamical systems using auto-encoder technology. *Phys. Fluids* **2020**, *32*, 067103. [CrossRef]

152. Thornber, B.; Mosedale, A.; Drikakis, D. On the implicit large eddy simulations of homogeneous decaying turbulence. *J. Comput. Phys.* **2007**, *226*, 1902–1929. [CrossRef]
153. Jiménez, J. Near-wall turbulence. *Phys. Fluids* **2013**, *25*, 1–28. [CrossRef]
154. Drikakis, D. Advances in turbulent flow computations using high-resolution methods. *Prog. Aerosp. Sci.* **2003**, *39*, 405–424. [CrossRef]
155. Kobayashi, H.; Matsumoto, E.; Fukushima, N.; Tanahashi, M.; Miyauchi, T. Statistical properties of the local structure of homogeneous isotropic turbulence and turbulent channel flows. *J. Turbul.* **2011**, *12*, N12. [CrossRef]
156. Duraisamy, K.; Iaccarino, G.; Xiao, H. Turbulence modeling in the age of data. *Annu. Rev. Fluid Mech.* **2019**, *51*, 357–377. [CrossRef]
157. Giralt, F.; Arenas, A.; Ferre-Gine, J.; Rallo, R.; Kopp, G. The simulation and interpretation of free turbulence with a cognitive neural system. *Phys. Fluids* **2000**, *12*, 1826–1835. [CrossRef]
158. Milano, M.; Koumoutsakos, P. Neural network modeling for near wall turbulent flow. *J. Comput. Phys.* **2002**, *182*, 1–26. [CrossRef]
159. Chang, F.J.; Yang, H.C.; Lu, J.Y.; Hong, J.H. Neural network modelling for mean velocity and turbulence intensities of steep channel flows. *Hydrol. Process. Int. J.* **2008**, *22*, 265–274. [CrossRef]
160. McConkey, R.; Yee, E.; Lien, F.S. A curated dataset for data-driven turbulence modelling. *Sci. Data* **2021**, *8*, 255. [CrossRef] [PubMed]
161. Bonnet, F.; Mazari, A.J.; Cinnella, P.; Gallinari, P. AirfRANS: High Fidelity Computational Fluid Dynamics Dataset for Approximating Reynolds-Averaged Navier-Stokes Solutions. *arXiv* **2023**, arXiv:2212.07564.
162. Ribeiro, M.D.; Rehman, A.; Ahmed, S.; Dengel, A. DeepCFD: Efficient Steady-State Laminar Flow Approximation with Deep Convolutional Neural Networks. *arXiv* **2021**, arXiv:2004.08826.
163. Xiao, H.; Wu, J.L.; Wang, J.X.; Sun, R.; Roy, C. Quantifying and reducing model-form uncertainties in Reynolds-averaged Navier–Stokes simulations: A data-driven, physics-informed Bayesian approach. *J. Comput. Phys.* **2016**, *324*, 115–136. [CrossRef]
164. Tracey, B.D.; Duraisamy, K.; Alonso, J.J. A machine learning strategy to assist turbulence model development. In Proceedings of the 53rd AIAA Aerospace Sciences Meeting, Kissimmee, FL, USA, 5–9 January 2015; p. 1287.
165. Ling, J.; Kurzawski, A.; Templeton, J. Reynolds averaged turbulence modelling using deep neural networks with embedded invariance. *J. Fluid Mech.* **2016**, *807*, 155–166. [CrossRef]
166. Kutz, J.N. Deep learning in fluid dynamics. *J. Fluid Mech.* **2017**, *814*, 1–4. [CrossRef]
167. Cheung, S.H.; Oliver, T.A.; Prudencio, E.E.; Prudhomme, S.; Moser, R.D. Bayesian uncertainty analysis with applications to turbulence modeling. *Reliab. Eng. Syst. Saf.* **2011**, *96*, 1137–1149. [CrossRef]
168. Edeling, W.; Cinnella, P.; Dwight, R.P.; Bijl, H. Bayesian estimates of parameter variability in the k–ε turbulence model. *J. Comput. Phys.* **2014**, *258*, 73–94. [CrossRef]
169. Zhang, Z.J.; Duraisamy, K. Machine learning methods for data-driven turbulence modeling. In Proceedings of the 22nd AIAA Computational Fluid Dynamics Conference, Dallas, TX, USA, 22–26 June 2015; p. 2460.
170. Duraisamy, K.; Zhang, Z.J.; Singh, A.P. New approaches in turbulence and transition modeling using data-driven techniques. In Proceedings of the 53rd AIAA Aerospace Sciences Meeting, Kissimmee, FL, USA, 5–9 January 2015; p. 1284.
171. Parish, E.J.; Duraisamy, K. A paradigm for data-driven predictive modeling using field inversion and machine learning. *J. Comput. Phys.* **2016**, *305*, 758–774. [CrossRef]
172. Geneva, N.; Zabaras, N. Quantifying model form uncertainty in Reynolds-averaged turbulence models with Bayesian deep neural networks. *J. Comput. Phys.* **2019**, *383*, 125–147. [CrossRef]
173. Sarghini, F.; De Felice, G.; Santini, S. Neural networks based subgrid scale modeling in large eddy simulations. *Comput. Fluids* **2003**, *32*, 97–108. [CrossRef]
174. Moreau, A.; Teytaud, O.; Bertoglio, J.P. Optimal estimation for large-eddy simulation of turbulence and application to the analysis of subgrid models. *Phys. Fluids* **2006**, *18*, 105101. [CrossRef]
175. Beck, A.D.; Flad, D.G.; Munz, C.D. Deep neural networks for data-driven turbulence models. *arXiv* **2018**, arXiv:1806.04482.
176. Maulik, R.; San, O.; Rasheed, A.; Vedula, P. Subgrid modelling for two-dimensional turbulence using neural networks. *J. Fluid Mech.* **2019**, *858*, 122–144. [CrossRef]
177. Bao, T.; Chen, S.; Johnson, T.T.; Givi, P.; Sammak, S.; Jia, X. Physics guided neural networks for spatio-temporal super-resolution of turbulent flows. In Proceedings of the Thirty-Eighth Conference on Uncertainty in Artificial Intelligence, PMLR, Eindhoven, The Netherlands, 1–5 August 2022; pp. 118–128. ISSN 2640-3498.
178. Fukami, K.; Fukagata, K.; Taira, K. Super-resolution reconstruction of turbulent flows with machine learning. *J. Fluid Mech.* **2019**, *870*, 106–120. [CrossRef]
179. Fukami, K.; Hasegawa, K.; Nakamura, T.; Morimoto, M.; Fukagata, K. Model Order Reduction with Neural Networks: Application to Laminar and Turbulent Flows. *Comput. Sci.* **2021**, *2*, 467. [CrossRef]
180. Fukami, K.; Fukagata, K.; Taira, K. Super-resolution analysis via machine learning: A survey for fluid flows. *Theor. Comput. Fluid Dyn.* **2023**. [CrossRef]
181. Khan, M.I.H.; Sablani, S.S.; Nayak, R.; Gu, Y. Machine learning-based modeling in food processing applications: State of the art. *Compr. Rev. Food Sci. Food Saf.* **2022**, *21*, 1409–1438. [CrossRef] [PubMed]
182. Marcato, A.; Boccardo, G.; Marchisio, D. From Computational Fluid Dynamics to Structure Interpretation via Neural Networks: An Application to Flow and Transport in Porous Media. *Ind. Eng. Chem. Res.* **2022**, *61*, 8530–8541. [CrossRef]

183. Dinesh Kumar, M.; Ameer Ahammad, N.; Raju, C.; Yook, S.J.; Shah, N.A.; Tag, S.M. Response surface methodology optimization of dynamical solutions of Lie group analysis for nonlinear radiated magnetized unsteady wedge: Machine learning approach (gradient descent). *Alex. Eng. J.* **2023**, *74*, 29–50. [CrossRef]
184. Priyadharshini, P.; Archana, M.V.; Ahammad, N.A.; Raju, C.; jin Yook, S.; Shah, N.A. Gradient descent machine learning regression for MHD flow: Metallurgy process. *Int. Commun. Heat Mass Transf.* **2022**, *138*, 106307. . [CrossRef]
185. Deng, Z.; He, C.; Liu, Y.; Kim, K.C. Super-resolution reconstruction of turbulent velocity fields using a generative adversarial network-based artificial intelligence framework. *Phys. Fluids* **2019**, *31*, 125111. [CrossRef]
186. Yousif, M.Z.; Yu, L.; Lim, H.C. Super-resolution reconstruction of turbulent flow fields at various Reynolds numbers based on generative adversarial networks. *Phys. Fluids* **2022**, *34*. 015130. [CrossRef]
187. Brenner, M.P.; Eldredge, J.D.; Freund, J.B. Perspective on machine learning for advancing fluid mechanics. *Phys. Rev. Fluids* **2019**, *4*, 100501. [CrossRef]
188. Dubey, V.; Sharma, A.K.; Pimenov, D.Y. Prediction of Surface Roughness Using Machine Learning Approach in MQL Turning of AISI 304 Steel by Varying Nanoparticle Size in the Cutting Fluid. *Lubricants* **2022**, *10*, 81. [CrossRef]
189. Cuomo, S.; Di Cola, V.S.; Giampaolo, F.; Rozza, G.; Raissi, M.; Piccialli, F. Scientific Machine Learning Through Physics—Informed Neural Networks: Where we are and What's Next. *J. Sci. Comput.* **2022**, *92*, 88. [CrossRef]
190. Schran, C.; Thiemann, F.L.; Rowe, P.; Muller, E.A.; Marsalek, O.; Michaelides, A. Machine learning potentials for complex aqueous systems made simple. *Proc. Natl. Acad. Sci. USA* **2021**, *118*, e2110077118. [CrossRef]
191. Alam, T.M.; Allers, J.P.; Leverant, C.J.; Harvey, J.A. Symbolic regression development of empirical equations for diffusion in Lennard-Jones fluids. *J. Chem. Phys.* **2022**, *157*, 014503. [CrossRef] [PubMed]
192. L'Heureux, A.; Grolinger, K.; Elyamany, H.F.; Capretz, M.A.M. Machine Learning with Big Data: Challenges and Approaches. *IEEE Access* **2017**, *5*, 7776–7797. [CrossRef]
193. Vinuesa, R.; Brunton, S.L. The Potential of Machine Learning to Enhance Computational Fluid Dynamics. *arXiv* **2021**, arXiv:2110.02085.
194. Poulinakis, K.; Drikakis, D.; Kokkinakis, I.W.; Spottswood, S.M. Machine-Learning Methods on Noisy and Sparse Data. *Mathematics* **2023**, *11*, 236. [CrossRef]

fluids

MDPI

Review

Machine Learning Reshaping Computational Fluid Dynamics: A Paradigm Shift in Accuracy and Speed

Aly Mousaad Aly

Windstorm Impact, Science, and Engineering (WISE) Research Lab, Louisiana State University, Baton Rouge, LA 70803, USA; aly@lsu.edu

Abstract

Accurate and efficient CFD simulations are essential for a wide range of engineering and scientific applications, from resilient structural design to environmental analysis. Traditional methods such as RANS simulations often face challenges in capturing complex flow phenomena like separation, while high-fidelity approaches including Large Eddy Simulations and Direct Numerical Simulations demand significant computational resources, thereby limiting their practical applicability. This paper provides an in-depth synthesis of recent advancements in integrating artificial intelligence and machine learning techniques with CFD to enhance simulation accuracy, computational efficiency, and modeling capabilities, including data-driven surrogate models, physics-informed methods, and ML-assisted numerical solvers. This integration marks a crucial paradigm shift, transcending incremental improvements to fundamentally redefine the possibilities of fluid dynamics research and engineering design. Key themes discussed include data-driven surrogate models, physics-informed methods, ML-assisted numerical solvers, inverse design, and advanced turbulence modeling. Practical applications, such as wind load design for solar panels and deep learning approaches for eddy viscosity prediction in bluff body flows, illustrate the substantial impact of ML integration. The findings demonstrate that ML techniques can accelerate simulations by up to 10,000 times in certain cases while maintaining or improving the accuracy, particularly in challenging flow regimes. For instance, models employing learned interpolation can achieve 40- to 80-fold computational speedups while matching the accuracy of baseline solvers with a resolution 8 to 10 times finer. Other approaches, like Fourier Neural Operators, can achieve inference times three orders of magnitude faster than conventional PDE solvers for the Navier–Stokes equations. Such advancements not only accelerate critical engineering workflows but also open unprecedented avenues for scientific discovery in complex, nonlinear systems that were previously intractable with traditional computational methods. Furthermore, ML enables unprecedented advances in turbulence modeling, improving predictions within complex separated flow zones. This integration is reshaping fluid mechanics, offering pathways toward more reliable, efficient, and resilient engineering solutions necessary for addressing contemporary challenges.

Keywords: machine learning; computational fluid dynamics; artificial intelligence; turbulence modeling; neural networks; data-driven fluid mechanics; CFD

1. Introduction

The escalating demand for clean energy and resilient infrastructure—particularly in urban environments vulnerable to climate change [1,2]—has underscored the critical importance of accurate and efficient FD simulations [3]. This necessitates robust predictive capabilities for phenomena like aerodynamic forces on civil structures, atmospheric dispersion in cities, and optimal designs for energy systems, areas where CFD plays a foundational role. CFD serves as a powerful tool for quantifying wind loads on structures and elucidating complex fluid phenomena [4,5]. Nonetheless, traditional CFD methodologies face significant limitations. Reynolds-averaged Navier–Stokes (RANS) simulations, while computationally affordable, often struggle with accuracy in predicting wind pressures, particularly within complex flow separation zones [4]. In contrast, high-fidelity techniques such as LES (Large Eddy Simulation) and DNS (Direct Numerical Simulation) offer superior accuracy but incur a prohibitive computational expense, thereby limiting their practical application in engineering workflows [4,6]. Wind structural engineering within the lower atmospheric boundary layer encapsulates a myriad of complex fluid dynamic phenomena that challenge both computational and experimental investigations [3,4,7].

These inherent challenges in traditional CFD have motivated the exploration of innovative approaches [8]. Overcoming the computational bottlenecks and accuracy limitations of conventional methods represents a critical frontier for advancing engineering and scientific discovery, particularly through the synergistic integration of data-driven intelligence with fundamental physical laws [9]. Recently, artificial intelligence (AI) and machine learning (ML) have emerged as groundbreaking technologies capable of addressing these limitations [10,11]. AI encompasses broader intelligence demonstrated by machines, whereas ML is a subset focused on algorithms that learn from data for prediction or decision-making without explicit programming [12,13]. A particularly promising advancement is the development of Physics-Informed Neural Networks (PINNs), which integrate physical laws directly into deep learning frameworks to ensure solutions adhere to fundamental physical principles [9,14,15]. This inherent physical consistency provides a crucial advantage over purely data-driven black-box models, enhancing interpretability and robustness, especially when extrapolating to unseen conditions. ML comprises data-driven methodologies that improve task performance by learning from data [12,13]. Integrating ML with CFD has demonstrated remarkable potential—accelerating simulations by factors of up to 10,000 without compromising accuracy [5], paving the way toward more efficient and reliable design processes. ML techniques have been leveraged to enhance CFD workflows by initializing solutions, approximating complex solution operators, and developing reduced-order models for physical phenomena such as turbulence modeling through eddy viscosity prediction [4]. This ML-driven paradigm unlocks new frontiers in fluid mechanics by harnessing the vast data generated from experiments and simulations alike, fundamentally reshaping the field and enabling more efficient, reliable, and resilient engineering designs critical to addressing contemporary global challenges, such as climate change impacts on infrastructure, the increasing demand for sustainable energy, and the need for advanced predictive tools in complex environmental systems. This is particularly relevant for sectors like aeronautics, where NASA's strategic implementation plan emphasizes the development of advanced computational and experimental tools for sustainable aviation, advanced air mobility, and virtual testing to meet future needs for safe, efficient, and environmentally sustainable air transportation. Specifically, NASA's vision for 2045 and beyond highlights the need for revolutionary advancements in areas such as ultra-efficient subsonic transport and innovative commercial high-speed aircraft, directly supported by accelerated CFD capabilities to develop sustainable and safe air travel solutions [16].

This paper presents an in-depth synthesis of recent advances in applications of ML to CFD, with particular emphasis on how AI integration enhances simulation accuracy, computational efficiency, and modeling capabilities [9,17]. Key themes covered include data-driven and physics-informed surrogate models, ML-assisted numerical solvers, and inverse design and control strategies, as well as emerging challenges such as multi-scale representation and the incorporation of physical knowledge within ML frameworks. Practical examples from the recent literature illustrate the tangible impact of these innovations.

The primary contributions and key focus areas addressed in this paper include the following:

- Advances in modeling capabilities, including data-driven surrogates, physics-informed surrogates [18], and ML-enhanced numerical solvers.
- Applications in inverse design and control [19–21].
- The identification of emerging challenges such as multi-scale modeling and encoding physical laws within ML models [22–24].
- Improvements in computational efficiency, highlighting speedups and reduced resource utilization achievable through ML integration [4,5].
- Enhancements in simulation accuracy provided by ML-driven approaches, particularly in regimes where traditional methods like RANS exhibit limitations (e.g., flow separation zones) [4].

2. Recent Advances and Applications of ML in CFD

The integration of machine learning (ML) techniques in computational fluid dynamics (CFD) has revolutionized the landscape of fluid simulations by significantly enhancing computational efficiency, accuracy, and physical consistency [10,17,25]. As summarized in Table 1, diverse ML approaches target key challenges in CFD, ranging from speedups in numerical predictions to improved turbulence modeling and optimal design [9,22]. This interdisciplinary convergence is not merely optimizing existing workflows but also fundamentally transforming the research ecosystem, pushing the boundaries of what is computationally feasible and scientifically discoverable [17]. Convolutional neural networks (CNNs) notably achieve up to 10,000-fold acceleration in predicting wind pressure and velocity fields, enabling rapid assessments critical for applications like solar panel optimization [5,26]. This speed allows for rapid iteration in design optimization, moving beyond expensive traditional wind tunnel testing or computationally intensive high-fidelity simulations for parameter sweeps. The work by Rana et al. [27] introduces a scalable CNN architecture based on a gated residual U-Net to predict fluid velocity fields in complex domains with multiple obstacles. The authors propose a domain decomposition strategy coupled with smoothness constraints to extend model applicability to larger, more intricate flow environments, achieving computational efficiency without retraining on extensive CFD datasets. This method notably advances the practicality of CNNs for real-world fluid dynamics problems by addressing error propagation and domain scalability. Physics-Informed Neural Networks (PINNs) advance surrogate modeling by embedding physical laws, ensuring solutions that respect underlying fluid dynamics, thus enhancing accuracy in drag reduction and wave dynamics [18,28–30]. Deep learning approaches to eddy viscosity modeling have improved the fidelity of turbulence representation in Reynolds-Averaged Navier–Stokes (RANS) simulations, which are essential for capturing complex bluff body separated flows [4]. This is particularly crucial for robust structural design in civil engineering, where accurate prediction of pressure loads in separated flow regions is paramount for safety and efficiency. Data-driven surrogate models facilitate efficient representation of multifaceted fluid phenomena, contributing to sustainable urban planning and rheological predictions of nanofluids [1,31,32]. Furthermore, ML-enabled inverse design and flow control empower the optimization of engineering geometries, such as breakwaters and ventilation systems, by leveraging data-driven feedback loops [19,20]. Emerging generative

AI techniques, including generative adversarial networks (GANs), offer capabilities for super-resolution reconstruction and augmentation of turbulent flow datasets, thus increasing the availability and quality of training data [24,33]. Such data-driven augmentation strategies are vital for addressing the inherent scarcity of high-fidelity simulation and experimental data, which often limits the training of robust ML models in fluid mechanics. In recent efforts to improve turbulent flow estimation from wall measurements, Cuéllar et al. (2024) introduced a novel three-dimensional generative adversarial network (3-D GAN) capable of reconstructing the entire volumetric flow field with a single unified model. This approach surpasses previous methods that independently estimated wall-parallel planes by directly providing full 3-D reconstructions with comparable error metrics and lower computational training costs. The study further reveals that reconstruction accuracy strongly correlates with the wall footprint of turbulent structures, with wall-attached structures predicted most accurately, offering new insights into coherent flow features near the wall [34]. Collectively, these advances illustrate how ML acts as a transformative tool accelerating CFD simulations while preserving or augmenting physical fidelity, propelling the field towards high-throughput and physically coherent fluid dynamics modeling. The ongoing integration promises a future where complex fluid phenomena can be modeled with unprecedented speed and accuracy, fundamentally altering engineering design cycles and scientific discovery across numerous disciplines.

Table 1. A summary of key machine learning techniques and their impacts on computational fluid dynamics.

ML Technique	Primary Impact	Example Applications/Context
Convolutional Neural Networks (CNNs)	Computational boosting, efficiency (e.g., 10,000× speedup)	Rapid prediction of wind pressure/velocity fields for solar panels [5,26,27]
Physics-Informed Neural Networks (PINNs)	Enhanced accuracy, physical consistency	Surrogate modeling for drag reduction [18], data assimilation for water wave dynamics [28], and mean-flow reconstruction [35,36]; applied to indoor airflow simulation [37] and parametric turbulence problems [38]
Deep Learning for Eddy Viscosity	Improved turbulence modeling, RANS accuracy	More precise prediction of pressure/velocity in bluff body separated flows [4]
Data-Driven Surrogate Models	Representing complex fluid dynamics, efficiency	Urban heat island mitigation, nanofluid rheology prediction [1,31,32]
ML for Inverse Design and Control	Optimal geometry/parameters, advanced control	Breakwater design optimization, room ventilation systems [19,20]
Generative AI (e.g., GANs)	Super-resolution reconstruction, data augmentation	High-fidelity turbulent flow field reconstruction [24,33,34]

2.1. Computational Boosting and Efficiency

One of the most impactful contributions of ML to CFD is the drastic reduction in computational time. Conventional CFD simulations, especially those involving high-fidelity methods like LES and DNS, are computationally expensive and time-consuming, hindering their widespread application in practical engineering scenarios [4]. For instance, a single high-fidelity simulation can demand thousands of CPU hours, translating into days or weeks of computation for complex industrial problems [39]. DNS, for example, is presently limited to low-Reynolds-number flows due to the significant increase in computational expense associated with higher Reynolds numbers, where the cost escalates approximately with the cube of the Reynolds number [6]. LES, while less expensive than DNS, still incurs high computational costs when resolving small spatiotemporal features [40]. For example, the grid-point requirements for wall-modeled Large Eddy Simulation (WMLES) typically scale linearly with an increasing Reynolds number [41,42].

Machine learning models, especially convolutional neural networks (CNNs), have proven their capability to predict outcomes with high accuracy while achieving computation speeds far exceeding those of conventional approaches. For example, research focusing on the aerodynamic design of solar panel installations has demonstrated that integrating machine learning with computational fluid dynamics (CFD) can accelerate the prediction of wind pressure and velocity fields by factors up to 10,000 without compromising precision [5,26]. This dramatic acceleration is critical for enabling real-time control, rapid design optimization, and large-scale parametric studies that are infeasible with traditional solvers, thereby expanding the Pareto frontier of efficient simulation [39]. In the broader context of machine-learning-accelerated CFD, end-to-end deep learning has been shown to improve approximations for modeling two-dimensional turbulent flows, achieving an accuracy comparable to that of baseline solvers with a resolution 8 to 10 times finer in each spatial dimension, leading to 40- to 80-fold computational speedups [39]. The approach, which involves learned interpolation and correction within standard numerical methods, remains stable and generalizes well to different forcing functions and Reynolds numbers, unlike black-box ML methods [39,43,44]. In another application, ML-enhanced RANS simulations achieved a speedup of over 50 times compared to LES for predicting the aerodynamic pressures on bluff bodies [4]. This notable improvement in efficiency is essential for applications that demand quick analysis, including design optimization and real-time control [45]. Examples include flow-informed aerodynamic control for unmanned aerial vehicles (UAVs) in turbulent conditions or active gust mitigation systems for wind turbines, where instantaneous predictions enable critical safety and performance adjustments [46]. Neural operators, particularly Fourier Neural Operators (FNOs), offer a powerful generalization of neural networks to learn mappings between infinite-dimensional function spaces. These operators are discretization-invariant, allowing them to be applied across different grid resolutions without retraining, and can achieve inference times orders of magnitude faster than conventional PDE solvers for problems like the Navier–Stokes equation [47]. ML methods can initialize CFD solutions or approximate solution operators, further contributing to computational acceleration [48]. Tools like TensorFlowFoam enable the direct deployment of trained neural networks within OpenFOAM, facilitating end-to-end acceleration of RANS simulations [45]. Recent empirical studies demonstrate that the application of machine learning methods can substantially accelerate CFD simulations without compromising accuracy and detail [49–52]. This includes applications like physics-informed graph neural networks (PIGNNs) for fluid and heat equations, which leverage deep neural networks to accelerate the solutions [53]. The ability of these models to learn complex, nonlinear mappings directly from data or governing equations fundamentally bypasses the iterative and computationally intensive nature of traditional numerical discretization schemes.

2.2. Advanced Turbulence Modeling

Turbulence modeling remains a critical challenge in CFD, particularly for RANS simulations, which often yield inaccurate results in complex flow scenarios such as separation zones due to limitations in their underlying turbulence closures. These inaccuracies can lead to significant errors in predicting forces, heat transfer, and mixing, which are crucial for reliable engineering design in aerospace, civil, and mechanical applications. These inaccuracies stem from the fundamental inability of conventional RANS models to accurately predict eddy viscosity [4].

ML has become a valuable asset in improving the precision of these closure models. Researchers are developing deep learning methodologies to adjust eddy viscosity within RANS equations, leading to improved predictions for bluff body aerodynamics. Unlike traditional turbulence models, which rely on empirical assumptions or simplified physics, ML offers a data-driven approach to discovering and refining complex constitutive relationships directly from high-fidelity simulation or experimental data [17,54]. A fundamental limitation of conventional RANS turbulence models, such as the k-ϵ [55] and k-ω SST [56] formulations, lies in their inadequate prediction of eddy viscosity. This shortcoming results in notable inaccuracies in the velocity and pressure fields, particularly within separation bubbles and shear layers [4]. For instance, the k-ϵ model typically generates elevated negative pressure regions near the windward side of a surface due to a small, confined separation bubble while exhibiting considerably lower negative pressures across the remainder of the roof, attributed to premature flow reattachment. Conversely, the k-ω SST model tends to predict more extensive flow separation on the roof, which produces a comparatively uniform negative pressure distribution.

ML models developed using high-fidelity LES datasets offer improved accuracy and a more balanced representation of eddy viscosity. For instance, one deep learning framework utilized a neural network architecture comprising four hidden layers with 32 neurons each, activated by tanh functions, and trained via the Adam optimization algorithm [57] at a learning rate of 0.001. The training dataset included LES results for a variety of bluff body geometries, such as forward- and backward-facing steps with width-to-height ratios between 0.2 and 6, as well as a flat-roofed structure known as the Silsoe cube. This machine learning approach achieves a refined balance in eddy viscosity, effectively delaying flow reattachment and yielding more precise predictions of pressure and velocity fields when compared to conventional closure models. Additionally, sensitivity analyses have underscored the critical influence of eddy viscosity on flow separation, reattachment dynamics, and pressure distribution around bluff bodies. Collectively, these methods represent a significant progression in turbulence modeling for complex flow scenarios [23,58–61]. This paradigm shift enhances the predictive capabilities of RANS, making it a more dependable tool for designing structures in environments dominated by complex turbulent flows while still maintaining computational tractability [4].

2.3. *Data-Driven and Physics-Informed Surrogates*

ML facilitates the development of data-driven surrogate models capable of accurately representing complex fluid dynamics or uncovering simplified representations of physical processes. Supervised learning, a primary ML category, is used to map labeled datasets to known outputs, typically dealing with regression and classification tasks [12]. A neural network architecture consists of input, hidden, and output layers. The input nodes receive and preprocess data, which then propagates through the hidden layers until an output is generated. Each neuron is characterized by associated weights and thresholds and utilizes an activation function to transform the weighted sum of inputs. CNNs have demonstrated a strong performance in forecasting pressure and velocity fields derived from CFD data [26].

In addition to purely data-centric methods, Physics-Informed Neural Networks (PINNs) are emerging as a powerful approach. By embedding physical laws and constraints directly into the neural network framework, PINNs merge machine learning techniques with established principles of physics. This integration not only improves predictive accuracy but also ensures that outcomes adhere to physical realism, which is essential for dependable results in engineering contexts [18]. For example, enforcing mass, momentum, and energy conservation laws during training prevents unphysical solutions and improves generalization, particularly when data is sparse or noisy [14,15]. PINNs achieve this by encoding governing differential equations to provide a hybrid physics-based and data-driven deep learning framework [14,37]. Research has examined how the availability of labeled data influences the effectiveness of parameterized PINNs, particularly in surrogate modeling and design optimization tasks such as drag reduction [18]. Furthermore, this approach has been utilized for data assimilation and parameter estimation related to water wave dynamics [28], as well as for comparative analyses of various RANS models within turbulence modeling using PINNs [23]. For instance, PINNs have been shown to significantly improve the prediction accuracy for indoor turbulent airflow when assimilating high-fidelity experimental data and embedding Reynolds-averaged Navier–Stokes (RANS) equations with turbulence models like the k-ϵ model [37]. PINNs also show promising potential in parametric engineering turbulence optimization problems, where specialized PT-PINN frameworks enhance accuracy and robustness through soft constraints and pre-training methods. These PT-PINNs can produce predictions closely matching experimental and traditional CFD results, offering a significant computational efficiency advantage with total model construction times 16 times faster and inference times 0.5% of a single CFD computation [38]. Broadly, data-driven methods in fluid mechanics have been investigated for applications including urban heat island mitigation [1], the prediction of nanofluid rheology [31], and fluid classification in non-contact dispensing processes [62]. Additionally, the fusion of artificial intelligence and machine learning holds significant promise for advancing solar energy deployment in smart city infrastructures [2] and residential developments in metropolitan environments [63]. These diverse applications highlight the versatility of ML in addressing complex, multi-physics problems across various scales, driving innovation toward more sustainable and technologically advanced solutions [32].

2.4. ML-Assisted Numerical Solvers and Bridging Scales

ML can directly assist numerical solvers by, for instance, providing improved initial conditions or approximating complex solution operators [48]. The ability of ML to accurately predict eddy viscosity, as discussed, is a form of ML-assisted numerical solving, directly improving RANS computations [4]. This also aids in bridging the gap between different computational scales. This capability is instrumental in constructing efficient hierarchical simulation frameworks that combine the strengths of different modeling approaches while overcoming their individual limitations [17,25]. While DNS is the most precise CFD method, its computational cost makes it unfeasible for high-Reynolds-number flows [6]. LES is less expensive than DNS but still requires significant resources [40]. A primary source of inaccuracies in LES stems from its reliance on sub-grid-scale (SGS) modeling. To address this, machine learning has emerged as a powerful tool to enhance existing turbulence modeling approaches in the fluid community [17,25,64]. Efforts have been devoted to developing SGS and wall models using ML, often relying on supervised learning to predict perfect SGS terms from filtered Direct Numerical Simulation (DNS) data [65–73]. The Building-block Flow Model (BFM), for instance, introduces a closure model for wall-modeled large-eddy simulation (WMLES) that leverages three feedforward Artificial Neural Networks (ANNs) to predict wall-shear stress and the SGS stress tensor. The BFM is designed to predict multiple flow regimes, unify closure models at solid boundaries, ensure consistency with numerical schemes, and be applicable to arbitrary complex geometries, demonstrating superior capabilities for external aerodynamic applications, including aircraft in landing configuration [74]. Additionally, PINNs augmented with turbulence models have proven effective for mean-flow reconstruction, achieving lower reconstruction error across various data resolutions by avoiding the discretization errors inherent in variational methodologies [35].

ML offers a pathway to achieve higher-fidelity results at lower computational costs by enhancing RANS models or by developing surrogate models for sub-grid-scale physics, thereby enabling more realistic solutions for complex turbulent flows [22,48,60,75]. This is particularly relevant given the computational efficiency challenges in general CFD modeling [76]. ML methods are also being used for super-resolution reconstruction of turbulent flows [24,49]. Integrating RANS models within sub-grid-scale frameworks enhances the precision of LES by refining the prediction of the resolved scales.

2.5. Inverse Design and Control

ML's predictive capabilities extend to inverse design and control problems in fluid mechanics. Instead of predicting the flow behavior from a given design, inverse design aims to find the optimal geometry or parameters to achieve the desired flow characteristics [77]. A key application is aerodynamics-guided machine learning for design optimization, especially for electric vehicles. This involves analyzing datasets of industry-quality automobile geometries and their aerodynamic performance obtained from high-fidelity Large Eddy Simulation (LES). Nonlinear autoencoders are used to extract low-dimensional relationships between geometries and aerodynamics, allowing for design optimization by estimating drag coefficients from latent variables. This approach has demonstrated agreement between optimized geometries and validation simulations, showcasing its potential for accelerating industrial design in a production environment [78]. AI-driven optimization has been applied to breakwater design to predict the optimal structural dimensions and wave reflection [19]. In control systems, machine learning models enable the forecasting of system dynamics, facilitating the development of advanced control strategies. Examples include AI-based surrogate models for room ventilation [20] and neural network approaches for estimating engine mass flow under semi-transient conditions [21]. These applications leverage ML to iteratively refine designs or control strategies based on predicted fluid dynamic responses. Other applications include predicting dynamic pressure coefficients in vertical water jets [79] and the electrical conductivity of ionic liquids [80]. The broader field also sees ML being used for wake modeling of wind turbines [81].

2.6. Emerging Challenges

Despite the remarkable progress, several challenges remain in the widespread adoption and further development of ML in CFD. A key challenge is generalization and extrapolation, ensuring that ML models perform accurately beyond the specific range of training data, especially for unseen geometries or flow conditions. This entails creating resilient normalization techniques and frameworks that preserve predictive accuracy, even when Reynolds numbers exceed the scope of the training dataset [75,82].

Another area of ongoing research involves multi-scale representation and physical knowledge encoding [23,24]. Effectively incorporating fundamental physical laws into ML models (as in PINNs) is critical for ensuring interpretability, robustness, and physical consistency. Additionally, the need for improved turbulence closures in both RANS simulations and LES persists, with ML playing a crucial role in developing more accurate eddy viscosity models [55,56]. Rather than replacing fundamental physics, ML-assisted approaches augment traditional models by learning residual errors or developing adaptive components that capture unresolved physics with greater fidelity and robustness [17,54]. Conventional RANS models exhibit limitations in accurately estimating eddy viscosity when benchmarked against LES reference data, resulting in notable deviations in velocity and pressure distributions [4]. Subsequent research will focus on formulating a dimensionless eddy viscosity parameter to capture the effects of inflow velocity variations, especially in regimes characterized by high Reynolds numbers.

The rising occurrence and severity of major wind-related disasters in tropical and subtropical areas, potentially influenced by climate change [83–85], present considerable threats to infrastructure, particularly renewable energy installations [5,86]. Addressing and mitigating the aerodynamic challenges posed by these wind events—especially those related to bluff body flow separation phenomena [87–89]—demand ongoing investigation [3,8,90–92]. Specific concerns such as the effects of wind on photovoltaic panels, including scale-dependent impacts [7,93,94] and the influence of panel height and spacing [95,96], as well as dynamic behaviors like torsional galloping [97–100], underscore the complexity of the problem. Furthermore, existing building codes have notable gaps; for instance, standards like ASCE 7-10/16 do not incorporate requirements tailored to solar tracking systems, ASCE 7-22 continues to overlook such systems, and the classification of rigid structures in ASCE 7 is inadequate for SATs, highlighting the necessity for further refinement and codification improvements [5,101]. Achieving reliable aerodynamic data from small-scale testing—which underpins many parametric design models—remains challenging [102,103]. To overcome these limitations, developments in large-scale open-jet wind tunnel testing facilities are underway [104,105].

3. Application Example 1: Wind-Resilient Design Strategies for Solar Panel Installations in Sustainable Communities

A key application of combining ML with CFD lies in the wind-resistant design of solar panels, which plays a vital role in fostering sustainable and resilient communities. Large-scale photovoltaic (PV) installations are increasingly recognized as essential clean and renewable energy solutions. However, their structural resilience is challenged by the heightened risk of windstorm damage, a predominant factor behind most weather-related losses and insurance claims at solar facilities. For instance, Hurricane Maria in 2017 destroyed a solar farm in Puerto Rico, with Phase 2 of the project suffering 75% destruction of its PV panels, in part because panels failed at a low angle (10.8°), an observation consistent with findings on resultant wind loads at low inclination angles. Hurricane Ida also caused significant damage to a solar farm in Louisiana, further highlighting the need for improved design protocols.

Accurately quantifying wind loads is essential. Traditional methods, even experimentally validated CFD, can yield results that differ significantly from standard practice. The ASCE 7-10 and ASCE 7-16 standards lacked solar-specific wind pressure calculations for ground-mounted single-axis trackers (SATs) and were often considered insufficient, frequently underestimating loads. Although ASCE 7-22 includes provisions for fixed-tilt ground-mounted solar panels, it still neglects tracking systems. Furthermore, the ASCE 7 standard's definition of a rigid structure, based on a 1 Hz threshold for dynamic insensitivity, is unsuitable for SATs, as their vortex shedding frequencies can significantly exceed this threshold. SATs are particularly vulnerable to dynamic mechanisms like torsional galloping, resonant vibration, and torsional flutter, especially when stowed at 0° tilt.

To address these challenges, a two-phase investigation was undertaken. The initial phase involved RANS simulations, rigorously validated through experimental data, to analyze wind flow behavior over solar panels and establish design-specific wind load parameters. In the subsequent phase, the insights derived from the RANS simulations were utilized to train a CNN. This model enables rapid and precise estimation of the pressure and velocity fields on solar panels across varied aspect ratios and angles of attack.

3.1. CFD-Based Assessment of Wind Loads (Study 1)

In the initial investigation, wind loads on an inclined solar panel were determined using a rigorously validated and optimized RANS atmospheric model, simulating an unsteady airflow in both two-dimensional and three-dimensional setups [5]. The 3D computational domain, established as the validation benchmark, featured dimensions proportional to the panel's mounting height (H): an upstream inflow length of 6.8H, a downstream outflow length of 13.6H, a vertical clearance above the panel of 7.5H, and lateral boundaries set at 3.3W, where W denotes the panel width [5]. Validation was conducted against a laboratory experiment employing a full-scale solar panel with dimensions of length L = 1.334 m, width W = 9.144 m, thickness t = 0.04 m, and mounting height H = 1.039 m. The mesh quality criteria were met by maintaining a non-dimensional wall distance (y^+) below one, with the finalized 3D mesh comprising 3.6 million cells. The k-ω SST turbulence model was chosen due to its superior correlation with empirical data relative to alternative RANS options. This rigorous 3D setup with the k-ω SST model provided a high-fidelity baseline for validating the CFD methodology against experimental observations. This established a benchmark for accuracy against which the efficiency and fidelity of ML-accelerated methods could be rigorously assessed. For extended parametric studies, a 2D model retained analogous geometric parameters, applied a single cell thickness of 200 mm in the z-direction, and consisted of approximately 1.6×10^4 cells with $y^+ < 1$.

The validation process involved comparing the mean pressure coefficients on the solar panel's centerline from the 3D model with experimental data. As shown in Figure 1, the CFD findings aligned well with the experimental results.

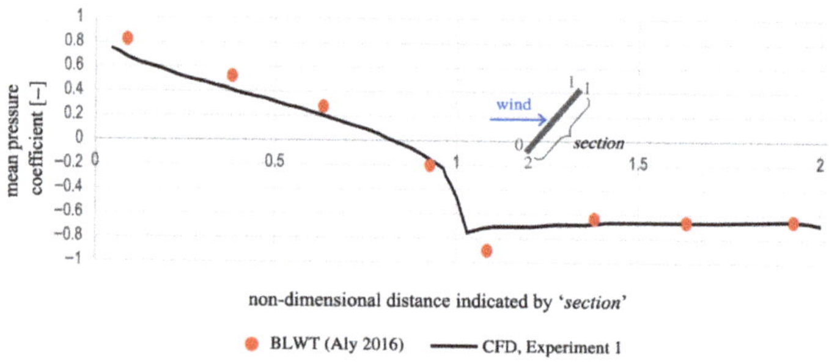

Figure 1. A comparison of the mean pressure coefficient distributions along the centerline of the solar panel's top and bottom surfaces at an angle of incidence $\theta = 40°$, contrasting boundary-layer wind tunnel benchmark measurements with the results obtained from the 3D model in Study 1 [93].

To identify the tilt angles corresponding to the maximum aerodynamic loads on solar panels, the aerodynamic drag coefficient (C_D), lift coefficient (C_L), and torsional moment coefficient (C_M) were computed using the subsequent expressions:

$$C_D = \frac{F_D}{0.5\rho U_S^2 A} \quad (1)$$

$$C_L = \frac{F_L}{0.5\rho U_S^2 A} \quad (2)$$

$$C_M = \frac{M_z}{0.5\rho U_S^2 AL} \quad (3)$$

where F_D denotes the drag force, F_L represents the lift force, M_z signifies the torsional moment, ρ corresponds to the air density, U_S is the wind velocity averaged across the surface of the solar panel, A indicates the solar panel's area, and L refers to its length.

Additionally, the Strouhal number (St), describing the frequency of vortex shedding, was calculated as

$$St = \frac{f_w d}{U} \quad (4)$$

where f_w denotes the vortex shedding frequency, d represents the projected width perpendicular to the flow direction, and U corresponds to the incoming velocity.

3.2. ML-Based Prediction of Wind Loads (Study 2)

The machine learning component of this research focused on developing a model capable of rapidly and precisely predicting the pressure and velocity fields on solar panels with novel geometries or orientations. Referred to as Study 2, the investigation was conducted in three distinct stages: generation of the dataset, model training, and evaluation of its predictive performance.

Figure 2 provides a comprehensive overview of Study 2, detailing the ML model and its process through three distinct parts. Part (a), "Overall Process", outlines the high level workflow, encompassing data generation from RANS-based simulations in OpenFOAM 5.0 (with an 80:20 training-to-testing ratio), the training phase using a modified U-Net architecture with the Adam optimizer in PyTorch, and subsequent performance analysis to evaluate accuracy and time. Part (b), "ML Model Overview", visually represents the conceptual flow of information within the ML model, showing the input (freestream conditions in the x and y directions and a mask of solar panel geometry), the neural network itself (specifically a U-Net architecture), and the output, which consists of predictions for pressure, velocity in the x and y directions, and their pre-computed target values. Finally, part (c), "Modified U-Net Architecture", elaborates on the specific details of the U-Net architecture, adapted from Deep Flow Prediction [26], describing it as an encoder–decoder architecture where the encoding portion progressively downsamples the image size through convolutional blocks to extract features, and the decoding portion uses seven symmetric layers to reconstruct the target functions, with ReLU and Leaky ReLU activation functions used in the encoding and decoding parts, respectively.

The RANS-based simulation dataset was generated using OpenFOAM 5.0 on Ubuntu 18.04 to maintain compatibility with PyTorch 1.10.2. Mesh creation was performed with Gmsh, producing a body-fitted triangular mesh featuring increased refinement near the solar panel. The solar panel geometry, consistent with Study 1 and scaled 1:1 (1.334 m length, 0.04 m thickness), was used. Unlike Study 1, where the k-ω SST model was used for high-fidelity validation, the Spalart–Allmaras turbulence model was chosen for its simplicity, lower memory requirements, and adequate accuracy under freestream flow conditions specifically for the purpose of generating a large and diverse dataset for machine learning training with acceptable computational cost. Computations ran on a Lenovo Yoga laptop equipped with an 8-core CPU, 16 GB of RAM, and an Intel 11th Gen i7 Evo processor [5].

The machine learning architecture implemented a tailored U-Net convolutional neural network designed for efficient and accurate image segmentation with minimal memory consumption. The encoder progressively compresses input images through convolutional blocks to a bottleneck of 512 features, while the decoder reconstructs the target outputs dimensionally through seven symmetrical layers. Input features included the freestream velocities in both the x and y directions, alongside a binary mask representing the solar panel's geometry. The network outputs predicted fields for the pressure and velocity components in the x and y directions, with target data pre-encoded to include Reynolds number information via scaled freestream velocity vectors.

The activation functions predominantly employed were the Rectified Linear Unit (ReLU) and Leaky ReLU. Leaky ReLU activations were used within encoding layers, whereas ReLU activations were applied in decoding layers (refer to Figure 3).

The training process employed the Adam optimizer implemented through PyTorch. The baseline configuration consisted of 100 training and testing samples, panel thickness ratios ranging from 1.5:1 to 4:1, an angle of attack (AoA) spanning ±22.5°, 10,000 iterations, and a learning rate of 0.0006 with decay applied. A parametric study was conducted by varying the sample size, learning rate, decay parameters, and AoA range to assess their impact on the model's predictive accuracy.

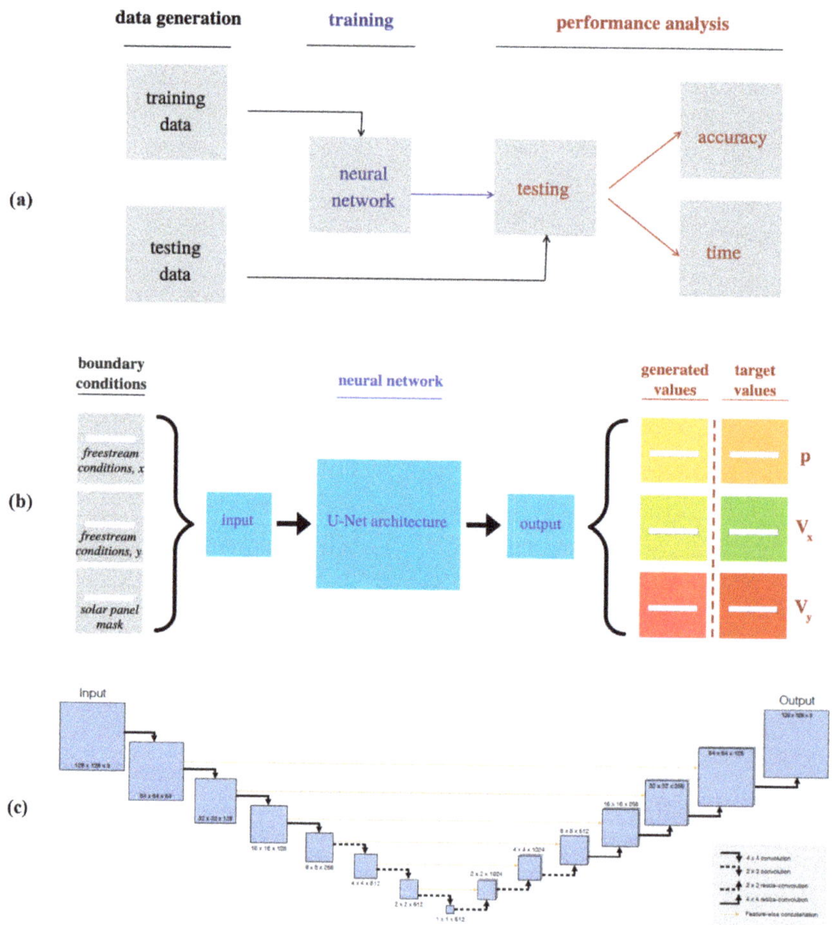

Figure 2. Overview of the Study 2 workflow, the machine learning model, and the adapted U-Net architecture: (**a**) overall process; (**b**) ML model overview; and (**c**) modified U-Net architecture called in the ML model.

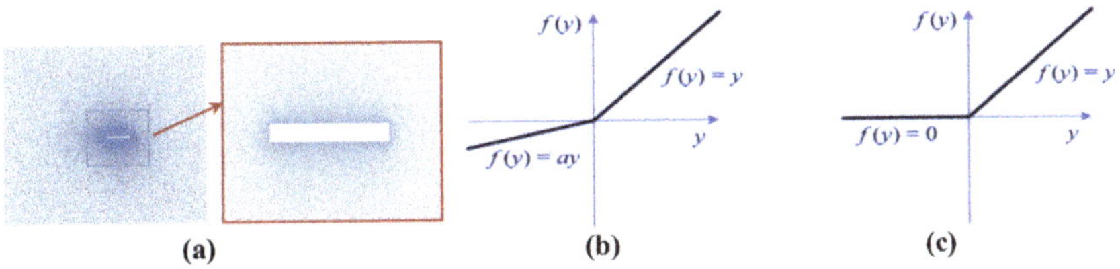

Figure 3. The solar panel simulation mesh indicating the CNN inference region (**a**), and illustrations of the Leaky ReLU (**b**) and ReLU (**c**) activation functions.

3.3. *Results and Key Findings*

3.3.1. Flow Characteristics and Aerodynamic Wind Loading

At a low wind velocity of 10 m/s, the airflow over the solar panel demonstrated laminar characteristics, marked by a consistent linear progression of separation bubble development across tilt angles ranging from −75° to +75° (refer to Figure 4a,b). Stagnation points were observed on the surfaces facing the wind, with the flow accelerating smoothly from the leading edge toward the trailing edge. Corresponding pressure profiles exhibited positive peaks near the windward leading edges and negative peaks near the leeward leading edges, as illustrated in Figure 4c,d.

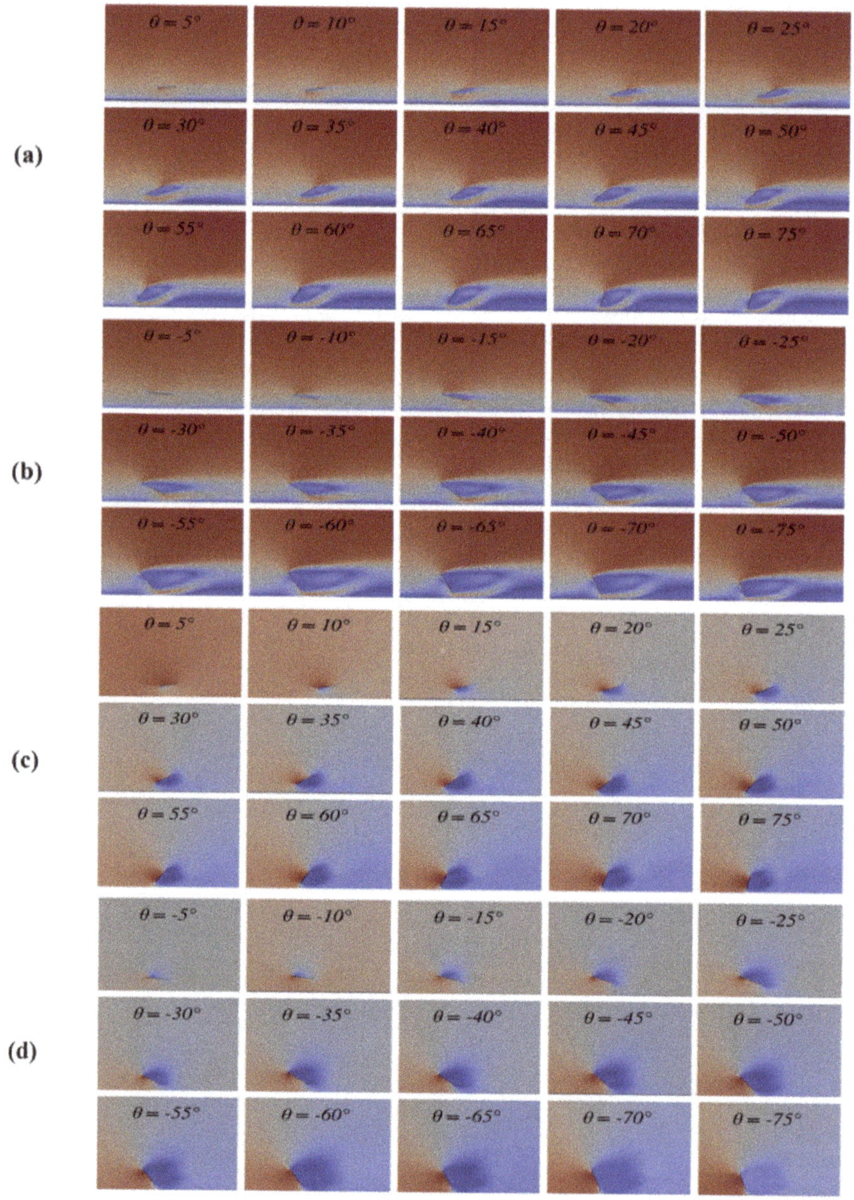

Figure 4. Velocity contours for positive (**a**) and negative (**b**) tilt angles ±5° to ±75°; pressure contours for positive (**c**) and negative (**d**) tilt angles.

At low velocity, the aerodynamic coefficients behaved distinctively (Figure 5):

- Drag coefficients peaked at steep tilt angles, approaching zero near 0°;
- Lift coefficients showed nearly parabolic behavior with the maximum lift at −45° and the maximum (negative) lift (downforce) at 40°;
- Torsional moment coefficients followed similar trends with extremes near ±40°.

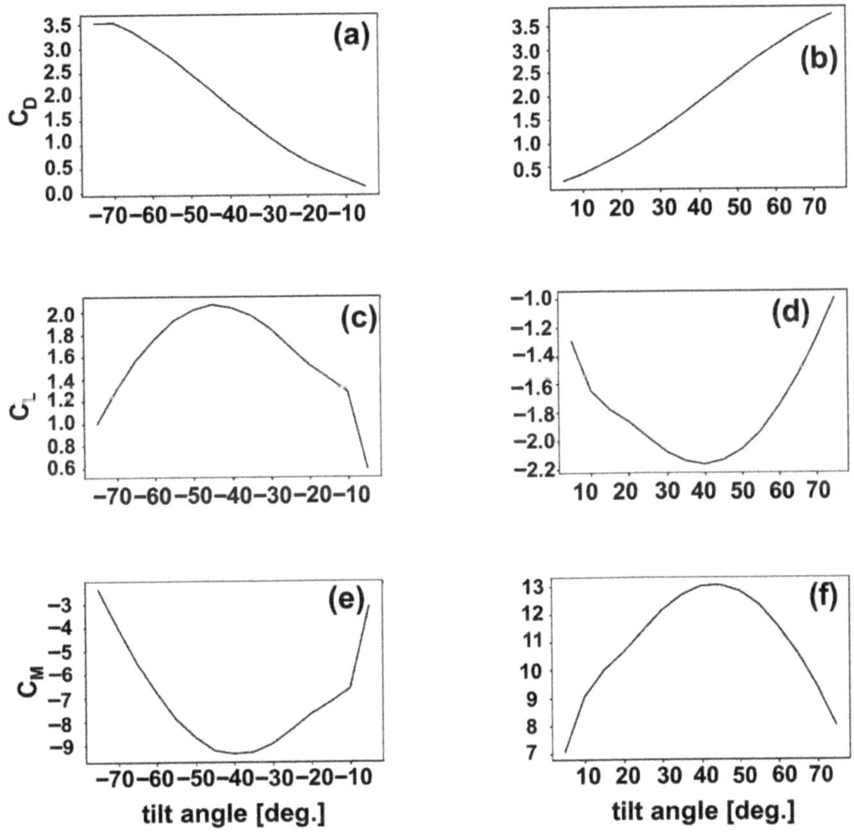

Figure 5. Drag (**a**,**b**), lift (**c**,**d**), and moment (**e**,**f**) coefficients across tilt angles ranging from −75° to +75°, sampled at 5° increments.

At the design wind speed (50.52 m/s), the aerodynamic loads showed transient variability. Tilt angles common for SATs (−45° to 45°) exhibited high leading-edge vorticity consistent with torsional galloping mechanisms.

The maximum absolute aerodynamic coefficients by tilt angle (Figure 6) identified:

- Drag peaks at steep angles;
- Lift peaks including lower tilt angles and 0°;
- Resultant drag–lift minima at −15° and 45°;
- Minimum torsional moments at −15° and 30°.

The vortex shedding details from transient simulation at −45° (Figure 7) reveal periodic vortex formation and shedding correlated with force fluctuations. Steeper tilts correspond to higher shedding frequencies and Strouhal numbers (Figure 8a).

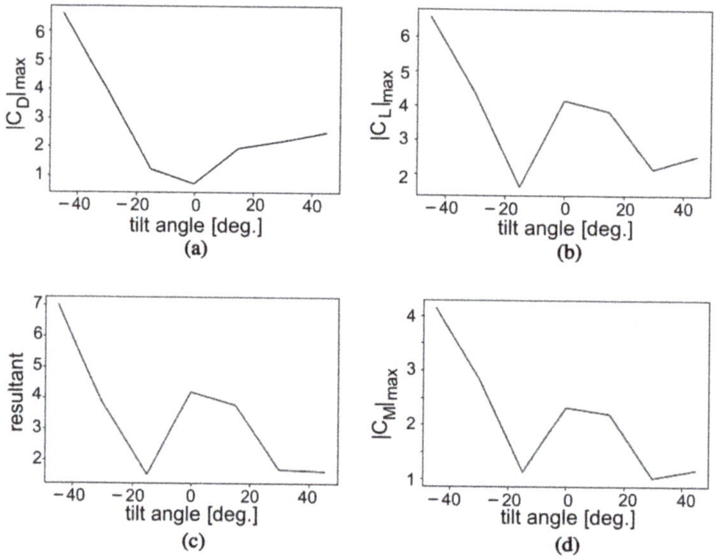

Figure 6. Maximum absolute drag (**a**), lift (**b**), resultant drag–lift (**c**), and hinge moment (**d**) coefficients over transient simulations at various tilt angles.

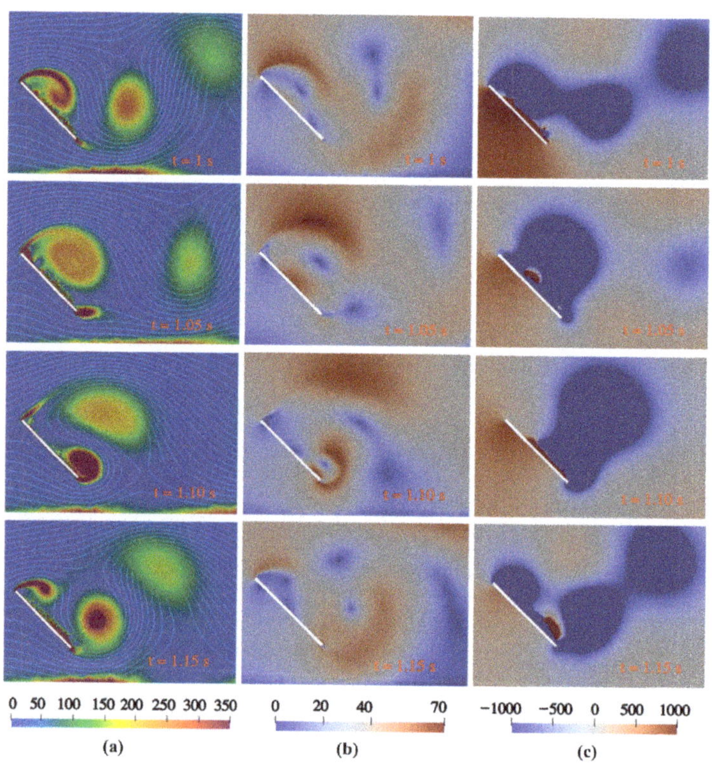

Figure 7. The vortex shedding period at an angle of −45°. (**a**) Streamlines overlaid onto vorticity contours; (**b**) instantaneous flow velocity distribution; and (**c**) snapshot of flow pressure fields.

(a)

(b)

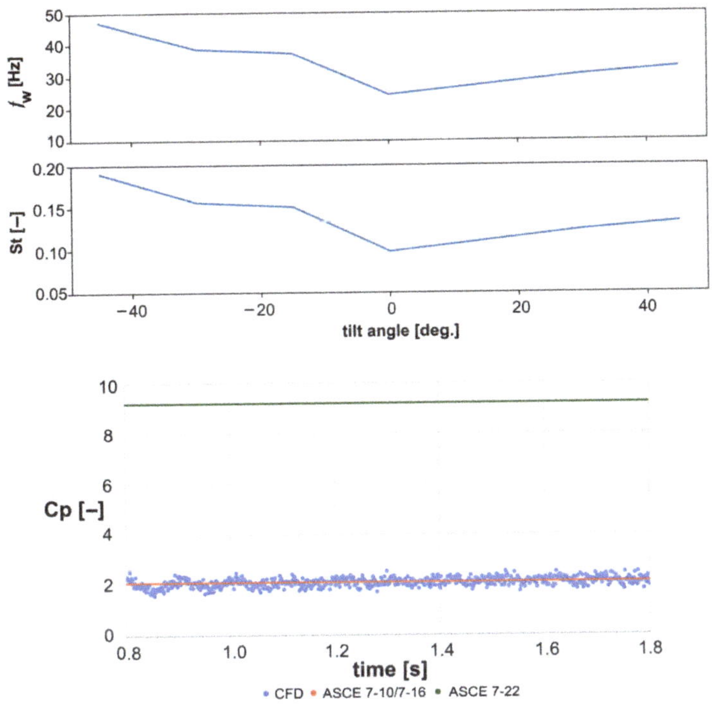

Figure 8. (**a**) Vortex shedding frequency and Strouhal number vs. tilt angle; (**b**) a comparison of pressure coefficients from ASCE standards and transient CFD simulations.

3.3.2. Comparison with ASCE 7 Standards

The peak pressure coefficients from transient CFD differed notably from the ASCE 7 estimates. The CFD maximum peak was 2.53 (mean: 1.30). ASCE 7-10 and 7-16 underestimated the loads 47% of the time; ASCE 7-22 overestimated with a coefficient of 9.2 vs. a CFD maximum of 2.53. These discrepancies support that the Louisiana solar farm failure during Hurricane Ida was likely linked to neglected dynamic wind loading in prior versions of ASCE 7.

3.3.3. ML Model Performance

ML-CFD integration yielded drastic speedups: the traditional CFD runtime averaged 45.55 s versus ML's 0.0048 s—a ≈10,000× improvement without losing accuracy, critical for expedited design.

ML predictions matched the targets well at low angles of attack (error <3%), but the errors increased past 20% for higher angles (see Figure 9). This observation is particularly significant as highly separated flows, often encountered at steep angles of attack, represent a major challenge for traditional CFD and remain a critical area for ML enhancement. The increased error at these angles highlights the importance of incorporating diverse and challenging flow regimes into training datasets and developing more robust ML architectures capable of generalizing to such complex phenomena. Future work will focus on improving the model performance in these challenging, highly separated flow conditions, potentially by integrating more physics-informed constraints or advanced data augmentation strategies.

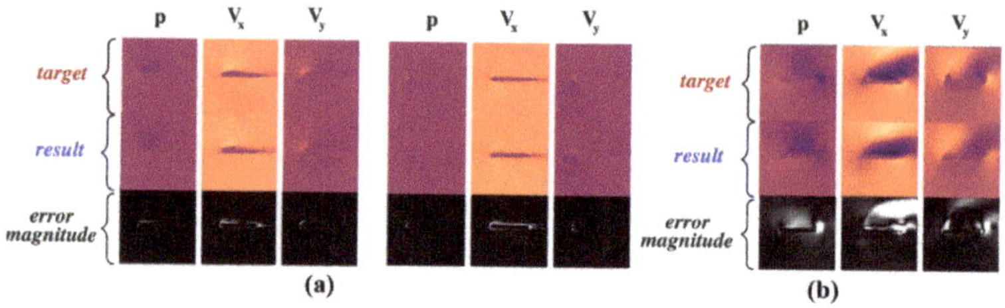

Figure 9. Base case examples: (**a**) small-magnitude error; (**b**) significant-magnitude error in pressure and velocity predictions.

Parameter studies showed:

- A learning rate that was too high (0.6) prevented convergence; the optimal was found at 0.0006 (Figure 10a).
- Learning rate decay reduced the validation loss variance (Figure 10b).
- Sample size variation (100–500) minimally impacted accuracy, except a spike at 400 samples without decay.
- Expanding the AoA range to ±75° increased the combined error to 27.51% vs. 8.5% for base (±22.5°), expected given the fixed sample size.

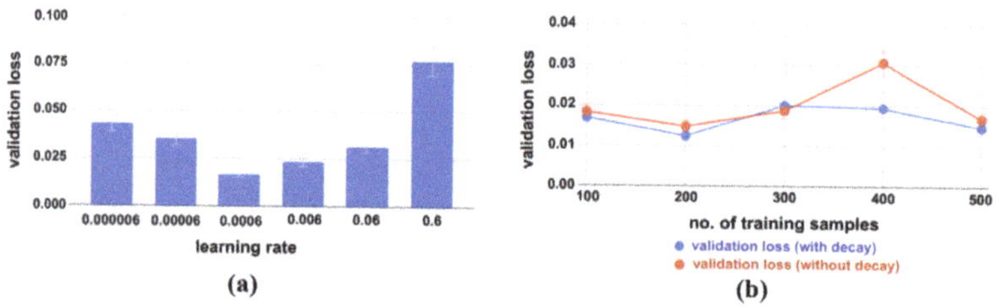

Figure 10. Validation loss: (**a**) learning rate variation; (**b**) training sample size variation with/without learning rate decay.

3.4. *Conclusion for Solar Panel Wind Design*

Experimentally validated CFD can produce results differing from the standards: older ASCE versions underestimate loads, while the latest version (ASCE 7-22) overestimates them. Importantly, stowing solar panels at −15° during wind events minimizes the resultant drag, lift, and torsional moment coefficients, making it an optimal, resilient position. This contrasts with the problematic 0° stow prone to torsional galloping. This example demonstrates ML's potential to improve the efficiency and safety of clean energy infrastructure, contributing significantly to climate adaptation by providing up to 10,000× faster time-to-solution at RANS-level accuracy. Ultimately, such ML-accelerated design methodologies pave the way for more widespread and reliable deployment of solar energy, a cornerstone of sustainable development [5].

4. Application Example 2: Deep-Learning-Based Eddy Viscosity Modeling for Bluff Bodies

An additional notable application involves improving the precision of RANS simulations in forecasting wind pressures on bluff bodies, especially within intricate flow separation regions, through the use of deep learning techniques for eddy viscosity modeling [4]. Accurate prediction of wind pressures on buildings is crucial for designing safe and efficient structures. Traditional computational methods, such as RANS simulations, often face challenges in accurately predicting complex flow behaviors in separation zones on bluff bodies, like flat roofs. In contrast, high-fidelity Large Eddy Simulation (LES) offer greater accuracy but require a substantially increased computational time and resources, owing to the necessity of resolving fine spatiotemporal scales. This section presents a novel deep learning approach designed to improve both the accuracy and efficiency of eddy viscosity modeling in RANS turbulence closures, with a focus on enhancing the predictive capabilities for bluff body aerodynamics.

4.1. Computational Fluid Dynamics Background

The study of fluid flow, particularly in turbulent regimes, is predominantly conducted using the Navier–Stokes equations. RANS simulations, a prevalent CFD technique in wind engineering, solve the time-averaged form of the Navier–Stokes (NS) equations. The RANS equations, combined with the continuity equation, are represented as follows:

$$\frac{\partial \rho \mathbf{U}}{\partial t} + \nabla \cdot (\rho \mathbf{U}\mathbf{U}^T) = -\nabla p + \nabla \cdot (2\mu_m \mathbf{S} + \rho \mathbf{R}) + \rho \mathbf{f} \tag{5}$$

where $\mathbf{U}$ represents the fluid velocity vector, ρ is the fluid density, $\mathbf{S}$ denotes the tensor corresponding to the mean strain rate, μ_m is the dynamic viscosity, p indicates the pressure, $\mathbf{R}$ is the tensor comprising the Reynolds stresses, and $\mathbf{f}$ signifies the externally applied body force. The Reynolds stresses arise from the advection term in the momentum conservation equation and characterize the averaged stresses due to velocity fluctuations. Typically, these stresses dominate over viscous stresses, except in the vicinity of solid boundaries where the fluctuation intensity diminishes.

To solve the RANS equations, it is essential to employ a turbulence closure model to account for the unknown Reynolds stress terms. According to the Boussinesq hypothesis, the effect of turbulent eddies on the momentum transfer is modeled through an eddy viscosity concept. Within the RANS closure framework, the parameter $\mathbf{R}$ is defined as

$$\mathbf{R} = -\overline{u'u'}^T = \frac{2\mu_t}{\rho}\mathbf{S} - \frac{2}{3}k\delta_{ij} \tag{6}$$

where k denotes the turbulent kinetic energy, μ_t represents the eddy viscosity, u' indicates the fluctuating component of velocity, and δ_{ij} is the Kronecker delta.

Two-equation turbulence models are widely used as closure approaches in RANS simulations. These models generally involve transport equations for the turbulent kinetic energy and a characteristic turbulence length scale. Among the most prevalent closures are the k-ϵ and the k-ω SST models. For instance, within the k-ϵ framework, the eddy viscosity, μ_t, is defined by

$$\mu_t = C_\mu \rho \frac{k^2}{\epsilon} \tag{7}$$

where ϵ represents the dissipation rate, and C_μ is a model constant. Nevertheless, conventional RANS turbulence closures frequently produce imprecise outcomes, especially in intricate flow regions such as separation zones. This shortfall arises from the inherent assumptions within these models and their limited capacity to accurately estimate eddy viscosity. Consequently, the discrepancy in the eddy viscosity predictions between RANS and LES approaches emphasizes the necessity for more advanced turbulence modeling techniques.

4.2. Deep Learning Model Description

This research introduces an innovative deep learning approach aimed at improving the accuracy and effectiveness of eddy viscosity modeling in RANS turbulence closures.

The employed deep neural network architecture is depicted schematically in Figure 11. This feedforward neural network consists of an input layer that receives the data, one or more hidden layers that apply transformations, and an output layer that produces the final predictions. Each layer consists of interconnected neurons, with weighted connections between them. These weights are iteratively updated during training to enhance the model's performance. Information flows in a single direction, from the input layer to the output layer, without any recurrent loops or feedback pathways. This architecture is particularly well suited to learning complex, nonlinear mappings between flow features and eddy viscosity, leveraging its ability to extract hierarchical representations from input data [57].

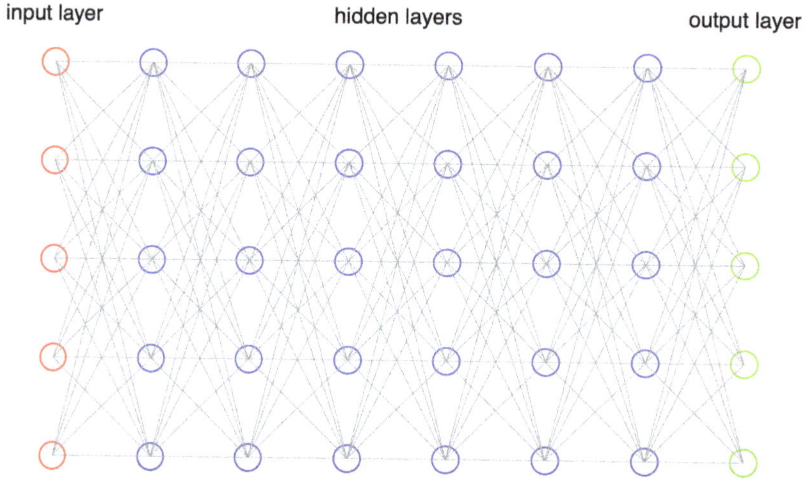

Figure 11. A diagram of a standard feedforward neural network architecture. The model comprises an input layer that accepts the input data, multiple hidden layers that perform nonlinear transformations, and an output layer that generates the final predictions. Each layer contains neurons interconnected by weighted links, which are updated throughout training to enhance the network's accuracy. Information flows in a single direction from the input layer through the hidden layers to the output layer, without any cycles or feedback loops.

In this implementation, the neural network architecture comprises four hidden layers, each containing 32 neurons. Every neuron is characterized by an associated weight and bias. The network computes a weighted sum of the inputs at each node, which is then processed by an activation function. This activation function governs whether the node's output is transmitted to the subsequent layer. By incorporating nonlinear activation functions, the network gains the capacity to capture complex patterns within the data and approximate sophisticated functions. For a given layer l, the output of the i-th neuron, denoted by $a_{i,l}$, is calculated as follows:

$$a_{i,l} = g\left(\sum_{j=0}^{n_{l-1}} w_{ij,l-1}\, a_{j,l-1}\right), \tag{8}$$

where n_{l-1} denotes the total number of nodes in the $(l-1)$-th layer, and $w_{ij,l-1}$ represents the corresponding weight. The hyperbolic tangent (tanh) activation function was utilized for both the input and hidden layers during training, as it yielded a superior performance relative to that of alternative activation functions such as ReLU and Leaky ReLU.

4.3. Data Generation and Training Approach

High-fidelity LES data were generated for a variety of bluff body configurations to facilitate the training of the deep learning model. The selected geometries encompassed a flat-roof structure—namely, the Silsoe cube, a well-established benchmark in wind pressure studies, measuring 6 m × 6 m × 6 m—and forward- and backward-facing steps with multiple width-to-height (w/h) ratios.

LES analyses were performed across a range of w/h ratios, specifically 0.2, 0.5, 0.75, 1, 1.25, 1.75, 2, 3, 4, and 6. For each case, high-quality prismatic mesh elements were employed to ensure computational accuracy. Figure 12 illustrates representative LES outputs used in the machine learning model development. The contours depict instantaneous velocity fields as well as the time-averaged total turbulent kinetic energy for various w/h ratios, both of which contributed to the computation of eddy viscosity. The dataset derived from these simulations—including velocity profiles, eddy viscosity distributions, and step depths—served as the input features to train the ML model for predicting eddy viscosity based on velocity and geometric parameters.

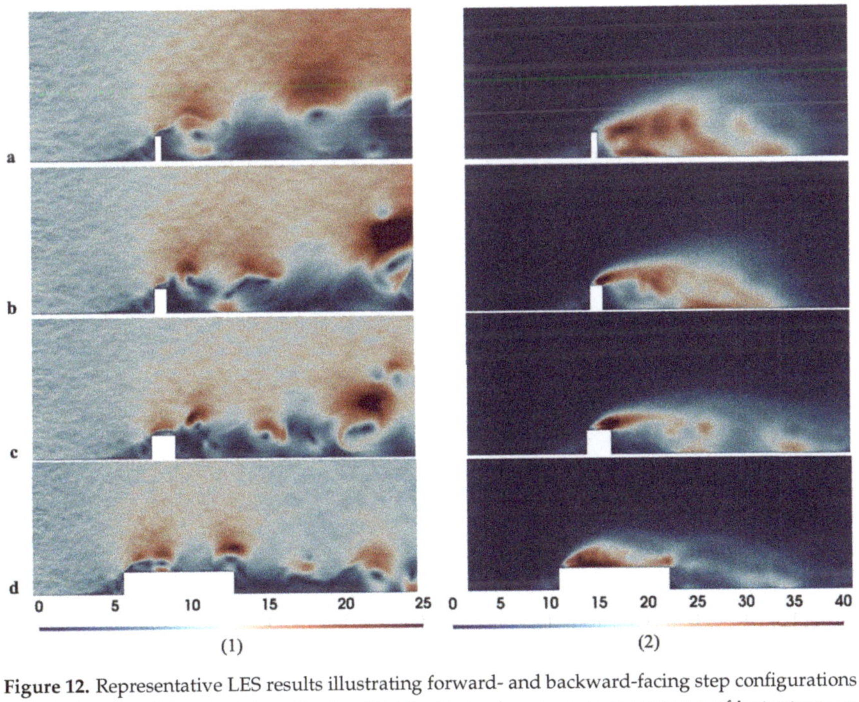

Figure 12. Representative LES results illustrating forward- and backward-facing step configurations with different width-to-height ratios (w/h). The first column presents contours of instantaneous velocity (m/s), while the second column displays contours of time-averaged total turbulent kinetic energy (m^2/s^2) corresponding to step geometries with (**a**) w = 0.2h, (**b**) w = 0.5h, (**c**) w = h, and (**d**) w = 6h. These LES datasets served as training inputs for the deep learning model aimed at predicting eddy viscosity based on the given step geometry.

The training dataset comprised the computed values of eddy viscosity, velocity, and spatial coordinates derived from ten large LES. Of these, eight datasets were allocated to training the ML model, while the remaining two served as testing samples. The deep learning framework was implemented using TensorFlow, seamlessly integrated with OpenFOAM by installing the TensorFlow library and appropriately configuring the OpenFOAM environment. Model training was conducted utilizing the Adam optimizer, adopting a learning rate of 0.001. The neural network architecture was designed to predict the fluid flow characteristics based on input parameters including the flow conditions, boundary constraints, and geometric features. The model produced predictions of eddy viscosity corresponding to specified step widths, which were subsequently employed in solving the RANS equations.

4.4. Results: Improved Accuracy and Computational Efficiency

The findings indicate that integrating the machine-learning-predicted eddy viscosity notably enhanced the congruence with the LES outcomes and experimental observations, especially within the separation bubble and shear layer regions.

4.4.1. Accuracy of Eddy Viscosity Predictions

The training phase revealed a strong correlation between the eddy viscosity values predicted by the machine learning model and the LES-derived ground-truth data. Figure 13 presents this comparison, highlighting the model's capability to precisely estimate the eddy viscosity distribution for flow around bluff bodies. This high fidelity in reconstructing the eddy viscosity field is critical for accurately capturing intricate turbulent flow structures that directly influence pressure and velocity distributions [4].

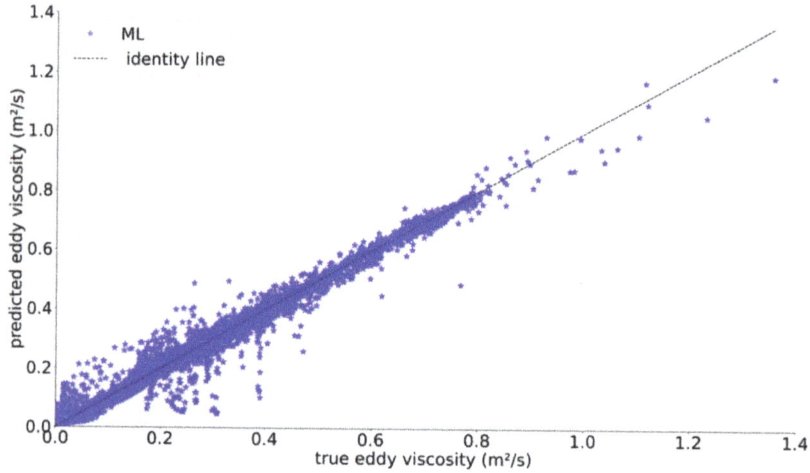

Figure 13. A comparison between the eddy viscosity values (m^2/s) predicted by the machine learning model and those obtained from LES data. The neural network was trained using LES results for forward- and backward-facing steps with varying width-to-height ratios. The strong agreement confirms the model's effectiveness in accurately reconstructing the eddy viscosity field in bluff body flows.

Notable differences were identified when comparing the eddy viscosity predictions from the machine-learning-enhanced model with those from conventional RANS turbulence closures. Figure 14 depicts contour maps of eddy viscosity for RANS and machine-learning-assisted simulations. The ML-based model predicted elevated eddy viscosity values along the roof and in the wake region of the bluff body, ranging approximately between 0.05 and 0.5 m^2/s. In contrast, classical RANS models such as k-ϵ and k-ω SST exhibited lower and more spatially uniform eddy viscosity values in these areas (0.2 m^2/s). Conversely, RANS closures predicted higher eddy viscosity levels near the upper boundary of the domain compared to the ML model. This more balanced magnitude and spatial distribution of eddy viscosity, as captured by the machine learning approach, contribute to a more faithful representation of the underlying flow physics.

(a)
(b)
(c)

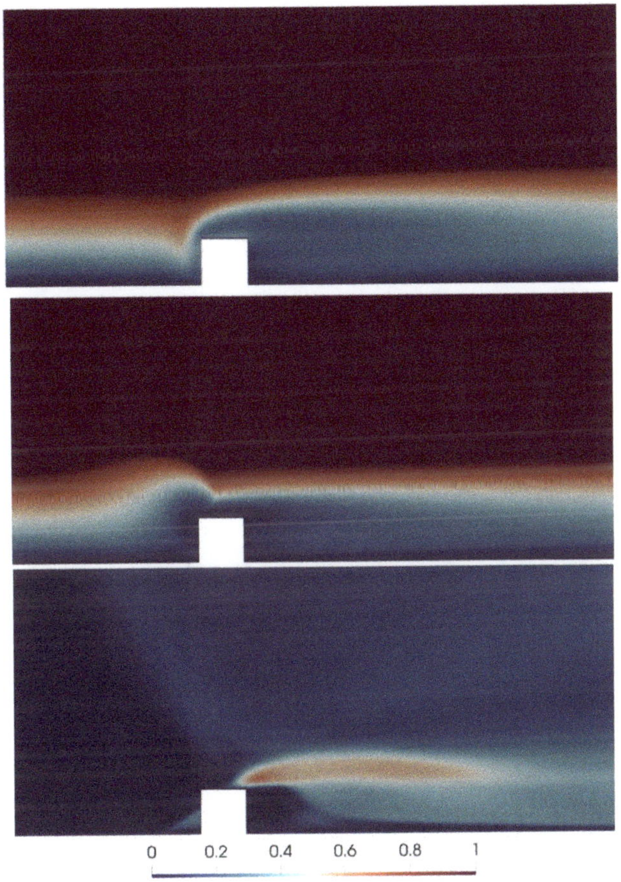

Figure 14. Eddy viscosity contours (m^2/s) obtained from (**a**) k-ϵ RANS, (**b**) k-ω SST RANS, and (**c**) machine-learning-enhanced RANS simulations. The ML approach predicts increased eddy viscosity values on the leeward side of the bluff body, particularly within the flow separation and wake regions. In comparison, the k-ϵ and k-ω SST models exhibit lower eddy viscosity in these areas. The RANS models also show an elevated eddy viscosity near the upper boundary, a feature less pronounced in the ML simulation. The balanced eddy viscosity distribution predicted by the ML model results in a more accurate depiction of flow behavior than traditional linear eddy viscosity turbulence models.

4.4.2. Enhanced Predictions of Pressure and Velocity

The machine learning (ML) model demonstrated a superior performance compared to that of the conventional k-ϵ and k-ω SST turbulence closures in forecasting the mean roof pressures, particularly within the flow separation region. As illustrated in Figure 15, the pressure coefficient predicted by the ML approach exhibits close agreement with the LES and the experimental data reported in the literature. The k-ϵ model tends to predict elevated negative pressures near the windward edge but underestimates negative pressures over the remaining roof area, a consequence of premature flow reattachment. Conversely, the k-ω SST closure shows extensive flow separation across the roof's surface, resulting in a more uniform distribution of negative pressure. The ML model accurately captures a flow separation zone near the windward side with slight reattachment towards the leeward side, aligning well with observed pressure patterns.

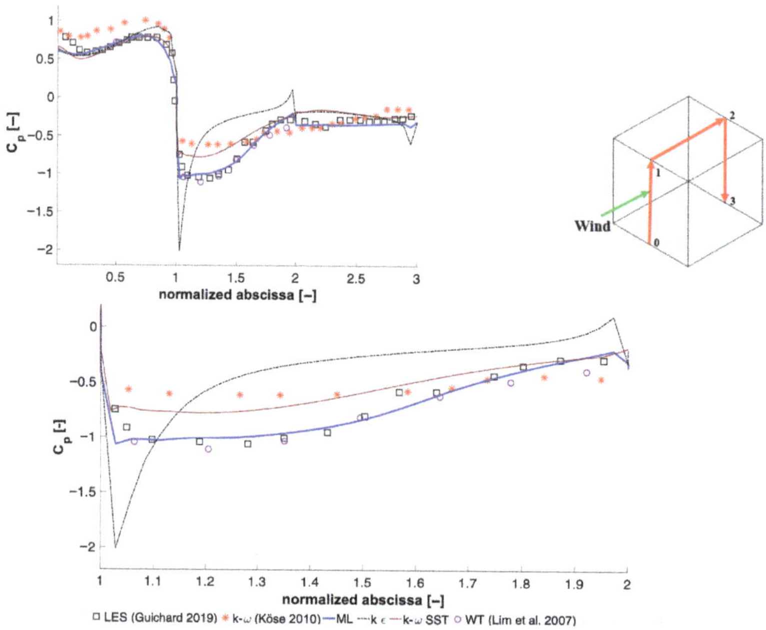

Figure 15. Comparison of roof pressure coefficients (C_p) obtained via ML, LES, RANS simulations utilizing the k-ϵ and k-ω SST turbulence models, and experimental measurements. The ML model closely replicates the LES results and experimental observations, especially within the flow separation zone on the roof. In contrast, the k-ϵ model predicts higher negative pressures near the windward edge owing to a small, compact separation bubble, whereas the k-ω SST model generally underestimates negative pressures on the roof. The ML model's accurate representation of the pressure field in the separated flow region underscores its capability to enhance RANS-based bluff body aerodynamic simulations compared to traditional turbulence closures [4,106–108].

Velocity contour analyses further substantiate these observations. The k-ϵ closure produces a short separation bubble with rapid flow reattachment, while the k-ω SST model forms an extended separation bubble persisting near the leeward side. The ML model predicts a separation bubble longer than that of the k-ϵ closure but with reattachment proximal to the leeward roof edge, thus providing an accurate depiction of the velocity field, particularly in regions influenced by separated flow.

4.4.3. Sensitivity Analysis of Eddy Viscosity

A comprehensive sensitivity analysis highlighted the critical influence of eddy viscosity on flow separation, reattachment, and pressure distribution patterns. By systematically adjusting the eddy viscosity values within the machine learning (ML) model—scaling them by factors ranging from 0.2 to 10—it was observed that increasing eddy viscosity promotes greater flow reattachment and results in less negative pressure predominately on the leeward surface. This trend aligns with the characteristics observed in the k-ϵ RANS model. Conversely, reducing the eddy viscosity caused the reattachment point to shift upstream, thereby increasing negative pressure across the leeward region, consistent with the trends predicted by the k-ω Shear Stress Transport (SST) model. The ML-based method achieves a balanced prediction of eddy viscosity, yielding a more accurate depiction of the underlying flow physics compared to conventional RANS turbulence models.

4.4.4. Computational Efficiency

In addition to an enhanced predictive accuracy, the ML approach substantially reduces the computational time. On a notebook PC equipped with 12 CPU cores and 40 GB of memory, the k-ϵ turbulence model required approximately 21.75 min to converge, whereas the k-ω SST model converged in 44.3 min. The LES, by comparison, demanded around 5.6 h. Notably, the ML-augmented RANS simulation converged in roughly 6.27 min, representing a speedup exceeding fiftyfold relative to the LES, while maintaining a comparable accuracy in aerodynamic pressure estimations. This efficiency gain enables seamless integration of the trained neural network directly within the OpenFOAM framework, facilitating an end-to-end acceleration of RANS computations.

4.5. Conclusion on Eddy Viscosity Modeling

The presented deep learning framework for eddy viscosity modeling effectively overcomes the shortcomings inherent to traditional RANS closures when predicting wind-induced pressures on bluff bodies. By generating a more precise and balanced eddy viscosity field, the ML model significantly enhances the fidelity of the pressure and velocity predictions, particularly within zones characterized by flow separation and shear layers. Furthermore, the dramatic reduction in computational expense positions high-fidelity RANS simulations as viable tools for practical engineering use, offering valuable insights poised to advance more robust computational strategies for structural design and the analysis of unsteady aerodynamic phenomena. Accurately capturing eddy viscosity within separation and shear regions is essential, and the proposed ML-based method achieves this with improved accuracy, alongside a considerable reduction in computational cost.

5. Discussion

The comprehensive investigation of recent advances underscores the transformative impact of machine learning and artificial intelligence on computational fluid dynamics. The integration of ML methods effectively addresses the long-standing challenges of computational cost and accuracy that have historically limited the application of CFD, particularly for high-fidelity simulations of turbulent flows [4]. This represents a fundamental paradigm shift, moving beyond incremental improvements to enable previously unattainable levels of efficiency and fidelity across the entire spectrum of fluid dynamics research and application [17,25].

A key contribution of ML is its ability to drastically reduce simulation times, exemplified by speedups of 10,000 times for solar panel design [5] and over 50 times for bluff body aerodynamics [4], all while maintaining or improving accuracy. Such breakthroughs accelerate not only engineering design but also fundamental scientific discovery, allowing researchers to explore vast parameter spaces and uncover new physical insights in complex systems that were previously computationally prohibitive [39,47]. This efficiency is critical for accelerating the design process, enabling engineers to perform more iterations and explore a wider range of design parameters in a fraction of the time previously required. The move towards ML-assisted numerical solvers and the deployment of neural networks directly within CFD software frameworks are further streamlining this process [45].

Furthermore, ML's capacity to enhance turbulence modeling, particularly by providing more accurate eddy viscosity predictions for RANS simulations, is revolutionizing how complex flow phenomena are captured [4]. The demonstrated improvement in predicting wind pressures on bluff bodies, especially in separation zones, highlights ML's potential to overcome the inherent limitations of traditional turbulence closures. This improved fidelity is crucial for designing resilient infrastructure that can withstand extreme environmental stressors like windstorms, contributing significantly to climate change adaptation and mitigation efforts.

The emergence of data-driven and physics-informed surrogates represents a powerful paradigm shift, moving beyond mere data fitting to models that inherently respect physical laws [18]. This dual approach ensures both computational efficiency and physical consistency, fostering greater trust in ML-generated simulations. The ability to perform inverse design and optimize control strategies based on ML predictions opens new avenues for material and structural engineering, allowing for proactive design solutions rather than reactive adjustments [19–21].

Despite these significant advancements, several challenges and areas for future work persist. Both application examples, solar panel wind design and eddy viscosity modeling for bluff bodies, underscore the immense potential of ML to accelerate CFD simulations and enhance accuracy in challenging flow regimes. Common insights reveal that ML models, when trained on high-fidelity data, can achieve remarkable speedups (e.g., $10,000\times$ for solar panels, $50\times$ for bluff bodies) while improving predictions in complex separated flows. However, both also highlight the need for robust generalization, especially for conditions beyond the training data, such as high angles of attack in solar panels or diverse bluff body geometries. The generalization capability of ML models, particularly when extrapolating beyond training data ranges or adapting to vastly different geometries, requires continuous research [39,75,82]. Quantitatively exploring the sensitivity of ML models to variations in the flow parameters, such as the Reynolds number, is crucial for assessing extrapolation risks and building models that reliably perform across a broad spectrum of real-world conditions [38]. Developing more robust and universally applicable ML-based turbulence closures is also a priority to fully replace or augment existing models across diverse flow regimes [35,55,56,74]. The integration of multi-scale representation and more sophisticated encoding of physical knowledge into ML models will push the boundaries of accuracy and applicability further [9,22,48,60]. Furthermore, the emergence of generative AI models, such as generative adversarial networks (GANs), shows promise for tasks like the super-resolution reconstruction of turbulent flows, potentially addressing data scarcity and enhancing simulation fidelity beyond the direct output of traditional solvers [24]. While this paper primarily focuses on wind engineering applications and turbulence modeling, other broad areas of CFD, such as multiphase flows and combustion, also stand to benefit significantly from ML integration and represent important avenues for future research [9]. Research into flow-informed aerodynamic control in turbulent wind conditions, leveraging reinforcement learning and recurrent neural networks, also highlights the need for effective disturbance rejection capabilities to ensure the safety and performance of technologies like UAVs [46]. Additionally, the aerodynamics of morphing drones, which adapt wing and tail configurations, present opportunities for ML to optimize the flight performance in complex scenarios. Future research could also investigate utility-scale PV plant modeling to explore the effects of winds across several acres of panels, low-cost damping mechanisms for affordable installation on SAT systems [5], and the development of dimensionless eddy viscosity numbers to refine turbulence adjustments for high-Reynolds-number flows. Additionally, the effect of sand accumulation on PV system efficiency is an area beyond the current scope that merits further analytical study and mitigation techniques.

In conclusion, the synergistic integration of ML and CFD is not merely an incremental improvement but a fundamental reshaping of the field of fluid mechanics. By providing unprecedented speed, enhanced accuracy, and novel modeling capabilities, ML is enabling more efficient, reliable, and resilient engineering designs crucial for addressing contemporary global challenges.

6. Conclusions

The integration of machine learning (ML) and AI into CFD represents a pivotal advancement, fundamentally enhancing the simulation accuracy, computational efficiency, and modeling capabilities. Traditional CFD methods have long struggled to balance high fidelity with practical computational costs, especially in predicting wind loads and modeling turbulent flows. This paper has demonstrated how ML techniques effectively overcome these challenges. This synergy marks a critical step towards democratizing access to high-fidelity fluid simulations, enabling a new era of predictive science and data-driven engineering design.

Key findings include:

- **Unprecedented Computational Efficiency**: ML-driven approaches have achieved dramatic speedups—up to 10,000 times faster in solar panel design and over 50 times faster in bluff body aerodynamics—while maintaining or improving accuracy compared to traditional high-fidelity simulations;
- **Enhanced Accuracy and Turbulence Modeling**: ML models improve Reynolds-averaged Navier–Stokes simulations by providing more accurate eddy viscosity predictions in complex flow regimes such as separation zones, yielding better representations of flow reattachment and pressure distributions;
- **Expanded Modeling Capabilities**: The development of data-driven and physics-informed surrogates, together with ML-assisted solvers, extends CFD applications into inverse design, control, and multi-scale modeling, enabling novel methodologies that were previously unfeasible.

The synergy of ML and CFD is fostering more efficient, reliable, and resilient engineering designs essential for addressing global challenges like climate change and sustainable infrastructure. Despite ongoing challenges—including generalization, extrapolation, and encoding physical knowledge—research continues to advance these frontiers. Future work will refine these methods, broaden practical applications, and enhance turbulence closures. The evolving integration of ML and AI promises to unlock unparalleled capabilities in modeling, understanding, and controlling fluid phenomena.

Funding: The author's research and expertise, which inform this paper, have been supported by (1) LA Board of Regents (BOR), Originating Sponsor: National Aeronautics & Space Administration (NASA) EPSCoR RID-RAP (Grant No. LEQSF-EPS(2025)-RAP-59); and (2) the Institute for Energy Innovation (Grant No. Phase II Energy Innovation). The findings in this paper are those of the author and do not necessarily reflect the opinion of the supporting agencies.

Institutional Review Board Statement: Not applicable.

Informed Consent Statement: Not applicable.

Data Availability Statement: Not applicable.

Conflicts of Interest: The author declares no conflicts of interest.

Abbreviations

The following abbreviations are used in this manuscript:

AI	Artificial Intelligence
AoA	Angle of Attack
ASCE	American Society of Civil Engineers
CFD	Computational Fluid Dynamics
CNN	Convolutional Neural Network
DNS	Direct Numerical Simulation
FD	Fluid Dynamics
LES	Large Eddy Simulation
ML	Machine Learning
PINN	Physics-Informed Neural Network
PV	Photovoltaic
RANS	Reynolds-Averaged Navier–Stokes
ReLU	Rectified Linear Unit
SAT	Single-Axis Tracker
SGS	Sub-Grid-Scale
SST	Shear Stress Transport
w/h	Width-to-Height
WISE	Windstorm Impact, Science, and Engineering

References

1. Aly, A.; Goodlow, B.; Conerly, J. Urban heat island mitigation via geometric configuration. *Theor. Appl. Climatol.* **2022**, *149*, 1329–1355. [CrossRef]
2. Ghadami, N.; Gheibi, M.; Kian, Z.; Faramarz, M.; Naghedi, R.; Eftekhari, M.; et al. Implementation of solar energy in smart cities using an integration of artificial neural network, photovoltaic system and classical Delphi methods. *Sustain. Cities Soc.* **2021**, *74*, 103149. [CrossRef]
3. Aly, A. Atmospheric boundary-layer simulation for the built environment: Past, present and future. *Build. Environ.* **2014**, *75*, 206–221. [CrossRef]
4. Aly, A. Deep Learning-Based Eddy Viscosity Modeling for Improved RANS Simulations of Wind Pressures on Bluff Bodies. *J. Appl. Fluid Mech.* **2024**, *17*, 2514–2532. [CrossRef]
5. Aly, A.M.; Clarke, J. Wind design of solar panels for resilient and green communities: CFD with machine learning. *Sustain. Cities Soc.* **2023**, *94*, 104529. [CrossRef]
6. Orszag, S. Analytical theories of turbulence. *J. Fluid Mech.* **1970**, *41*, 363–386. [CrossRef]
7. Aly, A.M.; Bitsuamlak, G. Aerodynamics of ground-mounted solar panels: Test model scale effects. *J. Wind. Eng. Ind. Aerodyn.* **2013**, *123*, 250–260. [CrossRef]
8. Baker, C. Wind engineering-past, present and future. *J. Wind. Eng. Ind. Aerodyn.* **2007**, *95*, 843–870. [CrossRef]
9. Zhao, C.; Zhang, F.; Lou, W.; Wang, X.; Yang, J. A comprehensive review of advances in physics-informed neural networks and their applications in complex fluid dynamics. *Phys. Fluids* **2024**, *36*, 101301. [CrossRef]
10. Drikakis, D.; Sofos, F. Can Artificial Intelligence Accelerate Fluid Mechanics Research? *Fluids* **2023**, *8*, 212. [CrossRef]
11. Samuel, A. Some Studies in Machine Learning Using the Game of Checkers. *IBM J. Res. Dev.* **1959**, *3*, 210–229. [CrossRef]
12. Mitchell, T.; Mitchell, T. *Machine Learning*; McGraw-Hill: New York, NY, USA, 1997.
13. Samuel, A.L. Artificial Intelligence: A Frontier of Automation. *Ann. Am. Acad. Political Soc. Sci.* **1962**, *340*, 10–20. [CrossRef]
14. Raissi, M.; Perdikaris, P.; Karniadakis, G.E. Physics-informed neural networks: A deep learning framework for solving forward and inverse problems involving nonlinear partial differential equations. *J. Comput. Phys.* **2019**, *378*, 686–707. [CrossRef]
15. Karniadakis, G.E.; Kevrekidis, I.G.; Lu, L.; Perdikaris, P.; Wang, S.; Yang, L. Physics-informed machine learning. *Nat. Rev. Phys.* **2021**, *3*, 422–440. [CrossRef]
16. National Aeronautics and Space Administration. *NASA Aeronautics Strategic Implementation Plan 2023*; Technical Report; National Aeronautics and Space Administration: Washington, DC, USA, 2023.
17. Brunton, S.L.; Noack, B.R.; Koumoutsakos, P. Machine learning for fluid mechanics. *Annu. Rev. Fluid Mech.* **2020**, *52*, 477–508. [CrossRef]
18. Gustafsson, E.; Andersson, M. Investigating the Effects of Labeled Data on Parameterized Physics-Informed Neural Networks for Surrogate Modeling: Design Optimization for Drag Reduction over a Forward-Facing Step. *Fluids* **2024**, *9*, 296. [CrossRef]

19. Loukili, M.; El Moumni, S.; Kotrasova, K. AI-Driven Optimization of Breakwater Design: Predicting Wave Reflection and Structural Dimensions. *Fluids* **2025**, *10*, 34. [CrossRef]
20. Luis-Gómez, J.; Martínez, F.; González-Barberá, A.; Mascarós, J.; Monrós-Andreu, G.; Chiva, S.; Borrás, E.; Martínez-Cuenca, R. AI-Driven Surrogate Model for Room Ventilation. *Fluids* **2025**, *10*, 163. [CrossRef]
21. Savioli, T.; Pampanini, M.; Visani, G.; Esposito, L.; Rinaldini, C. Engine Mass Flow Estimation through Neural Network Modeling in Semi-Transient Conditions: A New Calibration Approach. *Fluids* **2024**, *9*, 239. [CrossRef]
22. Majchrzak, M.; Marciniak-Lukasiak, K.; Lukasiak, P. A Survey on the Application of Machine Learning in Turbulent Flow Simulations. *Energies* **2023**, *16*, 1755. [CrossRef]
23. Pioch, F.; Harmening, J.; Müller, A.; Peitzmann, F.J.; Schramm, D.; el Moctar, O. Turbulence Modeling for Physics-Informed Neural Networks: Comparison of Different RANS Models for the Backward-Facing Step Flow. *Fluids* **2023**, *8*, 43. [CrossRef]
24. Ward, N. Physics-Informed Super-Resolution of Turbulent Channel Flows via Three-Dimensional Generative Adversarial Networks. *Fluids* **2023**, *8*, 195. [CrossRef]
25. Vinuesa, R.; Brunton, S.L. Enhancing computational fluid dynamics with machine learning. *Nat. Comput. Sci.* **2022**, *2*, 358–366. [CrossRef] [PubMed]
26. Thuerey, N.; Weissenow, K.; Prantl, L.; Hu, X. Deep Learning Methods for Reynolds-Averaged Navier–Stokes Simulations of Airfoil Flows. *AIAA J.* **2020**, *58*, 25–36. [CrossRef]
27. Rana, P.; Weigand, T.M.; Pilkiewicz, K.R.; Mayo, M.L. A scalable convolutional neural network approach to fluid flow prediction in complex environments. *Sci. Rep.* **2024**, *14*, 23080. [CrossRef]
28. Ehlers, S.; Wagner, N.; Scherzl, A.; Klein, M.; Hoffmann, N.; Stender, M. Data Assimilation and Parameter Identification for Water Waves Using the Nonlinear Schrödinger Equation and Physics-Informed Neural Networks. *Fluids* **2024**, *9*, 231. [CrossRef]
29. Lee, J.; Shin, S.; Kim, T.; Park, B.; Choi, H.; Lee, A.; Choi, M.; Lee, S. Physics informed neural networks for fluid flow analysis with repetitive parameter initialization. *Sci. Rep.* **2025**, *15*, 16740. [CrossRef]
30. Neelan, A.G.; Bosco, F.S.D.; Jarugumalli, N.S.; Vedarethinam, S.B. Physics-Informed Neural Networks: Bridging the Divide Between Conservative and Non-Conservative Equations. *arXiv* **2025**. [CrossRef]
31. Ali, A.; Noshad, N.; Kumar, A.; Ilyas, S.; Phelan, P.; Alsaady, M.; Nasir, R.; Yan, Y. Application of Machine Learning Algorithms in Predicting Rheological Behavior of BN-diamond/Thermal Oil Hybrid Nanofluids. *Fluids* **2024**, *9*, 20. [CrossRef]
32. Yang, S.; Kim, H.; Hong, Y.; Yee, K.; Maulik, R.; Kang, N. Data-driven physics-informed neural networks: A digital twin perspective. *Comput. Methods Appl. Mech. Eng.* **2024**, *428*, 117075. [CrossRef]
33. Banerjee, C.; Lilian, C.; Reasor, D.; Pasiliao, E.; Mukherjee, T. An Application of Generative Adversarial Networks for Robust Inference in Computational Fluid Dynamics. In Proceedings of the 2021 5th International Conference on Information System and Data Mining (ICISDM '21), Silicon Valley, CA, USA, 27–29 May 2021; pp. 74–83. [CrossRef]
34. Cuellar, A.; Guemes, A.; Ianiro, A.; Flores, O.; Vinuesa, R.; Discetti, S. Three-dimensional generative adversarial networks for turbulent flow estimation from wall measurements. *J. Fluid Mech.* **2024**, *991*, A1. [CrossRef]
35. Patel, Y.; Mons, V.; Marquet, O.; Rigas, G. Turbulence model augmented physics-informed neural networks for mean-flow reconstruction. *Phys. Rev. Fluids* **2024**, *9*, 034605. [CrossRef]
36. Molnar, J.P.; Venkatakrishnan, L.; Schmidt, B.E.; Sipkens, T.A.; Grauer, S.J. Estimating density, velocity, and pressure fields in supersonic flows using physics-informed BOS. *Exp. Fluids* **2023**, *64*, 14. [CrossRef]
37. Zhang, C.; Wen, C.Y.; Jia, Y.; Juan, Y.H.; Lee, Y.T.; Chen, Z.; Yang, A.S.; Li, Z. Enhancing the accuracy of physics-informed neural networks for indoor airflow simulation with experimental data and Reynolds-averaged Navier–Stokes turbulence model. *Phys. Fluids* **2024**, *36*, 065161. [CrossRef]
38. Jiang, L.; Cheng, Y.; Luo, K.; Fan, J. PT-PINNs: A Parametric Engineering Turbulence Solver based on Physics-Informed Neural Networks. *arXiv* **2025**, arXiv:2503.17704 .
39. Kochkov, D.; Smith, J.A.; Alievaa, A.; Wanga, Q.; Brenner, M.P.; Hoyera, S. Machine learning–accelerated computational fluid dynamics. *Proc. Natl. Acad. Sci. USA* **2021**, *118*, e2101784118. [CrossRef]
40. Lesieur, M.; Metais, O. New trends in large-eddy simulations of turbulence. *Annu. Rev. Fluid Mech.* **1996**, *28*, 45–82. [CrossRef]
41. Choi, H.; Moin, P. Grid-point requirements for large eddy simulation: Chapman's estimates revisited. *Phys. Fluids* **2012**, *24*, 011702. [CrossRef]
42. Yang, X.I.A.; Griffin, K.P. Grid-point and time-step requirements for direct numerical simulation and large-eddy simulation. *Phys. Fluids* **2021**, *33*, 015108. [CrossRef]
43. Bar-Sinai, Y.; Hoyer, S.; Hickey, J.; Brenner, M.P. Learning data-driven discretizations for partial differential equations. *Proc. Natl. Acad. Sci. USA* **2019**, *116*, 15344–15349. [CrossRef] [PubMed]
44. Zhuang, J.; Kochkov, D.; Bar-Sinai, Y.; Brenner, M.P.; Hoyer, S. Learned discretizations for passive scalar advection in a two-dimensional turbulent flow. *Phys. Rev. Fluids* **2021**, *6*, 064605. [CrossRef]

45. Maulik, R.; Sharma, H.; Patel, S.; Lusch, B.; Jennings, E. Deploying Deep Learning in OpenFOAM with TensorFlow. In Proceedings of the AIAA SciTech 2021 Forum, Virtual, 11–15 & 19–21 January 2021; p. 1485. [CrossRef]
46. Renn, P.I.; Gharib, M. Machine learning for flow-informed aerodynamic control in turbulent wind conditions. *Commun. Eng.* **2022**, *1*, 45. [CrossRef]
47. Kovachki, N.; Li, Z.; Liu, B.; Azizzadenesheli, K.; Bhattacharya, K.; Stuart, A.; Anandkumar, A. Neural Operator: Learning Maps Between Function Spaces With Applications to PDEs. *J. Mach. Learn. Res.* **2023**, *24*, 1–97.
48. Pathak, J.; Mustafa, M.; Kashinath, K.; Motheau, E.; Kurth, T.; Day, M. Using Machine Learning to Augment Coarse-Grid Computational Fluid Dynamics Simulations. *arXiv* **2020**, 1–8. [CrossRef]
49. Bao, K.; Zhang, X.; Peng, W.; Yao, W. Deep learning method for super-resolution reconstruction of the spatio-temporal flow field. *Adv. Aerodyn.* **2023**, *5*, 19. [CrossRef]
50. Zan, B.W.; Han, Z.H.; Xu, C.Z.; Liu, M.Q.; Wang, W.Z. High-dimensional aerodynamic data modeling using a machine learning method based on a convolutional neural network. *Adv. Aerodyn.* **2022**, *4*, 39. [CrossRef]
51. Zhao, W.; Jiang, L.; Yao, S.; Chen, W. Data-driven nonlinear constitutive relations for rarefied flow computations. *Adv. Aerodyn.* **2021**, *3*, 30. [CrossRef]
52. Zhu, L.; Zhang, W.; Tu, G. Generalization enhancement of artificial neural network for turbulence closure by feature selection. *Adv. Aerodyn.* **2022**, *4*, 1. [CrossRef]
53. Wang, Y.; Guan, F.; Low, M.Y.H.; Wang, Z.; Ng, A.B.; See, S. Towards Accelerating Physics Informed Graph Neural Network for Fluid Simulation. In Proceedings of the SIGGRAPH Asia 2024 Posters, Tokyo, Japan, 3–6 December 2024; pp. 1–3. [CrossRef]
54. Duraisamy, K.; Iaccarino, G.; Xiao, H. Turbulence Modeling in the Age of Data. *Annu. Rev. Fluid Mech.* **2019**, *51*, 357–377. [CrossRef]
55. Jones, W.; Launder, B. The calculation of low-Reynolds-number phenomena with a two-equation model of turbulence. *Int. J. Heat Mass Transf.* **1973**, *16*, 119–1130. [CrossRef]
56. Menter, F.R. *Improved Two-Equation k-ω Turbulence Models for Aerodynamic Flows*; Technical Report; NASA: Washington, DC, USA, 1992.
57. Kingma, D.; Ba, J. Adam: A method for stochastic optimization. *arXiv* **2015**, arXiv:1412.6980.
58. Heinz, S. The Potential of Machine Learning Methods for Separated Turbulent Flow Simulations: Classical Versus Dynamic Methods. *Fluids* **2024**, *9*, 278. [CrossRef]
59. Sanhueza, R.; Smit, S.; Peeters, J.; Pecnik, R. Machine learning for RANS turbulence modeling of variable property flows. *Comput. Fluids* **2023**, *255*, 105835. [CrossRef]
60. Singh, A.; Medida, S.; Duraisamy, K. Machine-Learning-Augmented Predictive Modeling of Turbulent Separated Flows over Airfoils. *AIAA J.* **2017**. [CrossRef]
61. Zhao, Y.; Akolekar, H.D.; Weatheritt, J.; Michelassi, V.; Sandberg, R.D. RANS turbulence model development using CFD-driven machine learning. *J. Comput. Phys.* **2020**, *411*, 109413. [CrossRef]
62. Sardana, P.; Zolfaghari, M.; Miotto, G.; Zengerle, R.; Brox, T.; Koltay, P.; Kartmann, S. DropletAI: Deep Learning-Based Classification of Fluids with Different Ohnesorge Numbers during Non-Contact Dispensing. *Fluids* **2023**, *8*, 183. [CrossRef]
63. Fardi Asrami, R.; Sohani, A.; Saedpanah, E.; Sayyaadi, H. Towards achieving the best solution to utilize photovoltaic solar panels for residential buildings in urban areas. *Sustain. Cities Soc.* **2021**, *71*, 102968. [CrossRef]
64. Duraisamy, K. Perspectives on machine learning-augmented Reynolds-averaged and large eddy simulation models of turbulence. *Phys. Rev. Fluids* **2021**, *6*, 050504. [CrossRef]
65. Sarghini, F.; De Felice, G.; Santini, S. Neural networks based subgrid scale modeling in large eddy simulations. *Comput. Fluids* **2003**, *32*, 97–108. [CrossRef]
66. Gamahara, M.; Hattori, Y. Searching for turbulence models by artificial neural network. *Phys. Rev. Fluids* **2017**, *2*, 054604. [CrossRef]
67. Wang, Z.; Luo, K.; Li, D.; Tan, J.; Fan, J. Investigations of data-driven closure for subgrid-scale stress in large-eddy simulation. *Phys. Fluids* **2018**, *30*, 125101. [CrossRef]
68. Zhou, Z.; He, G.; Wang, S.; Jin, G. Subgrid-scale model for large-eddy simulation of isotropic turbulent flows using an artificial neural network. *Comput. Fluids* **2019**, *195*, 104319. [CrossRef]
69. Xie, C.; Wang, J.; Li, H.; Wan, M.; Chen, S. Artificial neural network mixed model for large eddy simulation of compressible isotropic turbulence. *Phys. Fluids* **2019**, *31*, 085112. [CrossRef]
70. Xie, C.; Yuan, Z.; Wang, J. Artificial neural network-based nonlinear algebraic models for large eddy simulation of turbulence. *Phys. Fluids* **2020**, *32*, 115101. [CrossRef]
71. Park, J.; Choi, H. Toward neural-network-based large eddy simulation: application to turbulent channel flow. *J. Fluid Mech.* **2021**, *914*, A16. [CrossRef]

72. Wang, Y.; Yuan, Z.; Xie, C.; Wang, J. Artificial neural network-based spatial gradient models for large-eddy simulation of turbulence. *AIP Adv.* **2021**, *11*, 055216. [CrossRef]
73. Kang, M.; Jeon, Y.; You, D. Neural-network-based mixed subgrid-scale model for turbulent flow. *J. Fluid Mech.* **2023**, *962*, A38. [CrossRef]
74. Arranz, G.; Ling, Y.; Costa, S.; Goc, K.; Lozano-Duran, A. Building-block-flow computational model for large-eddy simulation of external aerodynamic applications. *Commun. Eng.* **2024**, *3*, 127. [CrossRef] [PubMed]
75. Liu, W.; Fang, J.; Rolfo, S.; Moulinec, C.; Emerson, D. An iterative machine-learning framework for RANS turbulence modeling. *Int. J. Heat Fluid Flow* **2021**, *90*, 108822. [CrossRef]
76. Khaled, M.; Aly, A.; Elshaer, A. Computational efficiency of CFD modeling for building engineering: An empty domain study. *J. Build. Eng.* **2021**, *42*, 102792. [CrossRef]
77. Chen, S.; Qiu, J.; Yang, H.; Yuan, W.; Gao, Z. Deep learning for inverse design of low-boom supersonic configurations. *Adv. Aerodyn.* **2023**, *5*, 13. [CrossRef]
78. Tran, J.; Fukami, K.; Inada, K.; Umehara, D.; Ono, Y.; Ogawa, K.; Taira, K. Aerodynamics-guided machine learning for design optimization of electric vehicles. *Commun. Eng.* **2024**, *3*, 174. [CrossRef]
79. Salemnia, A.; Hosseini Boldaji, S.; Atashi, V.; Fathi-Moghadam, M. Machine Learning for Dynamic Pressure Coefficient Prediction in Vertical Water Jets. *Fluids* **2024**, *9*, 205. [CrossRef]
80. Karakasidis, T.; Sofos, F.; Tsonos, C. The Electrical Conductivity of Ionic Liquids: Numerical and Analytical Machine Learning Approaches. *Fluids* **2022**, *7*, 321. [CrossRef]
81. Ti, Z.; Deng, X.W.; Yang, H. Wake modeling of wind turbines using machine learning. *Appl. Energy* **2020**, *257*, 114025. [CrossRef]
82. Liu, W.; Fang, J.; Rolfo, S.; Moulinec, C.; Emerson, D.R. On the improvement of the extrapolation capability of an iterative machine-learning based RANS Framework. *Comput. Fluids* **2023**, *256*, 105864. [CrossRef]
83. Magnus, J.R.; Melenberg, B.; Muris, C. Global Warming and Local Dimming: The Statistical Evidence. *J. Am. Stat. Assoc.* **2011**, *106*, 452–464. [CrossRef]
84. Meehl, G.A.; Zwiers, F.; Evans, J.; Knutson, T.; Mearns, L.; Whetton, P. Trends in Extreme Weather and Climate Events: Issues Related to Modeling Extremes in Projections of Future Climate Change. *Bull. Am. Meteorol. Soc.* **2000**, *81*, 427–436. [CrossRef]
85. Strauss, B.H.; Orton, P.M.; Bittermann, K.; Buchanan, M.K.; Gilford, D.M.; Kopp, R.E.; Kulp, S.; Massey, C.; Moel, H.d.; Vinogradov, S. Economic damages from Hurricane Sandy attributable to sea level rise caused by anthropogenic climate change. *Nat. Commun.* **2021**, *12*, 2720. [CrossRef]
86. Broach, D. These Are the 10 Most Expensive Hurricanes in U.S. History; See Video, Storm Tracks. Available online: https://www.kiplinger.com/slideshow/business/t019-s001-most-expensive-natural-disasters-in-u-s-history/index.html (accessed on 18 September 2025).
87. Gutnikov, V.A.; Lifanov, I.K.; Setukha, A.V. Simulation of the aerodynamics of buildings and structures by means of the closed vortex loop method. *Fluid Dyn.* **2006**, *41*, 555–567. [CrossRef]
88. Ke, J. RANS and hybrid LES/RANS simulations of flow over a square cylinder. *Adv. Aerodyn.* **2019**, *1*, 10. [CrossRef]
89. Shademan, M.; Naghib-Lahouti, A. Effects of aspect ratio and inclination angle on aerodynamic loads of a flat plate. *Adv. Aerodyn.* **2020**, *2*, 14. [CrossRef]
90. Gnatowska, R. Wind-induced pressure loads on buildings in tandem arrangement in urban environment. *Environ. Fluid Mech.* **2019**, *19*, 699–718. [CrossRef]
91. Kopp, G.A.; Morrison, M.J. Component and Cladding Wind Loads for Low-Slope Roofs on Low-Rise Buildings. *J. Struct. Eng.* **2018**, *144*, 04018019. [CrossRef]
92. Aldoum, M.; Stathopoulos, T. Wind loads on low-slope roofs of buildings with large plan dimensions. *Eng. Struct.* **2020**, *225*, 111298. [CrossRef]
93. Aly, A.M. On the evaluation of wind loads on solar panels: The scale issue. *Sol. Energy* **2016**, *135*, 423–434. [CrossRef]
94. Abiola-Ogedengbe, A. Experimental Investigation of Wind Effect on Solar Panels. Ph.D. Thesis, Western University, London, ON, Canada, 2013.
95. Jubayer, C.; Hangan, H. A numerical approach to the investigation of wind loading on an array of ground mounted solar photovoltaic (PV) panels. *J. Wind. Eng. Ind. Aerodyn.* **2016**, *153*, 60–70. [CrossRef]
96. Shademan, M.; Balachandar, R.; Barron, R. Detached eddy simulation of flow past an isolated inclined solar panel. *J. Fluids Struct.* **2014**, *50*, 217–230. [CrossRef]
97. Young, E.; He, X.; King, R.; Corbus, D. A fluid-structure interaction solver for investigating torsional galloping in solar-tracking photovoltaic panel arrays. *J. Renew. Sustain. Energy* **2020**, *12*, 63503. [CrossRef]
98. Valentín, D.; Valero, C.; Egusquiza, M.; Presas, A. Failure investigation of a solar tracker due to wind-induced torsional galloping. *Eng. Fail. Anal.* **2022**, *135*, 106137. [CrossRef]

99. Rohr, C.; Bourke, P.A.; Banks, D. Torsional instability of single-axis solar tracking systems. In Proceedings of the 14th International Conference on Wind Engineering, Porto Alegre, Brazil, 21–26 June 2015; pp. 21–26.
100. Martínez-Garcia, E.; Blanco-Marigorta, E.; Parrondo Gayo, J.; Navarro-Manso, A. Influence of inertia and aspect ratio on the torsional galloping of single-axis solar trackers. *Eng. Struct.* **2021**, *243*, 112682. [CrossRef]
101. ICC. International Codes-Adoption by State. Louisiana. Louisiana State Uniform Construction Code (LSUCC). 2021. Available online: https://www.iccsafe.org/wp-content/uploads/Master-I-Code-Adoption-Chart-AUG-2021.pdf (accessed on 18 September 2025).
102. Cóstola, D.; Blocken, B.; Hensen, J. Overview of pressure coefficient data in building energy simulation and airflow network programs. *Build. Environ.* **2009**, *44*, 2027–2036. [CrossRef]
103. Muehleisen, R.T.; Patrizi, S. A new parametric equation for the wind pressure coefficient for low-rise buildings. *Energy Build.* **2013**, *57*, 245–249. [CrossRef]
104. Aly, A.M.; Khaled, M.F.; Clancy, R. Large-Scale Open-Jet Testing: A new frontier in structural wind Engineering. *Eng. Struct.* **2022**, *266*, 114567. [CrossRef]
105. Lei, J.; Huang, P.; Zhang, L.; Yuan, Y.; Deng, W.; Mao, S.; Zhang, J. Experimental study on flow characteristics of a large-scale open jet wind tunnel for outdoor pool fire research. *J. Wind. Eng. Ind. Aerodyn.* **2021**, *210*, 104522. [CrossRef]
106. Lim, H.C.; Castro, I.P.; Hoxey, R.P. Bluff Bodies in Deep Turbulent Boundary Layers: Reynolds Number Issues. *J. Fluid Mech.* **2007**, *571*, 97–118. [CrossRef]
107. Guichard, R. Assessment of an improved Random Flow Generation method to predict unsteady wind pressures on an isolated building using Large-Eddy Simulation. *J. Wind. Eng. Ind. Aerodyn.* **2019**, *189*, 304–313. [CrossRef]
108. Kose, D.A.; Dick, E. Prediction of the pressure distribution on a cubical building with implicit LES. *J. Wind Eng. Ind. Aerodyn.* **2010**, *98*, 628–649. [CrossRef]

fluids

MDPI

Review

Machine Learning in Fluid Dynamics—Physics-Informed Neural Networks (PINNs) Using Sparse Data: A Review

Mouhammad El Hassan [1,*], Ali Mjalled [2], Philippe Miron [3], Martin Mönnigmann [2] and Nikolay Bukharin [4]

1 Department of Mechanical Engineering, Prince Mohammad Bin Fahd University, Al Khobar 31952, Saudi Arabia
2 Automatic Control and Systems Theory, Ruhr-Universität Bochum, Universitätstrasse 150, 44801 Bochum, Germany
3 Center for Ocean-Atmospheric Prediction Studies, Florida State University, Tallahassee, FL 32306, USA
4 The School of Manufacturing and Automation, The Southern Alberta Institute of Technology, Calgary, AB T2M 0L4, Canada; nikolay.bukharin@sait.ca
* Correspondence: melhassan@pmu.edu.sa

Abstract

Fluid mechanics often involves complex systems characterized by a large number of physical parameters, which are usually described by experimental and numerical sparse data (temporal or spatial). The difficulty of obtaining complete spatio-temporal datasets is a common issue with conventional approaches, such as computational fluid dynamics (CFDs) and various experimental methods, particularly when evaluating and modeling turbulent flows. This review paper focuses on the integration of machine learning (ML), specifically physics-informed neural networks (PINNs), as a means to address this challenge. By directly incorporating governing physical equations into neural network training, PINNs present a novel method that allows for the reconstruction of flow from sparse and noisy data. This review examines various applications in fluid mechanics where sparse data is a common problem and evaluates the effectiveness of PINNs in enhancing flow prediction accuracy. An overview of diverse PINNs methods, their applications, and outcomes is discussed, demonstrating their flexibility and effectiveness in addressing challenges related to sparse data and illustrating that the future of fluid mechanics lies in the synergy between data-driven approaches and established physical theories.

Keywords: physics informed neural networks (PINNs); sparse data; data-driven modeling; PIV; CFD; machine learning in fluids; sparse data reconstruction

1. Introduction

Over the past few years, significant progress has been made in solving partial differential equations (PDEs), which describe the physics of fluid flows in various fluid mechanics applications. Numerous numerical and analytical techniques, along with effective approximations and simplifications for each particular case, were developed. Moreover, significant progress was made in fluid mechanics experimental data post-processing methods. Despite this progress, various problems with current approaches to fluid flow prediction and analysis remain. Fluid mechanics problems often involve complex systems that are governed by a wide range of physical parameters. While some computational fluid dynamics (CFDs) and experimental techniques provide detailed flow dynamics insight, they are often costly, both in terms of computational resources and experimental setup. These methods are frequently limited by physical constraints and uncertainties, such as simplifications in modeling assumptions, numerical errors in computational simulations, and instrumentation inaccuracies. In response to these challenges, machine learning has been increasingly applied in the field of fluid mechanics, addressing tasks related to data processing [1,2], flow control [3–6] etc. While some studies used ML for applications such as super resolution [7–9] and flow classification [10,11], the main attention has been directed to data-driven modeling of flow fields from data. Specifically, artificial neural networks have been used for this purpose due to their ability to capture nonlinear relationships. A variety of neural network architectures have been used for modeling flow dynamics, including fully connected [12–14], convolutional [15–18], recurrent [19–21], and transformer networks [22–24], to name just a few. These models often aim to predict the dynamics either in a discrete manner, i.e., the model predicts future flow states at a fixed time interval based on the current flow state, or using a continuous formulation such as neural ordinary differential equations [25].

Recently, machine learning (ML) has become a promising option that can be implemented in a wide range of fluid mechanics problems. However, the proper training of ML neural networks requires a significant amount of high-quality data, which is not always available for a given application. Despite this challenge, these neural networks can be trained from supplementary data acquired by enforcing different laws of physics. This approach, known as physics-informed learning, combines noisy data with mathematical models, executing them using neural networks. Furthermore, there is a potential to develop specialized network structures that naturally fulfill certain physical principles, enhancing accuracy and generalization capabilities and optimizing training, [26]

In general, ML falls into three main categories: supervised, unsupervised, and semi-supervised learning. Supervised learning unites methods that derive input-output relationships from labeled data; in contrast, unsupervised learning extracts patterns from unlabeled data, while semi-supervised learning blends elements of both unsupervised and supervised learning. Supervised learning is commonly used for tasks such as classification and regression [27–31]. Reinforcement Learning (RL) is typically classified among general machine learning types; however, it will only be briefly addressed in this paper. It is based on structured learning through interaction, where an algorithm is given a set of actions, goals, and parameters. By exploring various strategies within these defined rules, the algorithm learns through trial and error. While reinforcement learning (RL) generally relies on a clearly defined reward function and is predominantly used for optimal control and decision-making tasks, emerging hybrid methods are beginning to incorporate physical constraints and considerations of data sparsity into the training process. However, these approaches tend to be less robust when it comes to addressing inverse problems or partial differential equations (PDEs) with sparse datasets, particularly in comparison to physics-informed neural networks (PINNs), which are explored in this review.

Deep learning (DL), a subset of ML and Artificial Intelligence (AI) (see Figure 1), can learn from data and has made it a popular subject in computing across diverse domains such as healthcare [32], visual recognition [33,34], cybersecurity [35], text analytics [36], etc. It is also recognized as a fundamental technology of Industry 4.0. Nonetheless, developing an effective DL model presents challenges due to the dynamic nature and variability of real-world problems and data [37].

Most DL models do not consider physical constraints, resulting in a poor fit with observational data. Introducing governing physical laws and information about the domain into model training can enhance its accuracy [38]. Therefore, PINNs were introduced in [39]. These frameworks, based on deep learning, can integrate both data and physical laws represented by governing PDEs to facilitate learning. PINNs have proven effective in addressing forward and inverse problems associated with various types of PDEs.

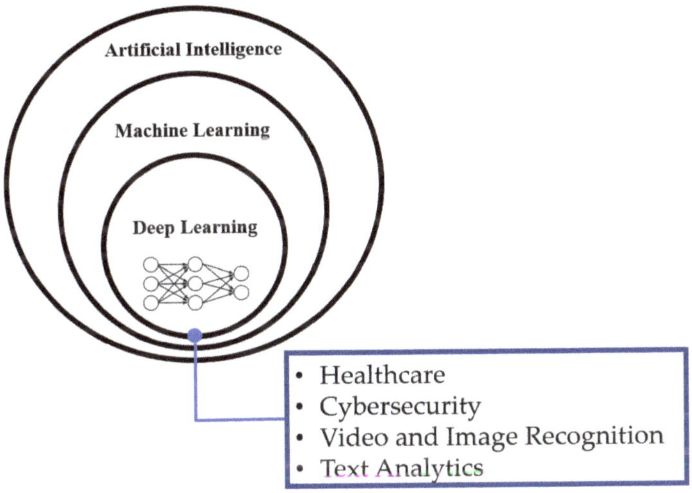

Figure 1. Deep learning applications.

The main requirement for developing an accurate data-driven model for capturing flow dynamics is access to high-dimensional spatio-temporal data. However, in many engineering applications, acquiring such data is challenging due to limitations such as experimental setup constraints, memory capacity, and sensor availability. These factors often result in sparse datasets. In the context of fluid mechanics, data sparsity refers to the condition where only a limited number of measurements are available in space and time relative to the high-dimensional nature of the flow field. Most likely, this occurs due to constraints in experimental resolution (limited sensors, optical access, probe locations, etc.) or computational resources (storage and post-processing limitations in high accuracy simulations). A dataset is considered sparse when the quantity, density, or coverage of the data points is insufficient to resolve the full flow structure or flow dynamics directly and requires the use of reconstruction or interpolation techniques.

Unlike general definitions where sparsity implies many missing or zero-valued entries, in fluid mechanics, sparsity more specifically reflects the incompleteness of physical observations relative to the complexity of the governing phenomena. Sparse data problems arise in various fluid mechanics applications [40–42], including turbulent flow analysis [38,43,44], multiphase flow modeling [45], and environmental fluid dynamics [46,47], where obtaining comprehensive datasets across the entire flow domain or time frame can be difficult and impractical.

Effective handling of sparse data in fluid mechanics is crucial for improving the accuracy of flow predictions, optimizing designs, and enhancing the understanding of physical phenomena. Traditional methods based on interpolating or extrapolating sparse data can introduce uncertainties or inaccuracies, especially in highly nonlinear systems. To mitigate these challenges, modern approaches such as data assimilation [48–51], machine learning techniques [52], and reduced-order modeling [53–56] have been adopted. These techniques leverage existing sparse datasets to reconstruct full-field flow information.

This review aims to explore various methods for handling sparse data in fluid mechanics, focusing on how recent advances in machine learning can overcome these challenges.

Key applications where sparse data is particularly problematic are discussed, and a review of the current state-of-the-art solutions based on physics-informed neural network (PINN) models is provided.

2. Physics-Informed Neural Network Methods Tailored for Navier–Stokes Equations

PINNs are commonly used to simulate fluid flows governed by Navier–Stokes (NS) equations [57,58]. The goal is to parameterize the solution using a neural network $\mathcal{N}$. For a given spatial location (x, y) defined in a 2D domain and time t, the neural network predicts the velocity components $\widetilde{u}$ and $\widetilde{v}$ in x and y directions, respectively, as well as the pressure $\widetilde{p}$, i.e.,

$$\widetilde{u},\ \widetilde{v},\ \widetilde{p} = \mathcal{N}(\theta, x, y, t), \tag{1}$$

where θ represents the parameters of the network. Training a PINN involves minimizing a loss function that is regularized by incorporating the physical laws as soft constraints. Depending on the specific characteristics of the underlying fluid problem, availability of data, and computational constraints, various PINN approaches can be utilized to model fluid flows effectively. Below, we provide an overview of the most commonly used PINN variants for solving Navier–Stokes equations (see Table 1 for a summary).

Table 1. PINN methods tailored for Navier–Stokes.

Method	Description	Usage	Limitation
Standard PINNs	Minimize the residuals of the continuity and momentum equations	Well-posed boundary and initial value problem	Struggle with high Reynolds numbers or complex geometries
Data-driven PINNs	Leverage labelled data to train the model in addition to physical constraints	Problems with available simulation or experimental data	Requires careful balance between physical loss and data-driven loss
Extended PINNs	The domain is decomposed into subregions	Large-scale problems	Requires handling on interfaces
Variational PINNs	The training considers an integral-based loss rather than a pointwise evaluation	Problems with discontinuity	Complex implementation
Stochastic PINNs	Incorporate probabilistic elements to model uncertainties	Fluid flows with uncertainty parameters	Suffer from instability during training
PINNs with turbulence	The turbulence equations are provided as additional constraints to the optimization problem	Highly turbulent flows	Rely on turbulence models, which are usually derived empirically

2.1. Standard PINNs

Standard PINNs are suitable for solving forward and inverse problems for the steady state or transient NS equations in well-defined geometries with known initial and boundary conditions, [39,59]. These models are particularly effective in canonical domains such as laminar flow over flat plates, flow in straight arteries, and benchmark lid-driven cavities. The loss function of the network $\mathcal{L}_{\text{PINN}}$ includes terms to satisfy the continuity and momentum equations, as well as boundary and initial conditions. For 2D incompressible NS equations, the loss function reads

$$\mathcal{L}_{\text{PINN}} = \mathcal{L}_{\text{Con}} + \mathcal{L}_{\text{Mom}-x} + \mathcal{L}_{\text{Mom}-y} + \mathcal{L}_{\text{BC}} + \mathcal{L}_{\text{IC}}, \tag{2}$$

where:

$$\mathcal{L}_{\text{Con}} = \frac{1}{N_f}\sum_{i=1}^{N_f} \epsilon^2_{\text{con}_i} = \frac{1}{N_f}\sum_{i=1}^{N_f}\left(\frac{\partial\widetilde{u}}{\partial x_i} + \frac{\partial\widetilde{v}}{\partial y_i}\right)^2, \tag{3}$$

$$\mathcal{L}_{\text{Mom}-x} = \frac{1}{N_f}\sum_{i=1}^{N_f} \epsilon^2_{x_i} = \frac{1}{N_f}\sum_{i=1}^{N_f}\left(\frac{\partial\widetilde{u}}{\partial t_i} + \widetilde{u}\frac{\partial\widetilde{u}}{\partial x_i} + \widetilde{v}\frac{\partial\widetilde{u}}{\partial y_i} + \frac{\partial\widetilde{p}}{\partial x_i} - \nu\left(\frac{\partial^2\widetilde{u}}{\partial x_i^2} + \frac{\partial^2\widetilde{u}}{\partial y_i^2}\right)\right)^2, \tag{4}$$

$$\mathcal{L}_{\text{Mom}-y} = \frac{1}{N_f}\sum_{i=1}^{N_f} \epsilon^2_{y_i} = \frac{1}{N_f}\sum_{i=1}^{N_f}\left(\frac{\partial\widetilde{v}}{\partial t_i} + \widetilde{u}\frac{\partial\widetilde{v}}{\partial x_i} + \widetilde{v}\frac{\partial\widetilde{v}}{\partial y_i} + \frac{\partial\widetilde{p}}{\partial y_i} - \nu\left(\frac{\partial^2\widetilde{v}}{\partial x_i^2} + \frac{\partial^2\widetilde{v}}{\partial y_i^2}\right)\right)^2, \tag{5}$$

with ρ and ν denoting the density and kinematic viscosity, respectively. The terms ϵ_{con_i}, and $\epsilon_{(x/y)_i}$ are the residuals for the continuity and momentum equations, respectively, evaluated at (x_i, y_i, t_i). Similarly, $\mathcal{L}_{\text{BC}}$ and $\mathcal{L}_{\text{IC}}$ are residual terms introduced to enforce boundary and initial conditions, respectively. During the training, N_f training points, usually denoted as collocation points, are used to evaluate the residuals and to backpropagate the errors to update the network parameters. Altogether, this loss formulation ensures the learned velocity and pressure fields simultaneously satisfy the governing PDEs and the initial-boundary conditions, integrating domain knowledge directly into the optimization process.

Although standard PINNs can handle well-posed boundary-value problems, they usually struggle when considering complex geometries and high Reynolds numbers. In such cases, the loss landscape becomes stiff [60], and the optimizer struggles to converge to a physically meaningful solution [61].

2.2. Data-Driven PINNs

Data-driven PINNs extend the standard framework by integrating partial simulation or experimental data, especially useful when observations are spatially or temporally sparse, as in biomedical flow imaging, environmental sensing, or wind tunnel measurements [62]. This is usually the case when simulation or experimental flow data are spatially or/and temporally sparse. Specifically, the neural network learns to interpolate between the available data points while ensuring that the predictions satisfy the Navier–Stokes (NS) equations. The loss function in this case incorporates an additional term penalizing deviation from the observed measurements, which reads

$$\mathcal{L}_{\text{data-driven}} = \mathcal{L}_{\text{PINN}} + \alpha\mathcal{L}_{\text{data}} = \mathcal{L}_{\text{PINN}} + \alpha\frac{1}{N_d}\sum_{i=1}^{N_d}(\widetilde{u}_i - u_i)^2 + (\widetilde{v}_i - v_i)^2 + (\widetilde{p}_i - p_i), \tag{6}$$

where α is a weighting factor for the multi-objective optimization problem, N_d is the number of observation points, and $u, v,$ and p are the observed velocity and pressure measurements. The data-driven PINN framework is outlined in Figure 2.

The total loss $\mathcal{L}_{\text{data-driven}}$ requires careful calibration between the physical loss term $\mathcal{L}_{\text{PINN}}$ and the data loss term $\mathcal{L}_{\text{data}}$. When α is set too low, the neural network overfits the physical equations and ignores the measurement data. In contrast, setting α too high reproduces the measurement data at the expense of violating the underlying governing equations. Accordingly, various methods have been proposed to mitigate this trade-off, e.g., gradient normalization [63], and self-adaptive learning [64].

2.3. Extended PINNs for Domain Decomposition

Extended PINNs (XPINNs) consist of decomposing the domain into multiple regions and solving the N-S equations in each subdomain, allowing for parallel computing [65]. Within each subdomain, a separate neural network is trained to satisfy the governing physical laws, such as the Navier–Stokes equations.

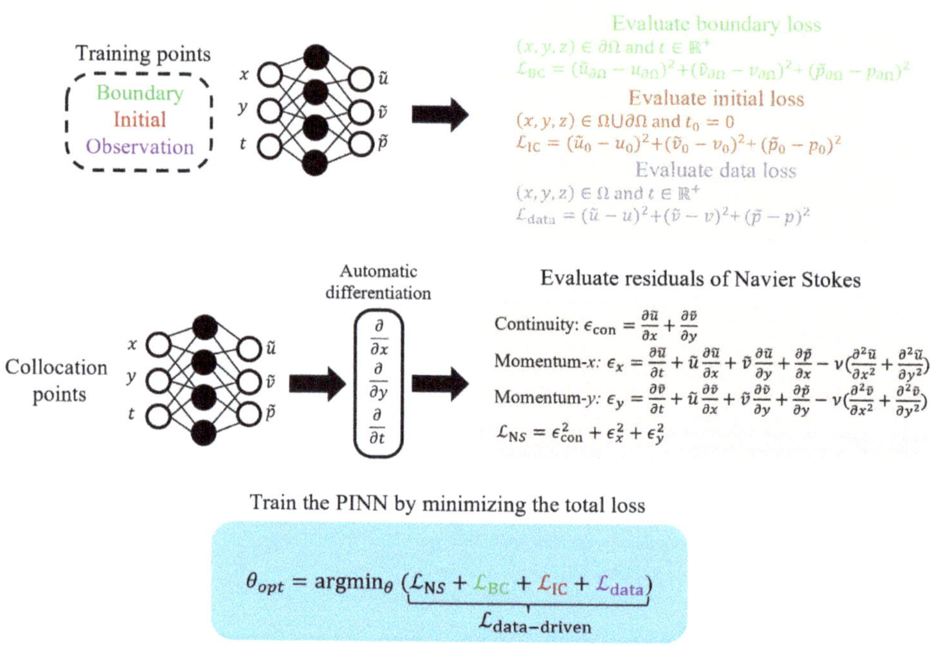

Figure 2. PINN framework for the 2D incompressible Navier–Stokes equations with observations. θ denotes the trainable parameters of the NN.

This domain decomposition approach brings several advantages:

- Parallelization: each subdomain can be trained independently, enabling efficient parallel computation.
- Scalability: XPINNs are particularly well-suited for large or complex geometries (such as vascular networks, combustor chambers, or porous geological formations), where a monolithic network would struggle with convergence or memory limitations.
- Flexibility: local variations in physical properties (e.g., heterogeneous media) can be modeled more effectively using specialized networks in different regions.

The loss function of XPINNs reads

$$\mathcal{L}_{\text{XPINN}} = \sum_k \mathcal{L}_{\text{PINN,k}} + \mathcal{L}_{\text{inter,k}}, \tag{7}$$

where $\mathcal{L}_{\text{PINN,k}}$ is the PINN loss (data, boundary, and PDE residuals) evaluated in each region k and $\mathcal{L}_{\text{inter,k}}$ denotes the interface loss, which enforces continuity of the solution and its derivatives across adjacent subdomain boundaries.

The residuals in this context refer to how well the neural network predictions satisfy the PDE and interface conditions. Specifically, for each subdomain, the residual of the PDE is minimized at a set of collocation points. At interfaces, additional residuals are defined to ensure the physical compatibility between subdomains—this typically includes continuity in both the solution and fluxes (e.g., pressure, velocity gradients).

The XPINN framework has been successfully applied in various applications, including flow simulations through highly heterogeneous media [66] and modeling of supersonic flows [67].

2.4. Variational PINNs

Variational PINNs (VPINNs) use a weak formulation of the NS equations, allowing better stability in irregular or highly curved domains, such as aneurysms, coronary arteries, or bioinspired microfluidic channels [68,69]. The training process consists of minimizing an integral-based loss, denoted as $\mathcal{L}_{\text{VPINN}}$, rather than pointwise evaluation, i.e.,

$$\mathcal{L}_{\text{VPINN}} = \sum_{i=1}^{N_b} \left(\int_{\Omega} \phi_i \cdot \left(\partial_t u + (u \cdot \nabla) u + \frac{1}{\rho} \nabla p - \nu \Delta u \right) d\Omega \right)^2, \tag{8}$$

where $\{\phi\}_{i=1}^{N_b}$ is a basis of test functions spanning the test space, and Ω denotes the fluid domain. In [68], a Petrov–Galerkin variational PINN was introduced by representing the trial functions—i.e., the unknowns being approximated using neural networks—while the test functions were taken as Legendre polynomials on each element. In this formulation, Legendre polynomials were chosen because they (i) are orthogonal on the reference element, (ii) offer good approximation properties for smooth functions, and (iii) permit efficient methods for computing integrals and derivatives.

2.5. Stochastic PINNs

Stochastic PINNs (SPINNs) address uncertainty in the fluid flow, such as random initial or boundary conditions [70,71]. Applications include weather prediction, turbulent combustion with input uncertainty, and patient-specific flows with uncertain measurements. The neural network treats the PDE constraints probabilistically, which is essential for modeling fluid flows with uncertain parameters or noisy observations. Within the SPINN framework, the individual terms of the loss function are enforced as expected values, with the option of incorporating an additional regularization term to minimize the variance of the predicted solution.

Let $\mathcal{L}_{\text{PINN}}(\xi)$ denote the standard PINN loss evaluated on a stochastic input sample ξ, e.g., drawn from uncertain boundary data, and similarly let $\mathcal{L}_{\text{data}}(\xi)$ be the corresponding data loss. The corresponding SPINN loss reads:

$$\mathcal{L}_{\text{SPINN}} = \mathbb{E}_{\xi}[\mathcal{L}_{\text{PINN}}(\xi)] + \alpha \mathbb{E}_{\xi}[\mathcal{L}_{\text{data}}(\xi)] + \lambda \sum_{\varphi \in \{u,v,p\}} \mathbb{V}_{\xi}[\varphi], \tag{9}$$

where $\mathbb{E}_{\xi}[\cdot]$ denotes the expectation over uncertain input samples, λ is a weighting factor controlling the regularization rate, $\mathbb{V}\xi[\cdot]$ denotes the pointwise variance of predictions, and α is a weighting factor that controls the relative importance of the data loss.

2.6. PINNs with Turbulence

In this family of methods, turbulence models are incorporated within the PINN framework (e.g., $\kappa - \epsilon$ [72,73] or Spalart–Allmaras [43] model). This extension helps model the effects of turbulence rather than directly solving them. The corresponding loss function is written as

$$\mathcal{L}_{\text{PINN,tr}} = \mathcal{L}_{\text{PINN}} + \mathcal{L}_{\text{data}} + \mathcal{L}_{\text{tur}}, \tag{10}$$

where $\mathcal{L}_{\text{tur}}$ denotes the residual of the turbulence model. These methods are used in aerodynamic simulations, jet mixing studies, and flow separation predictions.

2.7. B-PINNs: Bayesian Physics-Informed Neural Networks

In [70], the authors propose a Bayesian physics-informed neural network (B-PINN) framework to solve forward and inverse partial differential equation (PDE) problems under uncertainty. Unlike standard PINNs, B-PINNs incorporate Bayesian neural networks (BNNs) to enable uncertainty quantification and improve robustness against noisy data.

1. Problem setup

   Given a general PDE: $N_x(u;\lambda) = f(x),\ x \in D$ ($d-$dimensional physical domain), and boundary condition operator acting on the domain boundary $\Gamma:\ B_x(u;\lambda) = b(x),\ x \in \Gamma$ with noisy data:

$$\overline{u}^{(i)} = u\left(x_u^{(i)}\right) + \varepsilon_u^{(i)}, \tag{11}$$

$$\overline{f}^{(i)} = f\left(x_f^{(i)}\right) + \varepsilon_f^{(i)}, \tag{12}$$

$$\overline{b}^{(i)} = b\left(x_b^{(i)}\right) + \varepsilon_b^{(i)}, \tag{13}$$

   with $\varepsilon \sim N(0,\sigma^2)$. This noise reflects aleatoric uncertainty.

2. Bayesian Surrogate and Posterior Inference

   The surrogate solution $\widetilde{u}(x;\theta)$, parameterized by a BNN, satisfies:

$$\widetilde{f}(x;\theta) = N_x(\widetilde{u}(x;\theta)),\ \widetilde{b}(x;\theta) = B_x(\widetilde{u}(x;\theta)). \tag{14}$$

   The likelihood function over the noisy data is:

$$P(D|\theta) = P(D_u|\theta)P\left(D_f\middle|\theta\right)P(D_b|\theta), \tag{15}$$

   for example:

$$P(D_u|\theta) = \prod_{i=1}^{N_u} \frac{1}{\sqrt{2\pi\sigma_u^{(i)^2}}} \exp\left(-\frac{\left(\widetilde{u}\left(x_u^i;\theta\right) - \overline{u}^{(i)}\right)^2}{2\pi\sigma_u^{(i)^2}}\right). \tag{16}$$

   The posterior is obtained from Bayes' theorem:

$$P(\theta|D) = \frac{P(D|\theta)P(\theta)}{P(D)} \simeq P(D|\theta)P(\theta), \tag{17}$$

   which is approximated via:

   - Hamiltonian Monte Carlo (HMC)—accurate but computationally expensive
   - Variational Inference (VI)—efficient but less reliable due to mean-field assumptions
   - Dropout—used as a non-Bayesian baseline (performed poorly in uncertainty estimation)

3. Inverse Problem Extension

   For inverse problems, where PDE parameters λ are also unknown:

$$P(\theta, \lambda|\, D) \simeq P(D|\theta,\lambda)P(\theta,\lambda) = P(D|\theta,\lambda)P(\theta)P(\lambda). \tag{18}$$

   This enables joint inference of both the solution and the unknown physical parameters.

4. Main Results

- B-PINN-HMC yields superior predictive accuracy and reliable uncertainty estimates.
- B-PINN-VI underestimates uncertainty due to Gaussian factorization.
- Dropout-based PINNs fail to capture meaningful uncertainty.
- Replacing BNNs with a truncated Karhunen–Loève (KL) expansion yields similar accuracy at lower computational cost for low-dimensional problems.
- Deep Normalizing Flows (DNFs) are evaluated as alternative posterior estimators.

In general, it was demonstrated that B-PINNs offer a robust, probabilistically grounded alternative to deterministic PINNs, particularly in the presence of noise and limited data. The Bayesian formulation enables principled uncertainty quantification but requires careful consideration of prior choices and scalability for high-dimensional applications.

A similar approach was used in [74].

3. Physics-Informed Neural Networks in Experimental and Numerical Fluid Dynamics

Examples of PINNs in Laminar and Turbulent Flows

The physics-informed neural network (PINN) was applied in [44] to address certain limitations in simulating incompressible turbulent and laminar flows. Navier–Stokes flow nets (NSFnets) were introduced, which are designed based on the Navier–Stokes equations using two mathematical formulations: velocity-velocity (VV) and velocity-pressure (VP).

Databases from direct numerical simulation (DNS) and analytical solutions were used to provide boundary conditions and reference baselines for the NSFnets. The networks take spatiotemporal coordinates as input and generate instantaneous velocity as output. In the VP-NSFnet, the output includes vorticity fields, while the VV-NSFnet outputs both pressure fields and instantaneous velocity.

This approach eliminates the need for labeled data beyond the boundary conditions, properties of the fluid, and initial conditions. Residuals of the governing equations (VV or VP), together with boundary and initial conditions, are incorporated into the loss function of the NSFnets.

It is interesting to note that the VP-NSF-net does not receive direct pressure data; instead, this information is treated as a hidden state and is derived through the incompressibility constraint without any additional computational costs. It should be noted that this approach might not be directly applicable for compressible flows where pressure plays a crucial role in the governing equations. For compressible flows, pressure cannot be obtained from velocity and incompressibility constraints alone, since the flow's density and thermal energy also vary and influence the pressure field. Therefore, in models such as NSFnet for compressible flows, pressure must typically be treated as an explicit variable rather than a hidden state, and its evolution must be calculated directly based on the compressible Navier–Stokes equations, including the equation of energy and other thermodynamic relationships.

For laminar flow, it was demonstrated by [44] that both the VV and VP VV formulations achieve similar accuracy. In contrast to traditional numerical methods, NSFnets adopt neural network (NN) properties, leading to a total error composition comprising approximation, optimization, and generalization errors. In this study, an attempt was made to empirically quantify these errors by changing NN architectures, iterative solvers, and sampling ("residual") points. For the turbulent channel flow, NSFnets showed the ability to sustain turbulence at the skin-friction-velocity Reynolds number $Re_\tau \sim 1000$ [75]. However, due to intensive training requirements, the focus was made on enforcing velocity boundary conditions based on DNS data subdomain boundaries and a subset of the channel domain.

In [76], PINN was applied to study a flow with a low Reynolds number around a cylinder without simulation or experimental data input. Freestream conditions were used for PINN training and to define the boundaries of the computation domain.

The physics-informed neural network (PINN) consisted of layers (fully connected), where the biases and weights were adjusted according to specified boundary conditions and the governing physical equations. Every node in one layer of a fully connected network is connected to every other node in the next layer, which contains a number of nodes according to the features fed into the network and is known as the input layer. For problems related to fluid flow, these characteristics typically include time and spatial coordinates. The final layer, or output layer, provides the expected values for the input features, with the number of nodes matching the dimension of the prediction. The layers, which exist between the output and input layers, are known as hidden layers and contain a variable number of nodes.

This PINN effectively captures the overall flow field trends, with velocity fields generated by the PINN comparable to the fields obtained from computational fluid dynamics (CFDs). The Navier–Stokes equations solution was achieved numerically with simpleFoam, which utilizes the SIMPLE algorithm to iteratively update the pressure and velocity fields. Second-order upwind schemes were used for spatial discretization.

However, the PINN encountered challenges with pressure fields due to the absence of an explicit pressure term in the loss function. It was shown that an increased number of layers had the most significant impact on result accuracy, followed by expanding the point cloud size. The smallest performance improvement was observed when enhancing the nodes in each hidden layer. Furthermore, the PINN approach was more memory-efficient than CFD but did not necessarily outperform it in computational time. In simpler cases, where the CFD mesh contained a low number of cells, PINN took more time to compute results as compared to CFD. However, in scenarios involving high cell counts in CFD meshes, PINN demonstrated greater time efficiency.

Cases with different amounts of data demonstrated that the solution's resolution was significantly affected by the point cloud density. A low-density point cloud can fail to resolve near-wall flow conditions properly, while an overly dense point cloud can overcomplicate the neural network's training. To isolate the effect of point cloud density, each hidden layer was set to 50 nodes and kept constant for each of the 20 hidden layers of the network's architecture. PINN accuracy was tested with point clouds of 2000, 5000, 10,000, and 15,000 points.

It was shown that with only 2000 points, PINN failed to match CFD results, especially in capturing the leading-edge velocity profiles and pressure field magnitudes. This indicates that a 2000-point cloud is too scattered to resolve the pressure and velocity fields accurately for this scenario. When the point cloud density was increased to 5000 points, PINN began to show improved accuracy in both the field of pressure and velocity. The velocity profile became closer to CFD results; PINN still predicted a smaller flow acceleration region due to capturing a lower magnitude of velocity.

For the 10,000-point cloud, no significant improvements in the velocity components were observed compared to the 5000-point case. Although u-velocity components captured the wake profile, the magnitude of velocity near the leading edge was reduced in the high flow region. The v-velocity profile followed similar trends, with diminished high-velocity regions. However, the pressure profile from PINN showed increased symmetry, though the pressure magnitudes remained lower than those produced by CFD.

4. Physics-Informed Neural Networks Using Sparse Data

4.1. Introducing PINN Using Sparse Data

PINNs' approach was used in many applications [77–82], but in this review paper, the focus will be on applications of flow reconstruction using data from noisy and sparse data. Researchers are facing challenges in reconstructing a flow field from noisy and limited measurements, which are encountered across various engineering applications. To compensate for the sparsity and incompleteness of the data, supplementary information is required, which can be sourced from physics-based models or offline flow databases. With access to an extensive database of flow fields, correlation features, and coherent structures of fluid flow, it is possible to extract and utilize these features to reconstruct high-resolution flow fields using sparse data. Flow feature extraction techniques such as Proper Orthogonal Decomposition (POD) [83,84] or Dynamic Mode Decomposition (DMD) [85] are usually employed in this context. Recent progress in the development of deep learning algorithms for image super-resolution has opened up new possibilities for flow reconstruction from limited measurements. NNs were applied to either learn to directly construct end-to-end mapping between high-resolution flow fields and sparse measurements or to learn the POD coefficients.

However, these deep learning models heavily rely on sufficient training data (offline), which may not be available in many applications. To address the problem related to sparsity of data, a physics-constrained deep learning approach has been proposed for flow reconstruction [86].

4.2. PINNs Using Navier–Stokes Equations and Sparse Set of Video Frames

In [87], a novel method for reconstructing fluid flows utilizing the Navier–Stokes equations (governing physics) was introduced with an end-to-end optimization process. This approach bypasses the need for explicit inputs such as boundary conditions and geometry, using only a sparse set of video frames. This strategy establishes a smooth spatio-temporal representation of scenes, employing neural networks to model density and velocity solutions for fluid flows. A novel reconstruction of fluid interactions with static obstacles was achieved without requiring additional human labeling or geometry by segregating dynamic and static elements. This approach adheres to physical principles and benefits from image supervision.

To enhance optimization from sparse input views, a layer-by-layer progressive growth strategy was implemented, gradually increasing the network capacity of the resulting neural representation. Utilizing this progressively growing model alongside a newly introduced regularization term, density–color ambiguity in radiance fields was effectively addressed without encountering overfitting issues.

During training, a sliding window approach was used to progressively involve early-layer neurons, similar to shallow model training, and then to cover the whole architecture by slowly moving the window towards the deeper layers. For every hidden layer $m \in [0, N-1]$, the following weighting parameter governed the sliding window: $w_{lm} = \text{clamp}(1 + m_a - m, 0, 1) \cdot \text{clamp}\,(1 + m - m_a, 0, 1)$, where $m_a = 1 + (N-2)/S$ represents the current last hidden layer. Here, s refers to the present training iteration, and S represents the iteration step at the completion of progressive growth. The layer $m = 0$ (with $w_{l0} = 0$) is a permanent hidden layer. For $m = 1$, the layer fades out, whereas intermediate layers fade in and out, and the last layer ($m = N-1$) only fades in. A reasonable density profile was learned by the growing model in the second row after applying progressive growing.

However, at the 5000-iteration step, "ghost density" artifacts appeared. To address and penalize this ghost density, a regularization technique in 2D image space was proposed.

$$\mathcal{L}_{\text{ghost}}\left(C\left(r_{ijt}\right), B_{ijt}, a\left(r_{ijt}\right)\right) = \text{sigmoid}\left[-\left(C\left(r_{ijt}\right) - B_{ijt}\right)^2\right] \cdot A\left(r_{ijt}\right), \tag{19}$$

Background B_{ijt} is a known input, and $\mathcal{L}_{\text{ghost}}$ penalizes the opacity $t = k = 1KT(hk)\alpha(hk)$ when the pixel color $C\left(r_{ijt}\right)$ is close to the background color in the 2D image space. After applying $\mathcal{L}_{\text{ghost}}$ on the progressively growing model, it effectively refined the density profile during training. This approach, "ghost density" removal, proved effective across many scenes.

4.3. PINNs in Biomedical Applications Without Boundary Conditions

In [88], PINN was applied to enhance the quantification of the wall shear stress (WSS) in diseased arterial flow, particularly in scenarios where some boundary conditions, such as outlet and inlet boundaries, were not known. The WSS and blood flow near arterial walls play crucial roles in regulating major cardiovascular diseases. However, accurately quantifying them poses challenges due to issues such as uncertainty, low resolution, and noise in experimental and computational measurements of WSS for a specific patient. The PINN offers a flexible deep learning approach by combining mathematical equations of blood flow with experimental data. By incorporating the governing equations, particularly the Navier–Stokes equations, the substantial data requirements of the PINN can be dramatically reduced. In this study, an example using idealized aneurysm and stenosis models combined partial knowledge of flow physics with partial measurements, yielding a precise prediction of near-wall flow. The mathematical problem in [88] was presented with steady, incompressible, Newtonian Navier–Stokes equations and the continuity equation.

In cases where we have accurate knowledge of the parameters within the Navier–Stokes equations (represented by μ and ρ) and the boundary conditions within the computational domain, traditional computational fluid dynamics (CFDs) methods can effectively solve the equation mentioned above. However, in applications related to patient-specific cardiovascular fluid mechanics, there is often partial knowledge or uncertainty associated with these parameters and boundary conditions. In this scenario, the issue is that the problem lacks a clear definition, leading to a lack of a unique solution. However, it has been suggested that by using sparse measurements within the domain, a solution might be achievable. Suppose sparse set points were used to evaluate the velocity vector field. The main aim of PINN is to find the optimal solution $u(x)$ that not only conforms to these measured data points but also verifies the Navier–Stokes equations under unknown parameters or partially specified boundary conditions. In [88], fully connected neural networks (NNs) were specifically used; artificial neuron layers were stacked to form a composite function. An affine transformation was applied to every layer by the previous layer, introducing nonlinearity through an element-wise activation function. The biases and weights are determined by using Adam optimization, which is a variant of stochastic gradient descent optimization [89]. For all layers and cases, the Swish activation function was used.

With a proper definition of the boundary conditions (BC), a mathematically well-posed problem can be constructed by the two objectives mentioned above (BC and Navier–Stokes), which can be solved using a physics-informed neural network. However, in some cases of cardiovascular modeling of a specific patient, the boundary conditions are not fully known. Therefore, in [88], a boundary condition loss L_{BC} was only applied on certain parts of the boundary. While with conventional numerical techniques, solving partial differential equations (PDEs) is not usually feasible without well-defined BCs, PINNs can still converge to a solution by minimizing the residual of the field equation and handling fully unknown or partially known BCs. Despite this, the resulting solution may not be the true one, as many or infinite numbers of possibilities are present to find the solution of PDE when BCs are not fully defined. To address this challenge, sparse data from the measurements was introduced to define a new data loss term ($\mathcal{L}_{\text{data}}$):

$$\mathcal{L}_{\text{data}} = \|z - f\|_{G}, \tag{20}$$

where f is the function of velocity (it was assumed that the velocity field was measured in sparse data points $u(x_i) = f(x_i),\ x_i \in \Gamma$; with $\Gamma = \{x_i\},\ i = 1, \ldots, N$ representing discrete measurement locations where the velocity was obtained by the function f and z is the velocity measurement in sparse locations G. The total loss function was defined as

$$\mathcal{L}_{\text{tot}}(W_i, b_i) = \mathcal{L}_{\text{phys}} + \lambda_b \mathcal{L}_{\text{BC}} + \lambda_d \mathcal{L}_{\text{data}}, \tag{21}$$

$$W_i^*, b_i^* = \mathcal{L}_{\text{tot}}(W_i, b_i), \tag{22}$$

For each pressure network and velocity component b_i^* and W_i^*, the optimal weights and biases were obtained such that the total loss function $\mathcal{L}_{\text{tot}}$ was minimized.

The λ_d and λ_b are known as hyperparameters and were used to account for the contribution of the measurement data and boundary conditions (BCs), respectively. The framework presented in [88] was tested across various scenarios, including 2D blood flow simulation in both aneurysmal and stenotic conditions, 1D advection-diffusion transport, and blood flow simulations in 3D in an idealized aneurysm, and achieved highly accurate predictions of the wall shear stress (WSS). Specifically, two unknown parameters were required to solve the 1D problem analytically, but the boundary conditions were not specified. The coefficient was obtained using a regression problem that uses the sparse measurement data (three points in this case). In this example, a sharp boundary layer is applied to make the regression problem ill-conditioned, requiring at least one point within the boundary layer to address this issue. In the two-dimensional problems, the inlet boundary condition was not accurately identified by PINN; however, it correctly reconstructed the flow field where the sensors were placed and downstream of them. This suggests that the PINN framework in [88] should be viewed as an inverse modeling problem by which the inlet boundary condition is to be predicted. A common issue with many inverse problems is that for the same flow rate, different inlet velocity profiles can yield similar downstream solutions, and consequently, there may be no unique solution. Therefore, to accurately identify the inlet boundary conditions, sensors need to be placed near the inlet of the domain. It should be noted that the results in [88] demonstrated that accurately determining the WSS in the region of interest does not necessarily require knowing the inlet velocity profile. This framework illustrates the potential of integrating partial knowledge of cardiovascular flow physics, such as the Navier–Stokes equations, with sparse data gathered from localized regions without specific boundary conditions. This combination enables the extraction of clinically relevant hemodynamic parameters such as WSS.

4.4. PINNs Using Reynolds-Averaged Navier–Stokes Equation Without Closure Turbulence Model

In [90], PINNs incorporating RANS equations were used for two specific scenarios: the periodic hill problem ($5600 \leq Re_b \leq 37{,}000$; Re_b is the Reynolds number based on bulk velocity) and an adverse-pressure-gradient (APG) boundary layer ($1000 < Re_\theta < 3000$, where θ is the momentum thickness and Re_θ—Reynolds number based on momentum thickness).

In order to close the RANS equations, this approach does not rely on a turbulence model; instead, the neural network learns to infer the Reynolds stress fields during training as a means to obtain closure. Regarding the potential of PINNs to provide reliable estimations with sparse training data, this study indicates their capability to be trained using data typically collected in experiments. The goal is to determine the quantity and distribution of data required to model a time-averaged turbulent flow accurately with PINNs. For the APG boundary layer, the goal is to determine the amount and distribution of data needed to model a time-averaged turbulent flow accurately with PINNs. It was shown that PINNs can more effectively model wall pressure and WSS and show their performance when trained without crucial flow features such as a separation bubble. Additionally, it was demonstrated that the training cost of the network remains consistent across different Reynolds numbers. The PINN architecture overview is shown in Figure 2. The loss function in [90] was defined as

$$\mathcal{L} = \mathcal{L}_b + \lambda_r \mathcal{L}_r, \tag{23}$$

$$\mathcal{L}_b = \frac{1}{N_b} \sum_{j=1}^{N_d} \sum_{i=1}^{N_b} \left[T_{ij} - \hat{U}_{ij}\right]^2, \tag{24}$$

$$\mathcal{L}_r = \frac{1}{N_r} \sum_{j=1}^{N_e} \sum_{i=1}^{N_r} \epsilon_{ij}^2, \tag{25}$$

where $\mathcal{L}_b$ represents the mean squared error between training data T_{ij} and network prediction, it ensures that the network remains grounded in available empirical or simulated data, making it particularly valuable when measurements are sparse, noisy, or unevenly distributed. $\mathcal{L}_r$ is the residual loss. The weight λ_r was used to scale residual loss, N_d represents the count of network outputs, N_b is considered the number of training points provided to the network, N_e represents the residual equations to be solved, N_r is referred to as the number of residual points in the domain. The equation residual is denoted by ϵ_{ij}. Physical reasoning can be applied to impose Dirichlet conditions at specific points along the wall by introducing an additional term in the loss function in experimental campaigns. This was completed by using a set of spatial coordinates corresponding to the domain boundaries.

It was shown in [90] that PINNs could make accurate predictions for APG boundary layer problems with minimal data points. Accurate prediction of C_f (friction coefficient) and C_p (pressure coefficient) was achieved with just thirteen training points within δ_o (inside the BL) and six data points δ_{in} (velocity and Reynolds-stress), along with scattered pressure data on additional boundaries, including the wall and downstream locations.

However, the estimation of C_f can slightly degrade with less data availability. To test the network's performance without information on specific flow features such as separation, it was trained on a smaller dataset, excluding data from the separated flow region across different Re_b values. The network successfully predicted the recirculation zone extent but had issues with the estimation of the velocity magnitude within the separated region.

It was also shown that there is no significant increase in the computational cost of PINNs with increasing Reynolds numbers, making them suitable for modeling high Reynolds-number flows.

4.5. PINNs Using RANS with Reynolds Stresses; Solenoidal Forcing; Prediction Validation with Sparse Experimental Data

Sliwinski and Rigas in [91] focused on utilizing PINNs for reconstructing time-averaged quantities of an unsteady flow based on sparse velocity data. The schematic of physics-informed neural networks, which was used in [91], is shown in Figure 3. In the green box (Figure 4), network inputs are depicted, which can be either data point coordinates or collocation points. The flow variable modeled by the function and implemented through a neural network is represented by the blue box. The specific derivatives related to flow variables are denoted by the red box, computed through automatic differentiation. The residual operator is represented by the purple box, which takes the flow variable along with its derivatives to assess the physical residuals.

The RANS equations with the Reynolds stresses are incorporated in the top version of the network. On the other hand, RANS with solenoidal forcing is used by the bottom version of the network. When the network input consists of data points, the output layer is activated, and the residuals of the governing equations are not computed.

The data-assimilation framework aims to find a model function $g(x) = [g_1(x), g_2(x), \ldots, g_R(x)]^T$, represented as a neural network, by minimizing the objective function $\mathcal{L}$. This model maps the domain coordinates x and y to flow variables R.

$$\mathcal{L}(g) = \mathcal{L}_D(g) + \mathcal{L}_B(g) + \mathcal{L}_P(g), \tag{26}$$

where $\mathcal{L}_D$ represents data loss, the boundary condition loss is $\mathcal{L}_B$, and the physical loss by enforcing RANS equations is $\mathcal{L}_P$. The data loss is the mean squared error, computed at all points where flow variables are known.

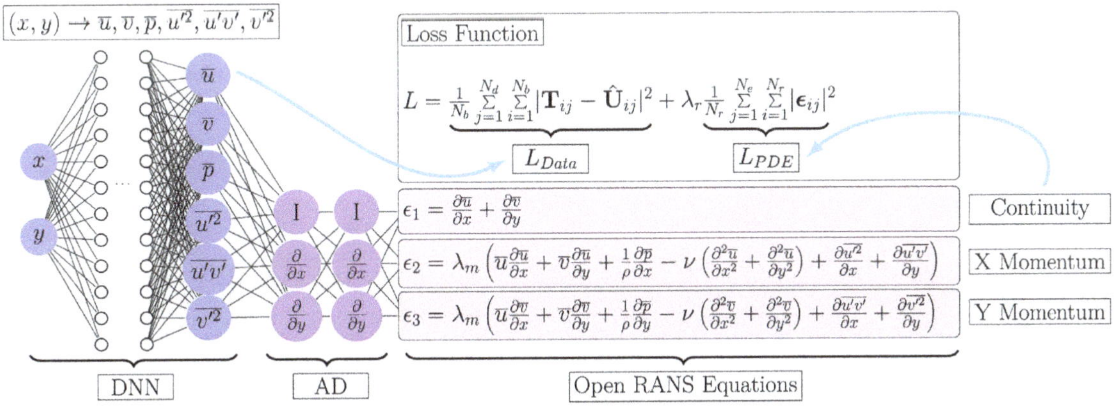

Figure 3. A PINN with embedded RANS equations [90], Network Quantities: $\mathcal{L}$—Loss, T_{ij}—Training data, $\hat{U}_{ij}$—Network predictions, ϵ_{ij}—Residual, N_b—No. data points, N_r—No. residual points, Flow Parameters: x, y—Residual coordinates, $\bar{u}$, $\bar{v}$—Velocity, $\bar{p}$—Pressure, $\overline{u'^2}$, $\overline{u'v'}$, $\overline{v'^2}$—Reynolds stresses, ρ—Density, ν-Viscosity. (Figure adopted under a Creative Commons license). The same figure was used in [92].

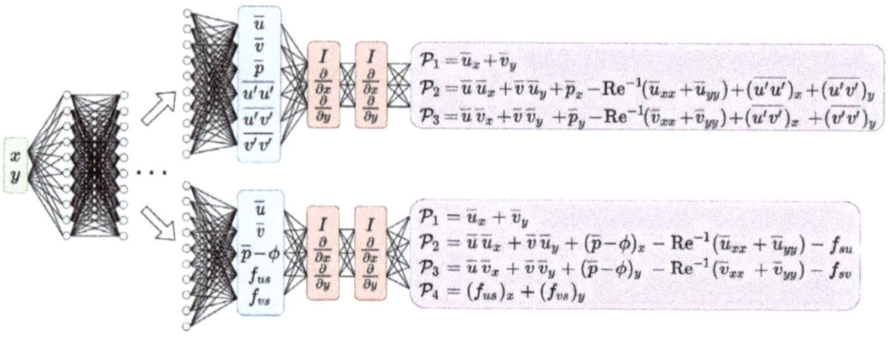

Figure 4. A diagram illustrating two different approaches of PINNs for flow reconstruction [91]. Figure adopted under a Creative Commons license.

To find some unknown closure quantities, such as the curl of unsteady RANS forcing, and to interpolate fields from sparse data [91] utilized the time-averaged velocity data. The capabilities of this technique were expanded to assimilate Reynolds stresses, while PINNs effectively incorporated the data to complete both velocity and stress fields. This process also provided insights into the pressure field.

The presented framework was validated in the context of unsteady cylinder flow at Re = 150, showing satisfactory predictions in the cylinder wake, while discrepancies were observed near the cylinder surface. An investigation into the impact of noise in the velocity data on PINN predictions revealed that it reduced the accuracy of the predicted forcing. However, PINN effectively filtered the noise from the input velocity data.

Additionally, PINNs demonstrated the ability to interpolate flows that depended on limited experimental velocity data, showcasing their potential for reconstructing flows from sparse measured data. The scope of PINN was expanded to explicitly incorporate Reynolds stresses (rather than implicitly through the forcing term), enabling the utilization of pressure measurement and second-order velocity statistics. Leveraging sparse measurements of mean Reynolds stresses and mean velocities, the method successfully interpolated velocity and Re-stress fields for the cylinder flow, although some discrepancies remained. Incorporating pressure measurements on the cylinder surface notably enhanced interpolation accuracy. The interpolated velocity fields also enabled accurate estimation of the skin friction distribution over the cylinder surface.

4.6. PINNs Using Experimental Data with Different Spatial Sparsities

In [93], an approach to reconstruct flow fields using PINN, which incorporates known data alongside physical principles to handle imperfect data, was introduced. The circular cylinder wake flow was used as a test scenario. The study explores two types of training CFD sets: one comprises velocity data with varying levels of sparsity, while the other includes data of velocity that is missing from different regions. Training convergence was improved using a learning rate schedule, specifically demonstrating the effectiveness of the cosine-annealing algorithm.

To replicate the conditions of a real experiment, data for continuous time were given to the physics-informed neural network at specific fixed points. Specifically, 1%, 5%, 20%, and 100% of the actual velocity data were used for fitting and prediction. It is important to note that the training set included only velocity information, as the nonlinearity in the flow is driven by velocity transport.

To quantitatively compare the deviation between the original and the predicted flow field across the whole time domain, the commonly used relative norm R_{L2} was introduced:

$$R_{L2} = \frac{||\hat{U} - U||}{||U||}, \tag{27}$$

where $||\hat{U} - U||$ represents the L2 norm of the deviation between the predicted quantity (e.g., u, v, p) at a particular time, and $||U||$ is the original quantity L2 norm at that specific time. R_{L2} metric provides a reliable measure of prediction accuracy at each time step. It was found that the average R_{L2} between the predicted and actual or original flow fields at a particular time, utilizing 100% of the velocity data for training was 3.5×10^{-2} for the pressure, 1.2×10^{-2} for the spanwise velocity, and 3.9×10^{-3} for the streamwise velocity. These indicate the successful wake flow fitting by the PINN.

It was also observed that the PINN had low dependency on the amount of data. With a suitable NN architecture and efficient training strategy, precise flow predictions were achieved using only 1% of the original data. Thus, PINNs can accurately reconstruct flow fields even with relatively sparse data.

After successfully fitting the momentum equation, the partial derivative of pressure was trained as a constraint, making the PINN-computed pressure gradient consistent with the original data. However, the pressure field differed by a constant value at each snapshot since pressure information was not directly included in the training set—only its spatial gradient was used.

The agreement between the predicted and original flow fields showed that by regularly selecting a few measurement points and using data from these points to build a training set, the PINN could reconstruct the flow field with high accuracy due to its adherence to physical principles. This makes the approach a promising data assimilation method for experimental fluid dynamics. It could potentially be used with particle image velocimetry (PIV) for velocity field reconstruction, allowing PINNs to predict flow in unmeasured regions and effectively expand the coverage of PIV measurements.

4.7. PINNs Compared to RANS and URANS

In [43], precise sparse mean velocity points along with the RANS equations with data assimilation techniques were integrated. The main goal was to bridge the gap between PINNS (data assimilation method) and the methods using traditional spatial discretization (variational methods).

A comparative analysis of both approaches was conducted using a turbulent flow case. By applying sparse data to constrain PINNs and solving the underdetermined RANS equations without full boundary conditions, superior accuracy in reconstructing the mean flow was demonstrated compared to a RANS solver employing the Spalart–Allmaras (SA) turbulence model. The following SA-augmented RANS equations were considered:

i. Continuity equation (mass conservation):

$$\frac{\partial U_i}{\partial x_i} = 0. \tag{28}$$

By penalizing deviations from zero divergence in the velocity field U_i, the model ensures that the volume of fluid is conserved throughout the domain.

ii. Momentum conservation:

$$U_j \frac{\partial U_i}{\partial x_j} + \frac{1}{\rho} \frac{\partial (P - \Phi)}{\partial x_i} - \frac{\partial \left[2(\nu + \nu_t) S_{ij} \right]}{\partial x_j} - f_{s,i} = 0. \tag{29}$$

iii. Forcing source term continuity:

$$\frac{\partial f_{s,i}}{\partial x_i} = 0. \tag{30}$$

This condition maintains solenoidal consistency for the applied forcing field $f_{s,i}$, ensuring that any synthetic or modeled body forces introduced into the system do not artificially create or destroy mass or momentum.

iv. Transport equation for turbulent viscosity:

$$U_j \frac{\partial \widetilde{\nu}}{\partial x_j} - S_p - S_{\text{diff}} - S_C - S_d = 0, \tag{31}$$

where P represents the mean pressure field, S_{ij} denotes the mean strain rate tensor, and the Reynolds forcing vector f_i was decomposed using Helmholtz decomposition into a divergence-free solenoidal, a potential part (Φ), a scalar, and a vector component ($f_{s,i}$); ν_t is an actual eddy viscosity and $\widetilde{\nu}$ is a pseudo eddy-viscosity variable. S_p represents the production, the diffusion is S_{diff}, the cross-diffusion is S_C, and S_d is the destruction term.

In PINN, the solution $\left(\hat{U}_i, \hat{P} - \hat{\Phi}, \hat{f}_{s,i}, \hat{\widetilde{\nu}}\right)(x;\theta)$ was determined by the biases and weights, θ, which characterized the neural network. The optimal set of biases and weights, $\hat{\theta}$ was obtained by minimizing a loss function $\mathcal{L}$,

$$\hat{\theta} = \mathcal{L}(U_i, P - \Phi, f_{s,i}, \widetilde{\nu})(x;\theta), \tag{32}$$

$\mathcal{L}$ represents the loss function used to enforce both the boundary conditions and the governing equations. These components were sufficient for optimizing the forward problem. For the inverse problem (data assimilation), however, a measurement error term—namely J—must be included, as follows:

$$\begin{aligned} \mathcal{L} = {} & \tfrac{\lambda^D}{N_m} J + \tfrac{\lambda^P}{N_C} \sum_{k=1}^{N_C} \text{RANS}[(U_i, P - \Phi, f_{s,i}, \widetilde{\nu})(x_k;\theta)]^2 + \tfrac{\lambda^P}{N_C} \sum_{k=1}^{N_C} \left(U_j \tfrac{\partial \widetilde{\nu}}{\partial x_j} - S_p - S_{diff} - S_C - S_d \right)(x_k;\theta)^2 \\ & + \tfrac{\lambda^B}{N_b} \sum_{k=1}^{N_C} \text{BC}[(U_i, P - \Phi, f_{s,i}, \widetilde{\nu})(x_k;\theta)]^2 + \tfrac{\lambda^R}{2} \tfrac{1}{N_C} \sum_{k=1}^{N_C} \sum_{i=1}^{3} f_{s,i}(x_k;\theta)^2. \end{aligned} \tag{33}$$

To simplify the equation, we can break it down into distinct terms

$$\mathcal{L} = \underbrace{\frac{\lambda^D}{N_m} J}_{\textit{Data term}} + \underbrace{\frac{\lambda^P}{N_C} \sum_{k=1}^{N_C} R_k^2}_{\textit{RANS}\text{ Residual}} + \underbrace{\frac{\lambda^P}{N_C} \sum_{k=1}^{N_C} T_k^2}_{\textit{Transport Residual}} + \underbrace{\frac{\lambda^B}{N_b} \sum_{k=1}^{N_C} B_k^2}_{\textit{Boundary conditions}} + \underbrace{\frac{\lambda^R}{2N_C} \sum_{k=1}^{N_C} \sum_{i=1}^{3} f_{s,i}(x_k;\theta)^2}_{\textit{Regularization}}, \tag{34}$$

where R_k, T_k, B_k denote the residuals of the Reynolds-averaged Navier–Stokes equations, transport, and boundary conditions at point x_k. The PINN architecture is illustrated in Figure 5.

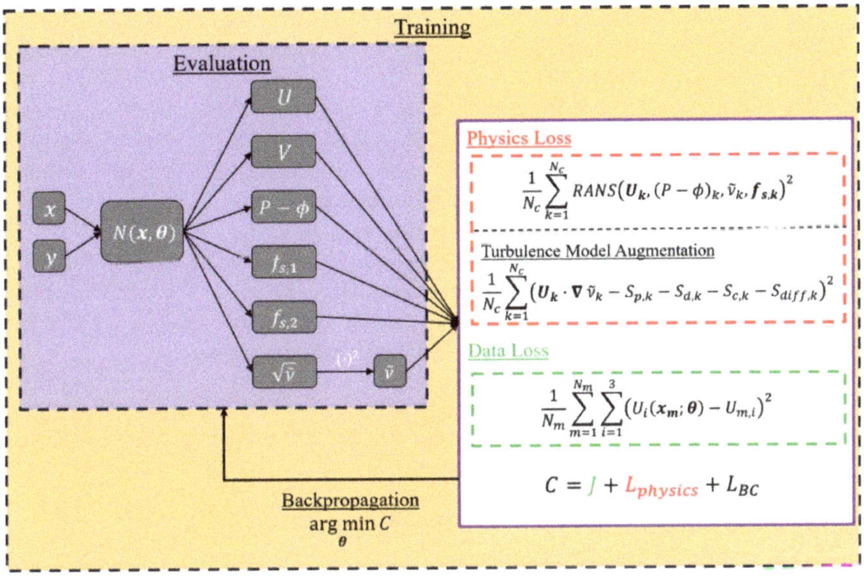

Figure 5. PINN's structure, which was used in [43]. Figure adopted under a Creative Commons license.

The governing equations were determined at N_C collocation points, while the boundary conditions were imposed at points N_b along the boundary. BC represents the error in the enforced boundary conditions, and *RANS* represents the residuals of the mean-flow equations. To compute the governing equations, the *RANS* equations gradients are precisely calculated by using automatic differentiation, which uses the chain rule to find the derivatives of all operations in the mapping $N(x;\theta)$. The physics term, the data term, and the boundary condition term are adjusted by the factors λ^P, λ^D, and λ^B, respectively. The purpose of applying additional regularization was to maintain the uniqueness of the distribution between the corrected and modeled forcing during optimization.

This hybrid approach significantly reduces mean velocity reconstruction errors by up to 73% when using coarse measurements. The incorporation of Spalart–Allmaras (SA) physics constraints prove particularly beneficial in improving reconstructions of flow in areas with pressure gradients, separation zones, and high-velocity.

The PINN-DA-SA methodology was compared to a variational data assimilation approach using the same physics constraints and sparse velocity measurements. It was shown that the PINN-DA-SA method yielded lower reconstruction errors across various data resolutions. This advantage is attributed to the avoidance of discretization errors—commonly observed in variational methodologies—which are effectively mitigated by PINNs.

The methodology was validated using high-fidelity measurements obtained directly from DNS simulations of turbulent periodic hill flow at Re = 5600.

A similar approach was used by [94]: 3D variational data assimilation of time-averaged sparse data into an unsteady Reynolds-averaged Navier–Stokes (URANS) simulation, using a stationary divergence-free forcing term within the URANS equations, was conducted. The presented approach enhances efficiency and speed by using coarse URANS simulations and the stationary discrete adjoint method for solving the time-averaged URANS equations. The data assimilation algorithms were produced utilizing OpenFOAM for URANS and adjoint problem solutions, alongside Python for gradient-based optimization.

The results in [93] have shown that assimilating sparse time-average velocity measurements enhanced both vortex dynamics prediction (shedding frequency) and facilitated correct reconstruction of the mean flow.

The effectiveness of this method was verified by applying it to turbulent flows around cylinders of different geometries for Reynolds numbers from 3000 to 22,000. The crucial role of including data points near the cylinder to improve the accuracy of vortex shedding frequency predictions was demonstrated, while additional downstream data points were essential for reconstructing the mean velocity field in the wake region.

A discrepancy term and a regularization term formed the cost function. In [94], total variation (TV) regularization was selected to decrease the ambiguity in the inverse problem by imposing a smoothness constraint on the parameter field. It was shown that in the optimization process, the regularization weight was a critical parameter.

Data-driven PINNs have rapidly emerged as a powerful approach for solving fluid dynamics problems, particularly when dealing with sparse data. In many examples, [76] and [68] found that PINNs models outperformed their CFD counterparts while requiring a fraction of the computation cost.

Similar to turbulence models, there is now a wide variety of PINN formulations available. The selection of an appropriate PINN method depends on several key factors: flow regime, boundary conditions, data availability, and computational constraints (see Table 2 for a non-exhaustive list of recent applications).

For the laminar regime [44] (NSFnets) and [76] show that PINNs can achieve accuracy comparable to traditional CFD while requiring significantly less computational memory.

Various promising approaches were also identified in the literature for situations involving turbulent flows. Eivazi et al. [38] highlight the excellent accuracy of PINN models in the turbulent case, even without providing a turbulent model.

Wang et al. [78] TF-Net leveraged multi-level spectral decomposition to predict the vortex dynamics at different scales, outperforming baseline models such as U-Net and ResNet.

Patel et al. [43] proposed a framework combining data measurements, RANS equations, and turbulence modeling, which offers greater accuracy for turbulent flows than traditional PINN methods.

Table 2. PINN examples for fluid flows.

Papers	PINN Methodology	Applications	Main Results
[38]	PINN neural network to solve the stationary two-dimensional RANS equations without any specific turbulent model. The inputs of the PINN are the spatial coordinates, and the outputs are the normal velocity, streamwise velocity, and Reynolds-stress components.	Falkner-Skan boundary layer, Adverse-pressure-gradient boundary layer, zero-pressure gradient boundary layer, turbulent flow over a periodic hill, and turbulent flow over a NACA4412 airfoil.	The PINN shows excellent ability in predicting laminar boundary layer flows (Falkner-Skan) and very good accuracy for the turbulent case.
[44]	Navier–Stokes flow nets (NSFnets) were introduced for simulating incompressible laminar and turbulent flows. Given the temporal and spatial coordinates as inputs, VP-NSFnet predicted the instantaneous velocity and pressure fields, while VV-NFSnet predicts instantaneous velocity and vorticity fields. The initial and boundary conditions are used as supervised loss, and the residuals of the Navier–Stokes equations are used for the unsupervised physics-informed loss.	2D steady Kovasznay flow, 2D unsteady cylinder wake, 3D unsteady Beltrami flow, and turbulent channel flow.	VP-NSFnet and VV-NSFnet show comparable accuracy for laminar flows.For turbulent flows, VP-NSFnet sustained turbulence at Re $\approx$ 1000 for a long period of time.
[76]	PINN for low Reynolds number flows. The network is trained without labeled data, i.e., unsupervised learning is used to minimize the residuals of Navier–Stokes equations and to satisfy initial and boundary conditions.	2D flow over a cylinder.	In terms of computational demands, the memory usage of PINN is 5–10 times smaller than the memory usage of CFD. For a small number of CFD cells, PINN takes longer to converge than the CFD solver. The number of layers is the most sensitive hyperparameter affecting the accuracy of the predictions.
[78]	Turbulent flow net (TF-Net) was introduced, which is a hybrid deep learning framework based on multi-level spectral decomposition for predicting the dynamics of different vortical structures The loss function is regularized by the continuity equation.	2D turbulent flow of the Rayleigh–Benard convection flow.	TF-net outperforms baseline models, such as U-net [95] and ResNet [34] in terms of prediction accuracy. Regularizing the loss function by penalizing the deviation from zero velocity divergence results in a more accurate model compared to the purely data-driven TF-net.

Table 2. *Cont.*

Papers	PINN Methodology	Applications	Main Results
[68]	PINN to solve forward and inverse problems for high-speed flows. The PINN predicts the density, velocity, and pressure fields for the Euler equations. The network is optimized to minimize a composite loss, integrating deviation between predicted states and ground truth values, and the residuals of the physical laws.	Forward problem: 1D Euler equation with moving contact discontinuity, 2D oblique shock wave problem Inverse problem: Sod's shock tube problem [96] and Lax's shock tube problem [97].	PINNs can approximate solutions for forward problems but are not as accurate as numerical methods for simulating high-speed flows. For inverse problems, PINNs show superior performance compared to classical methods. The clustering of collocation points around discontinuities improved the performance of the PINN.
[81]	PINN combined with the phase-field method for modeling two-phase incompressible flow. The Cahn-Hilliard equation and Navier–Stokes are encoded in the loss function of the neural network. A time-marching strategy, which consists of dividing the sampling domain into different parts, was introduced to help the network convergence.	Reversed single vortex case, bubble-rising problem in two-phase flow at a large density ratio.	The PINNs were able to capture the interface dynamics and velocity fields. The time-marching strategy employed for training the PINNs is required for obtaining accurate results.
[82]	PINN to predict 3D velocity and pressure fields using 3D temperature snapshots obtained via tomographic background-oriented Schlieren (Tomo-BOS) imaging. The loss function of the PINN contains data mismatch term as well as the residuals of the Navier–Stokes and heat transfer equations.	Buoyancy-driven flow over an espresso cup.	PINN results were validated using PIV experimental data It was shown that PINNs can handle sparse and noisy experimental data, yielding accurate and continuous descriptions of flow fields.
[86]	Physics-constrained Bayesian neural network (PC-BNN) for reconstruction of flow fields from noisy and sparse data. The violation of Navier–Stokes equations is penalized during the training of the neural network.	Vascular flow in stenosis and aneurysm geometries.	The PC-BNN model showed higher accuracy in reconstructing flow fields compared to purely data-driven methods, where the physics is not integrated.

Table 2. *Cont.*

Papers	PINN Methodology	Applications	Main Results
[87]	PINN combined with Neural Radiance Fields to handle complex fluid interactions from sparse Multi-view RGB videos. The loss function of the neural network is constrained with the residuals of Navier–Stokes equations and trained in an end-to-end optimization to learn the spatio-temporal fields without providing geometry information or boundary conditions as input.	Smoke flow involving fluid and static obstacles.	Accurate flow reconstruction for density and velocity fields reduces the need for detailed lighting or boundary conditions by using image sequences and PINN.
[88]	PINN incorporates Navier–Stokes equation and is trained using very sparse boundary data to estimate cardiovascular flow characteristics such as wall shear stress.	2D and 3D flow fields in diseased arteries (stenosed and aneurysmal) without full knowledge of boundary conditions.	PINNs can accurately predict wall shear stress from sparse velocity measurements, even in the case of unknown boundary conditions. The proposed method achieved high accuracy compared to CFD models for flow fields in stenosed and aneurysmal arteries.
[90]	PINN based on the RANS equations without using a turbulent closure model. The loss function contains a data mismatch term as well as residuals of the Navier–Stokes equations.	Adverse-pressure-gradient boundary layer and turbulent periodic hill flow.	Accurate predictions with limited training data, especially for wall shear stress and pressure predictions. Near-wall velocity gradients were poorly captured due to insufficient training data. PINN was able to predict reattachment points, even at high Reynolds numbers, for the periodic hill case.
[91]	A PINN methodology has been proposed to reconstruct time-averaged quantities of unsteady fluid flows using sparse velocity data. The training integrates sparse flow measurements with the governing RANS equations.	2D flow around a circular cylinder.	Accurate prediction of mean flow and force distribution, with small discrepancies near the cylinder. Errors were reduced by introducing inlet boundary conditions and pressure measurements. PINNs reduce noise in input velocity data and show good performance in interpolating missing data.

Table 2. *Cont.*

Papers	PINN Methodology	Applications	Main Results
[93]	PINN is employed to reconstruct high-dimensional flow fields with sparse or incomplete data. The residuals of the Navier–Stokes equations are integrated in the loss function of the network, alongside sparse velocity measurements.	2D flow around a circular cylinder.	PINNs can reconstruct velocity and pressure fields even if the data sparsity reaches 1%, where the average L2-norm error was shown to be as low as 0.0087 (compared to 0.0039 when sparsity reaches 100%). The cosine annealing learning rate scheduler reduced the number of training epochs required for convergence.
[43]	The PINN framework combined with the Spalart–Allmaras (SA) turbulence model to improve the accuracy of reconstructing mean flow fields from sparse high-fidelity measurements. The loss function of the network combines sparse data measurements, RANS equations, and the SA turbulence model.	Turbulent periodic hill flow.	Incorporating the SA turbulence model in the PINN framework improved the results by up to 73% compared to the baseline PINN model. The accuracy of PINN-RANS-SA degrades when using coarser data resolutions.

In [98], a reactive flow physics-informed neural network (RF-PINN) framework aimed at reconstruction of steady-state mean flow fields from limited particle image velocimetry (PIV) data was introduced without predicting the chemical species themselves. As illustrated in Figure 6, the method utilizes sparse data within the computational domain to accurately recover mean flow features. To reduce reliance on detailed species transport equations, an alternative strategy based on the progress variable (θ) equation was used, which included the reaction rate term. This approach streamlines the training process by minimizing the number of governing equations that act as physical constraints.

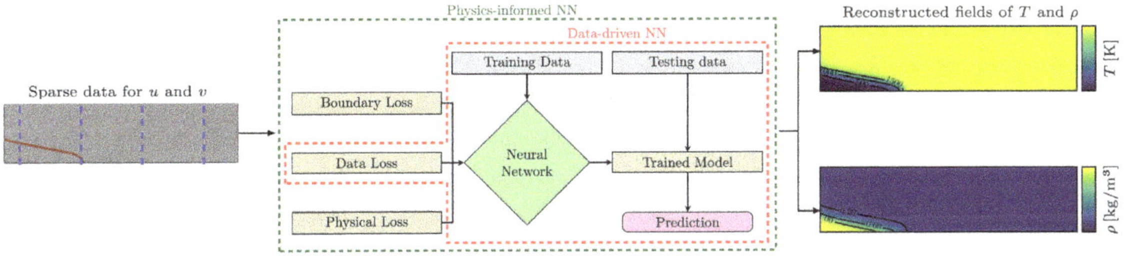

Figure 6. Reconstruction via RF-PINNs with sparse data. One way to use RF-PINNs is to reconstruct the T and ρ fields based on PIV sparse data [98]. Figure adopted under a Creative Commons license.

In cases where the boundary conditions are unknown or limited, models such as the one presented in [88] were able to predict wall shear in cardiovascular flow with higher accuracy than traditional CFD methods. Similarly, approach in [90] achieves a reliable wall shear stress approximation with limited training data. For even more challenging situations, Chu et al. [87] developed a methodology for complex fluid-structure interactions that eliminates the need for detailed boundary conditions and geometric information.

While most models perform well with large amounts of data, some specialize in handling sparse or noisy data. The model presented in [93] maintains a high level of accuracy with as little as 1% of available data across the domain. Furthermore, in [86], a physics-constrained Bayesian neural network, specifically addresses sparse as well as noisy datasets. Methodology in [91] extended these capabilities to time-averaged quantities, successfully reconstructing mean flows and forces despite the presence of noisy inlet data.

Recent studies have also focused on applying machine learning techniques to supersonic flow regimes [99], where traditional modeling approaches face significant challenges due to shock interactions, flow separation, and sparse measurements.

Deng et al. [100] introduced an intelligent reconstruction framework for combustion flow fields that integrates heuristic-guided learning (HGL) to address the challenge of sparse data environments. Rather than relying on conventional end-to-end learning approaches, the proposed HGL-MSFNN model adopts a two-stage learning strategy: in the first stage, it learns a broad range of unknown prior distributions; in the second stage, these priors guide the posterior reconstruction of the flow field, thereby enhancing both generality and generalization capacity.

The model was validated using supersonic wind tunnel experiments at free-stream Mach numbers of 2.5 and 3.0, where both flame and schlieren flow fields were captured. Notably, the system reduced pressure measurement points from 32 to just 14, demonstrating the model's effectiveness under extreme data sparsity. Comparative results indicate that the HGL-MSFNN significantly outperforms baseline models such as Sin-Transformer in reconstructing spatiotemporal dynamical fields with high accuracy and robustness.

Despite its promising performance, the model still exhibits limitations in capturing fine-scale turbulence structures, particularly those associated with multi-scale and chaotic vortex dynamics. The authors suggest that this limitation stems from a linear imbalance in the available training features and a lack of high-quality turbulence descriptors. To address this, they propose that data augmentation strategies could be employed to improve sample diversity and enhance the model's ability to resolve smaller-scale flow features.

5. Conclusions

Physics-informed neural networks (PINNs) have significantly advanced fluid dynamics in recent years by integrating physical laws (e.g., Navier–Stokes equations) with data-driven learning. Key successes include:

- Mesh-free solving: PINNs bypass complex meshing required in traditional CFD, enabling simulations on irregular geometries and high-dimensional parametric spaces.
- Inverse problem capability: they uniquely derive unknown parameters (such as boundary conditions and turbulence model coefficients) from sparse/noisy data, which is not possible with conventional solvers.
- Data assimilation: PINNs effectively deal with sparse experimental data (e.g., particle image velocimetry, video frames) to reconstruct flow fields, even with <1% spatial coverage.
- Hybrid approaches: techniques such as re-initialization escape local minima in stiff flows, while gradient-free PINNs and U-Net++ architectures capture vortex shedding in bluff-body flows.

Despite significant progress in the application of PINNs in different fields of fluid mechanics, this approach still has some important limitations:

- High-frequency/transient dynamics: PINNs struggle with vortex shedding, shocks, or turbulence due to spectral bias (neural networks favor low-frequency features) and inadequate periodic inductive bias.
- Optimization complexity: loss landscapes for multi-scale PDEs (e.g., RANS with Reynolds stresses) induce training instability, slow convergence, and local minima.
- Theoretical gaps: rigorous error bounds and convergence guarantees for PINNs remain underdeveloped, especially for turbulent flows.
- Computational costs: training deep PINNs for 3D flows demands extensive resources, offsetting gains in online prediction speed.

Ultimately, the future of fluid mechanics depends on how well data-driven approaches can be combined with traditional physical theories. As researchers continue to improve these methods and explore new applications, PINNs have the potential to deepen our understanding and improve how we model fluid systems, both in theory and in practice. Integrating machine learning into fluid dynamics not only unlocks numerous opportunities for obtaining more accurate results but also enables engineers to develop faster and more practical solutions for real-world problems.

Author Contributions: Conceptualization, M.E.H.; methodology, M.E.H., N.B. and A.M.; writing—original draft preparation, M.E.H., A.M., P.M., M.M. and N.B.; writing—review and editing, M.E.H., A.M., P.M., M.M. and N.B.; supervision, M.E.H. and M.M.; project administration, M.E.H. and N.B. All authors have read and agreed to the published version of the manuscript.

Funding: This research received no external funding.

Institutional Review Board Statement: Not applicable.

Informed Consent Statement: Not applicable.

Data Availability Statement: No original data generated.

Conflicts of Interest: The authors declare no conflict of interest.

References

1. Brunton, S.L. Applying machine learning to study fluid mechanics. *Acta Mech. Sin.* **2021**, *37*, 1718–1726. [CrossRef]
2. Stroeva, P.; Frolova, E.; Nichik, M.; Dulin, V.; Markovich, D. Flow Dynamics of an Axisymmetric Impinging Jet under Two-Frequency External Forcing. A Study by Time-Resolved PIV and DMD. *Int. J. Heat Fluid Flow* **2023**, *103*, 109196. [CrossRef]
3. Xie, Y.; Franz, A.; Chu, M.; Thuerey, N.; Franz, E. TempoGAN: A Temporally Coherent, Volumetric GAN for Super-Resolution Fluid Flow. *ACM Trans. Graph.* **2018**, *37*, 1–15. [CrossRef]
4. Ren, F.; Hu, H.-B.; Tang, H. Active flow control using machine learning: A brief review. *J. Hydrodyn.* **2020**, *32*, 247–253. [CrossRef]
5. Li, Y.; Chang, J.; Kong, C.; Bao, W. Recent progress of machine learning in flow modeling and active flow control. *Chin. J. Aeronaut.* **2022**, *35*, 14–44. [CrossRef]
6. Renn, P.I.; Gharib, M. Machine learning for flow-informed aerodynamic control in turbulent wind conditions. *Commun. Eng.* **2022**, *1*, 45. [CrossRef]
7. Fukami, K.; Fukagata, K.; Taira, K. Super-Resolution Reconstruction of Turbulent Flows with Machine Learning. *J. Fluid Mech.* **2019**, *870*, 106–120. [CrossRef]
8. Bright, I.; Lin, G.; Kutz, J.N. Compressive Sensing Based Machine Learning Strategy for Characterizing the Flow around a Cylinder with Limited Pressure Measurements. *Phys. Fluids* **2013**, *25*, 127102. [CrossRef]
9. Su, H.; Li, Y.; Xu, Y.; Fu, X.; Liu, S. A review of deep-learning-based super-resolution: From methods to applications. *Pattern Recognit.* **2025**, *157*, 110935. [CrossRef]
10. Colvert, B.; Alsalman, M.; Kanso, E. Classifying Vortex Wakes Using Neural Networks. *Bioinspiration Biomim.* **2018**, *13*, 025003. [CrossRef] [PubMed]
11. Roxas, R., II; Evangelista, M.A.; Sombillo, J.A.; Nnabuife, S.G.; Pilario, K.E. Machine Learning Based Flow Regime Identification using Ultrasonic Doppler Data and Feature Relevance Determination. *Digit. Chem. Eng.* **2022**, *3*, 100024. [CrossRef]
12. Lui, H.F.S.; Wolf, W.R. Construction of Reduced-Order Models for Fluid Flows Using Deep Feedforward Neural Networks. *J. Fluid Mech.* **2019**, *872*, 963–994. [CrossRef]
13. Zhu, L.; Sun, X.; Liu, Y.; Zhang, W. One neural network approach for the surrogate turbulence model in transonic flows. *Acta Mech. Sin.* **2022**, *38*, 321187. [CrossRef]
14. Cuomo, S.; Di Cola, V.S.; Giampaolo, F.; Rozza, G.; Raissi, M.; Piccialli, F. Scientific Machine Learning Through Physics–Informed Neural Networks: Where we are and What's Next. *J. Sci. Comput.* **2022**, *92*, 88. [CrossRef]
15. Wu, P.; Sun, J.; Chang, X.; Zhang, W.; Arcucci, R.; Guo, Y.; Pain, C.C. Data-Driven Reduced Order Model with Temporal Convolutional Neural Network. *Comput. Methods Appl. Mech. Eng.* **2020**, *360*, 112766. [CrossRef]
16. Guastoni, L.; Güemes, A.; Ianiro, A.; Discetti, S.; Schlatter, P.; Azizpour, H.; Vinuesa, R. Convolutional-network models to predict wall-bounded turbulence from wall quantities. *J. Fluid Mech.* **2021**, *928*, A27. [CrossRef]
17. Mjalled, A.; El Hassan, M.; Assoum, H.H.; Mönnigmann, M. Data-Driven Modeling Using Convolutional Neural Network for Experimental Velocity Fields of an Impinging Jet. In Proceedings of the 35th European Modeling and Simulation Symposium, Athens, Greece, 18–20 September 2023. [CrossRef]
18. Rana, P.; Weigand, T.M.; Pilkiewicz, K.R.; Mayo, M.L. A scalable convolutional neural network approach to fluid flow prediction in complex environments. *Sci. Rep.* **2024**, *14*, 23080. [CrossRef] [PubMed]
19. Maulik, R.; Lusch, B.; Balaprakash, P. Reduced-Order Modeling of Advection-Dominated Systems with Recurrent Neural Networks and Convolutional Autoencoders. *Phys. Fluids* **2021**, *33*, 037106. [CrossRef]
20. Hasegawa, K.; Fukami, K.; Murata, T.; Fukagata, K. CNN-LSTM Based Reduced Order Modeling of Two-Dimensional Unsteady Flows around a Circular Cylinder at Different Reynolds Numbers. *Fluid Dyn. Res.* **2020**, *52*, 065501. [CrossRef]
21. Xie, H.-R.; Hua, Y.; Li, Y.-B.; Aubry, N.; Wu, W.-T.; He, Y.; Peng, J.-Z. Estimation of sequential transient flow around cylinders using recurrent neural network coupled graph convolutional network. *Ocean. Eng.* **2024**, *293*, 116684. [CrossRef]
22. Fu, R.; Xiao, D.; Navon, I.; Fang, F.; Yang, L.; Wang, C.; Cheng, S. A Non-linear Non-intrusive Reduced Order Model of Fluid Flow by Auto-encoder and Self-attention Deep Learning Methods. *Int. J. Numer. Methods Eng.* **2023**, *124*, 3087–3111. [CrossRef]
23. Zhang, B. Airfoil-Based Convolutional Autoencoder and Long Short-Term Memory Neural Network for Predicting Coherent Structures Evolution around an Airfoil. *Comput. Fluids* **2023**, *258*, 105883. [CrossRef]

24. Jiang, J.; Li, G.; Jiang, Y.; Zhang, L.; Deng, X. TransCFD: A transformer-based decoder for flow field prediction. *Eng. Appl. Artif. Intell.* **2023**, *123*, 106340. [CrossRef]
25. Mjalled, A.; El Hassan, M.; Boldocky, J.; Gulan, M.; Mönnigmann, M. Improved Neural Ordinary Differential Equation-Based Reduced Model for Impinging Jet Using Wall Shear Stress. *Phys. Fluids* 2024, *in press*. [CrossRef]
26. Karniadakis, G.E.; Kevrekidis, I.G.; Lu, L.; Perdikaris, P.; Wang, S.; Yang, L. Physics-Informed Machine Learning. *Nat. Rev. Phys.* **2021**, *3*, 422–440. [CrossRef]
27. Bishop, C.M.; Nasrabadi, N.M. *Pattern Recognition and Machine Learning*; Springer: New York, NY, USA, 2006; Volume 4.
28. Sharma, P.; Chung, W.T.; Akoush, B.; Ihme, M. A Review of Physics-Informed Machine Learning in Fluid Mechanics. *Energies* **2023**, *16*, 2343. [CrossRef]
29. Li, X.; Sun, W.; Qin, C.; Yan, Y.; Zhang, L.; Tu, J. Evaluation of supervised machine learning regression models for CFD-based surrogate modelling in indoor airflow field reconstruction. *Build. Environ.* **2025**, *267*, 112173. [CrossRef]
30. Jagtap, A.D.; Shin, Y.; Kawaguchi, K.; Karniadakis, G.E. Deep Kronecker neural networks: A general framework for neural networks with adaptive activation functions. *Neurocomputing* **2022**, *468*, 165–180. [CrossRef]
31. Shukla, K.; Jagtap, A.D.; Karniadakis, G.E. Parallel physics-informed neural networks via domain decomposition. *J. Comput. Phys.* **2021**, *447*, 110683. [CrossRef]
32. Miotto, R.; Wang, F.; Wang, S.; Jiang, X.; Dudley, J.T. Deep learning for healthcare: Review, opportunities and challenges. *Brief. Bioinform.* **2018**, *19*, 1236–1246. [CrossRef]
33. Chai, J.; Zeng, H.; Li, A.; Ngai, E.W. Deep learning in computer vision: A critical review of emerging techniques and application scenarios. *Mach. Learn. Appl.* **2021**, *6*, 100134. [CrossRef]
34. He, K.; Zhang, X.; Ren, S.; Sun, J. Deep Residual Learning for Image Recognition. In Proceedings of the 2016 IEEE Conference on Computer Vision and Pattern Recognition (CVPR), Las Vegas, NV, USA, 27–30 June 2016; IEEE: New York, NY, USA, 2016; pp. 770–778. [CrossRef]
35. Dixit, P.; Silakari, S. Deep Learning Algorithms for Cybersecurity Applications: A Technological and Status Review. *Comput. Sci. Rev.* **2021**, *39*, 100317. [CrossRef]
36. Asudani, D.S.; Nagwani, N.K.; Singh, P. Impact of word embedding models on text analytics in deep learning environment: A review. *Artif. Intell. Rev.* **2023**, *56*, 10345–10425. [CrossRef] [PubMed]
37. Sarker, I.H. Deep Learning: A Comprehensive Overview on Techniques, Taxonomy, Applications and Research Directions. *SN Comput. Sci.* **2021**, *2*, 420. [CrossRef]
38. Eivazi, H.; Tahani, M.; Schlatter, P.; Vinuesa, R. Physics-Informed Neural Networks for Solving Reynolds-Averaged Navier–Stokes Equations. *Phys. Fluids* **2022**, *34*, 075117. [CrossRef]
39. Raissi, M.; Perdikaris, P.; Karniadakis, G.E. Physics-Informed Neural Networks: A Deep Learning Framework for Solving Forward and Inverse Problems Involving Nonlinear Partial Differential Equations. *J. Comput. Phys.* **2019**, *378*, 686–707. [CrossRef]
40. Jia, L.C.; Zhu, Y.D.; Jia, Y.X.; Yuan, H.J.; Lee, C.B. Image Pre-Processing Method for near-Wall PIV Measurements over Moving Curved Interfaces. *Meas. Sci. Technol.* **2017**, *28*, 035201. [CrossRef]
41. Kähler, C.J.; Scharnowski, S.; Cierpka, C. On the Uncertainty of Digital PIV and PTV near Walls. *Exp. Fluids* **2012**, *52*, 1641–1656. [CrossRef]
42. Schlueter-Kuck, K.L.; Dabiri, J.O. Coherent Structure Colouring: Identification of Coherent Structures from Sparse Data Using Graph Theory. *J. Fluid Mech.* **2017**, *811*, 468–486. [CrossRef]
43. Patel, Y.; Mons, V.; Marquet, O.; Rigas, G. Turbulence Model Augmented Physics-Informed Neural Networks for Mean-Flow Reconstruction. *Phys. Rev. Fluids* **2024**, *9*, 034605. [CrossRef]
44. Jin, X.; Cai, S.; Li, H.; Karniadakis, G.E. NSFnets (Navier-Stokes Flow Nets): Physics-Informed Neural Networks for the Incompressible Navier-Stokes Equations. *J. Comput. Phys.* **2021**, *426*, 109951. [CrossRef]
45. Zhang, J.; Braga-Neto, U.; Gildin, E. Physics-Informed Neural Networks for Multi-Phase Flow in Porous Media Considering Dual Shocks and Interphase Solubility. *Energy Fuels* **2024**, *38*, 17781–17795. [CrossRef]
46. Wei, C.; Ooka, R. Indoor airflow field reconstruction using physics-informed neural network. *Build. Environ.* **2023**, *242*, 110563. [CrossRef]
47. Ohara, Y.; Moteki, D.; Muramatsu, S.; Hayasaka, K.; Yasuda, H. Physics-informed neural networks for inversion of river flow and geometry with shallow water model. *Phys. Fluids* **2024**, *36*, 106633. [CrossRef]
48. Cotter, S.L.; Dashti, M.; Robinson, J.C.; Stuart, A.M. Bayesian Inverse Problems for Functions and Applications to Fluid Mechanics. *Inverse Probl.* **2009**, *25*, 115008. [CrossRef]
49. Gao, X.; Wang, Y.; Overton, N.; Zupanski, M.; Tu, X. Data-assimilated computational fluid dynamics modeling of convection-diffusion-reaction problems. *J. Comput. Sci.* **2017**, *21*, 38–59. [CrossRef]

50. Di Leoni, P.C.; Mazzino, A.; Biferale, L. Synchronization to Big Data: Nudging the Navier-Stokes Equations for Data Assimilation of Turbulent Flows. *Phys. Rev. X* **2020**, *10*, 011023. [CrossRef]
51. Angriman, S.; Cobelli, P.; Mininni, P.D.; Obligado, M.; Di Leoni, P.C. Assimilation of statistical data into turbulent flows using physics-informed neural networks. *Eur. Phys. J. E* **2023**, *46*, 13. [CrossRef] [PubMed]
52. Vinuesa, R.; Brunton, S.L. Enhancing Computational Fluid Dynamics with Machine Learning. *Nat. Comput. Sci.* **2022**, *2*, 358–366. [CrossRef]
53. Kaiser, E.; Noack, B.R.; Cordier, L.; Spohn, A.; Segond, M.; Abel, M.; Daviller, G.; Östh, J.; Krajnović, S.; Niven, R.K. Cluster-Based Reduced-Order Modelling of a Mixing Layer. *J. Fluid Mech.* **2014**, *754*, 365–414. [CrossRef]
54. Buoso, S.; Manzoni, A.; Alkadhi, H.; Kurtcuoglu, V. Stabilized Reduced-Order Models for Unsteady Incompressible Flows in Three-Dimensional Parametrized Domains. *Comput. Fluids* **2022**, *246*, 105604. [CrossRef]
55. Amsallem, D.; Zahr, M.J.; Farhat, C. Nonlinear Model Order Reduction Based on Local Reduced-order Bases. *Int. J. Numer. Methods Eng.* **2012**, *92*, 891–916. [CrossRef]
56. Kadeethum, T.; Ballarin, F.; O'mAlley, D.; Choi, Y.; Bouklas, N.; Yoon, H. Reduced Order Modeling for Flow and Transport Problems with Barlow Twins Self-Supervised Learning. *Sci. Rep.* **2022**, *12*, 20654. [CrossRef] [PubMed]
57. Khademi, A.; Dufour, S. A novel discretized physics-informed neural network model applied to the Navier–Stokes equations. *Phys. Scr.* **2024**, *99*, 076016. [CrossRef]
58. Khademi, A.; Dufour, S. Physics-informed neural networks with trainable sinusoidal activation functions for approximating the solutions of the Navier-Stokes equations. *Comput. Phys. Commun.* **2025**, *314*, 109672. [CrossRef]
59. Sun, L.; Gao, H.; Pan, S.; Wang, J.-X. Surrogate Modeling for Fluid Flows Based on Physics-Constrained Deep Learning without Simulation Data. *Comput. Methods Appl. Mech. Eng.* **2020**, *361*, 112732. [CrossRef]
60. Sharma, P.; Evans, L.; Tindall, M.; Nithiarasu, P. Stiff-PDEs and Physics-Informed Neural Networks. *Arch. Comput. Methods Eng.* **2023**, *30*, 2929–2958. [CrossRef]
61. Rathore, P.; Lei, W.; Frangella, Z.; Lu, L.; Udell, M. Challenges in Training PINNs: A Loss Landscape Perspective. *arXiv* **2024**, arXiv:2402.01868. [CrossRef]
62. Raissi, M.; Yazdani, A.; Karniadakis, G.E. Hidden Fluid Mechanics: Learning Velocity and Pressure Fields from Flow Visualizations. *Science* **2020**, *367*, 1026–1030. [CrossRef]
63. McClenny, L.D.; Braga-Neto, U.M. Self-adaptive physics-informed neural networks. *J. Comput. Phys.* **2023**, *474*, 111722. [CrossRef]
64. Chen, Z.; Badrinarayanan, V.; Lee, C.-Y.; Rabinovich, A. Gradnorm: Gradient normalization for adaptive loss balancing in deep multitask networks. In *PMLR, Proceedings of the 35th International Conference on Machine Learning, Stockholm, Sweden, 10–15 July 2018*; ML Research Press: Cambridge, MA, USA, 2018; pp. 794–803. [CrossRef]
65. Jagtap, A.D.; Karniadakis, G.E. Extended Physics-Informed Neural Networks (XPINNs): A Generalized Space-Time Domain Decomposition Based Deep Learning Framework for Nonlinear Partial Differential Equations. *Commun. Comput. Phys.* **2020**, *28*, 2002–2041. [CrossRef]
66. Alhubail, A.; He, X.; AlSinan, M.; Kwak, H.; Hoteit, H. Extended physics-informed neural networks for solving fluid flow problems in highly heterogeneous media. In Proceedings of the International Petroleum Technology Conference, Dhahran, Saudi Arabia, 21–23 February 2022; p. D031S073R001.
67. Jagtap, A.D.; Mao, Z.; Adams, N.; Karniadakis, G.E. Physics-informed neural networks for inverse problems in supersonic flows. *J. Comput. Phys.* **2022**, *466*, 111402. [CrossRef]
68. Kharazmi, E.; Zhang, Z.; Karniadakis, G.E. Variational Physics-Informed Neural Networks for Solving Partial Differential Equations. *arXiv* **2019**, arXiv:1912.00873. [CrossRef]
69. Kharazmi, E.; Zhang, Z.; Karniadakis, G.E. hp-vpinns: Variational physics-informed neural networks with domain decomposition. *Comput. Methods Appl. Mech. Eng.* **2021**, *374*, 113547. [CrossRef]
70. Yang, L.; Meng, X.; Karniadakis, G.E. B-PINNs: Bayesian Physics-Informed Neural Networks for Forward and Inverse PDE Problems with Noisy Data. *J. Comput. Phys.* **2021**, *425*, 109913. [CrossRef]
71. O'LEary, J.; Paulson, J.A.; Mesbah, A. Stochastic physics-informed neural ordinary differential equations. *J. Comput. Phys.* **2022**, *468*, 111466. [CrossRef]
72. Ghosh, S.; Chakraborty, A.; Brikis, G.O.; Dey, B. Using Parametric PINNs for Predicting Internal and External Turbulent Flows. *arXiv* **2024**, arXiv:2410.18917. [CrossRef]
73. Luo, S.; Vellakal, M.; Koric, S.; Kindratenko, V.; Cui, J. Parameter identification of rans turbulence model using physics-embedded neural network. In *International Conference on High Performance Computing*; Springer: Berlin/Heidelberg, Germany, 2020; pp. 137–149.

74. Dabrowski, J.J.; Pagendam, D.E.; Hilton, J.; Sanderson, C.; MacKinlay, D.; Huston, C.; Bolt, A.; Kuhnert, P. Bayesian Physics Informed Neural Networks for data assimilation and spatio-temporal modelling of wildfires. *Spat. Stat.* **2023**, *55*, 100746. [CrossRef]
75. Wenzel, C.; Selent, B.; Kloker, M.; Rist, U. DNS of Compressible Turbulent Boundary Layers and Assessment of Data/Scaling-Law Quality. *J. Fluid Mech.* **2018**, *842*, 428–468. [CrossRef]
76. Ang, E.H.W.; Wang, G.; Ng, B.F. Physics-Informed Neural Networks for Low Reynolds Number Flows over Cylinder. *Energies* **2023**, *16*, 4558. [CrossRef]
77. Zhang, E.; Dao, M.; Karniadakis, G.E.; Suresh, S. Analyses of Internal Structures and Defects in Materials Using Physics-Informed Neural Networks. *Sci. Adv.* **2022**, *8*, eabk0644. [CrossRef]
78. Wang, R.; Kashinath, K.; Mustafa, M.; Albert, A.; Yu, R. Towards Physics-Informed Deep Learning for Turbulent Flow Prediction. In Proceedings of the 26th ACM SIGKDD International Conference on Knowledge Discovery and Data Mining, New York, NY, USA, 6–10 July 2020; ACM: New York, NY, USA, 2020; pp. 1457–1466. [CrossRef]
79. Mao, Z.; Jagtap, A.D.; Karniadakis, G.E. Physics-Informed Neural Networks for High-Speed Flows. *Comput. Methods Appl. Mech. Eng.* **2020**, *360*, 112789. [CrossRef]
80. Cai, S.; Mao, Z.; Wang, Z.; Yin, M.; Karniadakis, G.E. Physics-Informed Neural Networks (PINNs) for Fluid Mechanics: A Review. *Acta Mech. Sin.* **2021**, *37*, 1727–1738. [CrossRef]
81. Qiu, R.; Huang, R.; Xiao, Y.; Wang, J.; Zhang, Z.; Yue, J.; Zeng, Z.; Wang, Y. Physics-Informed Neural Networks for Phase-Field Method in Two-Phase Flow. *Phys. Fluids* **2022**, *34*, 052109. [CrossRef]
82. Cai, S.; Wang, Z.; Fuest, F.; Jeon, Y.J.; Gray, C.; Karniadakis, G.E. Flow over an Espresso Cup: Inferring 3-D Velocity and Pressure Fields from Tomographic Background Oriented Schlieren via Physics-Informed Neural Networks. *J. Fluid Mech.* **2021**, *915*, A102. [CrossRef]
83. Berkooz, G.; Holmes, P.; Lumley, J.L. The Proper Orthogonal Decomposition in the Analysis of Turbulent Flows. *Annu. Rev. Fluid Mech.* **1993**, *25*, 539–575. [CrossRef]
84. Ye, M.; Xu, M.; Liu, H.; Yao, W. A Method of Flow Field Reconstruction via Proper Orthogonal Decomposition. In Proceedings of the 2010 2nd International Conference on Information Engineering and Computer Science, Ternopil, Ukraine, 25–26 December 2010; IEEE: New York, NY, USA, 2010; pp. 1–4. [CrossRef]
85. Schmid, P.J. Dynamic Mode Decomposition of Numerical and Experimental Data. *J. Fluid Mech.* **2010**, *656*, 5–28. [CrossRef]
86. Sun, L.; Wang, J.-X. Physics-Constrained Bayesian Neural Network for Fluid Flow Reconstruction with Sparse and Noisy Data. *Theor. Appl. Mech. Lett.* **2020**, *10*, 161–169. [CrossRef]
87. Chu, M.; Liu, L.; Zheng, Q.; Franz, A.; Seidel, H.-P.; Theobalt, C.; Zayer, R.; Franz, E. Physics Informed Neural Fields for Smoke Reconstruction with Sparse Data. *ACM Trans. Graph.* **2022**, *41*, 1–14. [CrossRef]
88. Arzani, A.; Wang, J.-X.; D'SOuza, R.M. Uncovering Near-Wall Blood Flow from Sparse Data with Physics-Informed Neural Networks. *Phys. Fluids* **2021**, *33*, 071905. [CrossRef]
89. Ilya, L.; Frank, H. SGDR: Stochastic Gradient Descent with Warm Restarts. *arXiv* **2016**, arXiv:1608.03983.
90. Hanrahan, S.; Kozul, M.; Sandberg, R. Studying Turbulent Flows with Physics-Informed Neural Networks and Sparse Data. *Int. J. Heat Fluid Flow* **2023**, *104*, 109232. [CrossRef]
91. Sliwinski, L.; Rigas, G. Mean Flow Reconstruction of Unsteady Flows Using Physics-Informed Neural Networks. *Data Centric Eng.* **2023**, *4*, e4. [CrossRef]
92. He, C.; Li, S.; Liu, Y. Data assimilation: New impetus in experimental fluid dynamics. *Exp. Fluids* **2025**, *66*, 94. [CrossRef]
93. Xu, S.; Sun, Z.; Huang, R.; Guo, D.; Yang, G.; Ju, S. A Practical Approach to Flow Field Reconstruction with Sparse or Incomplete Data through Physics Informed Neural Network. *Acta Mech. Sin.* **2023**, *39*, 322302. [CrossRef]
94. Plogmann, J.; Brenner, O.; Jenny, P. Variational Assimilation of Sparse Time-Averaged Data for Efficient Adjoint-Based Optimization of Unsteady RANS Simulations. *Comput. Methods Appl. Mech. Eng.* **2024**, *427*, 117052. [CrossRef]
95. Olaf, R.; Fischer, P.; Brox, T. U-Net: Convolutional Networks for Biomedical Image Segmentation. In Proceedings of the International Conference on Medical Image Computing and Computer-Assisted Intervention, Munich, Germany, 5–9 October 2015; pp. 234–241. [CrossRef]
96. Sod, G.A. A Survey of Several Finite Difference Methods for Systems of Nonlinear Hyperbolic Conservation Laws. *J. Comput. Phys.* **1978**, *27*, 1–31. [CrossRef]
97. Toro, E.F. *Riemann Solvers and Numerical Methods for Fluid Dynamics*; Springer: Berlin/Heidelberg, Germany, 2009. [CrossRef]
98. Yadav, V.; Casel, M.; Ghani, A. RF-PINNs: Reactive flow physics-informed neural networks for field reconstruction of laminar and turbulent flames using sparse data. *J. Comput. Phys.* **2025**, *524*, 113698. [CrossRef]

99. Deng, X.; Tian, Y.; Chen, E.; Yang, M.; Zhang, H.; Le, J. Research on intelligent prediction method of supersonic flow field in scramjet based on deep learning: A review. *Expert Syst. Appl.* **2025**, *279*, 127500. [CrossRef]
100. Deng, X.; Tian, Y.; Yang, M.; Chen, E.; Deng, J.; Zhang, H. Intelligent reconstruction of supersonic combustion flow fields using sparse sensor data and heuristic-guided learning. *Adv. Eng. Inform.* **2025**, *66*, 103486. [CrossRef]

fluids

MDPI

Review

The Potential of Machine Learning Methods for Separated Turbulent Flow Simulations: Classical Versus Dynamic Methods

Stefan Heinz

Department of Mathematics and Statistics, University of Wyoming, 1000 E. University Avenue, Laramie, WY 82071, USA; heinz@uwyo.edu

Abstract: Feasible and reliable predictions of separated turbulent flows are a requirement to successfully address the majority of aerospace and wind energy problems. Existing computational approaches such as large eddy simulation (LES) or Reynolds-averaged Navier–Stokes (RANS) methods have suffered for decades from well-known computational cost and reliability issues in this regard. One very popular approach to dealing with these questions is the use of machine learning (ML) methods to enable improved RANS predictions. An alternative is the use of minimal error simulation methods (continuous eddy simulation (CES), which may be seen as a dynamic ML method) in the framework of partially or fully resolving simulation methods. Characteristic features of the two approaches are presented here by considering a variety of complex separated flow simulations. The conclusion is that minimal error CES methods perform clearly better than ML-RANS methods. Most importantly and in contrast to ML-RANS methods, CES is demonstrated to be well applicable to cases not involved in the model development. The reason for such superior CES performance is identified here: it is the ability of CES to properly account for causal relationships induced by the structure of separated turbulent flows.

Keywords: computational fluid dynamics; machine learning; large eddy simulation (LES); Reynolds-averaged Navier-Stokes (RANS) methods; hybrid RANS-LES methods

1. Introduction

The ability to accurately predict wall-bounded turbulent flows usually involving flow separation is highly essential to most aerospace and wind energy problems. This concerns, for example, aircraft wing aerodynamics, hypersonic flow, aircraft stability and control, wind turbine blade performance, wind farm layout and interaction, and offshore wind turbines. The challenge of simulating such flows is the appearance of detached flow behind complex geometries: flow separates from the surface of an object due to adverse pressure gradients or sudden changes in geometry and a recirculation region is formed (see, e.g., the flows described in Sections 2 and 3). The latter usually has a significant influence on the overall flow behavior.

The advantages and disadvantages of basic computational methods usually applied to address such problems are well-known. The direct solution of Navier–Stokes equations, direct numerical simulation (DNS) [1], is computationally way too expensive for simulations of most flows of practical relevance. For example, comparisons with DNS are involved in the discussions presented below only in regard to periodic hill flows up to a Reynolds number (*Re*) of 5.6 K [2,3]. Stationary Reynolds-averaged Navier–Stokes (RANS) methods are applied with the motivation to fully model such flows. This means the turbulence equations applied in simulations are supposed to properly describe the structure of turbulent flows without the simulation of instationary turbulent motions. Technically, this can be accomplished by using a relatively large characteristic turbulence length scale (which implies a relatively large turbulent viscosity that damps turbulent fluctuations). Such simulations are computationally very efficient, but proved to be incapable of appropriately representing the structure of turbulent flows involving flow separation [1,4,5].

The large eddy simulation (LES), in particular wall-resolved LES (WRLES), is applied with the motivation to almost fully resolve turbulent flows. This means the equations applied are supposed to produce turbulent fluctuations (resolved flow) as seen in reality in separated turbulent flows. Technically, this can be accomplished by using a relatively small model length scale (which implies a relatively small turbulent viscosity). WRLES is able to accurately simulate separated turbulent flows, but unfortunately, the high computational cost of LES often does not allow simulations of practical relevance [1,6–8]. The search for appropriate ways to deal with these challenges has continued for several decades. Several strategies were applied to address these challenges [9,10].

A first basic approach, referred to as wall-modeled LES (WMLES), is to relax the resolution requirements of LES for wall-bounded turbulent flows by involving RANS components close to the wall. The basics of this approach can be found in Refs. [6,11–17], more recent reviews can be found in Refs. [18–25]. This approach is known to substantially reduce the computational cost of LES [23,24,26]. However, on top of simulation performance issues [27,28], this approach suffers from functionality issues. Such WMLES results depend on the use of different (equilibrium or non-equilibrium) wall models, definitions of regions where different models and grids are applied, different mesh distributions, and setup options to manage the information exchange between such different flow regions. Conceptual questions related to the use of alternative LES, Reynolds stress-constrained LES (RSC-LES) are discussed elsewhere [9,29–31]. The latter methods were applied to a large variety of cases like three-dimensional turbulent channel flow and flow past a circular cylinder [32], a full commercial aircraft simulation [33], periodic hill flows [34], compressible turbulent channel flows, and compressible flow past a circular cylinder [35,36], laminar-turbulent transition in a temporally developing channel flow [37], a U-duct flow [38], different airfoil type flows [39], and turbulent flow and heat transfer in a stationary ribbed duct [40].

A second basic approach, referred to as detached eddy simulation (DES), is to expand the capability of RANS equations by switching from the RANS turbulence length scale applied close to the wall to a much smaller LES-type length scale away from the wall. The basics of this approach can be found in Refs. [41–48], and more recent reviews can be found in Refs. [49–52]. The computational cost of this approach can very much exceed the cost of RANS simulations [53]. This approach also suffers from both simulation performance and functionality issues [27,28]. For each flow considered, there is the need to consider the validity of modifications of the basic approach. This involves, e.g., comparing the validity of zonal improved delayed DES (IDDES, which applies a synthetic turbulence generator at a certain flow-dependent RANS-LES interface) and non-zonal approaches, based, for example, on a shear layer-adapted (SLA) definition of the subgrid length scale (the filter size Δ) [28]. It is worth noting that corresponding issues do not only apply to aerospace and wind energy problems but to a variety of other problems, as, for example, mesoscale and microscale modeling in regard to atmospheric simulations and many technical applications [54–56].

Several alternative strategies were studied to overcome these issues [9,10], e.g., unified RANS-LES (UNI-LES) [57–63], partially averaged Navier–Stokes (PANS) [64–75], partially integrated transport modeling (PITM) [76–86], and scale-adaptive simulation (SAS) methods [52,87–92]. These methods aim at theory-supported modifications of two-equation RANS models to enable the generation of resolved motion. A typical feature of these methods is that the resolution imposed by the model is often poorly realized in flow simulations. The latter can significantly affect the model functioning, as, e.g., the accurate prediction of characteristic separation zone features [27,28,93].

Given this situation, what are the options to enable feasible and reliable computational predictions of separated turbulent flows? Observations reported above reveal that the implementation of RANS or LES features in different equations cannot be expected to result in methods that perform equally well (using the same model) for a variety of separated turbulent flow types. The most promising alternatives are mathematically consistent extensions of RANS equations that either enable the inclusion of resolved motion, or pure RANS modeling that includes characteristic features of flow separation, meaning the significant flow structure modifications due to nonequilibrium processes.

The first alternative mentioned in the preceding paragraph was recently realized via the introduction of a new computational approach: continuous eddy simulation (CES). The latter minimal error simulation methods (which minimize the hybridization error) [9,27,28,93–101] were presented as an alternative to existing hybrid RANS-LES. The methodological setup can be applied to all popular turbulence models and in several types of hybridization. The most relevant feature of these methods is their derivation based on strict mathematics. The latter has significant implications for the model mechanism. In contrast to existing computational methods, in minimal error simulation methods, the model receives information about the actual flow resolution degree, which enables the model to appropriately respond by increasing or decreasing its contribution to the simulation. The latter is the essential requirement for a functional swing between almost modeled (RANS) and almost resolved (LES) regimes. It is also the requirement for reliable predictions of very high *Re* flow regimes, which cannot be properly studied based on existing methods.

The second alternative is this use of popular machine learning (ML) methods for the extension of RANS methods (to enable RANS to properly deal with characteristic features of flow separation) [3,102–121]. The predominant approach is the modification of RANS equations considered (usually by introducing parameters that modify the production in turbulence equations). The latter parameters are determined first depending on physical space by using field inversion (FI) techniques in conjunction with high-fidelity reference data. In the second step, these spatial distributions of introduced model parameters are determined in dependence on model parameters available through the simulation by using ML techniques.

Given the two alternatives described in the preceding two paragraphs, the use of ML-RANS seems to be the more attractive approach because of their lower computational cost (ML-RANS do not attempt to resolve flow). However, such a preference should be accompanied by convincing evidence that ML-RANS is capable of providing reliable separated turbulent flow predictions even under conditions unseen in the model training. The latter question currently represents a question mark. In particular, there is a lack of clarity on what actually causes ML-RANS to be not convincing in this regard. Based on the latest developments, comparisons of ML-RANS with CES will be used to identify the reason for corresponding ML-RANS shortcomings. Simultaneously, these comparisons will be used to demonstrate the superior predictive power of CES in comparison to ML-RANS and other usually applied computational simulation methods. Characteristic features of these two alternatives considered will be presented in the following two sections including a discussion of differences. Conclusions will be presented finally.

2. CES Methods

2.1. Minimal Error Simulation Methods

CES methods [9,27,28,93–101] apply a sensibilization of RANS equations such that RANS equations become aware of the flow resolution using appropriate grids. More specifically, this approach enables the model to properly respond to the amount of flow resolution: the model contribution decreases (increases) if the degree of flow resolution is high (low). The latter is the essential requirement to properly transition from RANS regimes (dominated by the model contribution) and LES regimes (dominated by resolved flow). The CES approach can be seen as a dynamic ML approach: the model learns on the fly about the flow resolution amount and adjusts its contribution to the flow simulation properly. The mathematical approach, which is based on variational analysis, ensures that the model minimizes the hybridization error; see the discussion below related to Table 1. The approach can be applied in conjunction with all popular turbulence models, including Reynolds stress equation methods and probability density function (PDF) methods [100].

Table 1. Minimal error $k-\omega$ models: KOS and KOK hybridizations are considered in analysis options $\mathcal{O}_2$. Model errors λ, first variations, and resulting mode controls β^* are provided. Variations applied are given in brackets.

$\frac{Dk}{Dt} = P - \epsilon + D_k, \quad \frac{D\omega}{Dt} = C_{\omega_1}\omega^2\left(\frac{P}{\epsilon} - \beta^*\right) + D_\omega, \quad D_k = \frac{\partial}{\partial x_j}\nu_t\frac{\partial k}{\partial x_j}, \quad D_\omega = \frac{\partial}{\partial x_j}\frac{\nu_t}{\sigma_\omega}\frac{\partial \omega}{\partial x_j}$	(KOS)
$\frac{Dk}{Dt} = P - \psi_\beta\epsilon + D_k, \quad \frac{D\omega}{Dt} = C_{\omega_1}\omega^2\left(\frac{P}{\epsilon} - \beta\right) + D_\omega, \quad \beta^* = 1 + \beta - \psi_\beta$	(KOK)
• Analysis option $\mathcal{O}_2$ [without $Dk/Dt, D\omega/Dt$]: $\quad [\delta D_k/D_k = 3\delta k/k - \delta\epsilon/\epsilon,\ \delta D_\omega/D_\omega = \delta k/k]$	
$\lambda_2 = C_{\omega_1}\omega^2\left(\frac{P}{\epsilon} - \beta_2^*\right) + D_\omega, \quad \delta\left(\frac{\lambda_2}{k}\right) = \frac{C_{\omega_1}}{L^2}(\beta_2^* - 1)\left[\frac{\delta L^2}{L^2} - \frac{\delta\beta_2^*}{\beta_2^* - 1}\right], \quad \int_\beta^{\beta_2^*}\frac{dx}{x-1} = \int_{L_{tot}^2}^{L^2}\frac{dy}{y}, \quad \frac{\beta_2^* - 1}{\beta - 1} = L_+^2$	

To illustrate the structure of CES models, we consider incompressible flow. The incompressible continuity equation $\partial\widetilde{U}_i/\partial x_i = 0$ and momentum equation are considered,

$$\frac{D\widetilde{U}_i}{Dt} = -\frac{\partial(\widetilde{p}/\rho + 2k/3)}{\partial x_i} + 2\frac{\partial(\nu+\nu_t)\widetilde{S}_{ik}}{\partial x_k}. \tag{1}$$

Here, $D/Dt = \partial/\partial t + \widetilde{U}_k\partial/\partial x_k$ denotes the filtered Lagrangian time derivative and the sum convention is used throughout this paper. $\widetilde{U}_i$ refers to the i-th component of the spatially filtered velocity. We have here the filtered pressure $\widetilde{p}$, ρ is the constant mass density, k is the modeled energy, ν is the constant kinematic viscosity, and $\widetilde{S}_{ij} = (\partial\widetilde{U}_i/\partial x_j + \partial\widetilde{U}_j/\partial x_i)/2$ is the rate-of-strain tensor. The modeled viscosity is given by $\nu_t = C_\mu k^{1/2}L$. Here, C_μ is a model parameter with standard value $C_\mu = 0.09$, and L is a characteristic length scale. L can be calculated in different ways using $L = k^{3/2}/\epsilon = k^{1/2}\tau = k^{1/2}/\omega$, where the dissipation rate $\epsilon = k/\tau$ of modeled kinetic energy, the dissipation time scale τ, and the turbulence frequency $\omega = 1/\tau$ are involved.

Table 1 shows CES methods in regard to the use of the $k-\omega$ model. Here, $P = \nu_t S^2$ is the production of k, where $S = (2\widetilde{S}_{mn}\widetilde{S}_{nm})^{1/2}$, $C_{\omega_1} = 0.49$, and $\sigma_\omega = 1.8$. In contrast to RANS, where β^* is a constant, β^* is considered here an undetermined parameter that needs to be chosen to minimize the hybridization error. The abbreviations KOS and KOK refer to the consideration of the $k-\omega$ model where the hybridization is accomplished in the scale equation (involving β^* that needs to be determined) or k-equation (involving ψ_β that needs to be determined), respectively. Mathematically, both variants (KOS and KOK) are equivalent if the coefficient relation $\beta^* = 1 + \beta - \psi_\beta$ is honored. These models can be analyzed by the two options considered (option O_1 and option O_2). Option O_1 is an exact hybridization where total viscosities need to be applied in turbulent transport terms. Option O_2 is a hybridization where the usual model viscosities are involved in turbulent transport terms and an approximation is applied (given by the neglect of substantial derivatives in regard to the model coefficient calculation). Only the second option is presented in Table 1, which was found to work very well in all applications. The way to obtain β_2^* in option O_2 is explained in the lower part of Table 1. The hybridization error is given by λ_2. We consider the variation (δ) of the normalized error λ_2/k under consideration of variational properties of D_k and D_ω given in the bracket terms. This leads to the second relation in the last row of Table 1. An extremal state is given by setting the right-hand side of this relation equal to zero. The integration of the resulting $\delta L^2/L^2 = \delta\beta_2^*/(\beta_2^* - 1)$ from a RANS state (where $\beta_2^* = \beta$) to a state with a certain degree of flow resolution implies then $\beta_2^* = 1 + L_+^2(\beta - 1)$. The same analysis in regard to the determination of ψ_β provides the relationship $\beta_2^* = 1 + \beta - \psi_\beta$.

The difference between CES methods and other hybrid RANS-LES is the appearance of resolution indicators like $0 \le L_+ \le 1$ in model equations, which indicates the degree of flow resolution. The definition of $L_+ = L/L_{tot}$ follows the definition of the modeled-to-total kinetic energy ratio $k_+ = k/k_{tot}$: L is the modeled length scale contribution and L_{tot} is the total length scale contribution. $L_+ \approx 1$ indicates an almost completely modeled (RANS) regime, and $L_+ \approx 0$ indicates an almost completely resolved (LES) regime. The mechanism of the model response to the actual flow resolution is the following. A higher resolution

(L_+ becomes smaller) decreases β^*. Thus, there is less dissipation of ω, and ω increases, which decreases the model viscosity $\nu_t = C_\mu k/\omega$.

CES methods show significant differences from WMLES and DES methods in regard to their functionality and computational cost. The issues related to properly setting up WMLES and DES methods were described in the second and third paragraphs of the introduction. In contrast to that, CES methods are independent of such functionality requirements, the model can be used as is, and CES can be expected to enable reliable predictions under conditions where validation data are unavailable. The differences in CES computational cost to other usually applied methods are discussed elsewhere [27,28].

2.2. Periodic Hill Flow Simulations

One of the applications of CES methods is the simulation of periodic hill flows, as illustrated in Figure 1 [93]. This flow is a channel flow involving periodic restrictions. This flow, which is used a lot for the evaluation of turbulence models [9], involves features such as separation, recirculation, and natural reattachment [122,123]. A thorough evaluation of the performance of CES methods in regard to simulating periodic hill flows at the highest Re = 37 K (based on the hill height), for which experimental data for model evaluation are still available, can be found elsewhere, including evidence for the excellent CES prediction of mean velocities and stresses [93]. Figure 1 illustrates the remarkable differences to RANS predictions (see also the corresponding streamline results obtained by ML-RANS in Section 3.1. In contrast to RANS, the CES model provides a much more accurate characterization of the recirculation zone characteristics: the reattachment point is obtained as $x/h = 3.78$, which differs from the corresponding value $x/h = 3.76$ obtained in experiments only by 0.5%.

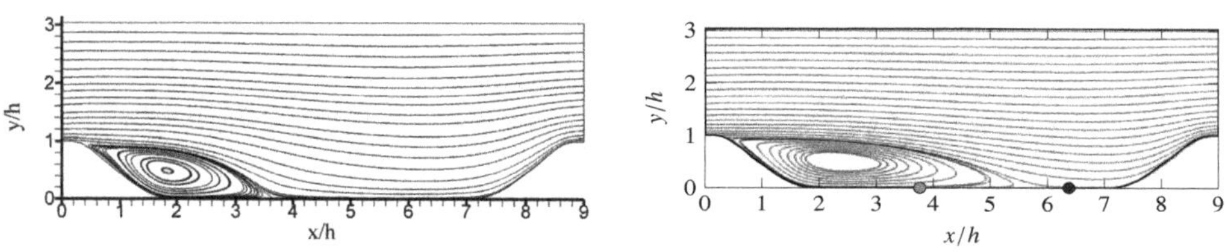

Figure 1. Periodic hill flow velocity streamlines for $Re = 37$ K. **Left**: CES results. Reprinted with permission from Ref. [93]. Copyright 2020 AIP Publishing. **Right**: Corresponding RANS results obtained by Köhler et al. [107]. Reprinted by permission of the American Institute of Aeronautics and Astronautics, Inc. (Pine Brook, NJ, USA). The red dot refers to the experimental measurement of the reattachment point, the black dot refers to the corresponding RANS result.

More specifically, the analyses presented in Ref. [93] involved four cases: $Re = 37$ K (G_{500}), which is almost fully resolved, $Re = 500$ K (G_{120}), which is almost fully modeled, and the transitional cases $Re = 37$ K (G_{120}) and $Re = 500$ K (G_{500}). Here, G_{500} and G_{120} refer to the use of grids with 500 K and 120 K grid points, respectively. Corresponding plots of velocity fluctuations are presented in Figure 2. These plots, too, underline the significant difference to steady RANS simulations, where fluctuations are excluded. The almost fully resolved $Re = 37$ K (G_{500}) case reveals a very good resolution of fine scale structures. It is remarkable to see that grid and Re variations imply smooth variations in these fluctuation structures, the merging of structures due to a coarser grid or higher Re. Interestingly, the almost fully modeled $Re = 500$ K (G_{120}) case still involves strong fluctuations, which are as strong as under resolving conditions. The latter clearly speaks for a stable generation mechanism of fluctuations, which is implied by the derivation of CES methods.

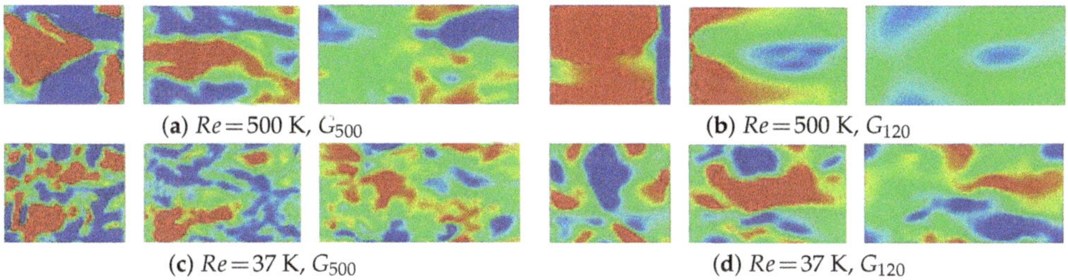

Figure 2. Periodic hill flows, CES predictions: velocity fluctuations u in xz planes at $y = (0.01, 0.5, 1.5)h$ (left/right) for (**a**) Re = 500 K (G_{500}), (**b**) Re = 500 K (G_{120}), (**c**) Re = 37 K (G_{500}), (**d**) Re = 37 K (G_{120}) cases. Red (blue) areas refer to positive (negative) fluctuations. The range of velocity fluctuations is given by $-0.1 \leq u \leq 0.1$ $[y/h = 0.01]$ and $-0.3 \leq u \leq 0.3$ $[y/h = (0.5, 1.5)]$. Reprinted with permission from Ref. [93]. Copyright 2020 AIP Publishing.

2.3. NASA Hump Flow Simulations

The NASA wall-mounted hump flow illustrated in Figure 3 was studied by Seifert and Pack [124], aiming at the investigation of unsteady flow separation, reattachment, and flow control at a high Reynolds number $Re = c\rho_{ref}U_{ref}/\mu \approx 936K$ based on the chord length c and freestream velocity U_{ref}. Here, μ is the dynamic viscosity and the abbreviation ref indicates the reference freestream conditions, which are determined at the axial point $x/c = -2.14$. The model reflects the upper surface of a 20-thick Glauert–Goldschmied airfoil that was originally designed for flow-control purposes in the early twentieth century. As a benchmark for comparison, we used the experiment conducted by Greenblatt et al. [125] without flow control. This benchmark case has been extensively documented on the NASA Langley Research Center's Turbulence Modeling Resource webpage and has been widely used for evaluating different turbulence modeling techniques, as discussed in the 2004 CFD Validation Workshop.

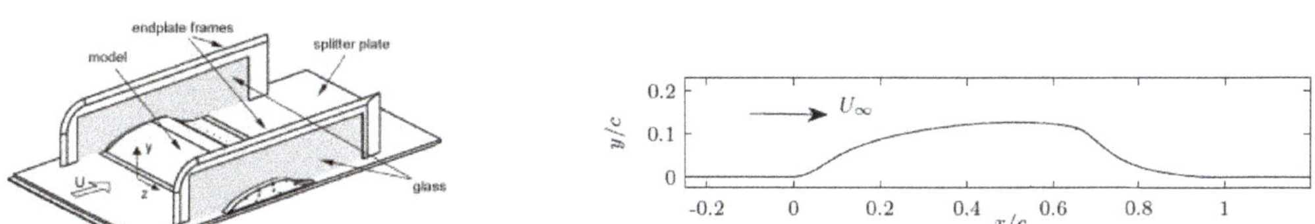

Figure 3. Wall-mounted NASA hump geometry. **Left**: experimental setup [124]; **right**: 2D computational layout [28].

Advantages of the CES-KOS model are illustrated in Figure 4. It may be seen that all methods involved in this comparison show a reasonable agreement with the pressure coefficient profile obtained via experiments. The WRLES predictions match the experimental measurement profile downstream and the model is capable of properly representing the dominant features of the flow. Regarding the reattachment region, we note that the second wall pressure peak is underpredicted by WRLES compared to CES-KOS and WMLES. Figure 4 also shows the mean skin friction coefficient obtained by CES-KOS, WMLES, and WRLES simulations, demonstrating their agreement with experimental values. In the separation zone, from $0 \leq x/c \leq 0.65$, WRLES underpredicts the skin friction coefficient, while WMLES overestimates the actual peak. In regard to the post-reattachment, the C_f profiles of WRLES and CES-KOS match relatively well, despite the very different computational setup of these methods.

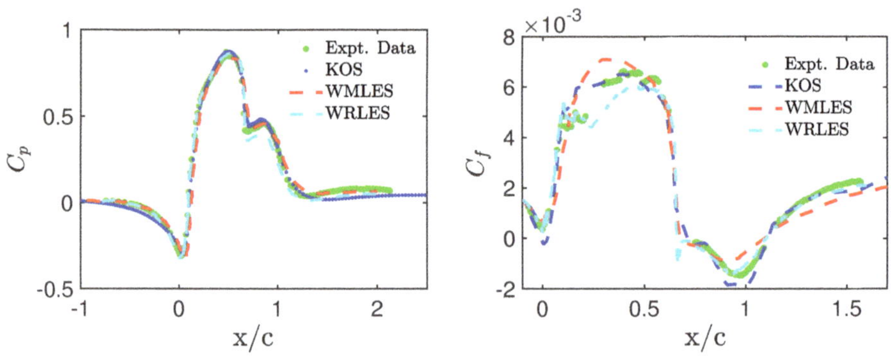

Figure 4. Wall-mounted NASA hump flow CES results [28]: CES-KOS, WMLES [126], and WR-LES [127,128] simulation results on the G_4 grid at $Re = 936$ K: pressure (C_p) and skin-friction (C_f) coefficients.

A demonstration of the difficulty of dealing with accurate predictions for this flow in the RANS framework is given by the following. In an attempt to design more generally applicable RANS models, Srivastava et al. [105] recently considered two augmentation strategies, one of them (the reattachment augmentation) focusing on the improvement of predictions for the point of reattachment after smooth-body separation. The modeling approach uses the Spalart–Allmaras (SA) turbulence model [129] in conjunction with an adjustable parameter β that modifies the production term. The motivation is illustrated in terms of Figure 5. The left-hand side shows the β distribution obtained from an FIML method. Three regions can be seen: region 1 with $\beta > 1$, region 2 with $\beta < 1$, and region 3 with $\beta > 1$. The focus is on region 3, leaving the modeling of regions 1 and 2 for future work. A very essential fact is that there is no use of ML in this approach, but an analytical model is provided for β. The latter includes a parameter η_r which roughly correlates to region 3. The authors then apply a smoothed version of the expression $\beta = 1$ if $\eta_r < c_{\eta_r}$ and $\beta = c_\beta^{max}$ otherwise. Here, the constants $c_{\eta_r} = 0.9$ and $c_\beta^{max} = 5$ are applied. A characteristic skin-friction result result is also provided in Figure 5 (SST+fix refers to the enhanced RANS model). It may be seen that this approach leads to improvements but significant model deficiencies remain; see the comparison with Figure 4.

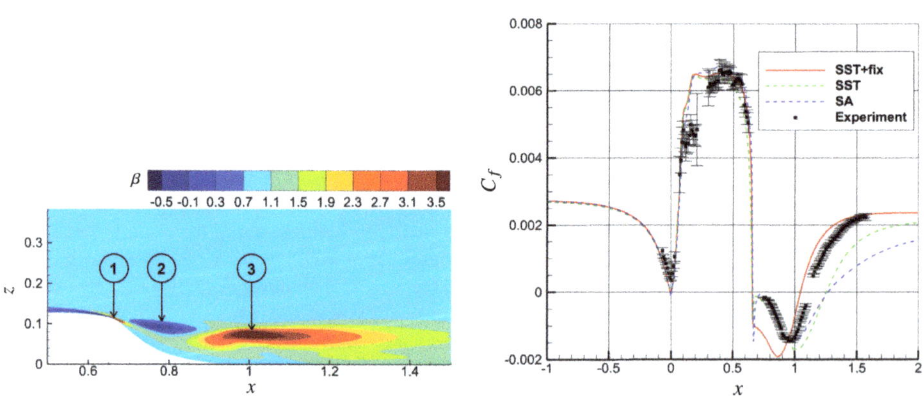

Figure 5. Field inversion results for the NASA wall-mounted hump flow (**left**) and corresponding skin-friction coefficient distributions (**right**) [105]. Reprinted with permission of the authors from Ref. [105].

2.4. Bachalo and Johnson Axisymmetric Transonic Bump Flow Simulations

The experimental configuration and the computational domain of the axisymmetric transonic bump considered by Bachalo and Johnson [130,131], along with the applied boundary conditions is illustrated in Figure 6. This case pertains to shock-triggered boundary layer separation induced by an axially symmetric bump mounted on a slim spherical cylinder, which extends 61cm upstream. The case reflects the upper surface of a transonic wing. It is characterized by a Mach number (M_∞) of 0.875 and a Reynolds number (Re) of 2.763 M relative to the airfoil's chord length c.

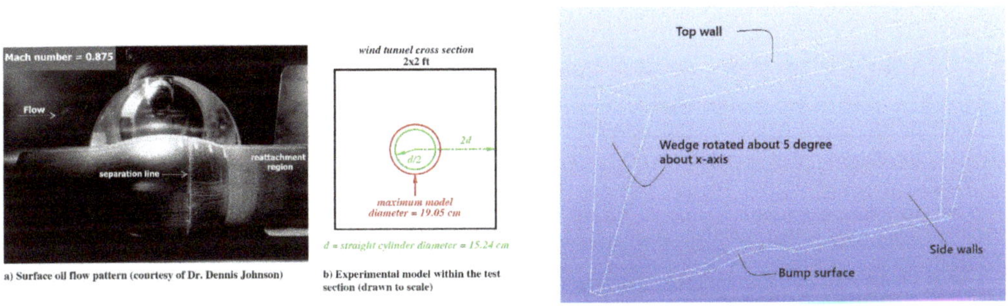

Figure 6. Axisymmetric transonic bump geometry [27]: experimental and computational configuration [132,133].

Streamwise velocity profiles obtained by CES-KOS, WMLES [132], and WRLES [133] are shown in the first row of Figure 7. The CES-KOS model predicts the streamwise velocity more accurately than WMLES and WRLES. In regard to turbulent shear stress profiles shown in the second row, we see a reasonable agreement of WMLES, WRLES, and CES-KOS with experimental data. Interestingly, CES-KOS and WRLES provide very similar results. In attached flow regions, WMLES overpredicts the turbulent shear stress. Due to its delayed reattachment point, WMLES predicts a faster separated shear layer growth and a higher maximum Reynolds stresses compared to CES-KOS. These observations are fully in line with past numerical studies using alternative models [134].

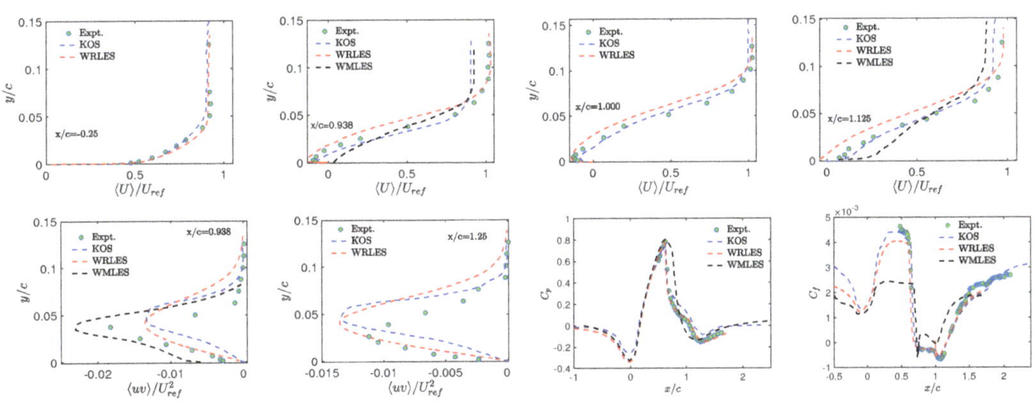

Figure 7. Axisymmetric transonic bump flow results [27]: CES-KOS vs. LES-type WRLES [133] and WMLES [132] models. Profiles of the normalized streamwise velocity $\langle U\rangle/U_{ref}$, Reynolds stress $\langle uv\rangle/U_{ref}^2$, pressure ($C_p$) and skin-friction ($C_f$) coefficients at different locations.

Figure 7 also shows pressure coefficient distributions obtained by CES-KOS, WMLES, and WRLES. Based on their sufficient flow resolution ability, CES-KOS and WRLES accurately predict pressure coefficient profiles. In contrast to that, WMLES fails to accurately capture the separation zone characteristics. Both CES-KOS and WRLES reasonably predict the shock location and post-shock pressure recovery. We note that WRLES results agree slightly better with the experimental data downstream of the bump ($1.1 \leq x/c \leq 1.3$) compared to the CES-KOS model. Skin-friction coefficient distributions obtained by CES-KOS, WMLES, and WRLES are also shown in this figure. We observe that WMLES significantly underestimates the skin-friction coefficient in the separation region and fails to accurately represent the post-separation flow characteristics. The predictions of CES-KOS and WRLES are very similar, although CES-KOS agrees better with the experimental data in the C_f plateau region upstream of separation. Hence, the most accurate flow predictions are provided by CES-KOS.

3. Ml Methods for Separated Turbulent Flows

Characteristic features of ML-RANS methods will be described next by focusing on hill-type simulations followed by an analysis of differences to CES methods. The usual methodological setup of such ML methods was described at the end of the introduction: with the exception of Ref. [106], a correction parameter β modifies the production in turbulence equations. Then, the spatial distribution of this parameter is found via FI techniques followed by an ML representation of β.

3.1. 2D Periodic Hill Flow Simulations

By focusing on 2D periodic hill flow simulations, three ML-based extensions of RANS equations [3,106,107] are presented in this subsection. Essential characteristics of these studies are presented in Table 2.

Table 2. ML-RANS simulations of hill flows: references and characteristics. Here, SA refers to the Spalart–Allmaras (SA) turbulence model [129]. The parameter α refers to different hill geometries.

Reference	Characteristics (Geometry, *Re*, Validation Data, Training and Validation, Output Variables)
Köhler et al. [107]	• Periodic hills, fixed geometry. • DNS/LES [2] [$Re = (0.7, 1.4, 2.8, 5.6, 10.6)$ K], Exp. [122] [$Re = 37$ K]. • Training: $Re = (0.7, 1.4, 2.8, 10.6)$ K, validation: $Re = (5.6, 37)$ K. • $k - \omega$ model, ML production [streamwise velocity profiles, streamlines, C_f, k, $\langle u'v' \rangle$].
Yan et al. [3]	• Periodic hills, variable geometry $\alpha = (0.5, 0.8, 1.0, 1.2, 1.5)$. • DNS [135] [$Re = 5.6$ K, α = (0.5–1.5)], LES [136] [$Re = 10.6$ K, $\alpha = 0.875$]. • Training: $Re = 5.6$ K, $\alpha = (0.5, 0.8)$, validation: $Re = 10.6$ K, $\alpha = 0.875$. • SA, ML production [streamwise velocity profiles, velocity contours, C_f].
Volpiani et al. [106]	• Periodic hills, variable geometry $\alpha = (0.8, 1.0, 1.2)$. • DNS [135] [$Re = 5.6$ K, α = (0.8–1.2)], LES [136] [$Re = (2.8, 10.6, 19)$ K, $\alpha = 1$]. • Training: $\alpha = 1$, $Re = (2.8, 10.6, 19)$ K (setup I), $Re = (2.8, 5.6, 10.6)$ K (setup II). • Validation: [$Re = 19$ K, $\alpha = 1$, setup II], [$Re = 5.6$ K, $\alpha = (0.8, 1.2)$, setup I and II]. • SA, ML momentum eq. [streamwise and vertical velocity profiles, velocity contours, C_f].
Yan et al. [108]	• Fixed geometry: 3D symmetric bump and 3D FAITH hill. • Exp. of sym. bump [137,138] [$Re = 130$ K], Exp. of FAITH hill [139] [$Re = 500$ K]. • Training: 3D sym. bump: full (ANN1), truncated (ANN2), sampled (ANN3) data. • Validation: 3D FAITH hill. • SA, ML production [velocity profiles, velocity contours, C_p].

One of these periodic hill flow simulations was presented by Köhler et al. [107]. The FIML model is referred to as an augmented $k - \omega$ model, the term inverse $k - \omega$ model refers to the model obtained by the FI method. The number of grid points was not reported. The $Re = 5600$ case simulations show that there is basically no difference between the predictions of the augmented $k - \omega$ model and the inverse $k - \omega$ model. Streamwise velocity profiles show a clear improvement over RANS results, but the comparison with LES results reveals deficiencies. Corresponding observations can be made in regard to modeled kinetic energy (k) and Reynolds shear stress ($\langle u'v' \rangle$) predictions; the agreement with LES is even less satisfactory. The skin-friction coefficient (C_f) predictions show significant improvements compared to RANS results. The most relevant conclusions may be seen by considering the model application to the $Re = 37$ K case. The $Re = 37$ K case simulations show substantial differences between the inverse and augmented $k - \omega$ model: the inverse $k - \omega$ model is performing better.

- In particular, Figure 8 shows separation zone characteristics for the $Re = 37$ K case unseen in the training. Although the augmented $k - \omega$ model is performing much better than the baseline $k - \omega$ model, there are clear disagreements with corresponding LES results: the separation zone characteristics are improperly described.

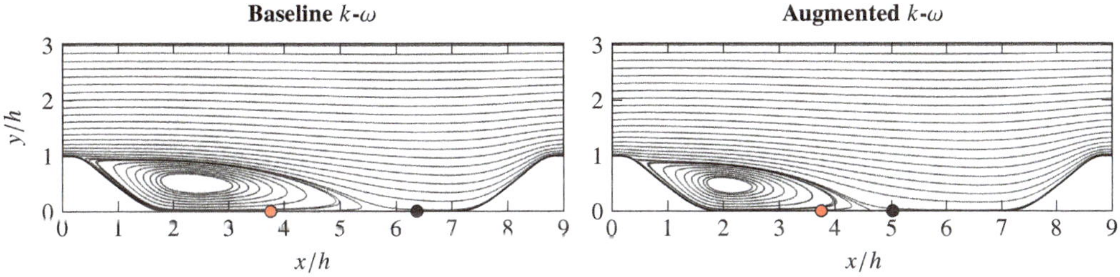

Figure 8. Periodic hill flow streamlines at $Re = 37$ K obtained by Köhler et al. [107] for the baseline $k - \omega$ model (**left**) and the augmented $k - \omega$ model (**right**) with the reattachment point at $x/h = 6.37$ and $x/h = 5.02$, respectively (black dot), compared to the reattachment point from the experiment at $x/h = 3.76$ (red dot). Reprinted by permission of the American Institute of Aeronautics and Astronautics, Inc.

Another ML-based analysis of periodic hill flows was presented by Yan et al. [3], which covers geometry variations (see the different hill sizes depending on α in Figure 9). The ML-supported model was referred to as the FIML method. It involves the adjustable parameter β, which modifies the production term in the Spalart–Allmaras (SA) model [129]. The results depend on finding an appropriate regularization (a weight used in the objective function to limit variations of β from the corresponding RANS value), which is important for guiding the FI to a physically realizable field [3]. One out of three options considered was found to be optimal. The grid applied used 81 points in the streamwise direction and 77 points in the normal direction. The $Re = 5600$ case simulations show a relatively good model performance for the $\alpha = (0.8, 1.2)$ cases involved in the training. There is an obvious improvement in RANS results, although the FIML model does not provide highly accurate predictions, in particular, close to the lower wall.

- The effect of geometry variations is shown in Figure 9. The FIML model performance is good for $\alpha = 0.5$ (involved in model training), reasonable for $\alpha = 1$ (slightly outside of the training range), and unsatisfactory for $\alpha = 1.5$ (clearly outside of the model training). In particular, there is hardly any improvement compared to the performance of the RANS model.

The effect of an increased Re is shown in Figure 10: the model is applied to the Re = 10,595 case in conjunction with $\alpha = 0.875$ instead of the $Re = 5600$ case considered so far. The FIML model obviously improves RANS predictions. Its performance is acceptable, although the FIML predictions are not highly accurate. The acceptable model performance may be related to the fact that the $\alpha = 0.875$ case considered is not too far from the range considered in the model training. The authors state that the results show a certain extent of generalization ability both on different geometry configurations and Reynolds numbers.

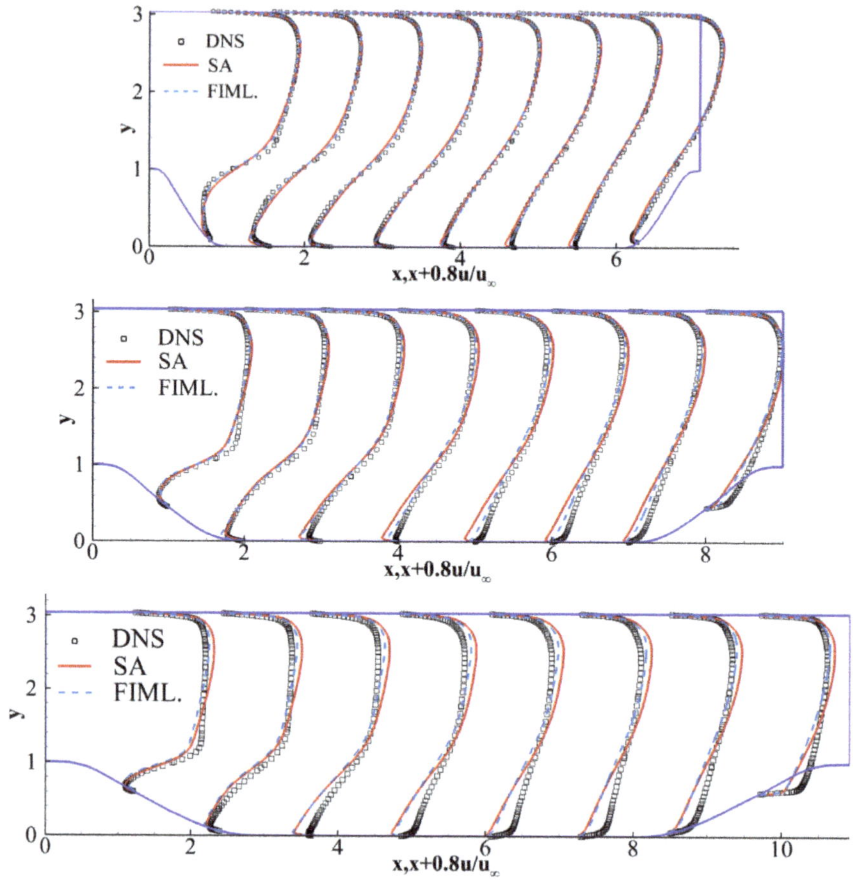

Figure 9. Periodic hill flow results obtained by Yan et al. [3] at $Re = 5600$, grid effect: streamline velocities at $\alpha = (0.5, 1, 1.5)$ are shown from above to below, respectively. Reprinted with permission from Ref. [3]. Copyright 2018 Elsevier.

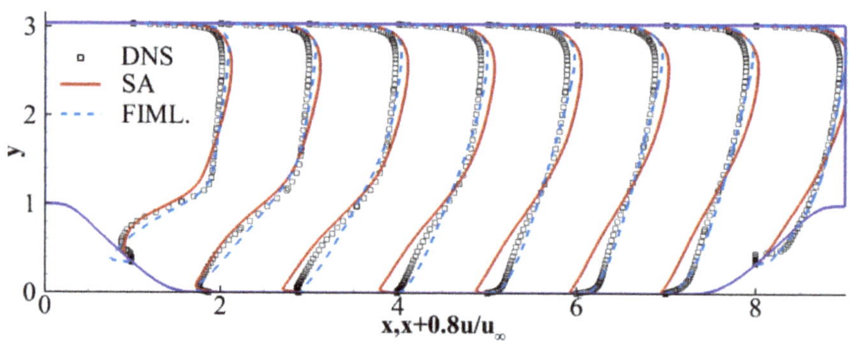

Figure 10. Periodic hill flow results obtained obtained by Yan et al. [3] at $Re = 10{,}595$ and $\alpha = 0.875$: Reynolds number effect. Reprinted with permission from Ref. [3]. Copyright 2018 Elsevier.

Another comprehensive analysis of periodic hill flows was presented by Volpiani et al. [106]. Using a neural network (NN) and high-fidelity numerical data, a vectorial source correction term in the momentum equations (output quantity) applied in conjunction with the SA turbulence model is determined. The corresponding model is referred to as NN-RANS. Two scenarios are considered: setup I applies model training based on Re = (2800, 10,595, 19,000) simulations, whereas setup II applies model training based on Re = (2800, 5600, 10,595) simulations. The number of grid points was not reported.

- Using the same $\alpha = 1$ geometry as in the training, the model used at a higher Re =19,000 than that considered in training leads to significant improvements compared to RANS, although NN-RANS velocity discrepancies can be found in the upper channel region, where the extrapolation overestimated the volume forces.
- Similar observations are found in regard to applying the model for the $Re = 5600$ case for different geometries than considered in the model training: there are significant improvements compared to RANS. It is of interest to note that setup I (trained using different geometries and Reynolds numbers) performs slightly better in the $\alpha = 0.8$ case, but slightly underperforms setup II in the $\alpha = 1.2$ case.
- We note that the range of modifications of unseen cases is smaller than in the approaches considered in the preceding paragraphs: the $Re = 37$ K and $\alpha = (0.5, 1.5)$ cases were not involved.

With respect to modifications of NN input parameter variations, the authors note that the choice of input features to improve NN-based RANS models is still a subject that needs further investigation.

3.2. 3D Hill-Type Flow Simulations

Next, let us consider the performance of 3D hill-type simulations based on ML extensions of RANS models, as presented by Yan et al. [108] and Ho and West [109]. Essential characteristics of these simulations are shown in Table 2 and Table 3, respectively. The grid applied for the FAITH hill simulation by Yan et al. was $165 \times 69 \times 93 = 1.06$ M, whereas Ho and West [109] applied a grid of 0.4 M cells for this simulation (and 374,400 cells for the 3D non-axisymmetric bump simulation).

Table 3. Ho and West analysis [109]: RF refers to random forest, OTC refers to one-time correction, ITER refers to the iterative approach, 2D refers to training on 2D cases, 3D refers to training on all 2D and 3D cases. An NN approach also considered by Ho and West [109] is excluded here because of its poor performance. ω equation with ML production.

Model and Training	Validation
RF-2D-OTC	
• 2D Channel [140] [DNS, $Re_\tau = 2$ K]	• 2D periodic hills [141] [LES, $Re = 10.6$ K]
• 2D NASA hump [125,142] [Exp. and LES, $Re = 936$ K]	• 3D FAITH hill [139] [Exp., $Re = 500$ K]
• 2D Curved backstep [143] [LES, $Re = 13.7$ K]	• 3D non-axisymmetric bump [144]
• 2D Cylinder [145,146] [Exp. and LES, $Re = 3.9$ K]	[Exp., $Re = 16$ K]
RF-2D-ITER	
• Same as above	• 2D periodic hills [141] [LES, $Re = 10.6$ K]
RF-3D-OTC	
• Same as above	• 3D non-axisymmetric bump [144]
• 3D FAITH hill [139] [Exp., $Re = 500$ K]	[Exp., $Re = 16$ K]

Yan et al. [108] report that the effect of the three ANN considered on the pressure coefficient distributions is very similar, which enables us to focus on the ANN1 results. Reference to the term artificial NN (ANN) is made here to provide a direct link to the cited work, bearing in mind that the terms ANN and NN have the same meaning. Characteristic results obtained by Yan et al. [108] for the unseen geometry of the 3D FAITH hill are shown in Figures 11 and 12. Figure 11 shows substantial differences between the flow field structures seen in experiments and using the FIML model, although the ANN-enhanced SA model shortens the separation region compared to the original SA model.

- Figure 12 shows that there are only very minor differences between the prediction of the original SA model and the ANN-enhanced SA model for the unseen geometry of the 3D FAITH hill, both the original and enhanced SA models underestimate the length of the separation region.

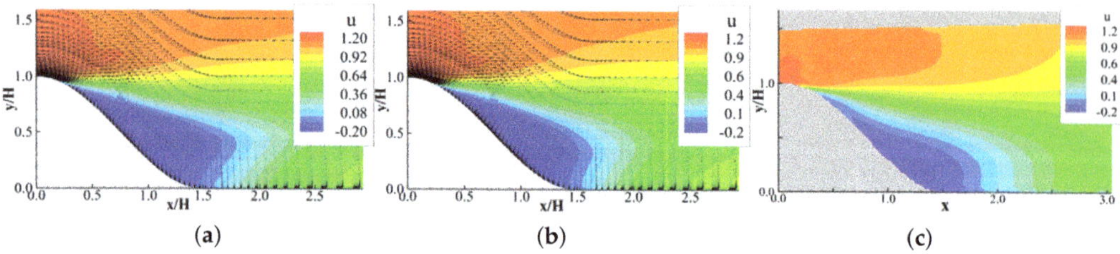

Figure 11. Comparison of the flow fields on the symmetry plane for the FAITH hill obtained from (**a**) the original SA model, (**b**) the FIML enhanced SA model, and (**c**) experimental results [108]. Reprinted with permission from Ref. [108]. Copyright 2020 AIP Publishing.

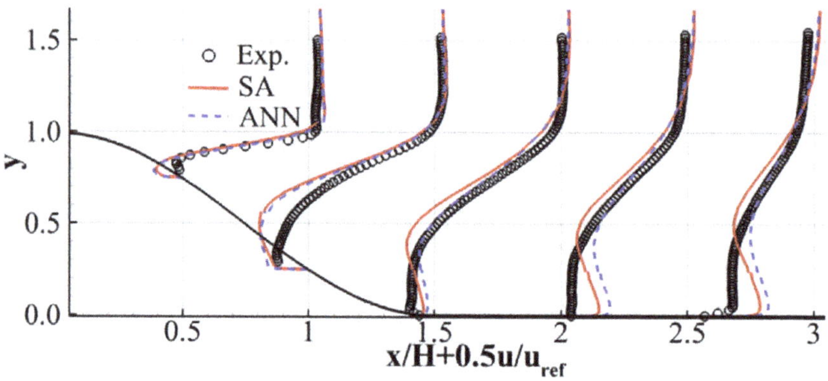

Figure 12. Comparison of the velocity profiles on the symmetry plane for the FAITH hill: the original SA model, the enhanced SA model, and experimental results [108]. Reprinted with permission from Ref. [108]. Copyright 2020 AIP Publishing.

Ho and West [109] applied the ML model called on individual cells to predict the correction factor during the simulation. Two options were considered: the correction factor calculation during every iteration in the iterative approach or just once in the one-time correction approach. The one-time correction (OTC) approach (having the lowest computational cost) was found to be most appropriate. The authors note that the iterative approach led to non-unique solutions based on the initialization of the flow field while the NN model produced large errors. In regard to 2D periodic hill flow simulations, the RF-2D-OTC model performed better than the RANS model as seen in mean velocities and k profiles, but the results are not very accurate in comparison to LES. The most interesting validations concern 3D FAITH hill and 3D non-axisymmetric bump applications.

- In regard to the 3D FAITH hill simulations, the results are very similar to the results of Yan et al. [108], shown in Figures 11 and 12. A minor difference is that the velocity profiles obtained by Ho and West using their RF-2D-OTC model agree a little better with the experimental results.
- The 3D non-axisymmetric bump flow simulations and streamwise velocity contour plots are shown in Figure 13. The conclusion is that there are improvements compared to the RANS result, but significant differences to the experimental results remain. It

may be seen that the inclusion of the 3D FAITH hill training data via the RF-3D-OTC model has a positive effect on the 3D non-axisymmetric bump results.

An interesting attempt to improve the ML approach was recently presented by Ho et al. [147]. This work is based on a probabilistic ML approach which reduces unwanted corrections, leading to more robust predictions. The latter is accomplished through only accepting predictions from the model above a certain confidence [147].

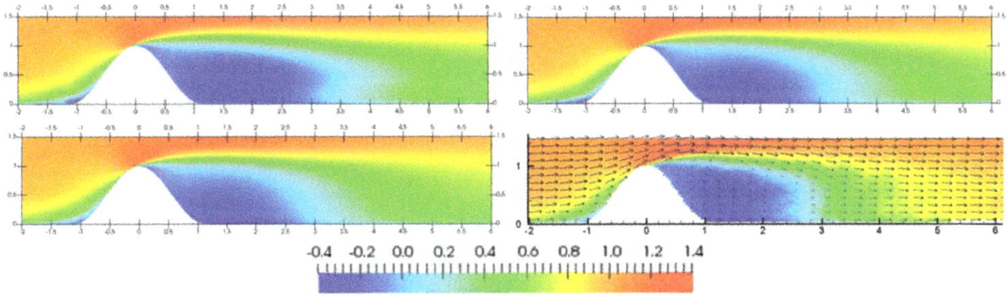

Figure 13. The 3D non-axisymmetric bump results obtained by Ho and West [109], streamwise velocity. First row: RANS model at $y/H = 0.5$ and symmetry plane (**left**) and RF-2D-OTC model results (**right**). Second row: RF-3D-OTC model results (**left**) and experimental results (**right**) [144]. The streamwise velocity color scale is shown in the last row. Reproduced with permission from Joel Ho and Alastair West.

3.3. CES vs. ML-RANS Methods

The general idea of developing ML-RANS methods is highly attractive: such well-functioning methods can be computationally highly efficient because no attempt is made to resolve flows. The core of this idea is the development of equations that represent well the physics of the flows considered. More specifically, such equations are supposed to represent causal relationships in close agreement with first principles. An illustration of such causal relationships is given in Figure 14: this figure illustrates regions of high and low turbulence production (compared to the underlying standard RANS model) implied by different hill shapes. The task to improve RANS predictions means to explain how such areas of high and low production are related to the specifics of the hill considered.

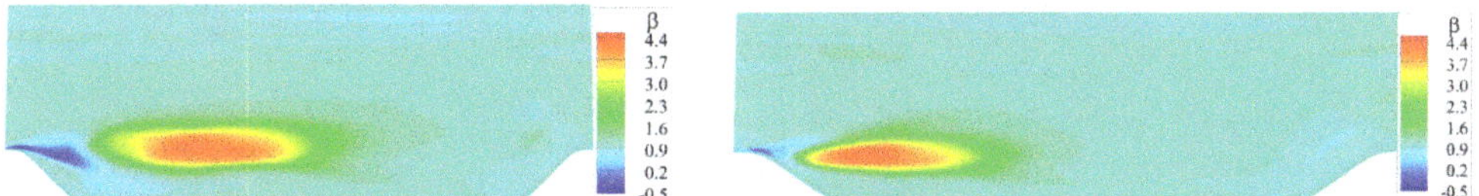

Figure 14. Periodic hill flow results at $Re = 5.6$ K [3]: characteristic model correction β distributions for different hill shapes [$\alpha = 0.8$ (**left**) and $\alpha = 1.2$ (**right**)]. Reprinted with permission from Ref. [3]. Copyright 2018 Elsevier.

In regard to the use of ML-RANS models, we observe the following:

1. The intellectual breeding ground of ML methods (the data applied in the model development) are reflections of causal relationships, i.e., not the causal relationships themselves. For example, by taking reference to Figure 14, ML methods are capable of explaining how the high turbulence production relates to flow variables (like velocity gradients and velocity differences) and other model variables. Such reflections of causal relationships have to be expected to vary (significantly) depending on the geometry and *Re* considered: see Figure 14 in regard to the geometry effect.
2. As a consequence, ML methods can be expected to improve RANS results, but they cannot be expected to reliably predict flows involving geometries and *Re* unseen

in model testing. The latter expectation is fully confirmed by the results presented above. As described in Sections 3.1 and 3.2, there may be hardly any improvements of RANS results.

3. From a general viewpoint, the development of equations for separated turbulent flows in the steady RANS framework involving all relevant causal relationships seems to be out of reach. At the end, this would require, e.g., to explain how the geometry considered implies the production increase in Figure 14. There is no mathematical basis for doing this.

This situation is different in the frame of CES methods (an illustration of differences between ML-RANS methods and CES methods can be found in Figure 15):

4. The geometry of separated turbulent flows implies the appearance of instantaneous detached flow (behind hills), which essentially determines the flow structure. In simulations, such flow corresponds to resolved flow, which is the instantaneous component of flow simulations. This leads to the task of explaining the causal relationship between the relative amount of resolved flow and the variation in production or dissipation in turbulence equations.
5. The latter is not doable in the RANS framework, which excludes resolved flow, but it can be conducted in the CES framework. This is the core idea of the CES approach, which uses exact mathematics to determine the variation in the dissipation in turbulence equations as a consequence of resolved flow implied, e.g., by obstacles in the flow. In particular, the generality of such causality relationships is confirmed via the general mathematical derivation.
6. From a general viewpoint, the use of CES methods may be seen as an experiment: causalities are involved to test whether the missing inclusion of causalities is indeed the origin of problems seen in regard to the use of ML-RANS methods for cases unseen in model testing. The results reported in Section 2 speak a clear language: the inclusion of causalities in CES methods enables much better flow predictions than given by ML-RANS models.

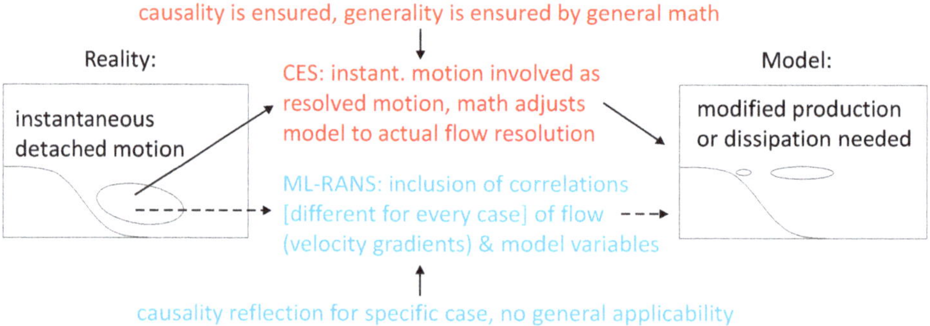

Figure 15. An illustration of conceptual differences between CES methods and ML-RANS methods in regard to hill-type separated turbulent flows. The illustration of required model modifications compared to standard RANS methods references Figure 14.

4. Summary

The issues of existing computational approaches related to the feasible and reliable simulation of high *Re*-separated turbulent flows led to the search for mathematically better-founded simulation methods. In a nonresolving framework, one way to address these questions is the very popular use of ML methods to enable improved RANS simulations. Another way is the use of minimal error CES simulation methods in the framework of methods being applicable under conditions ranging from resolving to almost modeled simulations.

The facts reported in Section 3, in regard to the performance of ML-RANS methods for the flows considered, speak a clear language. The setup of such ML methods usually involves a variety of adjustable model settings (see above) that need to be chosen to enhance the model performance. Such methods can provide improvements in RANS predictions (ML-RANS results are usually equivalent or better than RANS predictions), but in general, there is currently no indication that such methods are capable of providing reliable predictions for flow geometries or *Re* not involved in model training (it is worth noting that the results reported here were presented in the very last years). One may argue that this situation can be improved by involving larger amounts of data in model training, hoping that the range of model applicability can be enhanced in this way. As discussed in regard to Figure 13, the latter can lead to improvements, which agrees with observations made by Fang et al. [148]. But there is no indication so far that this approach can lead to systematic improvements (a gradual increase in the predictive power of ML-RANS). Table 4 shows facts of interest in this regard [103]. On average, this strategy simply fails. In line with Ref. [148] results, these results indicate that competing types of ML corrections are required for different cases.

Table 4. ML-RANS results presented by Rumsey et al. [103]. Upper part: cases used in training. ZPG refers to zero pressure gradient, APG refers to adverse pressure gradient, *Ma* refers to Mach number. Lower part: ML-RANS performance for cases unseen in training. All the lower part cases were trained by involving all upper part cases.

Case	Ma	Re	Unseen Case Predictions
ZPG-APG plate	0.1	0.18×10^6	
ZPG plate	0.2	5.0×10^6	
NACA 0012, $\alpha = 10°$	0.15	6.0×10^6	
NASA hump	0.1	0.936×10^6	
ZPG-APG plate, lower Re	0.1	0.08×10^6	worse than RANS
Channel at very high Re	0.2	80×10^6	worse than RANS
NACA 0012, other α	0.15	6.0×10^6	worse than RANS
Periodic hill	0.2	10,595	same as RANS
Axisymmetric jet	0.01	5601	much worse than RANS
Curved backstep	0.1	13,700	slightly better than RANS

The CES results reported in Section 2 also speak a clear language in regard to the three complex flow simulations considered so far. Such CES methods perform very well, better than WMLES or specifically designed DES methods, and at least as good or better than WRLES (at a little fraction of WRLES computational cost). These applications include evidence that these methods can work as resolving or partially resolving simulation methods. A remarkable feature of these CES methods is their well-balanced performance in contrast to other methods which usually provide appropriate predictions of some characteristics at the cost of other simulation characteristics. It is worth mentioning that such CES methods are independent of a variety of model setup options: they work on the basis of the RANS model considered, taking the mathematically determined hybridization into account.

The reason for the different features of ML-RANS and CES methods was identified in Section 3.3: it is given by the different way of accounting for causal relationships. The latter are a characteristic feature of separated turbulent flows, the flow geometry implies characteristic separation region features. The setup of ML-based methods enables the identification of correlations between model and flow variables, but the latter can vary a lot from flow to flow depending on the specific separation features. In contrast, CES methods can account for such causality via their inclusion of resolved motion implied by the type of flow separation. The generality of this inclusion of causality is confirmed via the general mathematical derivation and applications performed.

The conclusion of these facts is that the use of CES methods for separated turbulent flow simulations provides a much better alternative compared to the use of ML-RANS methods. Given that, what is the way to take maximum advantage of ML methods? First, in the frame of nonresolving methods, a more detailed investigation of the predictive power of ML methods in regard to relatively minor case variations would be of significant interest. This concerns, e.g., the prediction of *Re* effects for the same geometry under conditions where different *Re* cases are involved in the model training. The latter matters with respect to questions about asymptotic flow structures, and the known difficulty of resolving computational methods (DNS and LES) with extreme *Re* predictions [149,150]. Second, in the frame of partially resolving methods, a merging of ML methods as considered here with CES concepts would be highly beneficial to take advantage of both approaches. The current development of CES methods is based on two-equation turbulence equations, which include an (ω or ϵ) scale equation, or an equation for the turbulent viscosity [96]. However, there are several usually applied codes that do not involve dissipation transport equations, as is currently the case with the widely used Weather Research and Forecasting Model (WRF) dealing with atmospheric flow simulations. By following the CES concept, ML can help to identify corresponding causal relationships between parameters or source terms in simplified turbulence equations (without scale equation, or even without an k transport equation). The latter can significantly contribute to the availability of corresponding computational methods via available codes for a variety of equation structures.

Funding: This research received no external funding.

Data Availability Statement: The data that support the findings of this study are available within the article.

Acknowledgments: I would like to acknowledge support from the National Science Foundation (AGS, Grant No. 2137351, with N. Anderson as Technical Officer) and support from the Hanse-Wissen-schaftskolleg (Delmenhorst, Germany, with M. Kastner as Technical Officer). This work was supported by the Wyoming NASA Space Grant Consortium (NASA Grant No. 80NSSC20M0113) and the University of Wyoming School of Computing (Wyoming Innovation Partnership grant). I am thankful to J. Ho and A. West for their permission to reuse their Figure 13. I am also thankful to V. Srivastava, C. Rumsey, G. Coleman, and L. Wang for their permission to reuse Figure 5.

Conflicts of Interest: The author declares no conflicts of interest.

References

1. Pope, S.B. *Turbulent Flows*; Cambridge University Press: Cambridge, UK, 2000.
2. Breuer, M.; Peller, N.; Rapp, C.; Manhart, M. Flow over periodic hills—Numerical and experimental study in a wide range of Reynolds numbers. *Comput. Fluids* **2009**, *38*, 433–457. [CrossRef]
3. Yan, C.; Li, H.; Zhang, Y.; Chen, H. Data-driven turbulence modeling in separated flows considering physical mechanism analysis. *Int. J. Heat Fluid Flow* **2022**, *96*, 109004. [CrossRef]
4. Wilcox, D.C. *Turbulence Modeling for CFD*, 2nd ed.; DCW Industries: La Canada, CA, USA, 1998.
5. Hanjalić, K. Will RANS survive LES? A view of perspectives. *ASME J. Fluids Eng.* **2005**, *127*, 831–839. [CrossRef]
6. Sagaut, P. *Large Eddy Simulation for Incompressible Flows: An Introduction*; Springer: Berlin/Heidelberg, Germany, 2002.
7. Lesieur, M.; Metais, O.; Comte, P. *Large-Eddy Simulations of Turbulence*; Cambridge University Press: Cambridge, UK, 2005.
8. Piomelli, U. Large-eddy simulation: Achievements and challenges. *Prog. Aerosp. Sci.* **1999**, *35*, 335–362. [CrossRef]
9. Heinz, S. A review of hybrid RANS-LES methods for turbulent flows: Concepts and applications. *Prog. Aerosp. Sci.* **2020**, *114*, 100597. [CrossRef]
10. Menter, F.; Hüppe, A.; Matyushenko, A.; Kolmogorov, D. An overview of hybrid RANS–LES models developed for industrial CFD. *Appl. Sci.* **2021**, *11*, 2459. [CrossRef]
11. Deardorff, J.W. A numerical study of three-dimensional turbulent channel flow at large Reynolds numbers. *J. Fluid Mech.* **1970**, *41*, 453–480. [CrossRef]
12. Schumann, U. Subgrid scale model for finite difference simulations of turbulent flows in plane channels and annuli. *J. Comput. Phys.* **1975**, *18*, 376–404. [CrossRef]
13. Grötzbach, G. Direct numerical and large eddy simulation of turbulent channel flow. In *Encyclopedia of Fluid Mechanics*; Gulf: West Orange, NJ, USA, 1987; pp. 1337–1391.
14. Piomelli, U. Wall-layer models for large-eddy simulations. *Prog. Aerosp. Sci.* **2008**, *44*, 437–446. [CrossRef]
15. Piomelli, U.; Ferziger, J.; Moin, P. New approximate boundary conditions for large eddy simulations of wall-bounded flows. *Phys. Fluids A* **1989**, *6*, 1061–1068. [CrossRef]
16. Cabot, W.; Moin, P. Approximate wall boundary conditions in the large-eddy simulation of high Reynolds number flow. *Flow Turbul. Combust.* **1999**, *63*, 269–291. [CrossRef]
17. Piomelli, U.; Balaras, E. Wall-layer models for large-eddy simulations. *Annu. Rev. Fluid Mech.* **2002**, *34*, 349–374. [CrossRef]

18. Kawai, S.; Larsson, J. Wall-modeling in large eddy simulation: Length scales, grid resolution, and accuracy. *Phys. Fluids* **2012**, *24*, 015105. [CrossRef]
19. Bose, S.T.; Moin, P. A dynamic slip boundary condition for wall-modeled large-eddy simulation. *Phys. Fluids* **2014**, *26*, 015104. [CrossRef]
20. Park, G.I.; Moin, P. An improved dynamic non-equilibrium wall-model for large eddy simulation. *Phys. Fluids* **2014**, *26*, 015108. [CrossRef]
21. Piomelli, U. Large eddy simulations in 2030 and beyond. *Philos. Trans. R. Soc. Math. Phys. Eng. Sci.* **2014**, *372*, 20130320. [CrossRef] [PubMed]
22. Moin, P.; Bodart, J.; Bose, S.; Park, G.I. Wall-modeling in complex turbulent flows. In *Advances in Fluid-Structure Interaction, Notes on Numerical Fluid Mechanics and Multidisciplinary Design 133*; Braza, M., Bottaro, A., Thompson, M., Eds.; Springer International Publishing: Cham, Switzerland, 2016; pp. 207–219.
23. Larsson, J.; Kawai, S.; Bodart, J.; Bermejo-Moreno, I. Large eddy simulation with modeled wall-stress: Recent progress and future directions. *Mech. Eng. Rev.* **2016**, *3*, 15-00418. [CrossRef]
24. Bose, S.T.; Park, G.I. Wall-modeled large-eddy simulation for complex turbulent flows. *Annu. Rev. Fluid Mech.* **2018**, *50*, 535–561. [CrossRef]
25. Yang, X.I.A.; Sadique, J.; Mittal, R.; Meneveau, C. Integral wall model for large eddy simulations of wall-bounded turbulent flows. *Phys. Fluids* **2015**, *27*, 025112. [CrossRef]
26. Yang, X.I.A.; Griffin, K.P. Grid-point and time-step requirements for direct numerical simulation and large-eddy simulation. *Phys. Fluids* **2021**, *33*, 015108. [CrossRef]
27. Fagbade, A.; Heinz, S. Continuous eddy simulation (CES) of transonic shock-induced flow separation. *Appl. Sci.* **2024**, *14*, 2705. [CrossRef]
28. Fagbade, A.; Heinz, S. Continuous eddy simulation vs. resolution-imposing simulation methods for turbulent flows. *Fluids* **2024**, *9*, 22. [CrossRef]
29. Verma, A.; Park, N.; Mahesh, K. A hybrid subgrid-scale model constrained by Reynolds stress. *Phys. Fluids* **2013**, *25*, 110805. [CrossRef]
30. Xiao, Z.; Shi, Y.; Xia, Z.; Chen, S. Comment on 'A hybrid subgrid-scale model constrained by Reynolds stress' [Phys. Fluids 25, 110805 (2013)]. *Phys. Fluids* **2014**, *26*, 059101. [CrossRef]
31. Verma, A.; Park, N.; Mahesh, K. Response to "Comment on A hybrid subgrid-scale model constrained by Reynolds stress" [Phys. Fluids 26, 059101 (2014)]. *Phys. Fluids* **2014**, *26*, 059102. [CrossRef]
32. Chen, S.; Xia, Z.; Pei, S.; Wang, J.; Yang, Y.; Xiao, Z.; Shi, Y. Reynolds-stress-constrained large-eddy simulation of wall-bounded turbulent flows. *J. Fluid Mech.* **2012**, *703*, 1–28. [CrossRef]
33. Chen, S.; Chen, Y.; Xia, Z.; Qu, K.; Shi, Y.; Xiao, Z.; Liu, Q.; Cai, Q.; Liu, F.; Lee, C.; et al. Constrained large-eddy simulation and detached eddy simulation of flow past a commercial aircraft at 14 degrees angle of attack. *Sci. China Phys. Mech. Astron.* **2013**, *56*, 270–276. [CrossRef]
34. Xia, Z.; Shi, Y.; Hong, R.; Xiao, Z.; Chen, S. Constrained large-eddy simulation of separated flow in a channel with streamwise-periodic constrictions. *J. Turbul.* **2013**, *14*, 1–21. [CrossRef]
35. Jiang, Z.; Xiao, Z.; Shi, Y.; Chen, S. Constrained large-eddy simulation of wall-bounded compressible turbulent flows. *Phys. Fluids* **2013**, *25*, 106102. [CrossRef]
36. Hong, R.; Xia, Z.; Shi, Y.; Xiao, Z.; Chen, S. Constrained large-eddy simulation of compressible flow past a circular cylinder. *Commun. Comput. Phys.* **2014**, *15*, 388–421. [CrossRef]
37. Zhao, Y.; Xia, Z.; Shi, Y.; Xiao, Z.; Chen, S. Constrained large-eddy simulation of laminar-turbulent transition in channel flow. *Phys. Fluids* **2014**, *26*, 095103. [CrossRef]
38. Xua, Q.; Yang, Y. Reynolds stress constrained large eddy simulation of separation flows in a U-duct. *J. Propul. Power Res.* **2014**, *3*, 49–58. [CrossRef]
39. Xia, Z.; Xiao, Z.; Shi, Y.; Chen, S. Constrained large-eddy simulation for aerodynamics. In *Progress in Hybrid RANS-LES Modelling, Notes on Numerical Fluid Mechanics and Multidisciplinary Design 130*; Girimaji, S., Haase, W., Peng, S.H., Schwamborn, D., Eds.; Springer International Publishing: Cham, Switzerland, 2015; pp. 239–254.
40. Jiang, Z.; Xiao, Z.; Shi, Y.; Chen, S. Constrained large-eddy simulation of turbulent flow and heat transfer in a stationary ribbed duct. *Int. J. Numer. Methods Heat Fluid Flow* **2016**, *26*, 1069–1091. [CrossRef]
41. Spalart, P.R.; Jou, W.H.; Strelets, M.; Allmaras, S.R. Comments on the feasibility of LES for wings, and on a hybrid RANS/LES approach. In *Advances in DNS/LES*; Liu, C., Liu, Z., Eds.; Greyden Press: Columbus, OH, USA, 1997; pp. 137–147.
42. Travin, A.; Shur, M.L.; Strelets, M.; Spalart, P. Detached-eddy simulations past a circular cylinder. *Flow Turbul. Combust.* **1999**, *63*, 113–138.
43. Spalart, P.R. Strategies for turbulence modelling and simulations. *Int. J. Heat Fluid Flow* **2000**, *21*, 252–263. [CrossRef]
44. Strelets, M. Detached eddy simulation of massively separated flows. In Proceedings of the 39th AIAA Aerospace Sciences Meeting and Exhibit, Reno, NV, USA, 8–11 January 2001; AIAA Paper 01-0879; pp. 1–18.
45. Travin, A.; Shur, M.L. Physical and numerical upgrades in the detached-eddy simulation of complex turbulent flows. In *Advances in LES of Complex Flows*; Friedrich, R., Rodi, W., Eds.; Kluwer Academic Publishers: Dordrecht, The Netherlands, 2002; pp. 239–254.

46. Menter, F.R.; Kuntz, M.; Langtry, R. Ten years of industrial experience with SST turbulence model. *Turb. Heat Mass Transf.* **2003**, *4*, 625–632.
47. Spalart, P.R.; Deck, S.; Shur, M.L.; Squires, K.D.; Strelets, M.K.; Travin, A. A new version of detached-eddy simulation, resistant to ambiguous grid densities. *Theor. Comput. Fluid Dyn.* **2006**, *20*, 181–195. [CrossRef]
48. Shur, M.L.; Spalart, P.R.; Strelets, M.K.; Travin, A. A hybrid RANS-LES approach with delayed-DES and wall-modelled LES capabilities. *Int. J. Heat Fluid Flow* **2008**, *29*, 1638–1649. [CrossRef]
49. Spalart, P.R. Detached-eddy simulation. *Annu. Rev. Fluid Mech.* **2009**, *41*, 181–202. [CrossRef]
50. Mockett, C.; Fuchs, M.; Thiele, F. Progress in DES for wall-modelled LES of complex internal flows. *Comput. Fluids* **2012**, *65*, 44–55. [CrossRef]
51. Friess, C.; Manceau, R.; Gatski, T.B. Toward an equivalence criterion for hybrid RANS/LES methods. *Comput. Fluids* **2015**, *122*, 233–246. [CrossRef]
52. Chaouat, B. The state of the art of hybrid RANS/LES modeling for the simulation of turbulent flows. *Flow Turbul. Combust.* **2017**, *99*, 279–327. [CrossRef] [PubMed]
53. Dong, T.; Minelli, G.; Wang, J.; Liang, X.; Krajnović, S. Numerical investigation of a high-speed train underbody flows: Studying flow structures through large-eddy simulation and assessment of steady and unsteady Reynolds-averaged Navier–Stokes and improved delayed detached eddy simulation performance. *Phys. Fluids* **2022**, *34*, 015126. [CrossRef]
54. Wyngaard, J.C. Toward numerical modeling in the "Terra Incognita". *J. Atmos. Sci.* **2004**, *61*, 1816–1826. [CrossRef]
55. Juliano, T.W.; Kosović, B.; Jiménez, P.A.; Eghdami, M.; Haupt, S.E.; Martilli, A. "Gray Zone" simulations using a three-dimensional planetary boundary layer parameterization in the Weather Research and Forecasting Model. *Mon. Weather Rev.* **2022**, *150*, 1585–1619. [CrossRef]
56. Heinz, S.; Heinz, J.; Brant, J.A. Mass transport in membrane systems: Flow regime identification by Fourier analysis. *Fluids* **2022**, *7*, 369. [CrossRef]
57. Heinz, S. Unified turbulence models for LES and RANS, FDF and PDF simulations. *Theoret. Comput. Fluid Dynam.* **2007**, *21*, 99–118. [CrossRef]
58. Heinz, S. Realizability of dynamic subgrid-scale stress models via stochastic analysis. *Monte Carlo Methods Applic.* **2008**, *14*, 311–329. [CrossRef]
59. Heinz, S.; Gopalan, H. Realizable versus non-realizable dynamic subgrid-scale stress models. *Phys. Fluids* **2012**, *24*, 115105. [CrossRef]
60. Gopalan, H.; Heinz, S.; Stöllinger, M. A unified RANS-LES model: Computational development, accuracy and cost. *J. Comput. Phys.* **2013**, *249*, 249–279. [CrossRef]
61. Mokhtarpoor, R.; Heinz, S.; Stoellinger, M. Dynamic unified RANS-LES simulations of high Reynolds number separated flows. *Phys. Fluids* **2016**, *28*, 095101. [CrossRef]
62. Mokhtarpoor, R.; Heinz, S. Dynamic large eddy simulation: Stability via realizability. *Phys. Fluids* **2017**, *29*, 105104. [CrossRef]
63. Mokhtarpoor, R.; Heinz, S.; Stöllinger, M. Dynamic unified RANS-LES simulations of periodic hill flow. In *Direct and Large-Eddy Simulation XI (ERCOFTAC Series)*; Springer: Cham, Switzerland, 2019; pp. 489–496.
64. Girimaji, S.; Srinivasan, R.; Jeong, E. PANS turbulence for seamless transition between RANS and LES: Fixed-point analysis and preliminary results. In Proceedings of the ASME FEDSM03, Honolulu, HI, USA, 6–10 July 2003; ASME Paper FEDSM2003-45336; pp. 1–9.
65. Girimaji, S.; Abdol-Hamid, K. Partially averaged Navier Stokes model for turbulence: Implemantation and validation. In Proceedings of the 43rd AIAA Aerospace Sciences Meeting and Exhibit, Reno, NV, USA, 10–13 January 2005; AIAA Paper 05-0502; pp. 1–14.
66. Girimaji, S. Partially-averaged Navier-Stokes method for turbulence: A Reynolds-averaged Navier-Stokes to direct numerical simulation bridging method. *ASME J. Appl. Mech.* **2006**, *73*, 413–421. [CrossRef]
67. Girimaji, S.; Jeong, E.; Srinivasan, R. Partially averaged Navier-Stokes method for turbulence: Fixed point analysis and comparisons with unsteady partially averaged Navier-Stokes. *ASME J. Appl. Mech.* **2006**, *73*, 422–429. [CrossRef]
68. Lakshmipathy, S.; Girimaji, S.S. Extension of Boussinesq turbulence constitutive relation for bridging methods. *J. Turbul.* **2007**, *8*, 1–20. [CrossRef]
69. Frendi, A.; Tosh, A.; Girimaji, S. Flow past a backward-facing step: Comparison of PANS, DES and URANS results with experiments. *Int. J. Comput. Methods Eng. Sci. Mech.* **2007**, *8*, 23–38. [CrossRef]
70. Lakshmipathy, S.; Girimaji, S.S. Partially averaged Navier-Stokes (PANS) method for turbulence simulations: Flow past a circular cylinder. *ASME J. Fluids Eng.* **2010**, *132*, 121202. [CrossRef]
71. Jeong, E.; Girimaji, S.S. Partially averaged Navier–Stokes (PANS) method for turbulence simulations: Flow past a square cylinder. *ASME J. Fluids Eng.* **2010**, *132*, 121203. [CrossRef]
72. Basara, B.; Krajnovic, S.; Girimaji, S.S.; Pavlovic, Z. Near-wall formulation of the partially averaged Navier-Stokes turbulence model. *AIAA J.* **2011**, *42*, 2627–2636. [CrossRef]
73. Krajnovic, S.; Lárusson, R.; Basara, B. Superiority of PANS compared to LES in predicting a rudimentary landing gear flow with affordable meshes. *Int. J. Heat Fluid Flow* **2012**, *37*, 109–122. [CrossRef]
74. Foroutan, H.; Yavuzkurt, S. A partially averaged Navier Stokes model for the simulation of turbulent swirling flow with vortex breakdown. *Int. J. Heat Fluid Flow* **2014**, *50*, 402–416. [CrossRef]

75. Drikakis, D.; Sofos, F. Can artificial intelligence accelerate fluid mechanics research? *Fluids* **2023**, *8*, 212. [CrossRef]
76. Schiestel, R.; Dejoan, A. Towards a new partially integrated transport model for coarse grid and unsteady turbulent flow simulations. *Theor. Comput. Fluid Dyn.* **2005**, *18*, 443–468. [CrossRef]
77. Chaouat, B.; Schiestel, R. A new partially integrated transport model for subgrid-scale stresses and dissipation rate for turbulent developing flows. *Phys. Fluids* **2005**, *17*, 065106. [CrossRef]
78. Chaouat, B.; Schiestel, R. From single-scale turbulence models to multiple-scale and subgrid-scale models by Fourier transform. *Theor. Comput. Fluid Dyn.* **2007**, *21*, 201–229. [CrossRef]
79. Befeno, I.; Schiestel, R. Non-equilibrium mixing of turbulence scales using a continuous hybrid RANS/LES approach: Application to the shearless mixing layer. *Flow Turbul. Combust.* **2007**, *78*, 129–151. [CrossRef]
80. Chaouat, B.; Schiestel, R. Progress in subgrid-scale transport modelling for continuous hybrid nonzonal RANS/LES simulations. *Int. J. Heat Fluid Flow* **2009**, *30*, 602–616. [CrossRef]
81. Chaouat, B. Subfilter-scale transport model for hybrid RANS/LES simulations applied to a complex bounded flow. *J. Turbul.* **2010**, *11*, 1–30. [CrossRef]
82. Chaouat, B. Simulation of turbulent rotating flows using a subfilter scale stress model derived from the partially integrated transport modeling method. *Phys. Fluids* **2012**, *24*, 045108. [CrossRef]
83. Chaouat, B.; Schiestel, R. Analytical insights into the partially integrated transport modeling method for hybrid Reynolds averaged Navier-Stokes equations-large eddy simulations of turbulent flows. *Phys. Fluids* **2012**, *24*, 085106. [CrossRef]
84. Chaouat, B.; Schiestel, R. Partially integrated transport modeling method for turbulence simulation with variable filters. *Phys. Fluids* **2013**, *25*, 125102. [CrossRef]
85. Chaouat, B.; Schiestel, R. Hybrid RANS-LES simulations of the turbulent flow over periodic hills at high Reynolds number using the PITM method. *Comput. Fluids* **2013**, *84*, 279–300. [CrossRef]
86. Chaouat, B. Application of the PITM method using inlet synthetic turbulence generation for the simulation of the turbulent flow in a small axisymmetric contraction. *Flow Turbul. Combust.* **2017**, *98*, 987–1024. [CrossRef]
87. Menter, F.R.; Kuntz, M.; Bender, R. A scale-adaptive simulation model for turbulent flow predictions. In Proceedings of the 41st AIAA Aerospace Sciences Meeting and Exhibit, Reno, NV, USA, 6–9 January 2003; AIAA Paper 03-0767; pp. 1–12.
88. Menter, F.R.; Egorov, Y. A scale-adaptive simulation model using two-equation models. In Proceedings of the 43rd AIAA Aerospace Sciences Meeting and Exhibit, Reno, NV, USA, 10–13 January 2005; AIAA Paper 05-1095; pp. 1–13.
89. Menter, F.R.; Egorov, Y. Revisiting the turbulent scale equation. In *IUTAM Symposium on One Hundred Years of Boundary Layer Research*; Meier, G.E.A., Sreenivasan, K.R., Eds.; Springer: Dordrecht, The Netherlands, 2006; pp. 279–290.
90. Menter, F.R.; Egorov, Y. The scale-adaptive simulation method for unsteady turbulent flow prediction: Part 1: Theory and model description. *Flow Turbul. Combust.* **2010**, *78*, 113–138. [CrossRef]
91. Menter, F.R.; Egorov, Y. The scale-adaptive simulation method for unsteady turbulent flow prediction: Part 2: Application to complex flows. *Flow Turbul. Combust.* **2010**, *78*, 139–165.
92. Jakirlić, S.; Maduta, R. Extending the bounds of "steady" RANS closures: Toward an instability-sensitive Reynolds stress model. *Int. J. Heat Fluid Flow* **2015**, *51*, 175–194. [CrossRef]
93. Heinz, S.; Mokhtarpoor, R.; Stoellinger, M.K. Theory-based Reynolds-averaged Navier-Stokes equations with large eddy simulation capability for separated turbulent flow simulations. *Phys. Fluids* **2020**, *32*, 065102. [CrossRef]
94. Fagbade, A.I. Continuous Eddy Simulation for Turbulent Flows. Ph.D. Thesis, University of Wyoming, Laramie, WY, USA, 2024. Available online: https://www.proquest.com/docview/3058393461 (accessed on 1 June 2024).
95. Heinz, S. A mathematical solution to the Computational Fluid Dynamics (CFD) dilemma. *Mathematics* **2023**, *11*, 3199. [CrossRef]
96. Heinz, S. Minimal error partially resolving simulation methods for turbulent flows: A dynamic machine learning approach. *Phys. Fluids* **2022**, *34*, 051705. [CrossRef]
97. Heinz, S. Remarks on energy partitioning control in the PITM hybrid RANS/LES method for the simulation of turbulent flows. *Flow, Turb. Combust.* **2022**, *108*, 927–933. [CrossRef]
98. Heinz, S. From two-equation turbulence models to minimal error resolving simulation methods for complex turbulent flows. *Fluids* **2022**, *7*, 368. [CrossRef]
99. Heinz, S. Theory-based mesoscale to microscale coupling for wind energy applications. *Appl. Math. Model.* **2021**, *98*, 563–575. [CrossRef]
100. Heinz, S. The continuous eddy simulation capability of velocity and scalar probability density function equations for turbulent flows. *Phys. Fluids* **2021**, *33*, 025107. [CrossRef]
101. Heinz, S. The large eddy simulation capability of Reynolds-averaged Navier-Stokes equations: Analytical results. *Phys. Fluids* **2019**, *31*, 021702. [CrossRef]
102. Singh, A.P.; Medida, S.; Duraisamy, K. Machine-learning-augmented predictive modeling of turbulent separated flows over airfoils. *AIAA J.* **2017**, *55*, 2215–2227. [CrossRef]
103. Rumsey, C.L.; Coleman, G.N.; Wang, L. In search of data-driven improvements to RANS models applied to separated flows. In Proceedings of the AIAA SciTech Forum, San Diego, CA, USA, 3–7 January 2022; AIAA Paper 22-0937; pp. 1–21.
104. Rumsey, C.L.; Coleman, G.N. *NASA Symposium on Turbulence Modeling: Roadblocks, and the Potential for Machine Learning*; Technical Report; Langley Research Center: Hampton, VA, USA, 2022.

105. Srivastava, V.; Rumsey, C.L.; Coleman, G.N.; Wang, L. On generalizably improving RANS predictions of flow separation and reattachment. In Proceedings of the AIAA SciTech Forum, Orlando, FL, USA, 8–12 January 2024; AIAA Paper 24-2520; pp. 1–13.
106. Volpiani, P.S.; Meyer, M.; Franceschini, L.; Dandois, J.; Renac, F.; Martin, E.; Marquet, O.; Sipp, D. Machine learning-augmented turbulence modeling for RANS simulations of massively separated flows. *Phys. Rev. Fluids* **2021**, *6*, 064607. [CrossRef]
107. Köhler, F.; Munz, J.; Schäfer, M. Data-driven augmentation of rans turbulence models for improved prediction of separation in wall-bounded flows. In Proceedings of the AIAA SciTech 2020 Forum, Orlando, FL, USA, 6–10 January 2020; AIAA Paper 20-1586; pp. 1–17.
108. Yan, C.; Zhang, Y.; Chen, H. Data augmented turbulence modeling for three-dimensional separation flows. *Phys. Fluids* **2022**, *34*, 075101. [CrossRef]
109. Ho, J.; West, A. Field inversion and machine learning for turbulence modelling applied to three-dimensional separated flows. In Proceedings of the AIAA Aviation 2021 Forum, Virtual Event, 2–6 August 2021; AIAA Paper 21-2903; pp. 1–18.
110. Alhafiz, M.R.; Palar, P.S.; Lavi, Z. An efficient data-driven neural network turbulence model for flow over periodic hills and parametric bump. In Proceedings of the AIAA SciTech 2024 Forum, Orlando, FL, USA, 8–12 January 2024; AIAA Paper 24-1985; pp. 1–16.
111. Ching, D.S.; Banko, A.J.; Milani, P.M.; Eaton, J.K. Machine learning modeling for RANS turbulent kinetic energy transport in 3D separated flows. In Proceedings of the 11th International Symposium on Turbulence and Shear Flow Phenomena, Southampton, UK, 30 July–2 August 2019; pp. 1–6.
112. Giyats, A.F.; Yamin, M.; Mahandari, C.P. Applied machine learning to estimate length of separation and reattachment flows as parameter active flow control in backward facing step. *J. Mech. Eng.* **2023**, *20*, 131–154.
113. Hanrahan, S.; Kozul, M.; Sandberg, R.D. Studying turbulent flows with physics-informed neural networks and sparse data. *Int. J. Heat Fluid Flow* **2023**, *104*, 109232. [CrossRef]
114. Pioch, F.; Harmening, J.H.; Müller, A.M.; Peitzmann, F.J.; Schramm, D.; Moctar, O.E. Turbulence modeling for physics-informed neural networks: Comparison of different RANS models for the backward-facing step flow. *Fluids* **2023**, *8*, 43. [CrossRef]
115. Shan, X.; Liu, Y.; Cao, W.; Sun, X.; Zhang, W. Turbulence modeling via data assimilation and machine learning for separated flows over airfoils. *AIAA J.* **2023**, *61*, 3883–3899. [CrossRef]
116. Yazdani, S.; Tahani, M. Data-driven discovery of turbulent flow equations using physics-informed neural networks. *Phys. Fluids* **2024**, *36*, 035107. [CrossRef]
117. Yin, Y.; Yang, P.; Zhang, Y.; Chen, H.; Fu, S. Feature selection and processing of turbulence modeling based on an artificial neural network. *Phys. Fluids* **2020**, *32*, 105117. [CrossRef]
118. Eivazi, H.; Tahani, M.; Schlatter, P.; Vinuesa, R. Physics-informed neural networks for solving Reynolds-averaged Navier–Stokes equations. *Phys. Fluids* **2022**, *34*, 075117. [CrossRef]
119. Xu, H.; Zhang, W.; Wang, Y. Explore missing flow dynamics by physics-informed deep learning: The parameterized governing systems. *Phys. Fluids* **2021**, *33*, 095116. [CrossRef]
120. Xu, S.; Sun, Z.; Huang, R.; Guo, D.; Yang, G.; Ju, S. A practical approach to flow field reconstruction with sparse or incomplete data through physics informed neural network. *Acta Mech. Sin.* **2023**, *39*, 322302. [CrossRef]
121. Girimaji, S.S. Turbulence closure modeling with machine learning: A foundational physics perspective. *New J. Phys.* **2024**, *26*, 071201. [CrossRef]
122. Rapp, C.; Manhart, M. Flow over periodic hills—An experimental study. *Exp. Fluids* **2011**, *51*, 247–269. [CrossRef]
123. Kähler, C.J.; Scharnowski, S.; Cierpka, C. Highly resolved experimental results of the separated flow in a channel with streamwise periodic constrictions. *J. Fluid Mech.* **2016**, *796*, 257–284. [CrossRef]
124. Seifert, A.; Pack, L. Active flow separation control on wall-mounted hump at high Reynolds numbers. *AIAA J.* **2002**, *40*, 1362–1372. [CrossRef]
125. Greenblatt, D.; Paschal, K.B.; Yao, C.-S.; Harris, J.; Schaeffler, N.W.; Washburn, A.E. Experimental investigation of separation control Part 1: Baseline and steady suction. *AIAA J.* **2006**, *44*, 2820–2830. [CrossRef]
126. Iyer, P.S.; Malik, M.R. Wall-modeled large eddy simulation of flow over a wallmounted hump. In Proceedings of the 2016 AIAA Aerospace Sciences Meeting, AIAA SciTech Forum, San Diego, CA, USA, 4–8 January 2016; AIAA Paper 16-3186; pp. 1–22.
127. Uzun, A.; Malik, M.R. Wall-resolved large-eddy simulation of flow separation over NASA wall-mounted hump. In Proceedings of the 55th AIAA Aerospace Sciences Meeting, Grapevine, TX, USA, 9–13 January 2017; AIAA Paper 17-0538.
128. Uzun, A.; Malik, M. Large-Eddy Simulation of flow over a wall-mounted hump with separation and reattachment. *AIAA J.* **2018**, *56*, 715–730. [CrossRef]
129. Spalart, P.; Allmaras, S. A one-equation turbulence model for aerodynamic flows. *La Rech. Aérospatiale* **1994**, *1*, 5–21.
130. Bachalo, W.D.; Johnson, D.A. Transonic, turbulent boundary-layer separation generated on an axisymmetric flow model. *AIAA J.* **1986**, *24*, 437–443. [CrossRef]
131. Lynch, K.P.; Lance, B.; Lee, G.S.; Naughton, J.W.; Miller, N.E.; Barone, M.F.; Beresh, S.J.; Spillers, R.; Soehnel, M. A CFD validation challenge for transonic, shock-induced separated flow: Experimental characterization. In Proceedings of the AIAA SciTech 2020 Forum, Orlando, FL, USA, 6–10 January 2020; AIAA Paper 20-1309; pp. 1–19.
132. Ren, X.; Su, H.; Yu, H.H.; Yan, Z. Wall-modeled large eddy simulation and detached eddy simulation of wall-mounted separated flow via OpenFOAM. *Aerospace* **2022**, *9*, 759. [CrossRef]

133. Uzun, A.; Malik, M.R. Wall-resolved large-eddy simulations of transonic shock-induced flow separation. *AIAA J.* **2019**, *57*, 1955–1971. [CrossRef]
134. Sahu, J.; Danberg, J. Navier–Stokes computations of transonic flows with a two-equation turbulence model. *AIAA J.* **1986**, *24*, 1744–1751. [CrossRef]
135. Xiao, H.; Wu, J.L.; Laizet, S.; Duan, L. Flows over periodic hills of parameterized geometries: A dataset for data-driven turbulence modeling from direct simulations. *Comput. Fluids* **2020**, *200*, 104431. [CrossRef]
136. Gloerfelt, X.; Cinnella, P. Large eddy simulation requirements for the flow over periodic hills. *Flow Turbul. Combust.* **2019**, *103*, 55–91. [CrossRef]
137. Simpson, R.L.; Long, C.H.; Byun, G. Study of vortical separation from an axisymmetric hill. *Int. J. Heat Fluid Flow* **2002**, *23*, 582–591. [CrossRef]
138. Byun, G.; Simpson, R.L.; Long, C.H. Study of vortical separation from three-dimensional symmetric bumps. *AIAA J.* **2004**, *42*, 754–765. [CrossRef]
139. Bell, J.; Heineck, J.; Zilliac, G.; Mehta, R.; Long, K. Surface and flow field measurements on the faith hill model. In Proceedings of the 50th AIAA Aerospace Sciences Meeting, Nashville, TN, USA, 9–12 January 2012; AIAA Paper 12-0704; pp. 1–14.
140. Hoyas, S.; Jiménez, J. Reynolds number effects on the Reynolds-stress budgets in turbulent channels. *Phys. Fluids* **2008**, *20*, 101511. [CrossRef]
141. Fröhlich, J.; Mellen, C.P.; Rodi, W.; Temmerman, L.; Leschziner, M.A. Highly resolved large-eddy simulation of separated flow in a channel with streamwise periodic constrictions. *J. Fluid Mech.* **2005**, *526*, 19–66. [CrossRef]
142. Avdis, A.; Lardeau, S.; Leschziner, M. Large eddy simulation of separated flow over a two-dimensional hump with and without control by means of a synthetic slotjet. *Flow Turbul. Combust.* **2009**, *83*, 343–370. [CrossRef]
143. Bentaleb, Y.; Lardeau, S.; Leschziner, M.A. Large-eddy simulation of turbulent boundary layer separation from a rounded step. *J. Turbul.* **2012**, *13*, N4. [CrossRef]
144. Ching, D.S.; Elkins, C.J.; Eaton, J.K. Investigation of geometric sensitivity of a non-axisymmetric bump: 3D mean velocity measurements. *Exp. Fluids* **2018**, *59*, 143. [CrossRef]
145. Ong, L.; Wallace, J. The velocity field of the turbulent very near wake of a circular cylinder. *Exp. Fluids* **1996**, *20*, 441–453. [CrossRef]
146. Lehmkuhl, O.; Rodríguez, I.; Borrell, R.; Oliva, A. Low-frequency unsteadiness in the vortex formation region of a circular cylinder. *Phys. Fluids* **2013**, *25*, 085109. [CrossRef]
147. Ho, J.; Pepper, N.; Dodwell, T. Probabilistic machine learning to improve generalisation of data-driven turbulence modelling. *Comput. Fluids* **2024**, *284*, 106443. [CrossRef]
148. Fang, Y.; Zhao, Y.; Waschkowski, F.; Ooi, A.S.H.; Sandberg, R.D. Toward more general turbulence models via multicase computational-fluid-dynamics-driven training. *AIAA J.* **2023**, *61*, 2100–2115. [CrossRef]
149. Heinz, S. On mean flow universality of turbulent wall flows. I. High Reynolds number flow analysis. *J. Turbul.* **2018**, *19*, 929–958. [CrossRef]
150. Heinz, S. On mean flow universality of turbulent wall flows. II. Asymptotic flow analysis. *J. Turbul.* **2019**, *20*, 174–193. [CrossRef]

fluids

MDPI

Article

Turbulence Modeling for Physics-Informed Neural Networks: Comparison of Different RANS Models for the Backward-Facing Step Flow

Fabian Pioch [1,*,†], Jan Hauke Harmening [1,*,†], Andreas Maximilian Müller [1], Franz-Josef Peitzmann [1], Dieter Schramm [2] and Ould el Moctar [3]

[1] Department of Mechanical Engineering, Mechatronics Institute Bocholt, Westphalian University, Münsterstraße 265, 46397 Bocholt, Germany

[2] Department of Mechanical Engineering, University Duisburg-Essen, Lotharstraße 1, 47057 Duisburg, Germany

[3] Department of Mechanical Engineering, Institute for Ship Technology, Ocean Engineering and Transport Systems, University Duisburg-Essen, Bismarckstraße 69, 47057 Duisburg, Germany

* Correspondence: fabian.pioch@w-hs.de (F.P.); jan.harmening@w-hs.de (J.H.H.)

† These authors contributed equally to this work.

Abstract: Physics-informed neural networks (PINN) can be used to predict flow fields with a minimum of simulated or measured training data. As most technical flows are turbulent, PINNs based on the Reynolds-averaged Navier–Stokes (RANS) equations incorporating a turbulence model are needed. Several studies demonstrated the capability of PINNs to solve the Naver–Stokes equations for laminar flows. However, little work has been published concerning the application of PINNs to solve the RANS equations for turbulent flows. This study applied a RANS-based PINN approach to a backward-facing step flow at a Reynolds number of 5100. The standard k-ω model, the mixing length model, an equation-free ν_t and an equation-free pseudo-Reynolds stress model were applied. The results compared favorably to DNS data when provided with three vertical lines of labeled training data. For five lines of training data, all models predicted the separated shear layer and the associated vortex more accurately.

Keywords: physics-informed neural networks; RANS; turbulence model; flow separation; DeepXDE

1. Introduction

Several methods of artificial intelligence have been developed and applied to the field of fluid dynamics. Among them, physics-informed neural networks (PINNs) are becoming popular, as these networks reduce the requirements for large high-quality datasets [1]. PINNs can be used to solve partial differential Equations (PDEs) and predict nonlinear solution fields. A PINN is a neural network (NN) trained to satisfy given laws of physics. Lagaris et al. [2] were the first to employ a PINN to solve ordinary differential Equations (ODEs), coupled ODEs and PDEs. Later, Raissi et al. [3–5] used a PINN to solve the Burgers equation, the Schrödinger equation, and the Allen–Cahn equation for time-dependent one-dimensional problems. The reported predictions agreed well with the exact solutions. These publications showed that PINNs can be used to solve different equations accurately. Hence, PINNs could also be applied to fluid dynamics to solve the Navier–Stokes equations.

The application of PINNs to solve the Navier-Stokes equations was investigated in several publications. Raissi et al. [6] deployed a PINN to predict a two-dimensional (2D) flow past a cylinder, a 2D channel flow over a spherical obstacle, and a three-dimensional (3D) flow through the intercranial aneurysm. The predictions agreed favorably with exact solutions. Sun et al. [7] presented a PINN that was trained using boundary-condition enforcement and adaptive activation functions. They reported good agreement between PINN and CFD simulations for the aneurysmal flow. Zhu et al. [8] used a convolutional

PINN to predict the 2D Darcy flow. The results showed more accurate predictions for the PINN than for a data-driven NN. Jin et al. [9] trained a PINN to predict a 2D Kovasznay flow, a 2D cylinder wake, a 3D Beltrami flow, and a turbulent channel flow. The predicted channel flow agreed favorably with DNS results. Zhu et al. [10] applied a PINN to melt pool fluid dynamics using hard boundary-constraint enforcing. Their employed PINN accurately predicted temperature and dynamics using only a moderate number of labeled datasets. Wang et al. [11] deployed a convolutional PINN to predict 2D multiscale turbulent-flow fields based on numerical simulations. Laubscher and Rousseau [12] trained a PINN on an incompressible flow with heat transfer around two sequentially positioned cylinders. They obtained good agreement with CFD results. Wand et al. [13] used a PINN to reconstruct the flow field from PIV measurements and found good agreement with DNS results. In a recent study, Eivazi and Vinuesa [14] trained a PINN to reconstruct a transient flow field from low-resolution and noisy measurements, showing that PINNs can be used to solve unsteady turbulent problems.

In most of the research presented above, PINNs were trained to predict a solution for a single geometry. Wandel et al. [15] trained a convolutional PINN with different 3D shapes, such as boxes, cylinders, or spinning balls, and tested the neural net for geometries not contained in the training dataset.

High-Reynolds-number flows play a major role in many technical problems. For turbulent flows, the velocity and pressure fields are complex in space and time. This imposes challenges on any training attempt with PINNs. As for most engineering applications only stationary information is required, PINN predictions of the Reynolds-averaged Naver–Stokes (RANS) equations are sufficient. To our best knowledge, predictions of the RANS equations have only been presented by Hennigh et al. [16], Eivazi et al. [17] and Xu et al. [18]. Hennigh et al. [16] applied the mixing-length turbulence model to predict a turbulent thermal flow. Eivazi et al. [17] trained a PINN to statisfy the RANS equations without a transport-equation-based model to predict a turbulent boundary layer, the high Reynolds flow in a sub domain on an airfoil, as well as a flow over a periodic hill. Xu et al. [18] used a similar approach and trained a PINN to predict the turbulent viscosity without any imposed equation. They trained the model to CFD data and used it to explore missing flow data inside of a finite area downstream of a backward-facing step. Hence, a comparison of different turbulence modeling approaches for PINN and an extension to transport-equation-based models, such as the k-ω model, remain an open question.

In this work, we contribute to the question of how different turbulence modeling approaches could be applied and how the prediction accuracy of the different PINNs compares. Therefore, we applied different turbulence models for the flow over a backward-facing step at a Reynolds number of 5100. For our study, we deployed the mixing-length model as used in [16], as well as the two-equation k-ω model. Additionally, we applied two equation-free turbulence models for PINNs similar to the approaches in [17,18]. These models were developed to evade the training difficulties associated with the PDEs of traditional turbulence models. The predictions obtained with different amounts of labeled training data are compared to the DNS results of Le et al. [19]. Furthermore, we discuss the strengths and weaknesses of the turbulence models regarding the usage for PINN. The scope of our work was to compare the prediction quality of different turbulence models when applied with a fully connected feedfoward PINN that was trained with different amounts of labeled training data. The motivation for this work was to develop a method to use turbulence models for PINNs and to serve as a reference for application and selection of turbulence models in future work.

2. Backward-Facing Step Flow

Several PINNs with different turbulence models were trained to predict the flow around a backward-facing step at a Reynolds number of 5100. Figure 1 schematically illustrates the expected mean flow field. As seen, the flow field exhibits a boundary layer on the channel wall, a separated and reattached shear layer downstream of the step, a recirculative vortex inside of the separation bubble, and a secondary corner vortex in the lower corner of the step. A review of the literature concerning the flow field around the backward-facing step can be found in [20].

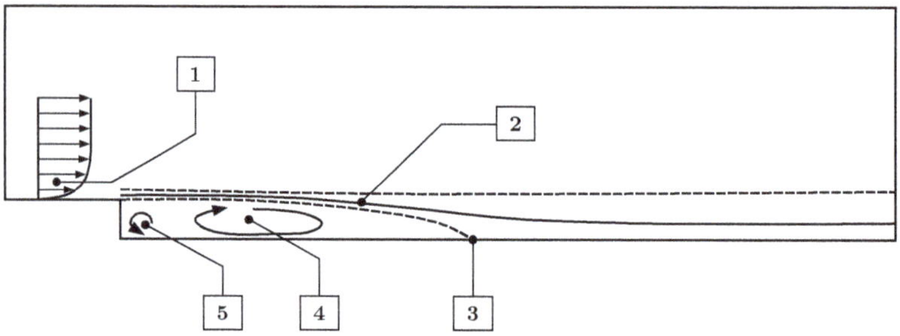

Figure 1. Backward-facing step flow features. 1: Boundary layer; 2: separated shear layer; 3: reattached flow; 4: recirculative vortex; 5: corner vortex.

The networks were trained with data and compared against DNS results for the backward-facing step flow of Le et al. [19]. They used the mean inlet velocity profile of Spalart [21] for the forced convection boundary condition. The velocity profile was superimposed with random velocity fluctuations. A convective boundary condition was applied for the outlet. In the spanwise dimension, they used a periodic boundary condition. The step had a contraction rate of 1.2. At the considered Reynolds number of 5100, the flow was turbulent [19]. Le et al. [19] reported that the DNS results agreed favorably with experiments.

3. Methodology

3.1. Flow Prediction with PINNs

A feedforward NN was trained to predict the two-dimensional stationary flow field. The network had three input neurons corresponding to coordinates x and y and wall distance d. An output layer displayed the variables of the flow field. In case the NN is trained to solve the Naver–Stokes equations for the variables u, v, and p, the network reads

$$NN_\theta \begin{pmatrix} x \\ y \\ d \end{pmatrix} = \begin{pmatrix} u \\ v \\ p \end{pmatrix} \quad (1)$$

where θ represents the trainable parameters (weights) of the NN. Figure 2 shows the corresponding network architecture of the PINN.

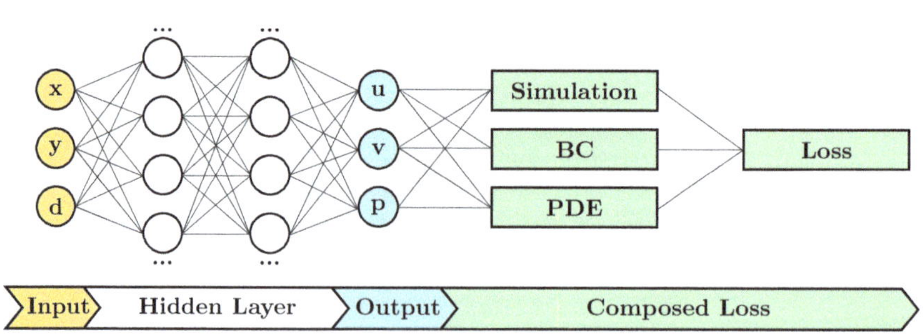

Figure 2. Neural network architecture of the PINN.

As shown in Figure 2, the output of the NN is fed into a so-called composed loss function which adds several individual losses.

3.2. Composed Loss Function

The pivotal part of a PINN is the composed loss function. Here, several individual losses are added. The composed loss function deployed here reads

$$Loss(\theta) = MSE_{BC}(\theta) + MSE_{Data}(\theta) + MSE_{PDE}(\theta) \tag{2}$$

where *BC* corresponds to the boundary conditions, *Data* to the simulated labeled training data, and *PDE* to the partial differential equation employed.

3.3. Governing Equations

The flow field considered here is governed by the two-dimensional Reynolds-averaged Navier–Stokes (RANS) equations and the continuity equation for an incompressible fluid. Hence, the PDEs considered in the composed loss function read

$$Loss_{continuity} = \frac{\partial \overline{u}}{\partial x} + \frac{\partial \overline{v}}{\partial y} \tag{3}$$

$$Loss_{momentum,x} = \overline{u}\frac{\partial \overline{u}}{\partial x} + \overline{v}\frac{\partial \overline{u}}{\partial y} + \overline{u'\frac{\partial u'}{\partial x}} + \overline{v'\frac{\partial u'}{\partial y}} + \frac{1}{\rho}\frac{\partial \overline{p}}{\partial x} - \nu(\frac{\partial^2 \overline{u}}{\partial x^2} + \frac{\partial^2 \overline{u}}{\partial y^2}) \tag{4}$$

$$Loss_{momentum,y} = \overline{u}\frac{\partial \overline{v}}{\partial x} + \overline{v}\frac{\partial \overline{v}}{\partial y} + \overline{u'\frac{\partial v'}{\partial x}} + \overline{v'\frac{\partial v'}{\partial y}} + \frac{1}{\rho}\frac{\partial \overline{p}}{\partial y} - \nu(\frac{\partial^2 \overline{v}}{\partial x^2} + \frac{\partial^2 \overline{v}}{\partial y^2}) \tag{5}$$

with averaged velocity components $\overline{u}$ and $\overline{v}$, averaged pressure $\overline{p}$, density ρ and kinematic viscosity ν. The RANS equations contain the following derivations of the Reynolds stress terms:

$$\frac{\partial \tau'_{xx}}{\partial x} + \frac{\partial \tau'_{xy}}{\partial y} = \overline{u'\frac{\partial u'}{\partial x}} + \overline{v'\frac{\partial u'}{\partial y}} \tag{6}$$

and

$$\frac{\partial \tau'_{yx}}{\partial x} + \frac{\partial \tau'_{yy}}{\partial y} = \overline{u'\frac{\partial v'}{\partial x}} + \overline{v'\frac{\partial v'}{\partial y}} \tag{7}$$

4. Experiments

4.1. Turbulence Modeling

We considered four different turbulence-modeling approaches. According to the Boussinesq hypothesis, for an incompressible flow the Reynolds stresses read as follows:

$$\tau'_{ij} = -\overline{u'_i u'_j} = \nu_t(\frac{\partial \overline{u_i}}{\partial x_j} + \frac{\partial \overline{u_j}}{\partial x_i}) - \frac{2}{3}k\delta_{ij} \tag{8}$$

where k is the turbulent kinetic energy, δ_{ij} is the Kronecker delta, and ν_t is the turbulent viscosity. The utilization of a turbulent viscosity is based on the gradient diffusion hypothesis. This hypothesis assumes that the turbulent transport is in negative gradient direction of the mean flow and depends on an isotropic turbulent viscosity serving as a diffusion coefficient.

The first model we considered is the standard k-ω model of Wilcox [22]. For this model, the turbulent viscosity is calculated as follows:

$$\nu_t = \frac{k}{\omega} \tag{9}$$

with the specific dissipation rate ω. The 2D stationary equations for k and ω read

$$\begin{aligned} \overline{u}\frac{\partial k}{\partial x} + \overline{v}\frac{\partial k}{\partial y} &= \tau'_{xx}\frac{\partial \overline{u}}{\partial x} + \tau'_{xy}\frac{\partial \overline{u}}{\partial y} + \tau'_{yx}\frac{\partial \overline{v}}{\partial x} + \tau'_{yy}\frac{\partial \overline{v}}{\partial y} - \beta^* k\omega \\ &+ \frac{\partial}{\partial x}\left((\nu + \sigma^*\nu_t)\frac{\partial k}{\partial x}\right) + \frac{\partial}{\partial y}\left((\nu + \sigma^*\nu_t)\frac{\partial k}{\partial y}\right) \end{aligned} \tag{10}$$

$$\overline{u}\frac{\partial \omega}{\partial x} + \overline{v}\frac{\partial \omega}{\partial y} = \alpha\frac{k}{\omega}\left(\tau'_{xx}\frac{\partial \overline{u}}{\partial x} + \tau'_{xy}\frac{\partial \overline{u}}{\partial y} + \tau'_{yx}\frac{\partial \overline{v}}{\partial x} + \tau'_{yy}\frac{\partial \overline{v}}{\partial y}\right) - \beta\omega^2 + \frac{\partial}{\partial x}\left((\nu + \sigma\nu_t)\frac{\partial \omega}{\partial x}\right) + \frac{\partial}{\partial y}\left((\nu + \sigma\nu_t)\frac{\partial \omega}{\partial y}\right) \tag{11}$$

with $\beta^* = 9/100$, $\sigma^* = \sigma = 0.5$, $\alpha = 5/9$, $\beta = 3/40$, as recommended by Wilcox [22]. Note that for the k-ω model, the transport equations for k and ω are added as additional PDEs into the composed loss function. The corresponding PINN reads

$$NN_\theta \begin{pmatrix} x \\ y \\ d \end{pmatrix} = \begin{pmatrix} u \\ v \\ p \\ k \\ \omega \end{pmatrix} \tag{12}$$

The second turbulence model we considered is the mixing-length model by Prandtl as applied in [16]. For this model, the turbulent viscosity is calculated as follows:

$$\nu_t = l_m^2\sqrt{G} \tag{13}$$

For a two-dimensional flow, the modulus of the mean rate of strain tensor G reads

$$G = 2\left(\frac{\partial \overline{u}}{\partial x}\right)^2 + 2\left(\frac{\partial \overline{v}}{\partial y}\right)^2 + \left(\frac{\partial \overline{u}}{\partial y} + \frac{\partial \overline{v}}{\partial x}\right)^2 \tag{14}$$

and the mixing length l_m is calculated as follows:

$$l_m = min(0.419d, 0.09d_{max}) \tag{15}$$

where d is the wall normal distance and d_{max} its maximal value. As no turbulent kinetic energy is calculated with the mixing-length model, the Reynolds stresses must be computed differently than in Equation (8). Since the 2D case is considered here, this reads

$$k = \frac{1}{2}\overline{u'_i u'_i} = \frac{1}{2}(\overline{u'u'} + \overline{v'v'}) \tag{16}$$

One can derive from Equation (8) the resulting terms of the Reynolds stresses, which read

$$\tau'_{xx} = 2\nu_t\left(\frac{\partial \overline{u}}{\partial x} + \frac{\partial \overline{u}}{\partial x} + \frac{\partial \overline{v}}{\partial y}\right) \tag{17}$$

$$\tau'_{xy} = \tau'_{yx} = \nu_t\left(\frac{\partial \overline{u}}{\partial y} + \frac{\partial \overline{v}}{\partial x}\right) \tag{18}$$

$$\tau'_{yy} = 2\nu_t\left(\frac{\partial \overline{v}}{\partial y} + \frac{\partial \overline{v}}{\partial y} + \frac{\partial \overline{u}}{\partial x}\right) \tag{19}$$

The PINN, then, reads

$$NN_\theta \begin{pmatrix} x \\ y \\ d \end{pmatrix} = \begin{pmatrix} u \\ v \\ p \end{pmatrix} \tag{20}$$

We applied two additional novel turbulence modeling approaches for PINNs, here referred to as the ν_t-model and the pseudo-Reynolds stress model. In both cases, no physics-based equation for the output is imposed.

For the ν_t-model, the Reynolds stresses are calculated in the same way as for the mixing length model. The deployed PINN then reads

$$NN_\theta \begin{pmatrix} x \\ y \\ d \end{pmatrix} = \begin{pmatrix} u \\ v \\ p \\ \nu_t \end{pmatrix} \tag{21}$$

In the training process, the weights are adjusted such that the predicted values of u, v, p, and ν_t satisfy the RANS equations.

In case of the pseudo-Reynolds stress model, the derivations of the Reynolds stresses are calculated directly from the pseudo-turbulent output of the PINN, as follows:

$$\frac{\partial \tau'_{xx}}{\partial x} + \frac{\partial \tau'_{xy}}{\partial y} = u'_p \frac{\partial u'_p}{\partial x} + v'_p \frac{\partial u'_p}{\partial y} \tag{22}$$

and

$$\frac{\partial \tau'_{yx}}{\partial x} + \frac{\partial \tau'_{yy}}{\partial y} = u'_p \frac{\partial v'_p}{\partial x} + v'_p \frac{\partial v'_p}{\partial y} \tag{23}$$

where u'_p and v'_p are pseudo turbulent velocity components. As u'_p and v'_p represent time-independent velocity components,

$$u'_p \frac{\partial u'_p}{\partial x} + v'_p \frac{\partial u'_p}{\partial y} \neq \overline{u' \frac{\partial u'}{\partial x}} + \overline{v' \frac{\partial u'}{\partial y}} \tag{24}$$

Therefore, the model is referred to as the pseudo-Reynolds stress model. The corresponding PINN reads

$$NN_\theta \begin{pmatrix} x \\ y \\ d \end{pmatrix} = \begin{pmatrix} u \\ v \\ p \\ u'_p \\ v'_p \end{pmatrix} \tag{25}$$

In the training process, the weights are adjusted such that the predicted u, v, p, u'_p and v'_p satisfy the RANS Equations (3)–(5).

4.2. Boundary Conditions

For the upper channel wall, a slip wall boundary condition was specified; for the lower channel walls a no-slip wall condition. For the no-slip wall, a smooth surface was considered. Additionally, a Neumann boundary condition was defined for the mean velocities u and v at the inlet edge of the training domain. The DNS data for the mean values of u and v were prescribed for six vertical lines. The lines were positioned at $x/h =$ 0.00, 7.00, 9.00, 12.98, 17.98, and 22.00. The associated boundary conditions were created using the DNS data of [19], which were taken from the ERCOFTAC database [23] (note the coordinate system was modified for our purposes). Data at $x/h = 0$ specified the inlet boundary condition of the domain. All other vertical lines were positioned inside the domain. For training, a point-wise boundary condition was used for the DNS data. Figure 3 displays the training setup and the associated boundary conditions.

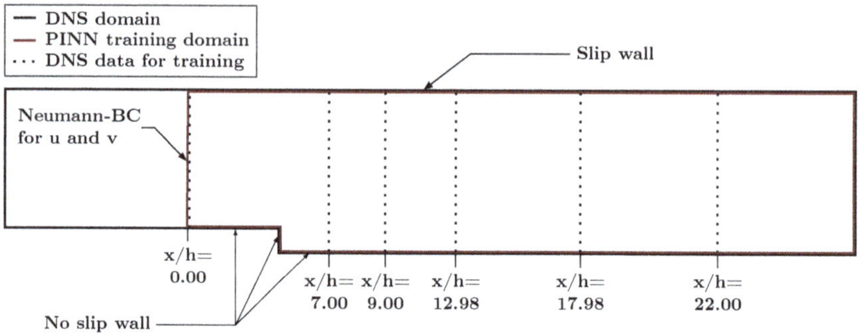

Figure 3. Boundaries for the PINN training process.

4.3. PINN Training Procedure

We used the Tensorflow-based library DeepXDE to set up and train the model [24]. As mentioned above, the PINN was trained to satisfy the RANS equations together with the turbulence models. Additionally, the model was trained to fulfill the boundary conditions and to match the DNS data. Figure 4 shows the implementation of the algorithm. The automatic differentiation allows to derive the outputs of the PINN depending on the turbulence model with respect to the inputs and to form the loss function of the PDE.

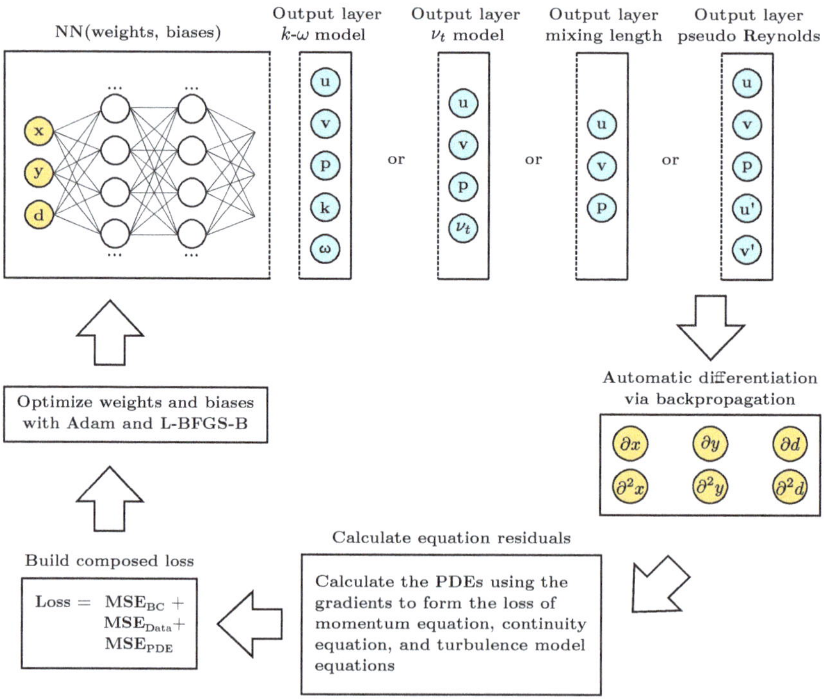

Figure 4. Flowchart of the applied training algorithm. For supplementary information on residuals of the specific turbulence models, see for the k-ω model Equations (8)–(11), the mixing-length model Equations (13)–(19), the ν_t model Equations (17)–(19), and the pseudo-Reynolds stress model Equations (22) and (23).

The PINN was trained for 30,000 epochs with the Adam optimizer and, afterwards, with the L-BFGS-B optimizer under predefined default settings. The boundary conditions and simulation data were trained at a total of 2000 training points. The PDEs were trained at 2000 training points inside of the domain. The applied number of points was based on a sensitivity analysis. Figure 5 displays the global normalized mean squared error (NMSE) for 1000, 2000, and 4000 points in the domain as well as on the boundaries. As seen, the error was minimal for the case of 2000 points. Consequently, this investigation was conducted with 2000 points. The training points were uniformly (equispaced) distributed.

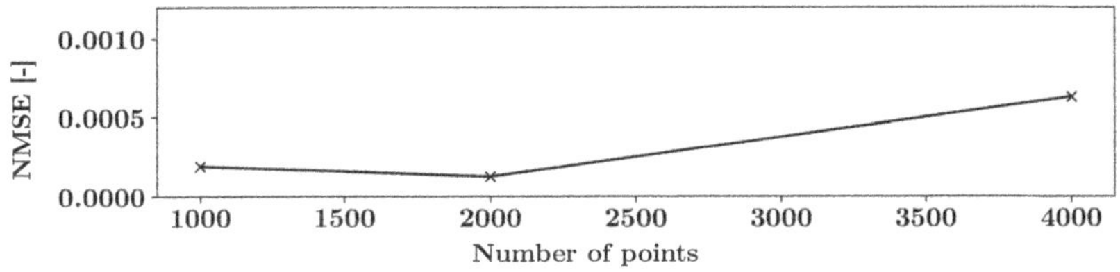

Figure 5. Global NMSE over the number of training points inside the domain as well as on the boundaries. The results were obtained for the pseudo-Reynolds stress model using three lines of labeled training data.

Table 1 lists the specified training parameters. Every neural net was reinitialized and trained ten times, as the initalization of the weights was shown to have a significant influence on the result [24]. If possible, all samples with physically unrealistic solutions were discarded, and the remaining sample with the lowest global NMSE was selected for further evaluation.

Table 1. PINN training parameters.

Parameter	Value/Setting
Architecture	Five hidden layers with 128 neurons each
Optimizer	Adam, L-BFGS-B
Epochs	30,000 (Adam)
Learning rate	10^{-3} (10,000 epochs) and 10^{-4} (20,000 epochs)
Activation function	tanh
Number of training points BC and data	2000
Number of training points PDE	2000

4.4. Comparison with DNS Data

To compare the turbulence models with the DNS data, we evaluated several error metrics. To measure systematic and random errors, the normalized mean squared error (NMSE) was used [25]. The factor of observations (FAC2) was utilized to asses extreme deviations [25]. Systematic errors were measured using the fractional bias (FB) [25]. Additionally, the validation metric V of Oberkampf and Trucano [26] was applied to assess the overall level of validation. Commonly used acceptance limits are: NMSE < 4, FAC2 > 0.5, and FB < 0.2 [27].

4.5. Analyses

An analysis was carried out to investigate the sensitivities of the predictions on turbulence modeling as well as on the quantity of labeled training data.

In the first step, the PINN was trained with DNS training data at $x/h = 0$ only. This training configuration investigated the training quality without any labeled training data inside of the flow domain. In a second step, the PINN was trained with DNS data at $x/h = 0.00, 7.00,$ 12.98, and 22.00. Both cases were compared with the DNS data to assess the accuracy of the predictions. In a third step, the PINN was trained with all DNS data available.

For all steps, the predictions obtained using the k-ω, the mixing length model, the proposed ν_t model, and the proposed pseudo-Reynolds stress model were compared.

5. Results

In this section, the predictions of the different PINNs are presented in comparison with the DNS data.

5.1. Training without Labeled Training Data Inside of the Domain

Figure 6 presents the magnitudes of fluid velocity predicted by the best-trained neural nets (lowest overal NMSE). The predictions are displayed together with DNS results and the deviations are shown. As seen, for all turbulence models, the training without labeled training data resulted in predictions which deviated significantly from the DNS results. Above the backward-facing step, all models underpredicted the fluid velocity; in the wake of the step, all models overpredicted the length of the recirculating vortex.

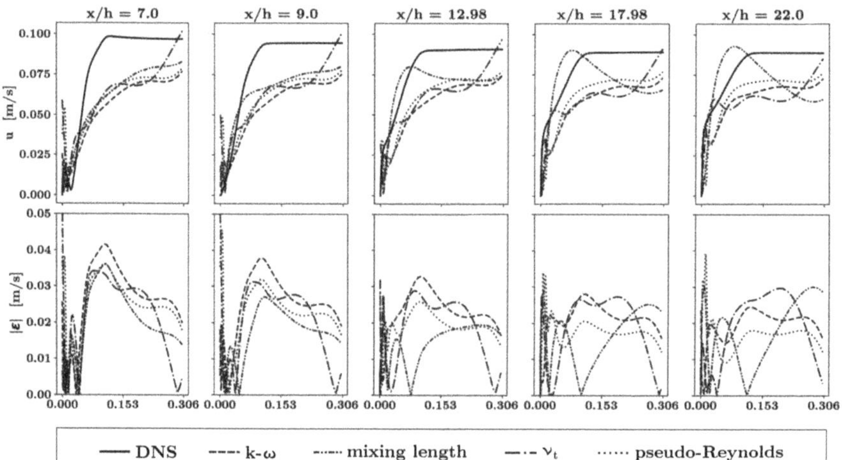

Figure 6. Comparative PINN predictions of velocity magnitudes (**top**) and their deviation from DNS results (**bottom**) for the training without any labeled training data inside of the flow domain. The results are shown at several longitudinal positions. The y-axis of the bottom row is clipped at 0.05.

5.2. Training with Three Lines of Labeled Training Data Inside of the Domain

Figure 7 displays the magnitudes of fluid velocity magnitudes predicted by the best-trained neural nets (lowest overal NMSE). The predictions are compared to the DNS results and the deviations are presented. As seen, all models favorably match the labeled training data. For data belonging to the vertical lines at $x/h = 9.00$ and $x/h = 17.98$, where no DNS data were used for the training, all models agreed favorably with the DNS data. The greatest deviations occured at $x/h = 9.00$ for the ν_t model. The mixing-length model and the ν_t model overpredicted the length of the separated shear layer. Above the step, the k-ω model as well as the mixing-length model exhibited deviations from the DNS results.

Table 2 lists error measures for the most accurate PINN sample for the five turbulence models considered. The average NMSEs for the k-ω model, the mixing length model, the ν_t model, and the pseudo-Reynolds-stress model were 2.0×10^{-3}, 1.2×10^{-3}, 2.2×10^{-3}, and 7.5×10^{-5}, respectively. The standard deviations of the NMSEs from ten reinitialized training runs were 66.9%, 48.8%, 37.2%, and 70.0%, which indicated significant variations in accuracy over the runs. The error measures for NMSE, FAC2, FB, and V, presented in Table 2, indicate low random errors, extreme deviations, bias errors and a good overall agreement, respectively.

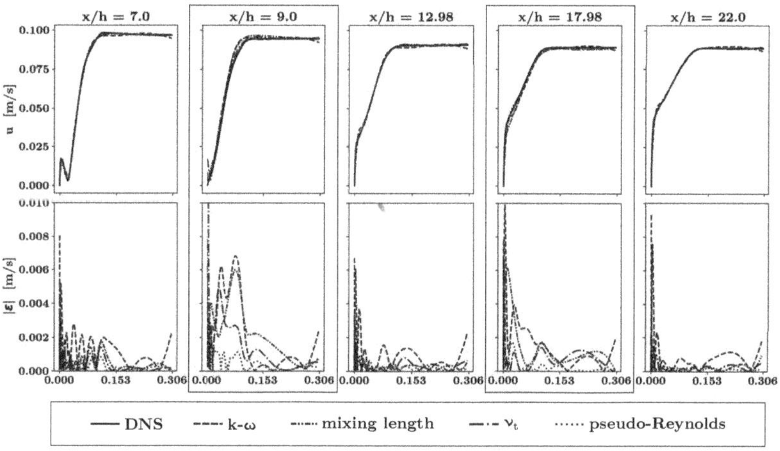

Figure 7. Comparative PINN predictions of velocity magnitudes (**top**) and their deviations from DNS results (**bottom**) for the training with three vertical lines of labeled training data inside of the flow domain. The y axis of the bottom row is clipped at 0.01. The red boxes indicate positions where no labeled training data was used for the training of the PINN.

Table 2. Error measures of the most accurate PINN sample for the five turbulence models.

Model	Metric	x/h = 7	x/h = 9	x/h = 12.98	x/h = 17.98	x/h = 22
k-*ω*	NMSE	1.3×10^{-3}	4.6×10^{-3}	7.5×10^{-4}	2.6×10^{-3}	1.1×10^{-3}
	FAC2	9.6×10^{-1}	8.8×10^{-1}	9.8×10^{-1}	9.8×10^{-1}	9.7×10^{-1}
	FB	4.5×10^{-3}	3.9×10^{-2}	9.0×10^{-4}	3.7×10^{-2}	2.5×10^{-4}
	V	9.1×10^{-1}	8.3×10^{-1}	9.5×10^{-1}	9.3×10^{-1}	9.5×10^{-1}
mixing length	NMSE	1.4×10^{-4}	4.1×10^{-3}	1.3×10^{-5}	1.9×10^{-3}	6.8×10^{-6}
	FAC2	9.8×10^{-1}	9.2×10^{-1}	9.9×10^{-1}	9.5×10^{-1}	10×10^{-1}
	FB	8.5×10^{-4}	4.0×10^{-2}	9.2×10^{-6}	3.0×10^{-2}	9.3×10^{-5}
	V	9.7×10^{-1}	8.4×10^{-1}	9.9×10^{-1}	9.3×10^{-1}	9.9×10^{-1}
ν_t	NMSE	3.4×10^{-4}	1.0×10^{-2}	1.3×10^{-4}	5.0×10^{-4}	6.6×10^{-5}
	FAC2	9.7×10^{-1}	8.6×10^{-1}	9.8×10^{-1}	9.8×10^{-1}	9.9×10^{-1}
	FB	3.3×10^{-4}	4.1×10^{-3}	3.2×10^{-4}	4.2×10^{-3}	2.3×10^{-4}
	V	9.5×10^{-1}	8.3×10^{-1}	9.7×10^{-1}	9.6×10^{-1}	9.8×10^{-1}
pseudo Reynolds stress	NMSE	1.3×10^{-4}	1.4×10^{-4}	4.0×10^{-5}	5.1×10^{-5}	9.0×10^{-6}
	FAC2	9.9×10^{-1}	9.7×10^{-1}	10×10^{-1}	10×10^{-1}	10×10^{-1}
	FB	8.6×10^{-4}	5.0×10^{-3}	4.9×10^{-4}	1.2×10^{-3}	4.4×10^{-5}
	V	9.6×10^{-1}	9.1×10^{-1}	9.8×10^{-1}	9.9×10^{-1}	10×10^{-1}

Figure 8 displays fluid velocities and streamlines of the flow predicted by the PINN using the *k*-*ω* model. As seen, the neural net predicted the separated shear layer as well as the vortex inside the separation bubble. At the vertical wall of the step, a small flux through the boundary is present (point 1 in Figure 8). The results did not exhibit a corner vortex (point 2 in Figure 8). Above the backward-facing step, a region of high-velocity fluid is present (point 3 in Figure 8). Downstream of the reattached shear layer, flux through the horizontal wall accompanied by an upward fluid motion is visible (compare denotation 4 in Figure 8).

Figure 9 shows the velocity as well as the streamlines predicted by the mixing-length model. The model predicted a separated shear layer that reattaches downstream of the step. Flux through the vertical and horizontal wall of the step is visible (point 1 in Figure 9). The PINN did not predict the vortex inside of the separation bubble nor the corner vortex (point 2 in Figure 9). The predicted shearlayer is thickened shortly downstream of the step (point 3 in Figure 9). Additionally, flux passing through the horizontal wall accompanied by an upward fluid motion is visible downstream of the step (point 4 in Figure 9).

Figure 10 exhibits the velocity field as well as the streamlines predicted by the PINN model using the ν_t model. The separated shear layer and the associated recirculative vortex are present in the results. As for the other models, the streamlines indicate a small flux through the vertical wall of the backward-facing step (point 1 in Figure 10). No corner vortex was predicted (point 2 in Figure 10). Above the separated shear layer and downstream of the step, an area of high-velocity fluid is present (point 3 in Figure 10).

Figure 11 displays the velocity as well as the streamlines predicted by the PINN using the pseudo-Reynolds-stress model. Similar to most other models, the pseudo-Reynolds-stress model predicted the separated shear layer and the associated vortex at the backward-facing step. At the corner of the backward facing step, flux passing through the vertical wall exists (point 1 in Figure 11). As for all other turbulence models, no corner vortex was predicted by the PINN using the pseudo-Reynolds-stress model (point 2 in Figure 11).

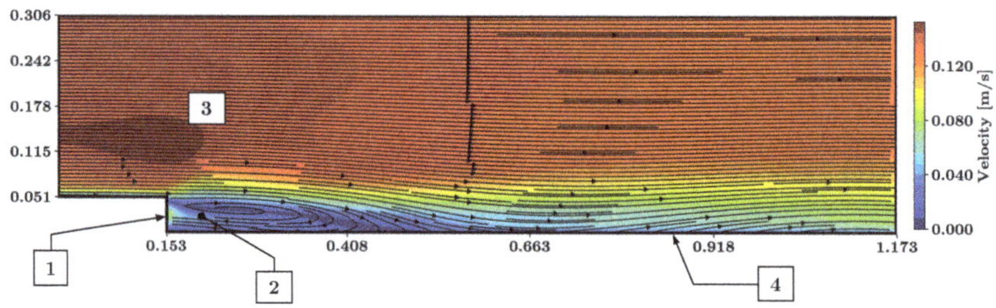

Figure 8. Predicted flow field around the backward-facing step using the k-ω model, based on training with three vertical lines of labeled training data. The magnitude of velocity is presented together with the streamlines. 1: Flux through the vertical wall; 2: no corner vortex predicted; 3: thickened shear layer; 4: flux through the horizontal wall accompanied by an upward motion.

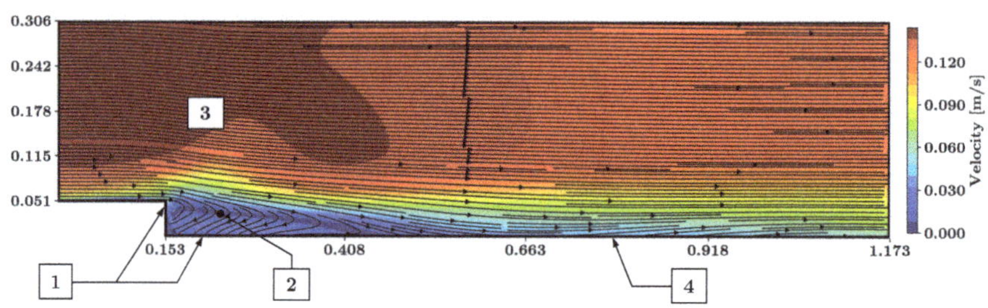

Figure 9. Predicted flow field around the backward-facing step using the mixing length model, based on training with three vertical lines of labeled training data. The magnitude of velocity is presented together with the streamlines. 1: Flux through the vertical wall; 2: no corner vortex predicted; 3: thickened shear layer; 4: flux through the horizontal wall accompanied by an upward motion.

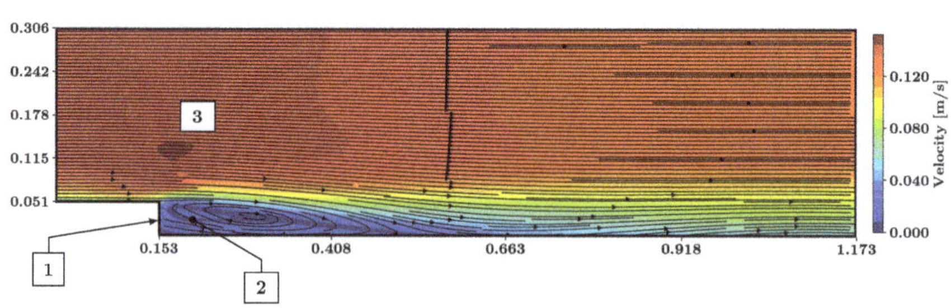

Figure 10. Predicted flow field around the backward-facing step using the ν_t model, based on training with three vertical lines of labeled training data. The magnitude of velocity is presented together with the streamlines. 1: Flux through the vertical wall; 2: no corner vortex predicted; 3: area of high velocity.

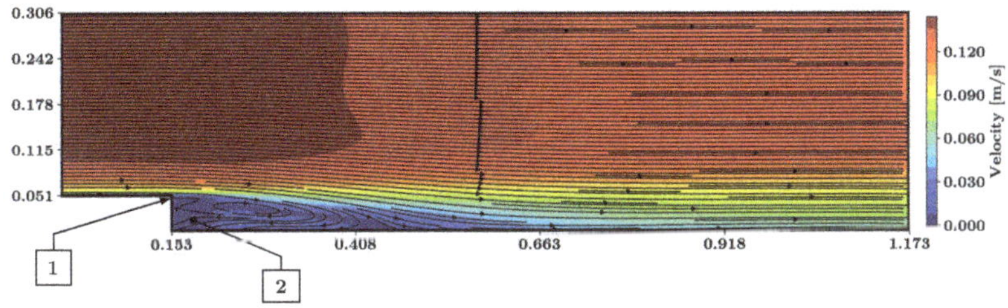

Figure 11. Predicted flow field around the backward-facing step using the pseudo Reynolds stress model, based on training with three vertical lines of labeled training data. The magnitude of velocity is presented together with the streamlines. 1: Flux through the vertical wall; 2: no corner vortex predicted.

5.3. Training with Five Lines of Labeled Training Data Inside of the Domain

The k-ω model, the mixing-length model, the ν_t model, and the pseudo-Reynolds model yielded global NMSE of 4.0×10^{-5}, 7.7×10^{-3}, 1.1×10^{-4}, and 5.6×10^{-5} as well as standard deviations of 1,196%, 61.8%, 46.2%, and 35.1%, respectively. This, again, indicated that the quality of the predictions obtained from the the ten reinitialized training sets varied significantly.

Figure 12 displays the fluid velocities and streamlines of the flow predicted by the PINN using the k-ω model. The neural network predicted the separated shear layer as well as the vortex inside of the separation bubble. At the vertical wall of the step, a small flux passing through the boundary is present (point 1 in Figure 12). The results do not exhibit a corner vortex (point 2 in Figure 12).

Figure 13 shows the velocity as well as the streamlines predicted by the PINN using the mixing-length model. The model predicted the separation bubble downstream of the backward-facing ramp as well as the vortex. A small flux through the vertical wall of the step is present (point 1 in Figure 13). The PINN did not predict the corner vortex (point 2 in Figure 13). The predicted shearlayer was thickened shortly downstream of the step and a high-velocity region was formed upstream (point 3 in Figure 13).

Figure 14 exhibits the velocity field as well as the streamlines predicted by the PINN model using the ν_t model for the PDE inside of the composed loss function. The separated shear layer and the associated recirculative vortex characterized the flow. As for the other models, the streamlines indicated a small flux passing through the vertical wall of the backward-facing step (point 1 in Figure 14). No corner vortex was predicted (point 2 in Figure 14). Above the separated shear layer and downstream of the step, an area of high-velocity fluid is present (point 3 in Figure 14).

Figure 15 displays the velocities as well as streamlines predicted by the best PINN using the pseudo-Reynolds-stress model. Predicted was the separated shear layer and the associated vortex at the backward-facing step. At the corner of the backward-facing step, a flux passing through the vertical wall is present (point 1 in Figure 15). As for all other turbulence models, no corner vortex was predicted by the PINN using the pseudo-Reynolds-stress model (point 2 in Figure 15).

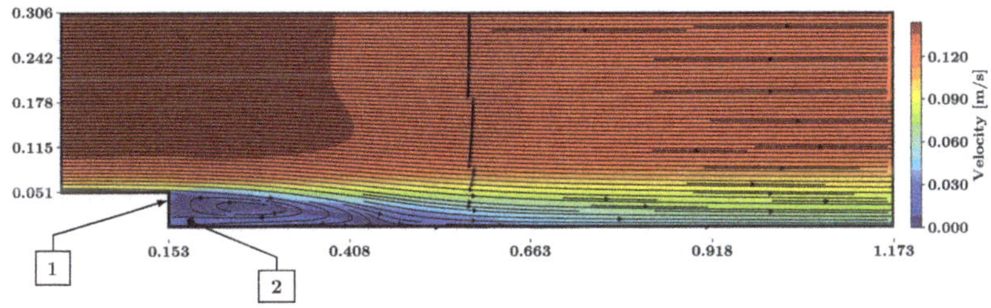

Figure 12. Predicted flow field around the backward-facing step using the k-ω model, based on training with five vertical lines of labeled training data. The magnitude of velocity is presented together with the streamlines. 1: Flux through the vertical wall; 2: no corner vortex predicted.

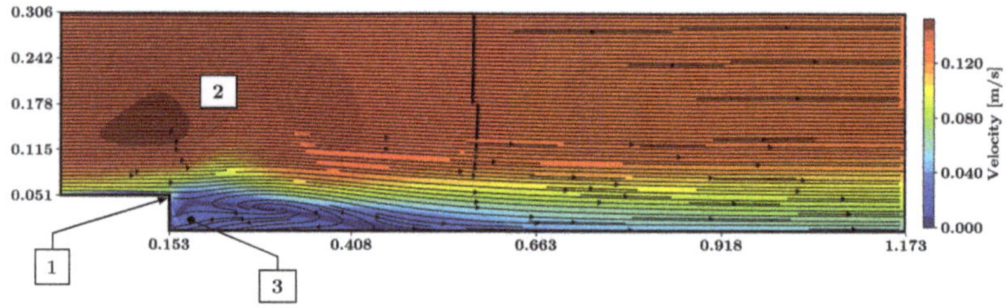

Figure 13. Predicted flow field around the backward-facing step using the mixing-length model, based on training with five vertical lines of labeled training data. The magnitude of velocity is presented together with the streamlines. 1: Flux through the vertical wall; 2: no corner vortex predicted; 3: thickened shear layer.

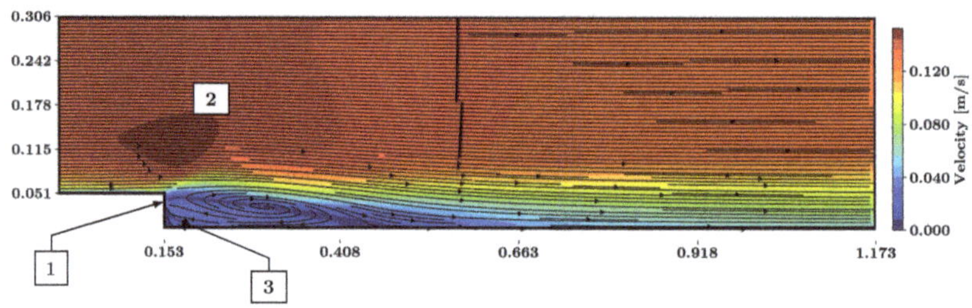

Figure 14. Predicted flow field around the backward-facing step using the ν_t model, based on training with five vertical lines of labeled training data. The magnitude of velocity is presented together with the streamlines. 1: Flux through the vertical wall; 2: no corner vortex predicted; 3: area of high velocity.

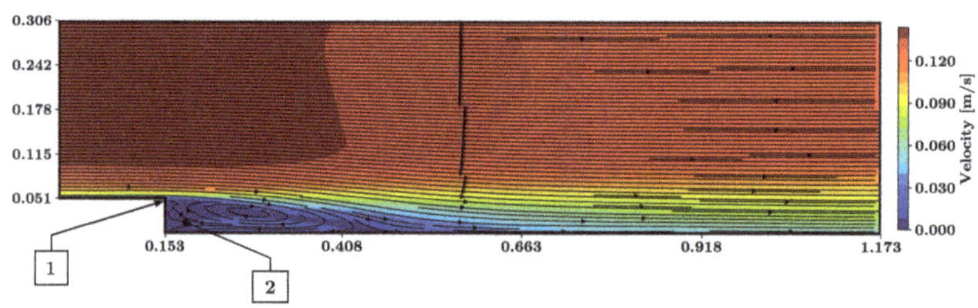

Figure 15. Predicted flow field around the backward-facing step using the pseudo-Reynolds-stress model, based on training with five vertical lines of labeled training data. The magnitude of velocity is presented together with the streamlines. 1: Flux through the vertical wall; 2: no corner vortex predicted.

6. Discussion

For training without labeled training data inside the domain, predictions of the fully connected feed-forward PINNs disagreed with corresponding DNS data. In part, these deviations were attributable to the spatial and temporal configuration. While the DNS results in [19] were obtained for a three-dimensional backward-facing step, we applied the PINN to a two-dimensional geometry.

Another explanation for the deviation from the DNS results could be the lack of labeled training data inside the domain. In contrast, several authors reported accurate predictions using PINNs that were trained without any labeled training data. However, the associated Reynolds numbers were below 300 and, hence, the flow corresponded to laminar conditions [6,7,9,12,28]. As the flow of our work had a Reynolds number of 5100, we attribute the deviation in the PINN when trained without labeled training data to the high Reynolds number.

For training with five vertical lines of labeled training data, excellent agreement was achieved between the predictions and the DNS results for all turbulence models. The reported error metrics turned out to be well below the acceptance limits for random and bias errors. These results support the claim that PINNs can be trained to yield accurate stationary predictions of turbulent flows.

The predictions were improved when using all available training data. This was especially so when using the k-w and the mixing-length models, which exhibited distinct flaws of the flow field when trained with only three lines (compare Figures 8 and 9 with Figures 12 and 13).

However, these results also demonstrated the limitations of the PINN approach deployed here, as small flux was formed through the boundaries, especially through the vertical step wall. This effect could be attributed to the loss reduction during the training process. Specifically, any violation of the imposed boundary conditions is penalized with an equivalent loss. As these losses cannot be eliminated by the optimizer, a small flux through the walls could remain after the training. These fluxes could accumulate over the boundaries, resulting in mass and momentum losses which deteriorate the flow field inside the domain. As a remedy, loss-specific weighting factors could be considered. Another potential solution is hard boundary constraint enforcement [7,8] or the direct-forcing immersed boundary method [29].

All models predicted the primary separation vortex of the backward-facing step when five vertical lines of training data were used. However, every applied turbulence model failed to predict the secondary corner vortex. Consequently, the shape of the primary corner vortex was flawed. Application of a residual adaptive refinement method (RAR) might have increased the accuracy of the PINN predictions for the backward-facing step flow. When training a PINN with a RAR, training points are added in areas where high residuals are present. This may yield more accurate solutions for the smaller areas, such as the secondary corner vortex. Lu et al. [24] applied this RAR method in the PINN framework DeepXDE.

Our results also indicate that the novel turbulence models considered here are capable of predicting turbulent flows when applied with a PINN. In fact, the PINN using the pseudo-Reynolds-stress model yielded more accurate predictions than the PINNs using the established turbulence models when trained with three lines of training data. In the following, we discuss the strengths and weaknesses of all employed models.

The advantage of the k-ω model is that two transport equations are employed: one equation to determine the kinetic turbulent energy; the other equation the specific dissipation rate. This model has been proven to yield accurate results for a wide variety of flows, including flows over a backward-facing step [30]. Here, we incorporated this model into a PINN obtaining favorable agreements with DNS results. However, introducing two additional complex PDEs might impede the optimization process in other cases. A too-complex optimization problem could then cause the optimizer failing to approach the global minimum and, hence, provide unrealistic predictions.

For the mixing-length model, no additional PDEs needed to be solved. Hence, the optimizer was expected to more easily solve the associated system of equations. However, as no transport equation for turbulent quantities was considered, the system of PDEs can be assumed to represent the real flow less accurately. This might have caused the overpredicted thickness of the shear layer behind the step, which could have been due to an overestimated turbulent viscosity. The mixing-length model was also applied by Hennigh et al. [16]. They reported good agreement with CFD results. However, no areas of separated flow seem to be present in [16].

For the ν_t model, the PINN predicted values for the velocity and pressure as well as the turbulent viscosity field. The Reynolds stress terms were then calculated from the output of the PINN using the Boussinesq hypothesis. In contrast to the established CFD turbulence models discussed above, no equation for ν_t needed to be satisfied. This is also the case for the proposed pseudo-Reynolds-stress model, where the Reynolds stresses were computed directly from the pseudo-turbulent output of the PINN. In the training process the weights of the PINN were adjusted such that the predicted u, v, and p, as well as the turbulent variables u'_p, and v'_p (of the pseudo-Reynolds-stress model) or ν_t (of the ν_t model) together satisfy the RANS equations. Presumably, as no physics-based equation was prescribed, this method might yield wrong predictions of u, v, and p.

However, in the example presented here, the predictions compared favorably to the DNS results. In fact, the pseudo-Reynolds-stress model and the ν_t model yielded more plausible flow fields than the traditional turbulence models, when trained with only three lines of labeled data (compare Figures 8–11). Harmening et al. [31] presented a PINN-based surrogate model which predicted the stationary flow and pressure field around a square within a range of high Reynolds numbers using the pseudo-Reynolds-stress model. Their predictions also agreed well with comparable RANS simulations. Eivazi et al. [17] presented a similar approach. They trained a PINN to predict several turbulent flows. The applied neural net was designed to predict velocity, pressure, and pseudo-Reynolds-stress terms. They also reported good agreements with simulations.

In many cases, experimental or simulative data are available. Therefore, the ν_t model and the pseudo-Reynolds-stress model could be a viable alternative to the k-ω model or the mixing-length model. This is especially the case for multidimensional surrogate models of variable geometry.

Data for the turbulent viscosity can be obtained from RANS simulations. Hence, although no equation was designated for the ν_t model, the corresponding PINN could be trained with labeled training data for all output variables. This is an advantage of the ν_t model when compared to the pseudo-Reynolds-stress model, where no training data can be derived for the pseudo turbulent output. The pseudo-Reynolds-stress model, on the other hand, provides more output variables that can be adjusted to satisfy the RANS equations. In fact, the pseudo-Reynolds-stress model yielded better results for the cases investigated here.

The presented RANS-based PINN method could be applied to a variety of turbulent engineering problems. This accounts for cases for which steady or unsteady solutions are required without resolving the turbulent motions. In this work, the training of the PINN took, on average, 1285 s for the pseudo-Reynolds-stress model and 2155 s for the k-ω model on a GeForce RTX 2070 GPU. This was 264% or 157% faster than a reference CFD simulation computed using a 120,000-cell computational grid on a i7-9700K CPU with 64 GB DDR4 RAM. However, the CFD simulations would be significantly faster for coarser grids. After training, the model required only 2.84 ms to predict the entire flow field (40,000 data points). This makes PINNs suitable for application as surrogate models for steering or optimization problems or for the generation of initial conditions for CFD calculations. As PINNs do not require a computational grid, they could also be applied as an alternative to CFD calculations to avoid expensive grid generation and errors associated with poor grid quality.

In future studies, we will investigate the capability of PINNs as multidimensional surrogate models for variable geometries. Furthermore, future work should concern improvements in the predicted flow field such that the corner vortex is also predicted by the PINN. The effect of different Reynolds numbers is a question for future studies as well. Additionally, an application of PINNs to the Lagrangian description of the Navier–Stokes equations using the smoothed particle hydrodynamics (SPH) method [32] would be interesting.

7. Conclusions

In this work, we compared four different RANS turbulence-modeling methods to predict an incompressible 2D turbulent backward-facing step flow using a PINN incorporating little labeled training data. The steady-state predictions were compared to averaged DNS results for a Reynolds number of 5100.

The models comprised the standard k-ω model and the mixing-length model. Additionally, we applied two equation-free models, namely, the ν_t model and a pseudo-Reynolds-stress model, which can be used to train a PINN. For the ν_t model and the pseudo-Reynolds-stress model, no additional equations needed to be solved. Instead, the PINN was trained to yield predictions that, together, satisfied the RANS equations.

Our predictions compared favorably to DNS results when three or five lines of labeled training data were used. For three lines of labeled training data, the pseudo-Reynolds-stress model produced the lowest errors. For this configuration, all models except the mixing-length model predicted the separated shear layer and the associated primary vortex. When five lines of labeled training data were used, the predicted flow fields were improved and the vortex behind the backward-facing step was predicted by all models. Every model failed to predict the secondary corner vortex.

Our work demonstrated the applicability of the k-ω model to PINNs. Furthermore, the pseudo-Reynolds-stress model was shown to be a viable alternative to the traditional turbulence models. For the high-Reynolds-number flow considered here, the fully connected feedforward PINN failed to yield acceptable predictions when trained without any labeled training data inside the domain. Our results demonstrated that the incorporation of few lines of training data could solve this issue.

Author Contributions: Conceptualization, F.P. and J.H.H.; methodology, F.P. and J.H.H.; software, F.P., J.H.H. and A.M.M.; validation, F.P. and J.H.H.; formal analysis, J.H.H. and F.P.; investigation, F.P., J.H.H. and A.M.M.; resources, F.-J.P.; data curation, A.M.M.; writing—original draft preparation, J.H.H. and F.P.; writing—review and editing, F.-J.P., O.e.M., D.S. and A.M.M.; visualization, F.P.; supervision, F.-J.P., O.e.M. and D.S.; project administration, F.P. and J.H.H. All authors have read and agreed to the published version of the manuscript.

Funding: We acknowledge support by the Open Access Publication Fund of the Westphalian University.

Data Availability Statement: Not applicable.

Conflicts of Interest: The authors declare no conflict of interest.

References

1. Sofos, F.; Stavrogiannis, C.; Exarchou-Kouveli, K.K.; Akabua, D.; Charilas, G.; Karakasidis, T.E. Current Trends in Fluid Research in the Era of Artificial Intelligence: A Review. *Fluids* **2022**, *7*, 116. [CrossRef]
2. Lagaris, I.E.; Likas, A.; Fotiadis, D.I. Artificial neural networks for solving ordinary and partial differential equations. *IEEE Trans. Neural Netw.* **1998**, *9*, 987–1000. [CrossRef] [PubMed]
3. Raissi, M. Physics Informed Deep Learning (Part I): Data-Driven Solutions of Nonlinear Partial Differential Equations. 2017. Available online: http://arxiv.org/pdf/1711.10561v1 (accessed on 29 December 2022).
4. Raissi, M. Physics Informed Deep Learning (Part II): Data-Driven Discovery of Nonlinear Partial Differential Equations. 2017. Available online: http://arxiv.org/pdf/1711.10566v1 (accessed on 29 December 2022).
5. Raissi, M.; Perdikaris, P.; Karniadakis, G.E. Physics-informed neural networks: A deep learning framework for solving forward and inverse problems involving nonlinear partial differential equations. *J. Comput. Phys.* **2019**, *378*, 686–707. [CrossRef]
6. Raissi, M.; Yazdani, A.; Karniadakis, G.E. Hidden fluid mechanics: Learning velocity and pressure fields from flow visualizations. *Science* **2020**, *367*, 1026–1030. [CrossRef] [PubMed]
7. Sun, L.; Gao, H.; Pan, S.; Wang, J.X. Surrogate modeling for fluid flows based on physics-constrained deep learning without simulation data. *Comput. Methods Appl. Mech. Eng.* **2020**, *361*, 112732. [CrossRef]
8. Zhu, Y.; Zabaras, N.; Koutsourelakis, P.S.; Perdikaris, P. Physics-constrained deep learning for high-dimensional surrogate modeling and uncertainty quantification without labeled data. *J. Comput. Phys.* **2019**, *394*, 56–81. [CrossRef]
9. Jin, X.; Cai, S.; Li, H.; Karniadakis, G.E. NSFnets (Navier-Stokes flow nets): Physics-informed neural networks for the incompressible Navier-Stokes equations. *J. Comput. Phys.* **2021**, *426*, 109951. [CrossRef]
10. Zhu, Q.; Liu, Z.; Yan, J. Machine learning for metal additive manufacturing: predicting temperature and melt pool fluid dynamics using physics-informed neural networks. *Comput. Mech.* **2021**, *67*, 619–635. [CrossRef]
11. Wang, R.; Kashinath, K.; Mustafa, M.; Albert, A.; Yu, R. Towards Physics-informed Deep Learning for Turbulent Flow Prediction. In *Proceedings of the 26th ACM SIGKDD International Conference on Knowledge Discovery & Data Mining*; Gupta, R., Liu, Y., Shah, M., Rajan, S., Tang, J., Prakash, B.A., Eds.; ACM: New York, NY, USA, 2020; pp. 1457–1466. [CrossRef]

12. Laubscher, R.; Rousseau, P. Application of a mixed variable physics-informed neural network to solve the incompressible steady-state and transient mass, momentum, and energy conservation equations for flow over in-line heated tubes. *Appl. Soft Comput.* **2022**, *114*, 108050. [CrossRef]
13. Wang, H.; Liu, Y.; Wang, S. Dense velocity reconstruction from particle image velocimetry/particle tracking velocimetry using a physics-informed neural network. *Phys. Fluids* **2022**, *34*, 017116. [CrossRef]
14. Eivazi, H.; Vinuesa, R. Physics-informed deep-learning applications to experimental fluid mechanics. *arXiv* **2022** , arXiv:2203.15402.
15. Wandel, N.; Weinmann, M.; Klein, R. Teaching the incompressible Navier–Stokes equations to fast neural surrogate models in three dimensions. *Phys. Fluids* **2021**, *33*, 047117. [CrossRef]
16. Hennigh, O.; Narasimhan, S.; Nabian, M.A.; Subramaniam, A.; Tangsali, K.; Fang, Z.; Rietmann, M.; Byeon, W.; Choudhry, S. NVIDIA SimNet™: An AI-Accelerated Multi-Physics Simulation Framework. In *Computational Science–ICCS 2021;* Lecture Notes in Computer Science; Paszynski, M., Kranzlmüller, D., Krzhizhanovskaya, V.V., Dongarra, J.J., Sloot, P.M., Eds.; Springer International Publishing: Cham, Switzerland, 2021; Volume 12746, pp. 447–461. [CrossRef]
17. Eivazi, H.; Tahani, M.; Schlatter, P.; Vinuesa, R. Physics-informed neural networks for solving Reynolds-averaged Navier–Stokes equations. *Phys. Fluids* **2022**, *34*, 075117. [CrossRef]
18. Xu, H.; Zhang, W.; Wang, Y. Explore missing flow dynamics by physics-informed deep learning: The parameterized governing systems. *Phys. Fluids* **2021**, *33*, 095116. [CrossRef]
19. Le, H.; Moin, P.; Kim, J. Direct numerical simulation of turbulent flow over a backward-facing step. *J. Fluid Mech.* **1997**, *330*, 349–374. [CrossRef]
20. Nadge, P.M.; Govardhan, R.N. High Reynolds number flow over a backward-facing step: structure of the mean separation bubble. *Exp. Fluids* **2014**, *55*, 1657. [CrossRef]
21. Spalart, P.R. Direct simulation of a turbulent boundary layer up to $R_\theta = 1410$. *J. Fluid Mech.* **1988**, *187*, 61–98. [CrossRef]
22. Wilcox, D.C. Reassessment of the scale-determining equation for advanced turbulence models. *AIAA J.* **1988**, *26*, 1299–1310. [CrossRef]
23. European Research Community on Flow, Turbulence and Combustion. Classic Collection Database: Case031 (Backward-Facing Step). Available online: http://cfd.mace.manchester.ac.uk/ercoftac/doku.php?id=cases:case031 (accessed on 9 September 2022).
24. Lu, L.; Meng, X.; Mao, Z.; Karniadakis, G.E. DeepXDE: A Deep Learning Library for Solving Differential Equations. *SIAM Rev.* **2021**, *63*, 208–228. [CrossRef]
25. Britter, R. (Ed.) *Background and Justification Document to Support the Model Evaluation Guidance and Protocol: COST Action 732 Quality Assurance and Improvement of Microscale Meteorological Models;* Meteorological Inst: Hamburg, Germany, 2007.
26. Oberkampf, W.L.; Trucano, T.G. Verification and validation in computational fluid dynamics. *Prog. Aerosp. Sci.* **2002**, *38*, 209–272. [CrossRef]
27. Meroney, R.; Ohba, R.; Leitl, B.; Kondo, H.; Grawe, D.; Tominaga, Y. Review of CFD Guidelines for Dispersion Modeling. *Fluids* **2016**, *1*, 14. [CrossRef]
28. Ma, H.; Zhang, Y.; Thuerey, N.; Hu, X.; Haidn, O.J. Physics-driven Learning of the Steady Navier-Stokes Equations using Deep Convolutional Neural Networks. 2021. Available online: http://arxiv.org/pdf/2106.09301v1 (accessed on 29 December 2022).
29. Huang, Y.; Zhang, Z.; Zhang, X. A Direct-Forcing Immersed Boundary Method for Incompressible Flows Based on Physics-Informed Neural Network. *Fluids* **2022**, *7*, 56. [CrossRef]
30. Kim, J.Y.; Ghajar, A.J.; Tang, C.; Foutch, G.L. Comparison of near-wall treatment methods for high Reynolds number backward-facing step flow. *Int. J. Comput. Fluid Dyn.* **2005**, *19*, 493–500. [CrossRef]
31. Harmening, J.H.; Pioch, F.; Schramm, D. Physics Informed Neural Networks as Multidimensional Surrogate Models of CFD Simulations. In Proceedings of the Machine Learning und Artificial Intelligence in Strömungsmechanik und Strukturanalyse, Wiesbaden, Germany, 16–17 May 2022; pp. 71–80.
32. Rakhsha, M.; Kees, C.E.; Negrut, D. Lagrangian vs. Eulerian: An Analysis of Two Solution Methods for Free-Surface Flows and Fluid Solid Interaction Problems. *Fluids* **2021**, *6*, 460. [CrossRef]

fluids

MDPI

Article

Physics-Informed Super-Resolution of Turbulent Channel Flows via Three-Dimensional Generative Adversarial Networks

Nicholas J. Ward

Department of Mechanical Engineering, Texas Tech University, Lubbock, TX 79409, USA; nicholas.ward@ttu.edu

Abstract: For a few decades, machine learning has been extensively utilized for turbulence research. The goal of this work is to investigate the reconstruction of turbulence from minimal or lower-resolution datasets as inputs using reduced-order models. This work seeks to effectively reconstruct high-resolution 3D turbulent flow fields using unsupervised physics-informed deep learning. The first objective of this study is to reconstruct turbulent channel flow fields and verify these with respect to the statistics. The second objective is to compare the turbulent flow structures generated from a GAN with a DNS. The proposed deep learning algorithm effectively replicated the first- and second-order statistics of turbulent channel flows of $Re_\tau = 180$ within a 2% and 5% error, respectively. Additionally, by incorporating physics-based corrections to the loss functions, the proposed algorithm was also able to reconstruct λ_2 structures. The results suggest that the proposed algorithm can be useful for reconstructing a range of 3D turbulent flows given computational and experimental efforts.

Keywords: turbulent channel flow; super-resolution; artificial intelligence; deep learning; generative adversarial network (GAN)

1. Introduction

Channel flow simulations have been utilized extensively in fluid mechanics research since the pioneering work performed by Kim et al. [1], Jiménez and Moin [2], Moser et al. [3], and Kim and Adrian [4]. Since these pioneering works were published, numerous different studies have been proposed. Lee and Moser [5] generated turbulent channel flow databases for viscous Reynolds numbers (Re_τ) up to 5200. In addition to the pioneering work in statistics and database generation, there have been significant advancements in near-wall turbulence. Jeong and Hussain [6] and Jeong et al. [7] discussed the identification, eduction, and dynamical inference of near-wall turbulence for turbulent channel flows. Schoppa and Hussain [8,9] enhanced the dynamical inference of near-wall turbulence via an inviscid streak transient growth phenomenon in the near-wall and buffer layer regions of a turbulent channel. Additionally, other studies have discussed large-scale structures in the near-wall [10,11] and buffer layers [4,12,13].

Over the last couple of decades, machine learning (ML) has been utilized for myriad problems in academic research. Among these problems are artificial intelligence, uncertainty quantification and propagation, algorithm development, and speech recognition. There is also interest in applying ML algorithms to fluid mechanics problems. These algorithms provided the foundation to deep learning, which applies a neural network (NN) with multiple layers to create hierarchical representations of data as a means of replacing a physical model. This means of substituting a physical model with a mathematical model (i.e., a machine-learning-based algorithm) is also known as surrogate modeling. For more information, Brunton et al. [14] reviewed each of the different ML algorithms utilized in fluid mechanics research in more detail.

Recently, there has been increasing interest in utilizing ML for problems in fluid mechanics and turbulence research. These include synthetic inlet flow generation, super-resolution of turbulent flows, and physics-informed neural networks for turbulent flows.

Synthetic inlet flow generation, a problem reviewed extensively by Tabor and Baba-Ahmadi [15], is a methodology that reconstructs 2D snapshots of the expected turbulence at the boundary condition in which flow enters the system; this is accomplished to reduce the dependence on precursor simulations, which can be computationally expensive. This study reviewed the pioneering works on inlet flow generation in large eddy simulations (LES). Wu [16] also reviewed the different methods utilized for inlet flow generation. This paper focused on the inflow–outflow turbulence generation for direct numerical simulation (DNS) and LES simulations. Because these simulations rely on non-periodic streamwise boundary conditions, the training of these generation methods requires some back-propagation or recycling technique. These techniques take the flow downstream of the inlet, and feed information back to the inlet to appropriately train the inlet to become and remain turbulent. These methods have been applied to homogeneous isotropic turbulence (HIT), incompressible and supersonic boundary layers, flows over rough surfaces, and mesoscale weather forecasting. Fukami et al. [17] discussed utilizing ML to generate time-dependent turbulent inflow generation for low-Reynolds-number channel flows. The authors utilized a convolutional neural network (CNN) encoder and decoder coupled with a multi-layer perceptron (MLP) classifier model to accomplish a priori and a posteriori studies of turbulence inflow generation. The a priori study was accomplished to train the model with respect to the DNS output. By using only one DNS snapshot at the initial timestep, the model was able to effectively predict the DNS results at subsequent timesteps. The a posteriori study took the results from the a priori study and applied a recursive input to investigate whether or not the turbulence was maintained in the simulation. While low-Reynolds-number flows were effectively predicted with the ML-based generator, the proposed method requires a higher-order resolution to train the model. This indicates that the proposed CNN-MLP generator needs to be significantly improved to tackle turbulent boundary layer and channel flow simulations in the future. Maulik et al. [18] discussed inlet flow generation using ML based on Reynolds-averaged Navier–Stokes (RANS) equations. The proposed framework utilizes a neural network as a surrogate model to interpolate the closure models (Spalart–Allmaras (SA), $k-\omega$, and $k-\varepsilon$) for RANS simulations. By applying this surrogate model, the authors were able to converge results up to sevenfold faster than using RANS with a closure model directly. The authors concluded that surrogate models of the closure problem in RANS garner accurate results in less time. However, this method is limited by the presence of sub-grid quantities in the simulation, meaning this surrogate model is good enough for lower-Reynolds-number simulations and the type of closure problem for the turbulence model. Kim and Lee [19] discussed synthetic inlet flow generation for a turbulent channel flow using neural networks. The study applied an ML technique utilizing a recurring neural network (RNN) coupled with a generative adversarial network (GAN) algorithm to generate inflow conditions of $Re_\tau = 180$–540. The model inputted temperature and velocity data to predict velocity, vorticity, and temperature. By applying this ML algorithm, the ML prediction accounts for the stochasticity of the DNS simulation. Additionally, the authors do not take into account that GANs and RNNs are semi-supervised and supervised models, meaning it was partially unsupervised. This shows that unsupervised models in turbulent flows still require supervision for optimization, regression, and classification purposes. These studies show that synthetic inflow generation is best served using multiple different types of NNs. However, because synthetic inflow generation depends on the type of simulation utilized, a CNN or GAN is recommended for DNS, but an ANN for LES and RANS simulations.

The super-resolution of turbulent flows involves using an ML algorithm as a surrogate model to recreate a high-resolution dataset to a similar quality as DNS from a low-resolution dataset. Fukami et al. [17] discussed the super-resolution of two-dimensional (2D) HIT. The authors utilized two different NNs to reconstruct the 2D HIT for laminar and turbulent flows over a cylinder as well as decaying isotropic turbulence. The study proposed a CNN and a hybrid downsampling connection and multi-scale (hDSC/MS) model to capture the physics of low-Reynolds-number DNS simulations. The authors concluded that utilizing the hDSC/MS model, which was a modification of a conventional CNN, enabled faster and more efficient convergence than DNS results in as little as 50 2D snapshots of data. The authors also claimed that the model was able to reconstruct the flow effectively without using a priori methods, which construct the flow using prior knowledge of the DNS simulations. However, the study showed that the deviation in the energy spectrum at higher wavenumbers indicates that the model is effective in capturing large-scale structures, and may need to be improved to capture small-scale structures. Liu et al. [20] improved on the work performed by Fukami et al. [17] by developing a CNN-based model that incorporates temporal information from consecutive snapshots to discuss the super-resolution of turbulent flows. Conventional CNNs are considered to be "static" NNs because they do not extract time-dependent information from the snapshots. The NN that the authors proposed extracted the features from each snapshot, and incorporated three parallel snapshot reconstruction modules in terms of forward-, central-, and backward-differencing. By applying weights to each parallel module, the outputs of each module were averaged to effectively produce a final output at the same timestep as the DNS simulation. However, like the study by Fukami et al. [17], the authors were unable to reproduce the kinetic energy at higher wavenumbers, meaning the small-scale structures also were not captured properly. Fukami et al. [21] discussed supervised machine learning techniques applied to super-resolution. The authors determined that utilizing basic ML algorithms such as MLP, random forest (RF), support vector regression (SVR), and extreme learning machine (ELM) (a simplified version of the MLP) can easily reproduce laminar flows. However, for turbulent flow data, the authors recommend that CNNs are effective in handling the large datasets that turbulent flows incur during DNS simulations. The study also suggests that CNNs are only effective within the distributions that the model trains and tests data from, meaning they should not be used for extrapolation. Because NNs are effective approximators of data resolution studies, future studies should focus on training data based on the available information. Kim et al. [22] accomplished this by applying unsupervised deep learning for super-resolution of paired and unpaired datasets for 2D snapshots of turbulent flows. Based on the results from these papers, it is clear that utilizing a CNN-based model that can capture the small-scale structures in turbulence from the DNS is of particular interest in future research.

Additionally, there has been increasing interest in physics-informed neural networks (PINNs), which incorporate the governing equations, statistics, and/or physical intuition into the neural network. Raissi and Karniadakis [23] and Raissi et al. [24] first introduced PINNs to solve different governing equations, such as Burger's equation and the Navier–Stokes equations. This has also been applied to super-resolution studies. Yousif et al. [25] applied PINNs to inlet flow generation. Sachin Venkatesh et al. [26] performed spatio-temporal super-resolution analysis on turbulent flow fields to reconstruct high-resolution flow fields from low-resolution flow field data. Bode et al. [27] used a physics-informed super-resolution GAN for sub-filter modeling of turbulent reactive flows. Yousif et al. [28] and Yousif et al. [29] applied physics-informed super-resolution to predict turbulent flows at different Reynolds numbers. Eivazi and Vinuesa [30] used PINNs to reconstruct turbulent flows from noisy datasets. Du et al. [31] also compared PINNs with four-dimensional adjoint-variational data assimilation in reconstructing turbulent channel flows from sparse data. Linqi et al. [32] then used PINNs in an enhanced super-resolution GAN to reconstruct turbulent channel flows from sparse data generated from upsampling via tri-linear interpolation. These studies indicate that there is a need for PINNs combined with the super-resolution capabilities of GANs to reconstruct turbulent flow fields and structures.

The novelty of this work lies in the combination of a GAN with physics-informed constraints to achieve the super-resolution of three-dimensional turbulence datasets. Previous work on super-resolution has typically relied on deep learning models that are trained on large datasets, but these models can struggle with capturing the complex physical processes that govern turbulent flows. By incorporating physics-informed constraints into the proposed GAN model, this work aims to improve the accuracy and reliability of the super-resolution results while maintaining computational efficiency. The physics-informed constraints are derived from first principles and are incorporated into the GAN loss function to ensure that the generated flow fields satisfy the laws of physics. This approach allows the model to capture important physical features of turbulence, such as vortices and energy spectra, while still achieving high-resolution results. The results show that the proposed physics-informed GAN approach achieves superior performance in terms of both accuracy and computational efficiency, making it a promising approach for super-resolution of three-dimensional turbulence datasets.

The goal of this work is to investigate the reconstruction of turbulence from minimal or lower-resolution datasets as inputs using reduced-order models. This work seeks to effectively reconstruct high-resolution 3D turbulent flow fields using unsupervised physics-informed deep learning. The first objective of this study is to reconstruct turbulent channel flow fields and verify these with respect to the statistics. The second objective is to compare the turbulent flow structures generated from the GAN with the DNS.

This paper is organized in the following structure. Section 2 describes the numerical set-up of the DNS data utilized for training the GAN, and explains the methodology of the proposed GAN. This section describes in detail the deep learning algorithm implemented, such as the generator and discriminator architecture, as well as the chosen values of the hyperparameters. Section 3.1 demonstrates the turbulent flow statistics generated from the GAN with respect to the DNS, and verifies the performance of the GAN compared to other models. Section 3.2 compares the reconstructed turbulent flow structures with respect to the DNS data. The conclusions from this study are stated in Section 4.

2. Methodology

DNS simulations were conducted to determine the training data utilized in this study, which used the POONPACK code developed by Lee and Moser [5]. For this study, the streamwise, wall-normal, and spanwise coordinates are denoted by x, y, and z, respectively. Additionally, the streamwise, wall-normal, and spanwise velocities are denoted by u, v, and w, respectively. The incompressible Navier–Stokes equations, which are determined as

$$\nabla \cdot \boldsymbol{u} = 0, \tag{1}$$

$$\frac{\partial \boldsymbol{u}}{\partial t} + \boldsymbol{u} \cdot \nabla \boldsymbol{u} = -\nabla p + \frac{1}{Re_\tau} \nabla^2 \boldsymbol{u}, \tag{2}$$

are solved using the methodology from Kim et al. [1].

For the turbulent channel flows used in this study, the diagram and the numerical discretization details are shown in Figure 1 and Table 1. In this study, the DNS simulation was conducted at a bulk Reynolds number Re_b of 2857. This corresponds to a viscous Reynolds number $Re_\tau = u_\tau \delta / \nu$ of 180, where u_τ, h, and ν are the friction velocity, channel half-height, and kinematic viscosity, respectively.

For the surrogate model used in this study, a machine learning algorithm was developed to effectively capture the flow physics in a turbulent channel flow without directly solving the Navier–Stokes equations. To effectively capture low-Reynolds-number turbulent channel flows, a GAN [33] was constructed. The GAN is comprised of a generator, which reconstructed the flow fields at each timestep using a random noise vector as an input, and a discriminator, which provided feedback to the generator to either accept or reject the outputted flow fields with respect to the real (DNS) data. Because the GAN is an unsupervised deep learning model [14,19,22], it takes a random input vector at each timestep and generates a three-dimensional (3D) flow field at an accuracy close to the DNS simulation.

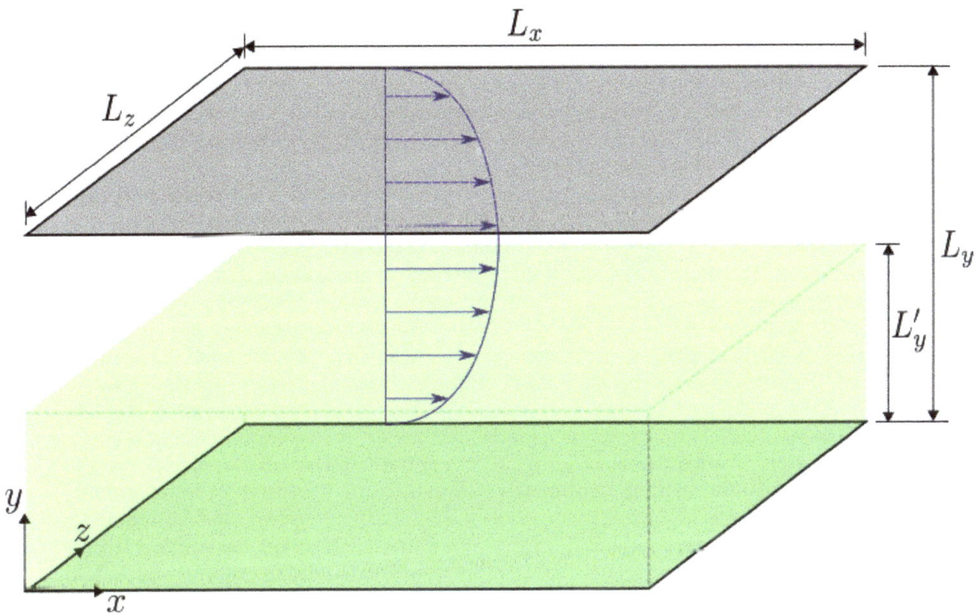

Figure 1. Sketch of the turbulent channel flow simulation used in this study. The DNS simulation encompasses the flow between both stationary walls. The proposed ML algorithm trains the bottom half of the channel, indicating that $L'_y = L_y/2$.

Table 1. Details of the numerical discretization employed for the simulations of the turbulent channel flow. The computational grid size is $N_x \times N_y \times N_z = 64 \times 128 \times 64$, and L_x, L_y, and L_z are the domain dimensions in each direction.

Re_τ	Re_b	$L_x \times L_y \times L_z$	Δx^+	Δy^+	Δz^+
180	2857	$\pi h \times 2.0h \times \pi h/2$	11.15	0.19–5.27	5.57

To ensure that it worked effectively, the GAN model required a significant number of samples, because increasing the number of samples enabled reduced errors between the generated samples and the DNS data. For the $Re_\tau = 180$ simulations, 10,000 samples of data were used to reconstruct the turbulent channel flow. The size of the training dataset does have an impact on the training results. Generally, a larger training set can lead to more accurate and reliable model predictions. With a smaller training set, there may be less information for the model to learn from, which could result in lower accuracy or overfitting. For the purposes of this study, 10,000 samples was considered to be an appropriate balance between having enough data to train the model effectively without overwhelming the model with too much data, which causes overfitting. By utilizing these DNS samples as inputs into the discriminator, the GAN model was able to generate the new data for each respective timestep. DNS simulation is not necessarily needed for the training of the GAN to reconstruct turbulence datasets. However, having access to DNS data can help to improve the accuracy and realism of the generated flow fields. The primary purpose of the GAN is to learn the underlying statistical distribution of a dataset, which can then be used to generate new data that is similar to the training data. In the case of turbulence datasets, the training data can come from a variety of sources, such as experimental measurements or LES. If DNS data are available, it can be used as part of the training dataset for the GAN. This can help to ensure that the generated flow fields accurately capture the complex physical processes that govern turbulent flows. However, if DNS data are not available, other sources of turbulence data can still be used for training the

GAN. It is worth noting that while DNS data can be useful for improving the accuracy and realism of GAN-generated flow fields, it is also computationally expensive and may not always be practical or feasible to obtain. Therefore, alternative sources of turbulence data may need to be used for training the GAN in some cases. The GAN model accomplished this by training for 40,000 training iterations. Additionally, the training and testing split for this study was 80% and 20%, respectively.

In addition to the number of samples and training iterations, the GAN model required loss functions for both the generator and the discriminator. These loss functions provided the GAN model with two functions to optimize during each training iteration. The selection of the loss function depends on the type of problem being solved. For example, for a regression problem, the mean squared error (MSE) loss function is commonly used, while for a binary classification problem, the binary cross-entropy loss function is often used. The choice of loss function can have a significant impact on the training process and the resulting model. Different loss functions have different properties and can optimize for different aspects of the model performance. For example, using a mean absolute error (MAE) loss function instead of MSE may result in a model that is more robust to outliers, but also less accurate overall. In addition to selecting an appropriate loss function, it is important to consider how the loss function is optimized during training. This is typically achieved by using an optimization algorithm such as stochastic gradient descent (SGD) or adaptive moment (Adam). The choice of optimization algorithm can also affect the training process and result in different model performance. In summary, selecting an appropriate loss function is important for achieving good model performance, and different loss functions can optimize for different aspects of model performance. However, it is also important to consider how the loss function is optimized during training and choose an appropriate optimization algorithm.

The loss functions were determined from the cycle-consistency GAN (CycleGAN) because this DL method is able to map between two unpaired datasets using two generators and two discriminators [34]. In this case, the two unpaired datasets are the low-resolution (LR) and high-resolution (HR) inputs to the two generators: G and F. The generator G takes an LR input and generates an HR output, and generator F does the opposite (HR input to LR output). The outputs from generators G and F are sent back through the other generators to create a cycle-consistency loss, such that $F(G(X_{LR})) \approx X_{LR}$ and $G(F(X_{HR})) \approx X_{HR}$ is satisfied. The two discriminators, D_Y and D_X, then process the generated outputs from G and F, respectively, to determine a probability of how real or fake the outputs are versus the real datasets. The loss functions for the discriminators ($\mathcal{L}_{D_Y}$ and $\mathcal{L}_{D_X}$) as well as for the generators ($\mathcal{L}_G$ and $\mathcal{L}_F$) are, respectively, defined as

$$\mathcal{L}_{D_Y} = \gamma_1[BCE(D_Y(X_G)) + BCE(D_Y(X))] + \lambda\mathbb{E}[(\|\nabla_{X_{int,G}} D(X_{int,G})\|_2 - 1)^2], \quad (3)$$

$$\mathcal{L}_{D_X} = \gamma_1[BCE(D_X(X_F)) + BCE(D_X(X))] + \lambda\mathbb{E}[(\|\nabla_{X_{int,F}} D(X_{int,F})\|_2 - 1)^2], \quad (4)$$

$$\mathcal{L}_G = BCE(D_Y(X_G)) + \lambda\mathcal{L}_{cycle} + \mathcal{L}_{div} + \mathcal{L}_{NSE} + \mathcal{L}_{SSIM} + \gamma_2\mathcal{L}_{down}, \quad (5)$$

$$\mathcal{L}_F = BCE(D_X(X_F)) + \lambda\mathcal{L}_{cycle} + \mathcal{L}_{div} + \mathcal{L}_{NSE} + \mathcal{L}_{SSIM} + \gamma_3\mathcal{L}_{up}, \quad (6)$$

where $BCE(\cdot)$ is the binary cross-entropy (BCE) loss, $\mathbb{E}[\cdot]$ is the expected value, X is the real data, X_G is the data outputted from generator G, X_F is the data outputted from generator F, λ is a coefficient of the gradient penalty, $X_{int} = \varepsilon X + (1-\varepsilon)X_{G/F}$ is the interpolated data between the real and generated data, ε is a uniform distribution of values between 0 and 1 for the purposes of determining the interpolated data, and $\|\cdot\|_2 = \sqrt{\sum_{i=1}^{n}\sum_{j=1}^{n}\sum_{k=1}^{n}|\cdot|^2}$ is the L2 matrix norm. The BCE loss is defined as

$$BCE(D(X)) = -\frac{1}{N}\sum_{i=1}^{N}\Big(D(X)\log(D(X)) + (1-D(X))\log(1-D(X))\Big), \quad (7)$$

where $D(X)$ is the discriminator output. A gradient penalty (GP) term was added to the discriminator loss functions because this loss term has proven to prevent mode collapse during the training, as well as to enable the discriminator to not overpower the generator [19,35]. For this study, λ was set to 10. To improve the predictions generated from the proposed algorithm, six additional loss function terms were introduced: a cycle-consistency loss function $\mathcal{L}_{cycle}$ [34], a divergence loss function $\mathcal{L}_{div}$, an NSE loss function $\mathcal{L}_{NSE}$ based on the material derivative, a structural similarity index measure (SSIM) loss $\mathcal{L}_{SSIM}$, a downsampling loss $\mathcal{L}_{down}$, and an upsampling loss $\mathcal{L}_{up}$ [36]. These loss functions are defined as

$$\mathcal{L}_{cycle}(X, X_{gen}) = MAE(X_{inp,LR}, X_F) + MAE(X_{inp,HR}, X_G), \tag{8}$$

$$\mathcal{L}_{div}(X, X_{gen}) = MSE(\nabla \cdot X, \nabla \cdot X_{gen}), \tag{9}$$

$$\mathcal{L}_{NSE}(X, X_{gen}) = MSE\left(\frac{\partial X}{\partial t} + X \cdot \nabla X, \frac{\partial X_{gen}}{\partial t} + X_{gen} \cdot \nabla X_{gen}\right), \tag{10}$$

$$\mathcal{L}_{SSIM}(X, X_{gen}) = \frac{1}{N}\sum_{i=1}^{N}(1 - SSIM(X, X_{gen})), \tag{11}$$

$$\mathcal{L}_{down}(X, X_{gen}) = MAE(X_{inp,LR}, f_{\downarrow}(X_G)), \tag{12}$$

$$\mathcal{L}_{up}(X, X_{gen}) = MAE(X_{inp,HR}, f_{\uparrow}(X_F)). \tag{13}$$

MSE and MAE represent the mean squared error and mean absolute error loss terms in the generator, which are defined as

$$MSE(X, X_{gen}) = \frac{1}{N}\sum_{i=1}^{N}(X^i - X^i_{gen})^2, \tag{14}$$

and

$$MAE(X, X_{gen}) = \frac{1}{N}\sum_{i=1}^{N}|X^i - X^i_{gen}|, \tag{15}$$

respectively. The parameter γ_1 is a constant that was set to 0.5 because that was determined to have the best fit to the data without over-powering the GP loss term in the discriminator loss functions. Equations (9)–(13) were added as additional regression losses to Equations (5) and (6) to penalize the generators more severely for larger errors between the generated samples and the DNS data. Each of these additional loss functions were able to significantly improve the training without increasing the computational expense. This is because the loss functions accounted for both the non-linearities in the governing equations, which is characteristic of the advection term in the material derivative, and the quality of the images reconstructed at both resolutions using the BCE, MSE, and MAE functions in the generator and discriminator. By utilizing both the governing equations as constraints to the GAN model and the data-driven image reconstruction techniques, the number of training iterations required to obtain visually similar datasets was reduced. Using the MSE or the MAE as an additional loss term to the generators has proven to be an effective constraint in previous studies [19,37]. The MAE is a better optimization function for image reconstruction purposes; however, unlike the MAE, the MSE enforces statistical constraints better because it accounts for any outliers in the predicted datasets.

For the CycleGAN, the conditions $F(G(X_{LR})) \approx X_{LR}$ and $G(F(X_{HR})) \approx X_{HR}$ are then satisfied using an identity loss [34]. This identity loss initially required the resolutions of both input datasets to be the same. Because the input datasets are not the same resolution, the identity losses for the generators are modified using the SSIM and either upsampling or downsampling losses. The SSIM [38] is a metric that measures the similarity between two datasets, and is defined as

$$SSIM(X_{LR}, X_{HR}) = \frac{(2\mu_{X_{LR}}\mu_{X_{HR}} + c_1)(2\sigma_{X_{LR}X_{HR}} + c_2)}{(\mu^2_{X_{LR}} + \mu^2_{X_{HR}} + c_1)(\sigma^2_{X_{LR}} + \sigma^2_{X_{HR}} + c_2)}. \quad (16)$$

As shown in Equation (16), $SSIM(X_{LR}, X_{HR})$ is a function of the mean μ, the standard deviation σ, and the covariance $\sigma_{X_{LR}X_{HR}}$. The constants c_1 and c_2 are set to small values to prevent the discriminator from going to zero. Additionally, it is recommended by Wang et al. [38] to evaluate the SSIM function on an 11×11 patch, but it was found that a 5×5 patch was a better choice for this study due to the relative size of the patch versus the final resolution size. The upsampling and downsampling losses incorporated the $f_{\uparrow}(\cdot)$ and $f_{\downarrow}(\cdot)$ functions, which are then defined as an upsampling and average sampling layer, as demonstrated in Figure 2 and Table 2.

After determining the appropriate loss functions, the GAN model then required objective functions to be utilized to provide optimized parameters to the loss functions during each training iteration. This indicated that an optimizer needed to be defined for the GAN model. Numerous optimizers have been proposed in previous works, such as RMSProp [39], Adam [40,41], etc. However, these optimizers suffer from warm-up constraints in learning rate optimization. To prevent this problem, the Rectified Adam (RAdam) optimizer was utilized for adaptive learning rate optimization [42]. This optimizer added an additional variance rectification method to enable faster convergence in less training iterations. By inputting the initial learning rate l, as well as the decay rates in calculating the first and second moments (i.e., the mean and variance) β_1 and β_2, the variance rectification and adaptive learning rate functions were determined at each timestep, as [42]:

$$\rho_\infty = \frac{2}{1-\beta_2} - 1, \quad (17)$$

$$g_t = \nabla_{\theta_t} f(\theta_t), \quad (18)$$

$$v_t = \beta_2 v_{t-1} + (1-\beta_2)g_t^2, \quad (19)$$

$$m_t = \beta_1 m_{t-1} + (1-\beta_1)g_t^2, \quad (20)$$

$$\widehat{m_t} = \frac{m_t}{1-\beta_1^t}, \quad (21)$$

$$\rho_t = \rho_\infty - \frac{2t\beta_2^t}{1-\beta_2^t}, \quad (22)$$

$$\theta_t = \begin{cases} \theta_{t-1} - \alpha_t r_t \widehat{m_t} l_t, & \text{if } \rho_t > 4 \\ \theta_{t-1} - \alpha_t \widehat{m_t}, & \text{otherwise} \end{cases}, \quad (23)$$

$$r_t = \sqrt{\frac{(\rho_t - 4)(\rho_t - 2)\rho_\infty}{(\rho_\infty - 4)(\rho_\infty - 2)\rho_t}}, \quad (24)$$

$$l_t = \sqrt{\frac{1-\beta_2^t}{v_t}}, \quad (25)$$

where ρ_∞ is the maximum length of the moving average, g_t is the gradient of the objective function f, v_t is the variance at time t, m_t is the mean, $\widehat{m_t}$ is the mean corrected for bias, ρ_t is the length of the moving average, r_t is the variance rectification, and l_t is the adaptive learning rate. For the purposes of this study, the input learning rates for the generator and discriminator were both set to 0.0001. The decay rates for both the generator and discriminator were set to 0.0 and 0.9 for β_1 and β_2.

The architecture that was utilized for this study is shown in Figure 2. Additionally, the description of each layer presented in Figure 2 is indicated in Table 2. The input that was used as an input into the generator was pre-processed to represent a normal distribution. This input was then inputted into the generator block, which is comprised of a fully connected layer with one activation layer, one reshape layer, and one normalization layer. The first activation layer was a leaky rectified linear unit (LeakyReLU), defined as

$$f(x) = \begin{cases} \alpha x, & \text{if } x \leq 0 \\ x, & \text{otherwise} \end{cases}, \tag{26}$$

with a slope coefficient α of 0.2. The LeakyReLU layer has been proven to perform well in both generators and discriminators for GAN models. The output from this layer was then sent to either an upsampling or average sampling layer, depending on the generator or discriminator, which is useful for preventing overfitting when using larger datasets.

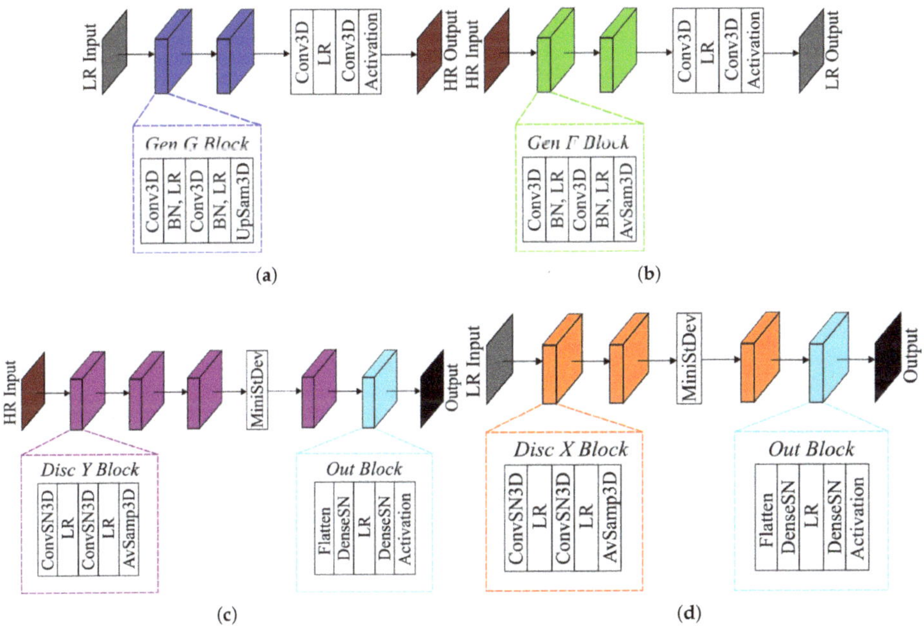

Figure 2. 3D GAN architecture utilized for this study at resolution ratio $R = 4$. The generator G and F are shown in (**a**,**b**), and the discriminator D_Y and D_X are shown in (**c**,**d**).

Table 2. Description of each layer used in the generators and discriminators of the proposed GAN model.

Generators G and F	*Description*
Conv3D	3D convolutional layer [32–128 features for all]
BN	Batch normalization (momentum = 0.8)
LR	Leaky rectified linear unit ($\alpha = 0.2$)
UpSam3D	3D upsampling layer
Activation	Activation function (*tanh*)
Discriminators D_X and D_Y	*Description*
ConvSN3D	3D convolutional layer with spectral normalization [16–256 features]
LR	Leaky rectified linear unit ($\alpha = 0.2$)
Dropout	Dropout ($r_{drop} = 0.2$)
MiniStDev	Mini-batch standard deviation layer
DenseSN	Fully connected dense layer with spectral normalization
Activation	Activation function (linear)

The output of the input block in the generator was then sent to a convolution block, which contained upsampling, 3D convolution, batch normalization, and LeakyReLU activation layers. The upsampling doubled the resolution of the input data, which was then sent to a transpose convolutional layer. The transpose convolutional layer applied an inverse convolution method to generate output feature maps that were greater than the inputs. To ensure that its output maintained a normal distribution, batch normalization, which is defined as

$$\mu = \frac{1}{n}\sum_i Z^{(i)}, \tag{27}$$

$$\sigma = \frac{1}{n}\sum_i (Z^{(i)} - \mu), \tag{28}$$

$$Z^{(i)}_{norm} = \frac{Z^{(i)} - \mu}{\sqrt{\sigma^2 - \epsilon}}, \tag{29}$$

$$\overline{Z} = \gamma Z^{(i)}_{norm} + \beta, \tag{30}$$

was applied, where μ is the mean, $Z^{(i)}$ is the input sample at each instance i, σ is the standard deviation, $Z^{(i)}_{norm}$ is the normalized input, ϵ is a constant (10^{-12}) to ensure the denominator does not go to 0, and $\overline{Z}$ is the modified input based on a constant multiplier γ and a bias term β, and averaged in each training batch. For this study, the batch size was set to 16. Each convolutional block was then repeated three more times. The output block applied the transpose convolutional layer and a hyperbolic tangent (*tanh*) activation function, defined as

$$f(x) = \frac{e^x - e^{-x}}{e^x + e^{-x}}. \tag{31}$$

For the discriminator, the activation and normalization layers remained the same, except the output activation function, which was chosen to be a linear activation function. The convolution layers applied in the discriminator were similar to previous studies; however, because of the computational expense and training instability associated with 3D GAN architectures, the convolutional layers required additional normalization to ensure faster convergence and more stabilized training. For this reason, spectral normalization [43] was applied to improve the performance of the discriminator without compromising accuracy and stability. The spectral normalization method applied a weight normalization to the discriminator by normalizing it with respect to the Lipschitz constraint ($\sigma(W) = 1$, where W is the weight matrix), such that

$$\overline{W}_{SN}(W) := W/\sigma(W), \tag{32}$$

$$\sigma(\overline{W}_{SN}(W)) = 1, \tag{33}$$

$$||f||_{Lip} = 1. \tag{34}$$

Additionally, a mini-batch standard deviation layer was applied to the last convolutional block of the discriminator. This layer, which was introduced by Karras et al. [44], calculated the standard deviation of each feature in each activation map, averaged the standard deviation over the mini-batch, and then added at the end of the discriminator network. This is calculated as

$$\overline{\sigma} = \sqrt{\frac{\sum_{i=1}^{n}(x_i - \overline{x})^2}{N-1}}. \tag{35}$$

The losses for the generator and discriminator at each iteration are displayed in Figure 3 for each resolution ratio R considered in this study. For the purposes of this study, generator G was optimized to be able to reconstruct a high-resolution dataset from a low-resolution input. This, therefore, meant that the discriminator D_Y was utilized to improve the generated datasets from G. The losses for each generator and discriminator are initialized to zero, but are then at their respective maxima within the first 500 iterations. This is because the GAN has not had enough iterations to minimize the regression problem associated with the combined losses. For both Reynolds numbers, Figure 3a,b indicate that the minimum number of iterations required for the GAN model to converge well-enough with the DNS results was 2000. However, this enables only the first-order statistics to converge to the DNS results within an error tolerance of 1%. To enable the higher-order statistics to converge to the DNS results, the number of iterations was, therefore, extended to 40,000 iterations for $Re_\tau = 180$.

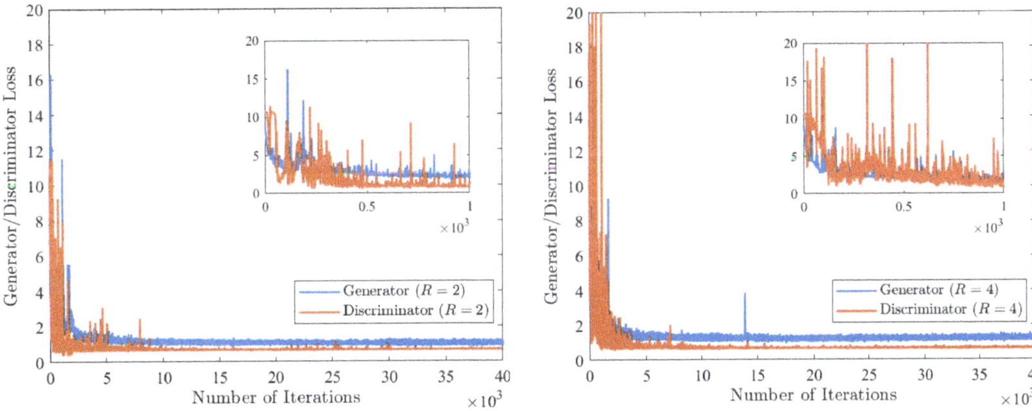

(a) (b)

Figure 3. Losses from the generator and discriminator as a function of the number of training iterations for the proposed GAN model at (**a**) $R = 2$ and (**b**) $R = 4$.

It is also important to note several reasons why testing a higher-turbulence dataset using a lower-turbulence training dataset could be problematic. If the GAN model is trained on a low-turbulence dataset, it may not generalize well to a higher-turbulence dataset. The model may have learned to generate flow fields that are specific to the low-turbulence conditions, and may not be able to accurately generate flow fields for higher-turbulence conditions. If the GAN model is only trained on low-turbulence data, it may have limited applicability to real-world scenarios that involve higher-turbulence conditions. This could limit the usefulness of the model in practical applications. Additionally, if the GAN model is tested on a higher-turbulence dataset that is significantly different from the training dataset, there is a risk of overfitting. The model may have learned to generate flow fields that closely match the training data, but may not be able to accurately generate flow fields for new and different conditions. Therefore, it is important to carefully consider the range of turbulence conditions that the GAN model will be expected to generate flow fields for, and to train and test the model on datasets that are representative of these conditions. This can help to ensure that the model is accurate and reliable across a range of different turbulence conditions.

3. Results and Discussion

3.1. *Super-Resolution Statistics*

This section discusses the super-resolution reconstruction statistics of the ML algorithm compared to DNS. In order for the data to be properly read by the GAN, the data were normalized to be between −1 and 1. Therefore, the data needed to be normalized to fit within the bounds of the GAN, which was accomplished via

$$X'_{train} = 2 * \frac{X_{train} - X_{train,min}}{X_{train,max} - X_{train,min}} - 1, \quad (36)$$

where X'_{train} is the normalized input suitable for the GAN training. After the training was completed and the GAN data were generated, the data were un-normalized to determine the velocity components at the appropriate scale for the DNS data.

Figure 4 indicates the velocity profiles for the u, v, and w components from the DNS or GAN models. The velocity profiles shown in Figure 4a,d,g correspond to the data from the DNS dataset. It is demonstrated that the GAN model is able to capture the flow in the bottom half of the channel within 5% of the DNS data instantaneously at the channel center. This also indicates that the GAN model is able to capture both the mean flow and the turbulent fluctuations for each velocity component. The differences in these snapshots based on visual inspection are attributed to the high turbulent fluctuations in the log-law region of the channel, as well as the information loss associated with higher resolution ratios. These differences are most present in the wall-normal and spanwise components v and w because the turbulent fluctuations for those components are not as strong or dominant as the streamwise velocity u.

Figure 5 indicates the mean velocity profile and the velocity fluctuations for each of the resolution ratios considered in the study, as well as the DNS data. The mean velocity profile shown in Figure 5a indicates that the GAN model is able to capture the mean flow in the bottom half of the channel within 2% of the DNS data. This also indicates that the GAN model is able to satisfy both the near-wall region, where $\bar{u} = y^+$, and the log-law region, where $\bar{u} = (1/\kappa)\ln y^+ + b$ (given that $\kappa = 0.40$ is the von Karman constant and $b = 5.5$ is the y-intercept for a smooth wall at $Re_\tau = 180$). Additionally, the velocity fluctuations are determined within a 5% error for a simulation with only 40,000 training iterations. The lowest variability was in the u' component of the velocity fluctuations in Figure 5b, whereas the highest variability was in w', from Figure 5d. This indicates that the GAN is able to recover the mean velocity profiles for the channel flow in less iterations than the velocity fluctuations at both resolution ratios. Likewise, the velocity fluctuations are more sensitive to the resolution ratios, meaning that increased resolution ratios reduce the ability of the GAN to capture the fluctuations effectively. Overall, the resolution ratio $R = 2$ performs the best because less information about the turbulent flow was lost. However, it was determined that the model was able to perform better at a lower resolution ratio for the wall-normal fluctuations v'. Despite that, the velocity fluctuations are more sensitive to the boundary conditions of the channel, especially at $y^+ = 180$.

Figure 6 demonstrates the Reynolds stresses in the channel. The Reynolds shear stress is shown in Figure 6a. The black line in the figure corresponds to $\overline{u'v'}/u_\tau^2 = 1 - (y^+/Re_\tau)$, which is the theoretical limit which the Reynolds shear stress cannot exceed [3]. For both resolution ratio cases, the predicted Reynolds stresses did not exceed that limit, and also collapsed within 5–10% of the DNS data. For the $R = 4$ case, the peak of the Reynolds shear stress is less than the peak for the DNS. This indicates that the GAN does not fully recover the total turbulent fluctuations in the fluid momentum, but captures generally the behavior of the Reynolds shear stress.

Likewise, the Reynolds normal stresses in the streamwise, wall-normal, and spanwise directions are displayed in Figure 6c,d. The peaks for each of the Reynolds normal stresses are $y^+ \approx 15$, $y^+ \approx 60$, and $y^+ \approx 40$ for the streamwise, wall-normal, and spanwise directions, respectively. Both the DNS and the predicted GAN normal stresses correspond to the same peak, indicating the consistence of the GAN with incompressible channel flow. Like the cases in Figure 5, the $R = 2$ cases generally outperform the $R = 4$ cases, with the exception of the wall-normal component of the Reynolds normal stresses, shown in Figure 6c.

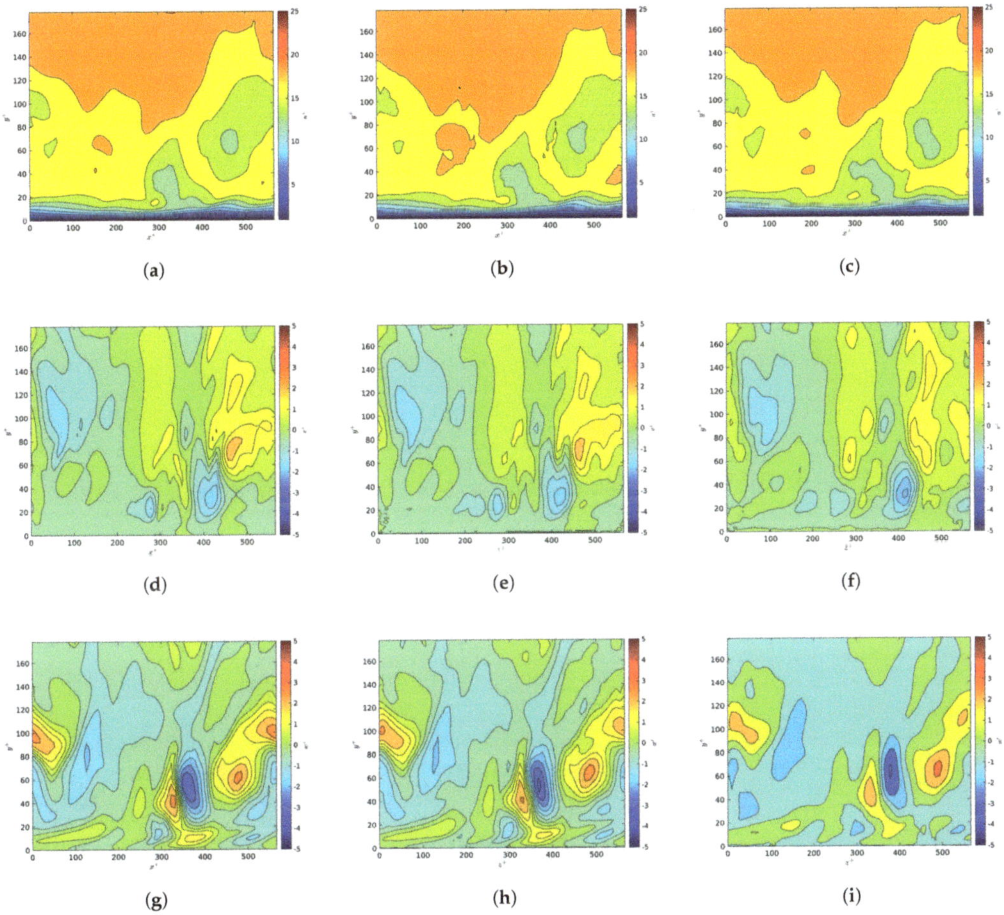

Figure 4. Instantaneous velocity components in the *yz*-plane at the channel center for $Re_\tau = 180$: DNS (**a**,**d**,**g**), GAN predictions for $R = 2$ (**b**,**e**,**h**), and GAN predictions for $R = 4$ (**c**,**f**,**i**). The top row (**a**–**c**), the middle row (**d**–**f**), and the bottom row (**g**–**i**) correspond to u, v, and w, respectively.

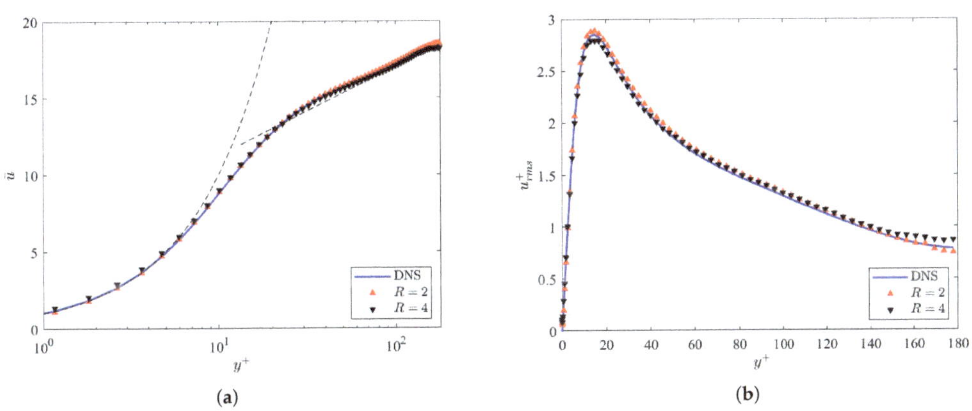

Figure 5. *Cont.*

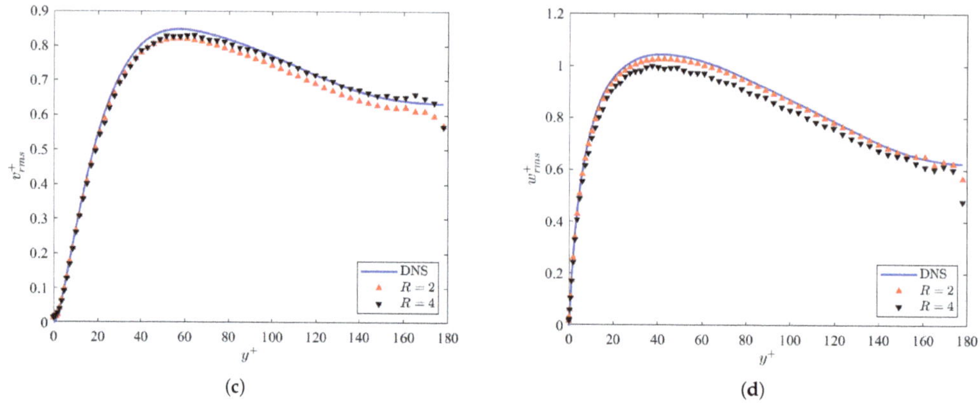

Figure 5. Mean velocity profile ($\overline{u}$) (**a**) and velocity fluctuations (**b–d**) as a function of the normalized wall-normal position y^+ for the DNS and ML models at $Re_\tau = 180$.

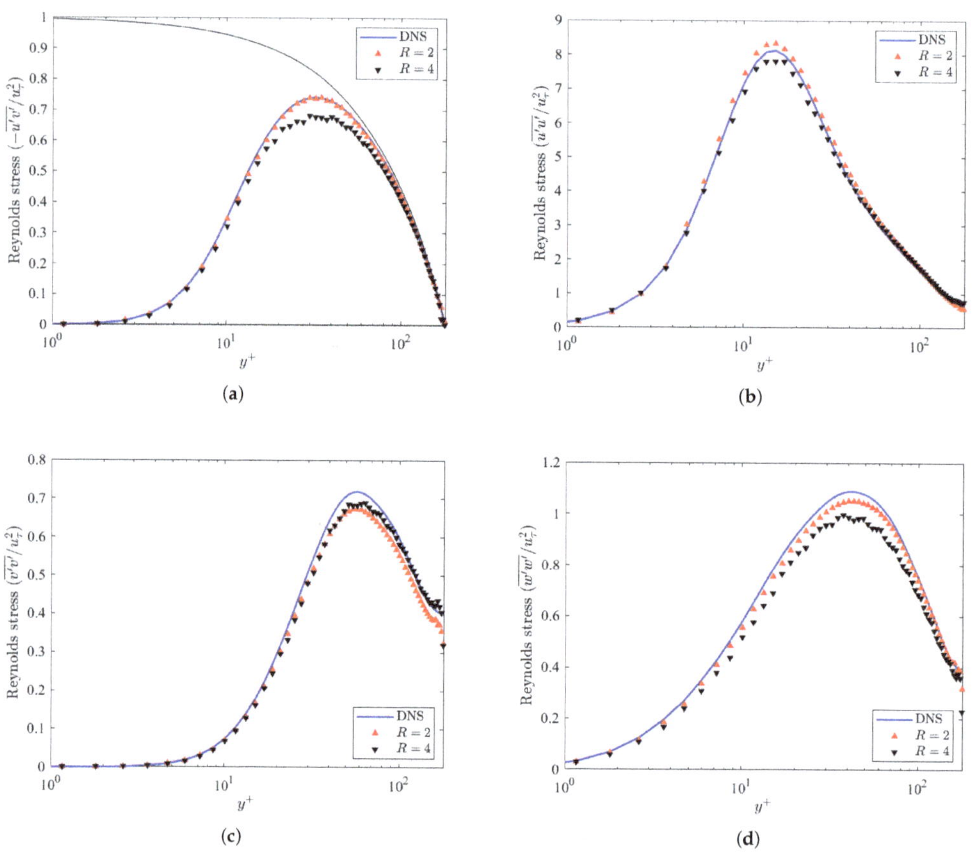

Figure 6. Reynolds shear stress fluctuations ($\overline{u'v'}$) (**a**) and Reynolds normal stress (**b–d**) as a function of the normalized wall-normal position y^+ for the DNS and ML models at $Re_\tau = 180$.

3.2. Turbulent Structures

This section discusses the turbulent structures from the data generated by the ML algorithm compared to DNS. Figure 7a–c display the vorticity fluctuations in the channel for the DNS and predicted GAN datasets. The vorticity fluctuations predicted by the GAN in Figure 7a,c are in good agreement with the DNS throughout the channel. However, the wall-normal vorticity fluctuations ω'_y are in good agreement until the peak at $y^+ \approx 30$. This is due to the fact that near the wall, ω'_y is dominated by the spanwise gradient of the streamwise velocity $\partial u/\partial z$. This can be confirmed visually via Figure 5a–c. The deviations in ω'_y at $y^+ > 30$ then demonstrate high sensitivity to the boundary conditions at the channel half-height $y^+ = Re_\tau$.

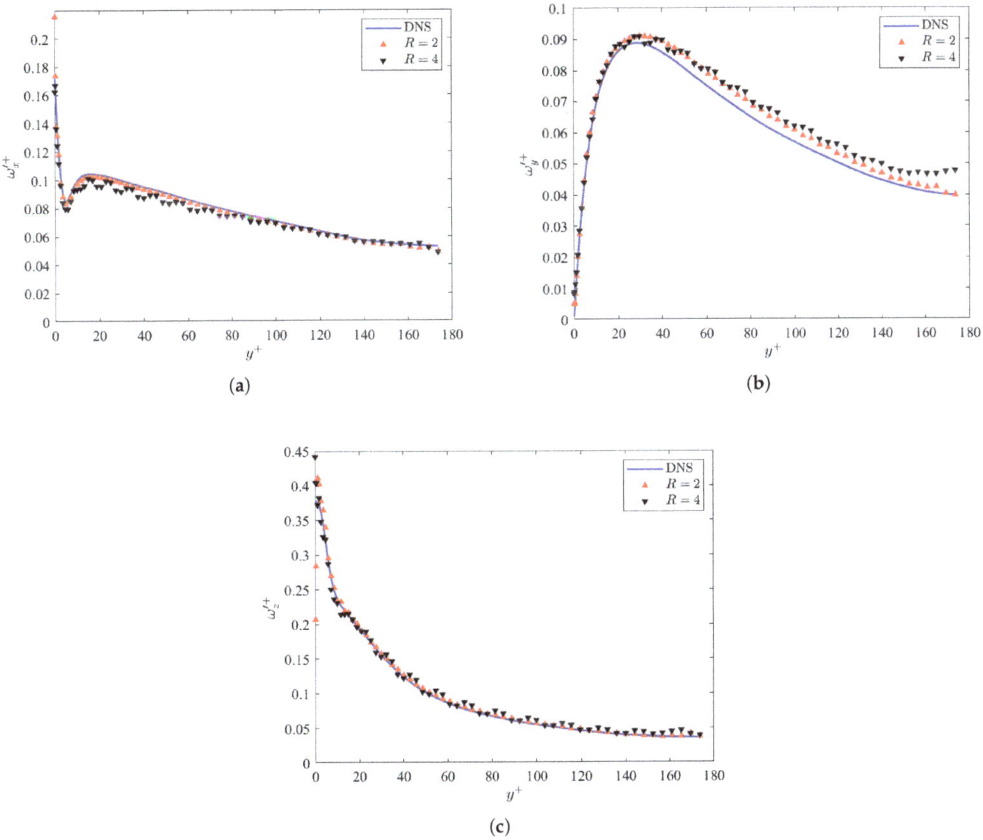

Figure 7. Vorticity fluctuations (ω') in the (**a**) streamwise, (**b**) wall-normal, and (**c**) spanwise directions as a function of the normalized wall-normal position y^+ for the DNS and ML models at $Re_\tau = 180$.

To determine the turbulent structures in the channel, a vortex identifier was needed [45,46]. Numerous methods have been proposed before, such as the Q-criterion [47], the Δ criterion [48], as well as the Omega and Rortex [49,50]. However, one of the most commonly accepted vortex identifiers is the λ_2 criterion, which was chosen for this analysis. The λ_2 vortex identification was defined in Jeong and Hussain [6] as the second eigenvalue of the tensor $S^2 + A^2$, which is a function of the symmetric $S = (\nabla \boldsymbol{u} + \nabla \boldsymbol{u}^T)/2$ and asymmetric $A = (\nabla \boldsymbol{u} - \nabla \boldsymbol{u}^T)/2$ components of the velocity gradient tensor $\nabla \boldsymbol{u}$, where T denotes the transpose of the tensor. A turbulent structure is, therefore, present when $\lambda_2 < 0$ [6,7]. It is demonstrated that the peak of the rms λ_2 plots occur at $y^+ \approx 23$, which confirms that the turbulent structures of interest are concentrated in the buffer region. The mean λ_2 is primarily negative, which indicates that, on average, there are vortical structures at each layer of the channel.

The streamwise energy spectra indicate the distribution of energy in the streamwise direction at different scales. For each of the resolution ratios considered in this analysis, the large- and small-scale structures would not be able to be effectively captured in the streamwise direction. This is because the peak of the turbulent kinetic energy in the streamwise direction is approximately 1000 wall units for a turbulent channel flow [10], which is greater than the domain size for the configuration in this study. The large-scale structures in the buffer and log-law layer region of the channel would, therefore, not be captured as effectively, so only the spanwise energy spectra were considered in this analysis.

Like the streamwise spectra, the spanwise energy spectra demonstrate the distribution of energy in the spanwise direction at different scales. The spanwise energy spectra for the streamwise, wall-normal, and spanwise velocities are displayed in Figure 8a–c. The large- and small-scale structures are effectively captured in the spanwise direction for the streamwise and wall-normal velocities, as shown in Figure 8a,b. Unlike the streamwise spectra, the spanwise energy spectra for the streamwise velocity in Figure 8a occurs at $y^+ \approx 15$, which corresponds to the peak of the turbulent energy production in a turbulent channel [1,4]. However, the large-scale structures in the buffer and log-law layer region of the channel are not as effectively captured in Figure 8 because of possible boundary artifacts affecting the energy distributions at the channel center.

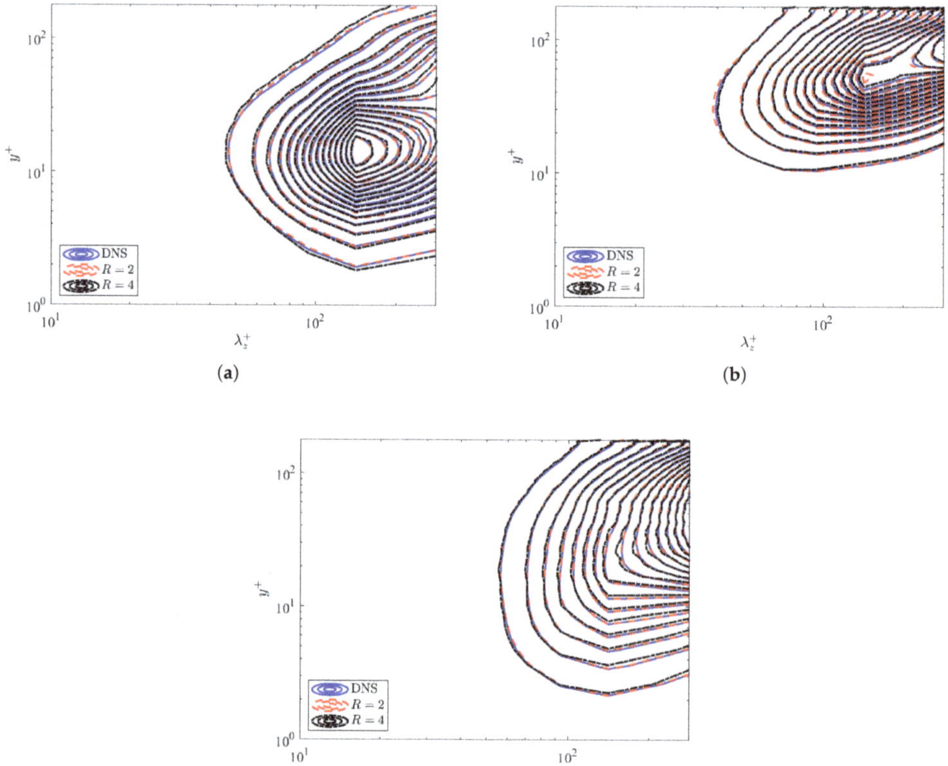

Figure 8. Premultiplied spanwise energy spectra as a function of wall-normal position and wavenumber for both the DNS and predicted data from the GAN: (**a**) $k_z E_{z,uu}$, (**b**) $k_z E_{z,vv}$, and (**c**) $k_z E_{z,ww}$.

The instantaneous λ_2 structures are displayed in Figure 9. The structures constructed by the GAN at $R = 2$ and $R = 4$ are shown in Figure 9a and 9b, respectively, while the DNS data are shown in Figure 9c. For reference, the turbulent structures are colored with respect to the streamwise vorticity ω_x. For each of the resolution ratios considered, the GAN is able to accurately reconstruct the large- and small-scale turbulent structures in the channel. Based on the high-resolution dataset size, it can be determined that the resolution ratio $R = 4$ is the lowest input resolution dataset that can be utilized in the GAN model in order to generate both the large and small scales of the turbulence in the channel. For resolution ratios $R > 4$, such as $R = 8$, too much information loss has occurred between the input and output datasets, so that accurate recovery for this channel flow configuration is not accomplished.

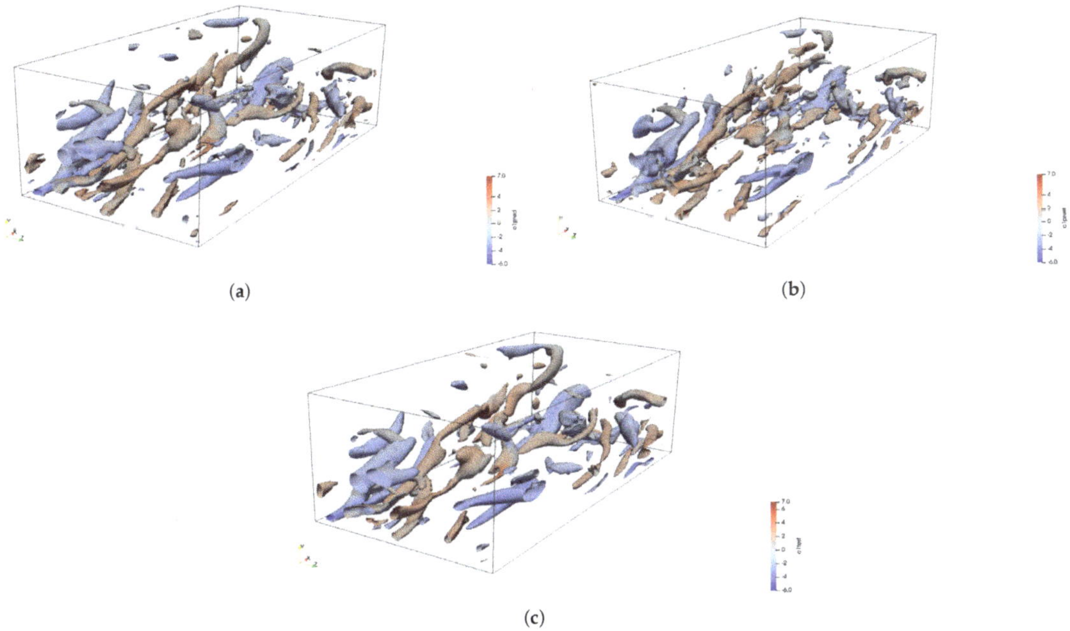

Figure 9. Instantaneous turbulent structures in the bottom half of the channel for $\lambda_2 = -1.0$: (**a**) GAN ($R = 2$), (**b**) GAN ($R = 4$), and (**c**) DNS.

4. Conclusions

This work sought to effectively reconstruct high-resolution 3D turbulent flow fields using unsupervised physics-informed deep learning. In particular, by combining physics-informed loss functions with other data-driven techniques, the GAN was able to reconstruct a turbulent channel flow dataset without the need for introducing any additional supervised learning, such as interpolation, to the input dataset. The first objective of this study was to reconstruct turbulent channel flow fields and verify these with respect to the statistics. It is demonstrated that the velocity profiles are able to be captured by the GAN within 5% of the DNS data instantaneously at the channel center. The GAN model is also able to capture each component of the velocity mean and fluctuations within 2% of the DNS data at the same Reynolds number. Additionally, for both resolution ratio cases, the predicted Reynolds stresses were predicted within 5–10% of the DNS data. This meant that the GAN was able to capture the general behavior of the turbulent channel flow, as well as recover most of the total turbulent fluctuations. The second objective was to compare the turbulent flow structures generated from the GAN with the DNS. In order to compare the turbulent flow structures, the general information about the vorticity, energy spectrum, and the

λ_2 vortex identifier needed to be determined. The GAN was able to accurately predict the vorticity fluctuations in the channel which indicated how effective the GAN was at determining the asymmetric components of the velocity gradient tensor. In addition, both the symmetric and asymmetric components of the velocity gradient were tested using the λ_2 vortex identifier, and the GAN was able to predict those well at both resolution ratios. Then, to determine whether or not the large- and small-scale structures were captured, the energy spectra and the instantaneous λ_2 structures were reconstructed. The GAN was able to effectively reconstruct both the large- and small-scale turbulent structures at each wall-normal position.

A GAN model trained on flow field data can potentially be used for other flow field calculations, such as HIT, turbulent boundary layers, and pipe flows. The primary purpose of GANs is to generate new data that is similar to the training data, so if the GAN has learned to accurately generate flow field data, then it could be used to generate new flow field data for other applications. However, it is important to note that the accuracy and generalizability of the GAN model will depend on the quality and diversity of the training data. If the training data does not represent a wide range of flow field scenarios, then the GAN may not be able to accurately generate new flow field data for other applications. Furthermore, it is important to carefully evaluate the performance of the GAN-generated flow field data for other applications, as it may not be as accurate or reliable as actual experimental or simulation data. Therefore, while a GAN model trained on flow field data may have potential uses in other applications, it should be used with caution and its performance should be carefully evaluated.

Funding: This research received no external funding.

Data Availability Statement: The data presented, as well as the code developed in this study are available on request from the corresponding author.

Acknowledgments: The author appreciates Prof. Stephen Ekwaro-Osire and Prof. Fazle Hussain for helpful discussion related to this study. The author thanks Prof. Ekwaro-Osire for reading the final version of the manuscript. Computational resources provided by the Texas Tech University High-Performance Computing Center are acknowledged.

Conflicts of Interest: The author declares no conflict of interest.

References

1. Kim, J.; Moin, P.; Moser, R. Turbulence statistics in fully developed channel flow at low Reynolds number. *J. Fluid Mech.* **1987**, *177*, 133–166. [CrossRef]
2. Jiménez, J.; Moin, P. The minimal flow unit in near-wall turbulence. *J. Fluid Mech.* **1991**, *225*, 213–240. [CrossRef]
3. Moser, R.D.; Kim, J.; Mansour, N.N. Direct numerical simulation of turbulent channel flow up to $Re_\tau = 590$. *Phys. Fluids* **1999**, *11*, 943–945. [CrossRef]
4. Kim, K.C.; Adrian, R.J. Very large-scale motion in the outer layer. *Phys. Fluids* **1999**, *11*, 417–422. [CrossRef]
5. Lee, M.; Moser, R.D. Direct numerical simulation of turbulent channel flow up to $Re_\tau \approx 5200$. *J. Fluid Mech.* **2015**, *774*, 395–415. [CrossRef]
6. Jeong, J.; Hussain, F. On the identification of a vortex. *J. Fluid Mech.* **1995**, *285*, 69–94. [CrossRef]
7. Jeong, J.; Hussain, F.; Schoppa, W.; Kim, J. Coherent structures near the wall in a turbulent channel flow. *J. Fluid Mech.* **1997**, *332*, 185–214. [CrossRef]
8. Schoppa, W.; Hussain, F. Coherent structure dynamics in near-wall turbulence. *Fluid Dyn. Res.* **2000**, *26*, 119–139. [CrossRef]
9. Schoppa, W.; Hussain, F. Coherent structure generation in near-wall turbulence. *J. Fluid Mech.* **2002**, *453*, 57–108. [CrossRef]
10. Hutchins, N.; Marusic, I. Large-scale influences in near-wall turbulence. *Philos. Trans. R. Soc. A Math. Phys. Eng. Sci.* **2007**, *365*, 647–664. [CrossRef]
11. Hutchins, N.; Marusic, I. Evidence of very long meandering features in the logarithmic region of turbulent boundary layers. *J. Fluid Mech.* **2007**, *579*, 1–28. [CrossRef]
12. Jiménez, J. Near-wall turbulence. *Phys. Fluids* **2013**, *25*, 101302. [CrossRef]
13. Jiménez, J. Coherent structures in wall-bounded turbulence. *J. Fluid Mech.* **2018**, *842*, P1.
14. Brunton, S.L.; Noack, B.R.; Koumoutsakos, P. Machine Learning for Fluid Mechanics. *Annu. Rev. Fluid Mech.* **2020**, *52*, 477–508. [CrossRef]

15. Tabor, G.R.; Baba-Ahmadi, M.H. Inlet conditions for large eddy simulation: A review. *Comput. Fluids* **2009**, *39*, 553–567. [CrossRef]
16. Wu, X. Inflow Turbulence Generation Methods. *Annu. Rev. Fluid Mech.* **2017**, *49*, 23–49. [CrossRef]
17. Fukami, K.; Fukagata, K.; Taira, K. Super-resolution reconstruction of turbulent flows with machine learning. *J. Fluid Mech.* **2019**, *870*, 106–120.
18. Maulik, R.; Sharma, H.; Patel, S.; Lusch, B.; Jennings, E. Accelerating RANS turbulence modeling using potential flow and machine learning. *arXiv* **2019**, arXiv:1910.10878.
19. Kim, J.; Lee, C. Prediction of turbulent heat transfer using convolutional neural networks. *J. Fluid Mech.* **2020**, *882*, 1–37. [CrossRef]
20. Liu, B.; Tang, J.; Huang, H.; Lu, X.Y. Deep learning methods for super-resolution reconstruction of turbulent flows. *Phys. Fluids* **2020**, *32*, 025105. [CrossRef]
21. Fukami, K.; Fukagata, K.; Taira, K. Assessment of supervised machine learning methods for fluid flows. *Theor. Comput. Fluid Dyn.* **2020**, *34*, 497–519. [CrossRef]
22. Kim, H.; Kim, J.; Won, S.; Lee, C. Unsupervised deep learning for super-resolution reconstruction of turbulence. *J. Fluid Mech.* **2021**, *910*, A29. [CrossRef]
23. Raissi, M.; Karniadakis, G.E. Hidden physics models: Machine learning of nonlinear partial differential equations. *J. Comput. Phys.* **2018**, *357*, 125–141. [CrossRef]
24. Raissi, M.; Perdikaris, P.; Karniadakis, G.E. Physics-informed neural networks: A deep learning framework for solving forward and inverse problems involving nonlinear partial differential equations. *J. Comput. Phys.* **2019**, *378*, 686–707. [CrossRef]
25. Yousif, M.Z.; Yu, L.; Lim, H.; Yousif, M.Z.; Yu, L.; Lim, H.C. Physics-guided deep learning for generating turbulent inflow conditions. *J. Fluid Mech.* **2022**, *936*, A21. [CrossRef]
26. Sachin Venkatesh, T.S.; Srivastava, R.; Bhatt, P.; Singh, R.K. A Comparative Study of Various Deep Learning Techniques For Spatio-Temporal Super-Resolution Reconstruction of Forced Isotropic Turbulent Flows. In Proceedings of the ASME 2021 International Mechanical Engineering Congress and Exposition, Online, 1–5 November 2021.
27. Bode, M.; Gauding, M.; Lian, Z.; Denker, D.; Davidovic, M.; Kleinheinz, K.; Jitsev, J.; Pitsch, H. Using physics-informed enhanced super-resolution generative adversarial networks for subfilter modeling in turbulent reactive flows. *Proc. Combust. Inst.* **2021**, *38*, 2617–2625. [CrossRef]
28. Yousif, M.Z.; Yu, L.; Lim, H.C. Super-resolution reconstruction of turbulent flow at various Reynolds numbers based on generative adversarial networks. *arXiv* **2021**, arXiv:2110.05047v1.
29. Yousif, M.Z.; Yu, L.; Lim, H.C. High-fidelity reconstruction of turbulent flow from spatially limited data using enhanced super-resolution generative adversarial network. *arXiv* **2021**, arXiv:2109.04250v2.
30. Eivazi, H.; Vinuesa, R. Physics-informed deep-learning applications to experimental fluid mechanics. *arXiv* **2022**, arXiv:2203.15402v1.
31. Du, Y.; Wang, M.; Zaki, T.A. State estimation in minimal turbulent channel flow: A comparative study of 4DVar and PINN. *arXiv* **2022**, arXiv:2210.09424.
32. Linqi, Y.; Yousif, M.Z.; Zhang, M.; Hoyas, S.; Vinuesa, R.; Lim, H.C. Three-dimensional ESRGAN for super-resolution reconstruction of turbulent flows with tricubic interpolation-based transfer learning. *Phys. Fluids* **2022**, *34*, 125126. [CrossRef]
33. Radford, A.; Metz, L.; Chintala, S. Unsupervised Representation Learning with Deep Convolutional Generative Adversarial Networks. *arXiv* **2016**, arXiv:1511.06434v2.
34. Zhu, J.Y.; Park, T.; Isola, P.; Efros, A.A. Unpaired Image-to-Image Translation using Cycle-Consistent Adversarial Networks. In Proceedings of the IEEE International Conference on Computer Vision, Venice, Italy, 22–29 October 2017; Institute of Electrical and Electronics Engineers Inc.: Piscataway, NJ, USA, 2017; pp. 2242–2251.
35. Gulrajani, I.; Ahmed, F.; Arjovsky, M.; Dumoulin, V.; Courville, A. Improved Training of Wasserstein GANs. *arXiv* **2017**, arXiv:1704.00028v3.
36. Zheng, T.; Oda, H.; Hayashi, Y.; Moriya, T.; Nakamura, S.; Mori, M.; Takabatake, H.; Natori, H.; Oda, M.; Mori, K. SR-CycleGAN: Super-resolution of clinical CT to micro-CT level with multi-modality super-resolution loss. *J. Med. Imaging* **2022**, *9*, 024003. [CrossRef] [PubMed]
37. Wu, J.L.; Kashinath, K.; Albert, A.; Chirila, D.; Xiao, H. Enforcing Statistical Constraints in Generative Adversarial Networks for Modeling Chaotic Dynamical Systems. *J. Comput. Phys.* **2020**, *406*, 109209. [CrossRef]
38. Wang, Z.; Bovik, A.C.; Sheikh, H.R.; Simoncelli, E.P. Image quality assessment: From error visibility to structural similarity. *IEEE Trans. Image Process.* **2004**, *13*, 600–612. [CrossRef] [PubMed]
39. Tieleman, T.; Hinton, G. RMSProp: Divide the Gradient by a Running Average of Its Recent Magnitude. *COURSERA Neural Netw. Mach. Learn.* **2012**, *4*, 26–31.
40. Kingma, D.P.; Lei Ba, J. Adam: A method for stochastic optimization. In Proceedings of the International Conference on Learning Representations, San Diego, CA, USA, 7–9 May 2015; pp. 1–15.
41. Reddi, S.J.; Kale, S.; Kumar, S. On the convergence of Adam and Beyond. In Proceedings of the International Conference on Learning Representations, Vancouver, BC, Canada, 30 April–3 May 2018; pp. 1–23.
42. Liu, L.; Jiang, H.; He, P.; Chen, W.; Liu, X.; Gao, J.; Han, J. On the Variance of the Adaptive Learning Rate and Beyond. In Proceedings of the Eighth International Conference on Learning Representations (ICLR), Addis Ababa, Ethiopia, 26–30 April 2020.

43. Miyato, T.; Kataoka, T.; Koyama, M.; Yoshida, Y. Spectral Normalization for Generative Adversarial Networks. In Proceedings of the International Conference on Learning Representations, Vancouver, BC, Canada, 30 April–3 May 2018; pp. 1–26.
44. Karras, T.; Aila, T.; Laine, S.; Lehtinen, J. Progressive Growing of GANs for Improved Quality, Stability, and Variation. In Proceedings of the 6th International Conference on Learning Representations, ICLR 2018—Conference Track Proceedings. International Conference on Learning Representations, ICLR, Toulon, France, 24–26 April 2017.
45. Chakraborty, P.; Balachandar, S.; Adrian, R.J. On the relationships between local vortex identification schemes. *J. Fluid Mech.* **2005**, *535*, 189–214. [CrossRef]
46. Liu, C.; Gao, Y.S.; Dong, X.R.; Wang, Y.Q.; Liu, J.M.; Zhang, Y.N.; Cai, X.S.; Gui, N. Third generation of vortex identification methods: Omega and Liutex/Rortex based systems. *J. Hydrodyn.* **2019**, *31*, 205–223. [CrossRef]
47. Hunt, J.C.R.; Wray, A.A.; Moin, P. Eddies, streams, and convergence zones in turbulent flows. In Proceedings of the Studying Turbulence Using Numerical Simulation Databases, 2. Proceedings of the 1988 Summer Program, Stanford, CA, USA, 1 December 1988; pp. 1–16.
48. Chong, M.S.; Perry, A.E.; Cantwell, B.J. A general classification of three-dimensional flow fields. *Phys. Fluids A Fluid Dyn.* **1990**, *2*, 777. [CrossRef]
49. Dong, X.; Gao, Y.; Liu, C. New normalized Rortex/vortex identification method. *Phys. Fluids* **2019**, *31*, 011701. [CrossRef]
50. Liu, J.; Liu, C. Modified normalized Rortex/vortex identification method. *Phys. Fluids* **2019**, *31*, 061704. [CrossRef]

fluids

MDPI

Article

Data Assimilation and Parameter Identification for Water Waves Using the Nonlinear Schrödinger Equation and Physics-Informed Neural Networks

Svenja Ehlers [1,*], Niklas A. Wagner [2], Annamaria Scherzl [3], Marco Klein [4], Norbert Hoffmann [1,5] and Merten Stender [6]

1 Dynamics Group, Hamburg University of Technology, 21073 Hamburg, Germany; norbert.hoffmann@tuhh.de
2 Communication Networks Institute, TU Dortmund University, 44227 Dortmund, Germany; niklas.wagner@tu-dortmund.de
3 Department of Mechanical Engineering, Technical University of Munich, 85748 Garching bei München, Germany
4 Ship Performance Department, Institute of Maritime Energy Systems, German Aerospace Center, 21052 Geesthacht, Germany; marco.klein@dlr.de
5 Department of Mechanical Engineering, Imperial College London, London SW7 2AZ, UK
6 Cyber-Physical Systems in Mechanical Engineering, Technische Universität Berlin, 10623 Berlin, Germany; merten.stender@tu-berlin.de
* Correspondence: svenja.ehlers@tuhh.de

Abstract: The measurement of deep water gravity wave elevations using in situ devices, such as wave gauges, typically yields spatially sparse data due to the deployment of a limited number of costly devices. This sparsity complicates the reconstruction of the spatio-temporal extent of surface elevation and presents an ill-posed data assimilation problem, which is challenging to solve with conventional numerical techniques. To address this issue, we propose the application of a physics-informed neural network (PINN) to reconstruct physically consistent wave fields between two elevation time series measured at distinct locations within a numerical wave tank. Our method ensures this physical consistency by integrating residuals of the hydrodynamic nonlinear Schrödinger equation (NLSE) into the PINN's loss function. We first showcase a data assimilation task by employing constant NLSE coefficients predetermined from spectral wave properties. However, due to the relatively short duration of these measurements and their possible deviation from the narrow-band assumptions inherent in the NLSE, using constant coefficients occasionally leads to poor reconstructions. To enhance this reconstruction quality, we introduce the base variables of frequency and wavenumber, from which the NLSE coefficients are determined, as additional neural network parameters that are fine tuned during PINN training. Overall, the results demonstrate the potential for real-world applications of the PINN method and represent a step toward improving the initialization of deterministic wave prediction methods.

Keywords: physics-informed neural network; hydrodynamic nonlinear Schrödinger equation; data assimilation; parameter identification; inverse problem; wave surface reconstruction

1. Introduction

The field of ocean engineering and experimental water wave research highly desires a deterministic description of wave quantities [1], which are usually described by partial differential equations (PDEs). Unlike statistical methods, deterministic prediction involves a phase-resolved tracing of wave fields, i.e., the wave elevation η as a function of space and time with high resolution. Unfortunately, acquiring such spatio-temporal data from experiments is often impractical: On the one hand, the reconstruction of wave surface elevations from real-world radar data is still an unresolved issue [2]. On the other hand, in situ measurement devices, such as wave gauges, measure the elevation time series at only a few selected locations due to the high operational costs. This sparsity of information leads to ill-posed inverse problems [3] when attempting to reconstruct the complete wave elevation $\eta(x,t)$ from gauge measurements $\eta_{\mathrm{m}}(t)$, which might be solved by numerical or machine learning methods.

Despite substantial advancements in common numerical PDE solvers, such as finite element, finite difference, and spectral methods, their applications to highly dynamic systems still incurs remarkable computational costs. These costs primarily stem from the need for fine-grained discretizations to ensure accurate solution approximations. Furthermore, conventional grid-based numerical solvers still face unresolved challenges, especially for addressing ill-posed inverse problems [4], such as wave reconstructions. In particular, inverse problems become computationally expensive due to the high number of forward evaluations required to estimate the inverse. Additionally, issues related to numerical stability and convergence necessitate the implementation of regularization techniques [5].

In the meantime, machine learning methods have revolutionized many scientific disciplines, including their applications in the fields of fast deterministic ocean wave prediction cf. [6–9] and reconstruction cf. [2,10]. However, in contrast to classical numerical solvers, these data-based approaches lack the incorporation of inherent knowledge about the physical laws that approximate the underlying system in the form of PDEs. Physical information is solely provided by observational training data for supervised learning, which are sourced from PDE simulations using classical numerical methods or from field measurements of real physical wave systems. Thus, the solutions generated by data-based approaches may not inherently ensure physical consistency, as the quality and quantity of training data limit their accuracy. This can be regarded as a neglect of established knowledge, especially when addressing the reconstruction problem outlined earlier, considering the centuries-long development of various model equations for describing surface gravity waves.

To overcome the limitations of both explicit numerical PDE solvers and neural network approaches, Raissi et al. [11] proposed physics-informed neural networks (PINNs). PINNs integrate observational data with physical laws by parameterizing the PDE solution as a neural network to solve forward, inverse parameter identification or inverse data assimilation problems. More precisely, the network's training process is constrained by a loss function that incorporates a PDE residual. To effectively calculate this residual, the PINN algorithm leverages the method of automatic differentiation [12]. In recent years, PINNs have gained attention across diverse scientific domains [4,13], including computational fluid dynamics [14–16], acoustic wave propagation [17], heat transfer [18,19], climate modeling [20], nano-optics [21], and the study of hyperelastic materials [22]. Notably, PINNs also employed in a few investigations related to water waves: For instance, Wang et al. [23] used a loss function based on the wave energy balance equation and the linear dispersion relation to reconstruct near-shore phase-averaged wave heights. The same equations also allow for solving sea bed depth inversion problems from statistical wave parameters in shallow-water regimes [24]. Furthermore, the Saint-Venant equations within the loss residuals of PINNs enable the downscaling of large-scale river models by assimilating remote sensing data in conjunction with in situ measurement data of the water surface [25]. Additionally, the research of Jagtap et al. [3] showcased the potential of PINNs in resolving ill-posed assimilation problems by leveraging analytical solitary surface measurements and the Serre–Green–Naghdi equations in shallow water.

However, in typical ocean engineering research, i.e., down-scaled testing in wave tanks, scenarios arise where the water depth significantly exceeds the wavelength. In such cases, it becomes essential to characterize the nonlinear behavior of water waves in the intermediate-to-deep water regime [26] rather than in shallow water. Zakharov [27] demonstrated that the amplitude envelope of slowly modulated wave groups approximately satisfies the nonlinear Schrödinger equation (NLSE). Consequently, various variants of the hydrodynamic nonlinear Schrödinger equation were investigated experimentally and numerically for deterministic wave prediction cf. [28–30] and rogue wave modeling cf. [31,32].

To the best of the authors' knowledge, there is no prior documentation of PINNs to solve the hydrodynamic form of the NLSE, despite the successful integration of other forms of the NLSE for different physical phenomena into the PINN framework cf. [11,33–38]. However, unifying most of this related work demonstrates the application of the NLSE-PINN methodology concerning the initial or boundary conditions derived from analytical soliton or breather solutions that correspond to their specific NLSE used in the loss function. While these approaches provide a theoretical proof of concept, they cannot inherently assess the practical viability of PINNs in real-world scenarios, as analytical solutions seldom align with the complexity and imperfections encountered in measurements.

Therefore, the objective of this study was to demonstrate the application of an NLSE-PINN framework in the more realistic scenario of irregular deep water gravity waves. Specifically, we aimed to leverage the capability of PINNs in solving data assimilation problems, which is crucial in ocean engineering to infer the dynamic system's state from partial or noisy surface elevation measurements $\eta_{\mathrm{m}}(t)$ cf. [17,25,39,40], thereby reducing the need for costly gauges while still being able to reconstruct the fully resolved wave fields $\eta(x,t)$ in between. For this purpose, we utilized irregular and nonlinear wave group data from gauge locations several meters apart within a numerical wave tank that was simulated using the high-order spectral method (HOSM) [41].

However, as our PINN loss function was constrained by the hydrodynamic NLSE associated with limitations in accommodating arbitrary irregular sea states due to the constraints in the bandwidth, steepness, and nonlinearity [1], we investigated the utilization of a slightly misspecified PINN for assimilating wave data derived from the more realistic HOSM; the HOSM also accounts for highly nonlinear, broad-banded, or directional sea states in non-breaking scenarios [42,43]. This misspecification acknowledges the common discrepancy between mathematical models (e.g., PDEs) and measurement data in complex, real-world physical systems [44], where perfect alignment between measurements and analytical solutions is uncommon, and exact knowledge of the physical equations that describe all phenomena encountered in measurements is rare. Moreover, an ongoing challenge in applying the NLSE to irregular waves with specific bandwidths lies in accurately determining its coefficients, which are derived from ratios of the base parameters of the carrier wave frequency ω_{p} and wavenumber k_{p}. In particular, relatively short measurement intervals and slight deviations in the measurement data from the narrow-band assumptions of the NLSE can introduce uncertainties in the determination of ω_{p} and k_{p}, which affect, for example, the propagation velocities, and thus, cause errors and offsets in the surface elevation reconstructions. Therefore, our hypothesis for this study was as follows:

- Alongside the assimilation of measurement data, the NLSE-PINN will facilitate fine tuning the base parameters (ω_{p} and k_{p}) of the NLSE coefficients. We expect this approach to enhance the reconstruction performance compared with using constant coefficients predetermined from spectral wave properties.

To address this, we first present the wave tank utilized for data generation in Section 2.1. Afterward, the hydrodynamic NLSE is introduced in Section 2.2 to subsequently develop the NLSE-PINN framework and training methodology in Section 2.3. Then, we first performed a pure data assimilation task with predetermined, constant NLSE coefficients to reconstruct wave surface envelopes from only two gauge measurements in Section 3.1. These results served as a benchmark to demonstrate the enhancement of reconstruction by incorporating the NLSE coefficient base parameters as additional tuneable PINN variables in Section 3.2. Finally, Section 4 summarizes the key findings, outlines the method limitations, and suggests potential directions for future research.

2. Method

The following subsection initiates a concise overview of the wave tank specifications and the numerical method to generate the synthetic wave elevation data. Next, the nonlinear Schrödinger equation for deep water gravity waves is introduced. Subsequently, this equation is integrated to develop a physics-informed neural network, with details on its architecture and training strategy provided in the last subsection.

2.1. Numerical Wave Tank and Measurement Data Generation

The numerical wave tank experiments were based on the dimensions of a real wave tank facility at Hamburg University of Technology. This wave tank possessed a cross-sectional area of 1.5 m × 1.5 m and extended over a length of 15 m (https://www.tuhh.de/mum/en/research/facilities (accessed on 11 September 2024)) and is visualized in Figure 1a. To generate waves, a flap-type board was installed on one side of the tank, while the opposite side had a beach element to minimize wave reflections. To simulate nonlinear wave propagation inside this numerical wave tank, we employed the high-order spectral method (HOSM) of order $M = 4$, as formulated by West et al. [41]. This method models highly realistic, irregular nonlinear water wave surfaces on a Cartesian coordinate system (x, z), with the mean free surface located at $z = 0$ m with z pointing upward. Assuming a Newtonian fluid that is incompressible, inviscid, and rotational, the HOSM solves the general initial boundary-value potential flow problem that is given by the Laplace equation:

$$\nabla^2\Phi = \Phi_{xx} + \Phi_{zz} = 0 \tag{1}$$

and the kinematic surface, dynamic surface, and bed boundary condition:

$$\eta_t + \eta_x\Phi_x - \Phi_z = 0 \quad \text{on } z = \eta(x,t) \tag{2}$$

$$\Phi_t + g\eta + \frac{1}{2}\left(\Phi_x^2 + \Phi_z^2\right) = 0 \quad \text{on } z = \eta(x,t) \tag{3}$$

$$\Phi_z = 0 \quad \text{on } z = -d. \tag{4}$$

Therein, $\Phi(x,z,t)$ is the velocity potential, $\eta(x,t)$ is the free surface elevation, and g is the gravity acceleration. The simulations were initialized using wave surfaces from the JONSWAP spectra [45] in the finite depth form [46].

For the data generation, we maintained a fixed water depth of $d = 1$ m and employed four virtual wave gauges to provide time-series measurements along the water surface at $x_g \in \{3,4,5,6\}$ m, which constituted a typical depth and measurement point spacing in educational wave tanks [47,48]. The outer gauges of the numerical wave tank were designated as locations of sparse measurements ($x_{g,\text{meas}} \in \{3,6\}$ m) for the PINN training, while measurements at the inner gauges ($x_{g,\text{test}} \in \{4,5\}$ m) were solely used to assess the PINNs' performance against a ground truth after the training process. We generated a temporal wave surface elevation series $\eta(x = x_g, t)$ over an interval that spanned t = 0–60 s to ensure that the wave packets fully passed all four gauges, while no significant wave reflections had yet occurred at the channel's end. A graphical representation of the resulting three-dimensional data structure is exemplified in Figure 1b.

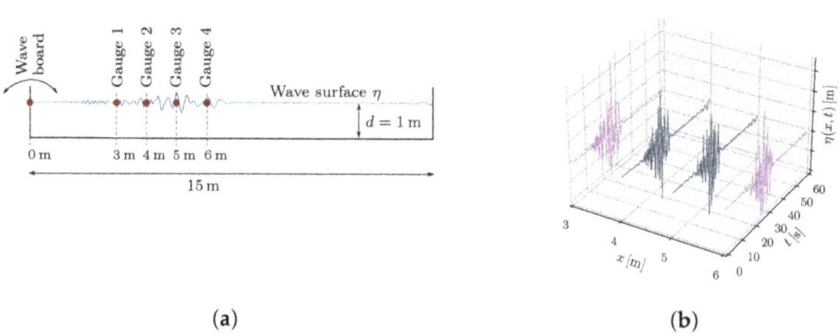

(a) (b)

Figure 1. In the numerical wave tank setup, wave gauges measured the elevations at four points $x_g = \{3,4,5,6\}$ m (**a**), which resulted in a sparse spatio-temporal data structure for each sample (**b**). During the PINN training, the violet elevation time series were utilized, while the grey series were reserved for the subsequent performance evaluation.

As we desired wave data that covered different wave conditions and significant wave heights H_s, the sea state parameters of the peak wave frequency $\omega_p = 2\pi/T_p$ (with T_p being the peak period), wave steepness $\epsilon = k_p H_s/2$ (with $k_p = 2\pi/L_p$ being the peak wavenumber and L_p being the peak wavelength), and peak enhancement factor γ were varied as follows:

$$\begin{aligned} \omega_p &\in \{3, 4, 5, 6, 7, 8, 9\} \ \tfrac{\text{rad}}{\text{s}} \\ \epsilon &\in \{0.0125, 0.0250, 0.0375, 0.0500, 0.0750, 0.1000\} \\ \gamma &\in \{1, 3, 6\}, \end{aligned}$$

where a higher γ indicates a narrow-banded spectra, as shown in Figure 2. By randomly selecting initial phase shifts of the component waves, five different elevations were generated for each ϵ–ω_p–γ combination, which resulted in 630 different wave samples in total, where each individual sample had the data structure shown in Figure 1b. For details on the nonlinear wave data generation in a numerical wave tank facility using the HOSM, the reader is referred to Klein et al. [7] and Lünser et al. [43].

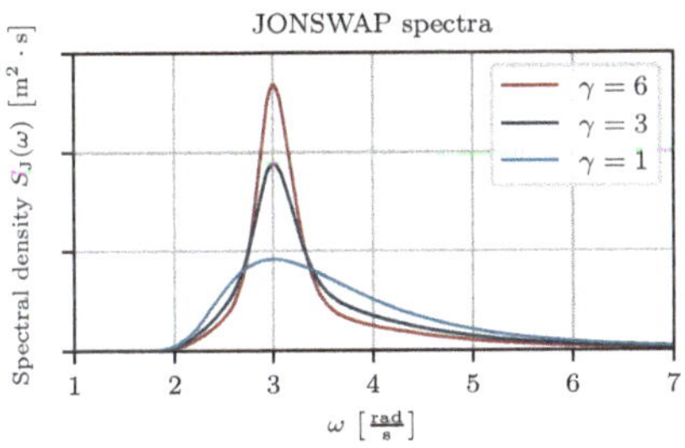

Figure 2. JONSWAP spectra for the different peak enhancement factors γ used to initialize the HOSM wave simulations. A higher value of γ resulted in a narrower spectra for the generated irregular waves.

2.2. Hydrodynamic Nonlinear Schrödinger Equation

In addition to employing numerical methods, such as HOSM, to approximate the potential flow Equations (1)–(4), an alternative approach is the derivation of simplified solutions. By utilizing perturbation theory around the parameters of wave steepness $\epsilon = k_p a$ and relative bandwidth $\mu = \Delta k/k_p << 1$, small-amplitude waves and a narrow bandwidth are assumed. Here, Δk denotes the width of the wavenumber spectrum around the peak wavenumber k_p. Moreover, the boundary value problem at the unknown free surface $z = \eta(x,t)$ can be approximated using a Taylor series expansion. By truncating the perturbation expansion at order $\mathcal{O}(\epsilon^3)$, the envelope equation of the nonlinear Schrödinger equation (NLSE) is derived [27,49,50] in terms of a complex wave envelope amplitude $A(x,t) = U(x,t) + iV(x,t)$ that varies slowly compared with the phase $\vartheta = k_p x - \omega_p t + \varphi$ of its underlying carrier wave $\eta(x,t)$, where φ is a phase shift. The hydrodynamic NLSE in the time-like form is as follows:

$$i\left(A_x + \frac{1}{c_g}A_t\right) + \delta A_{tt} + \nu|A|^2 A = 0 \quad (5)$$

which finds common application in boundary value wave tank problems [51], where

$$c_g = \frac{\omega_p}{2k_p}, \quad \delta = -\frac{k_p}{\omega_p^2}, \quad \nu = -k_p^3 \quad (6)$$

are the NLSE coefficients. The peak frequency ω_p of the carrier wave is related to the corresponding peak wavenumber k_p through the linear dispersion relation for deep water:

$$\omega_\text{p} = \sqrt{gk_\text{p}}. \tag{7}$$

The first term in Equation (5) characterizes the spatial variation of the amplitude, while the second term represents the wave propagation by the group velocity c_g. The third term introduces the dispersive effect, while the last term introduces the nonlinearity. In contrast to the potential flow equations, which require an approximation in the depth direction, the NLSE is regarded as the simplest nonlinear equation for deep-water wave dynamics, as it is uniquely formulated on the water surface [52]. However, while it captures the essential aspects of nonlinear water waves, the NLSE's limitations in nonlinearity magnitude and spectral bandwidth limit its general applicability in deterministic wave prediction [1].

In practical cases, where real-valued measurement series $\eta_\text{m}(x = x_\text{g}, t)$ are available, the values of ω_p and k_p are commonly derived from a spectral representation via a discrete Fourier transform $F(\omega)$ of the time domain surface elevation $\eta_\text{m}(t)$. We employed the advanced method of Sobey [53] and Mansard and Funke [54] by defining

$$\omega_\text{p} = \frac{\int \omega \cdot F(\omega)^5 d\omega}{\int F(\omega)^5 d\omega}, \tag{8}$$

which enhances the robustness compared with determining ω_p as the frequency where $F(\omega)$ attains its maximum. The corresponding k_p can be derived by the dispersion relation. The complex amplitudes for initializing the NLSE are computed as

$$A(x_\text{g}, t) = \left[\eta(x_\text{g}, t) + i\mathcal{H}(\eta(x_\text{g}, t))\right] \cdot \exp(-i\vartheta), \tag{9}$$

where $\mathcal{H}$ denotes the Hilbert transform of the measured signal [55,56] and $\vartheta = k_\text{p}x - \omega_\text{p} + \varphi$ is the carrier wave's phase. For the other way around, the equation

$$\eta(x_\text{g}, t) = \text{Re}\left[A(x_\text{g}, t) \cdot \exp(i\vartheta)\right] \tag{10}$$

instead allows for the reverse transformation from a complex envelope back to a carrier wave elevation after the reconstruction procedure, although this transformation is not presented in this work. The relationship between the carrier wave $\eta(x_\text{g}, t)$ and complex envelope $A(x_\text{g}, t) = U(x_\text{g}, t) + iV(x_\text{g}, t)$ is illustrated in Figure 3.

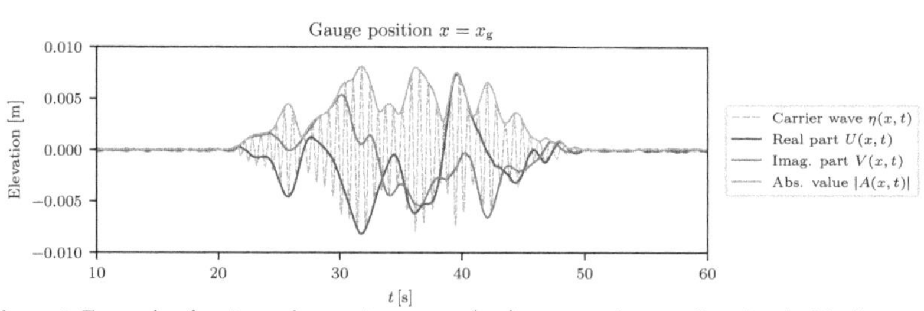

Figure 3. Example of an irregular carrier wave $\eta(x, t)$ measured at one location inside the wave tank. The real and imaginary parts of its corresponding complex envelope $A(x, t) = U(x, t) + iV(x, t)$ are visualized, along with its absolute value $|A(x, t)|$.

2.3. *Physics-Informed Neural Network for the NLSE*

In light of the simplifications inherent in the hydrodynamic NLSE concerning the generalized water wave problem discussed previously, we investigated the suitability of this equation within a PINN framework for analyzing surface waves that arose from sightly different wave physics by using a phenomenon that is termed model misspecification and commonly encountered in real-world physical systems. Our selection of this equation was motivated by the NLSE's simple definition (Equation (5)), which is limited to the (x, t) domain only. In contrast, for example, employing a PINN to solve the fully nonlinear potential flow equations (Equations (1)–(4)) would demand substantially greater human and computational effort. Hence, subsequent sections introduce the architecture, loss function, training methodology, and evaluation metrics adopted for our NLSE-PINN framework.

2.3.1. PINN Architecture

The neural network architecture of the PINN developed in our study is depicted on the lower-left side of Figure 4. It includes two input nodes for collocation points in space x and time t. Given the complex-valued solution $A(x,t) = U(x,t) + iV(x,t)$ of the NLSE, two output nodes are required [11], where U represents the real part and V represents the imaginary part. Moreover, the network consists of four intermediate hidden layers, with each comprising 200 nodes, which results in a total network depth of $D = 6$. The variables of each layer l, which are denoted as $\boldsymbol{\theta}^{(l)} = \{\mathbf{W}^{(l)}, \mathbf{b}^{(l)}\}_{1 \leq l \leq D}$, receive their initial values using the Xavier initialization technique for weights $\mathbf{W}^{(l)}$ [57], while the biases $\mathbf{b}^{(l)}$ are initialized to zero. The total amount of the network's weights and biases is denoted as $\boldsymbol{\theta} = \left[\boldsymbol{\theta}^{(1)}, \ldots, \boldsymbol{\theta}^{(D)}\right]$.

To enhance the convergence and accuracy during the training, we incorporated a strategy using layer-wise locally adaptive activation functions [58,59], which has shown promising applications, e.g., in Pu et al. [35], Jagtap et al. [3,60], Shukla et al. [61], and Guo et al. [62]. The input of a hidden layer $\mathbf{h}^{(l+1)}$ is derived from the output of the preceding layer $\mathbf{o}^{(l)}$ by following the rule

$$\mathbf{o}^{(l)} = \mathbf{W}^{(l)\mathrm{T}} \cdot \mathbf{h}^{(l)} + \mathbf{b}^{(l)} \tag{11}$$

$$\mathbf{h}^{(l+1)} = \tanh\left(s \cdot a^{(l)} \cdot \mathbf{o}^{(l)}\right) \tag{12}$$

for $l = 1, \ldots, D-1$. The last layer D has a linear activation function. In this context, $\mathbf{W}^{(l)}$ and $\mathbf{b}^{(l)}$ are the weight and bias matrices of the (l)-th layer, and $s = 10$ is a fixed scaling factor. Furthermore, $\boldsymbol{a} = \left[a^{(1)}, \ldots, a^{(D-1)}\right]$ represent additional variables of the network, which are fine tuned during the training process to modulate the slopes of the activations. In our case, we initialized the system with $a^{(l)} = 0.2$, which resulted in the initial slope of $s \cdot a^{(l)} = 2$ being slightly steeper than the usual tanh activation. Empirical evidence suggested that this initialization could be considered valid according to [58], as it accelerated the convergence rate in our setup without causing divergence or increasing oscillations in the loss function.

Moreover, over the course of our study (Section 3.2), we examined how the data-driven fine tuning of the NLSE coefficients (Equation (6)) while assimilating sparse measurement data affected the reconstruction quality. Given that all the coefficients in the NLSE were related to the peak wavenumber and frequency, we selected treating the coefficients' base parameters ω_p and k_p as additional PINN variables while sharing an optimizer with the neural network parameters $\boldsymbol{\theta}$ and $\boldsymbol{a}$. This is visualized in the lower-left box of Figure 4.

2.3.2. PINN Loss Function and Training

To enable the PINN's capability to approximate an NLSE surrogate model $\tilde{A}(x,t) = \tilde{U}(x,t) + i\tilde{V}$, we used wave elevation measurement data $\eta_\mathrm{m} \in \mathbb{R}^{N_\mathrm{d}}$ from the two outer gauge positions, $x_{\mathrm{g,meas}} \in \{3, 6\}$ m, within the numerical wave tank shown in Figure 1. The elevation data at the remaining locations, $x_{\mathrm{g,test}} \in \{4, 5\}$ m, were reserved for later evaluation and not incorporated in the PINN's training process. Each elevation measurement was transferred to a complex envelope $A_\mathrm{m} \in \mathbb{C}^{N_\mathrm{d}}$ using Equation (9), as depicted in the top boxes of Figure 4. In this work, $N_\mathrm{d} = 1200$ was the total number of measurement data points at the domain boundaries, which is denoted as $\{x_\mathrm{d} = x_{\mathrm{g,meas}}, t_\mathrm{d}^{(j)}\}_{j=1}^{N_\mathrm{d}}$. These points were obtained by sampling the temporal sequence t = 0–60 s with an increment of $\Delta t = 0.05$ s. In addition, we incorporated a set of $N_\mathrm{r} = 20,000$ randomly located collocation points $\{x_\mathrm{r}^{(j)}, t_\mathrm{r}^{(j)}\}_{j=1}^{N_\mathrm{r}}$ to enforce the NLSE solution across the entire computational (x,t)

domain. Using these measurements and sets of points, the multi-objective PINN loss function was defined by

$$\mathcal{L} = \mathcal{L}_{U,\text{data}} + \mathcal{L}_{V,\text{data}} + \mathcal{L}_{U,\text{res}} + \mathcal{L}_{V,\text{res}}, \tag{13}$$

where the loss components were composed of the mean-squared errors (MSEs):

$$\mathcal{L}_{U,\text{data}} = \underbrace{\frac{1}{N_\text{d}} \sum_{j=0}^{N_\text{d}} \left| \tilde{U}(x_\text{d}, t_\text{d}^{(j)}) - U_\text{m}^{(j)} \right|^2}_{\text{MSE}_{U,\text{data}}} \cdot \left(\lambda_\text{d}^{(j)}\right)^2 \tag{14}$$

$$\mathcal{L}_{V,\text{data}} = \underbrace{\frac{1}{N_\text{d}} \sum_{j=0}^{N_\text{d}} \left| \tilde{V}(x_\text{d}, t_\text{d}^{(j)}) - V_\text{m}^{(j)} \right|^2}_{\text{MSE}_{V,\text{data}}} \cdot \left(\mu_\text{d}^{(j)}\right)^2 \tag{15}$$

$$\mathcal{L}_{U,\text{res}} = \underbrace{\frac{1}{N_\text{r}} \sum_{j=0}^{N_\text{r}} \left| \tilde{R}_U(x_\text{r}^{(j)}, t_\text{r}^{(j)}) \right|^2}_{\text{MSE}_{U,\text{res}}} \cdot \left(\lambda_\text{r}^{(j)}\right)^2 \tag{16}$$

$$\mathcal{L}_{V,\text{res}} = \underbrace{\frac{1}{N_\text{r}} \sum_{j=0}^{N_\text{r}} \left| \tilde{R}_V(x_\text{r}^{(j)}, t_\text{r}^{(j)}) \right|^2}_{\text{MSE}_{V,\text{res}}} \cdot \left(\mu_\text{r}^{(j)}\right)^2. \tag{17}$$

The terms $\text{MSE}_{U,\text{data}}$ and $\text{MSE}_{V,\text{data}}$ served to quantify the error between PINN predictions $(\tilde{U}, \tilde{V})$ and measured envelope data (U_m, V_m) at the domain boundaries $x_{\text{g,meas}} \in \{3, 6\}$ m. In contrast, $\text{MSE}_{U,\text{res}}$ and $\text{MSE}_{V,\text{res}}$ quantified the degree to which the PINN solution conformed to the residuals of the NLSE, which are given by

$$\tilde{R}_U(x,t) := -\tilde{V}_x(x,t) - \frac{1}{c_g}\tilde{V}_t(x,t) + \delta\tilde{U}_{tt}(x,t) + \nu\left(\tilde{U}(x,t)^2 + \tilde{V}(x,t)^2\right)\tilde{U}(x,t) \tag{18}$$

$$\tilde{R}_V(x,t) := \tilde{U}_x(x,t) + \frac{1}{c_g}\tilde{U}_t(x,t) + \delta\tilde{V}_{tt}(x,t) + \nu\left(\tilde{U}(x,t)^2 + \tilde{V}(x,t)^2\right)\tilde{V}(x,t) \tag{19}$$

within the remaining computational domain. Thus, as the optimizer minimizes these residuals during training, the PINN solution increasingly aligns with the physical constraints imposed by the NLSE. The derivative terms in these residuals are calculated using automatic differentiation (AD) [12] of the neural network's outputs $\tilde{U}$ and $\tilde{V}$ with respect to the input variables x or t, as illustrated on the right of Figure 4. Unlike numerical differentiation methods, AD provides exact derivatives without any approximation error [13]. Furthermore, AD enhances the computational efficiency by decomposing complex functions into elementary operations and reusing intermediate results through the application of the chain rule, thus avoiding redundant calculations.

Furthermore, following the approach proposed by McClenny and Braga-Neto [63], trainable self-adaptation weights $\lambda_\text{d}^{(j)}$, $\mu_\text{d}^{(j)}$, $\lambda_\text{r}^{(j)}$, and $\mu_\text{r}^{(j)}$ were introduced in the loss components (Equations (14)–(17)). These variables, which are initially set to one, are associated with each measurement point $\{x_\text{d}, t_\text{d}^{(j)}\}_{j=1}^{N_\text{d}}$ or collocation point $\{x_\text{r}^{(j)}, t_\text{r}^{(j)}\}_{j=1}^{N_\text{r}}$. The PINN autonomously identifies challenging regions inside the solution domain characterized by high point-specific errors and increases the respective self-adaptation weights to emphasize the penalty, and thus, improve the approximation. This behavior is achieved through the concurrent minimization of the loss function $\mathcal{L} = \mathcal{L}(\boldsymbol{\theta}, \boldsymbol{a}, \boldsymbol{\lambda}_\text{d}, \boldsymbol{\mu}_\text{d}, \boldsymbol{\lambda}_\text{r}, \boldsymbol{\mu}_\text{r})$ by updating the network weights and biases $\boldsymbol{\theta}$, activation slopes $\boldsymbol{a}$, and coefficient parameters $\overline{\omega}_\text{p}$ and $\overline{k}_\text{p}$ alongside the maximization of the loss for the self-adaption weights $\boldsymbol{\lambda}_\text{d}$, $\boldsymbol{\mu}_\text{d}$, $\boldsymbol{\lambda}_\text{r}$, and $\boldsymbol{\mu}_\text{r}$ in each epoch of training.

In many instances, PINNs are trained using a two-step strategy that involves the Adam optimizer [64] for a defined number of epochs, followed by the L-BFGS optimizer [65]. This strategy is recognized as pivotal [66], with Adam initially preventing convergence to local minima, while L-BFGS refines small-scale solution components [13]. However, this approach, which is beneficial for smooth analytical solutions, proved less suitable for our studies: The L-BFGS optimizer tends to over-refine noisy elements within measurements. We found that the AMSGrad modification of the Adam optimizer [67] from the PyTorch library (Version 1.8.1, sourced from Hamburg, Germany) [68] yielded satisfactory results with a learning rate of $\alpha = 0.0001$ over 15,000 epochs of training.

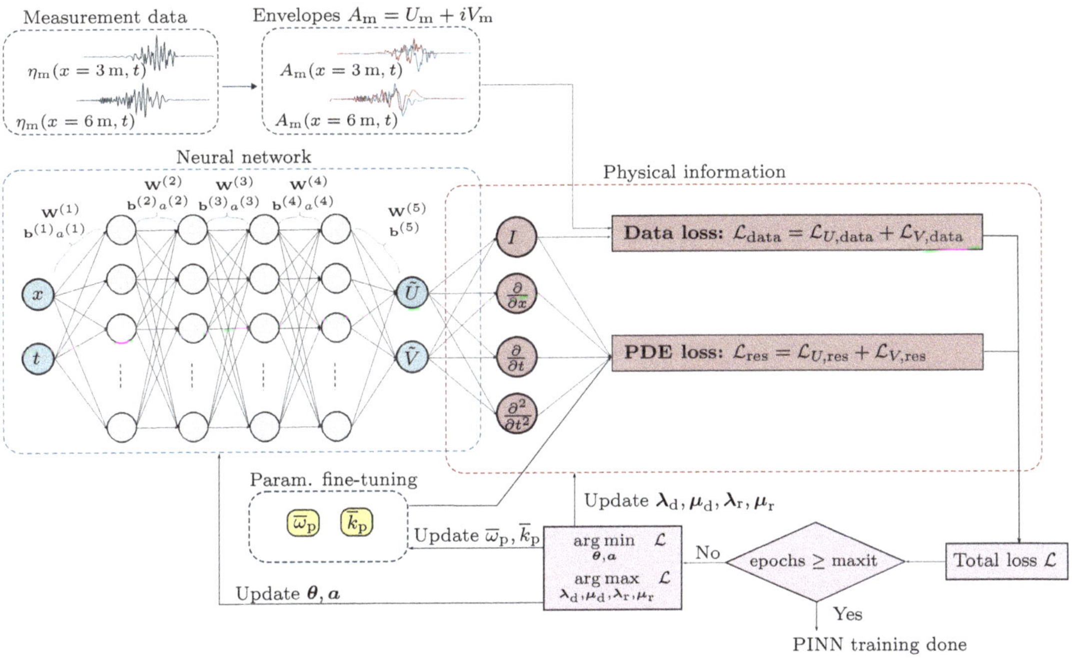

Figure 4. Schematic framework of the physics-informed neural network developed to solve the hydrodynamic nonlinear Schrödinger equation. The neural network architecture comprises two input nodes to insert points of the computational domain (x, t) and two output nodes to approximate the real and imaginary parts of the complex-valued NLSE solution $\tilde{A}(x,t) = \tilde{U}(x,t) + i\tilde{V}(x,t)$. Real wave measurement data η_m, which are obtained at the domain boundaries, are transformed into envelope representations A_m to guide the PINN's solution toward approximating these boundary values. This is achieved by the data loss component MSE_{data} (Equations (14) and (15)). To additionally guide the PINN solution toward ensuring physical consistency inside the entire computational domain, the PDE loss MSE_{res} incorporates NLSE residuals (Equations (16)–(19)). The PINN's variables $\boldsymbol{\theta}$ (weights $\mathbf{W}$ and biases $\mathbf{b}$); activation function slopes $\boldsymbol{a}$; and self-adaption weights λ_d, μ_d, λ_r, and μ_r (and, for the case of additional coefficient fine tuning during assimilation, the parameters $\overline{\omega}_p$ and $\overline{k}_p$) are updated iteratively in the training process to minimize the total loss $\mathcal{L}$, which is composed of data and PDE loss components.

2.3.3. Evaluation

After assimilating the measurements at the domain boundaries to reconstruct the envelopes in the entire computational domain, the evaluation of the PINNs' performance necessitated a metric that compared the measurement time series $A_m = A_m(x = x_g, t)$ with the reconstruction $\tilde{A} = \tilde{A}(x = x_g, t)$ at all gauge positions $x_g \in \{3, 4, 5, 6\}$ m. While metrics based on Euclidean distances, such as the mean-squared error (MSE), are scale dependent and treat deviations in frequency and phase as amplitude errors [69], the surface similarity parameter (SSP)

$$\mathrm{SSP}(A_{\mathrm{m}}, \tilde{A}) = \frac{\sqrt{\int |F_{A_{\mathrm{m}}}(\omega) - F_{\tilde{A}}(\omega)|^2 d\omega}}{\sqrt{\int |F_{A_{\mathrm{m}}}(\omega)|^2 d\omega} + \sqrt{\int |F_{\tilde{A}}(\omega)|^2 d\omega}} \in [0,1] \tag{20}$$

proposed by Perlin and Bustamante [70] combines phase , amplitude , and frequency errors into a scalar unified measure. In this metric, ω denotes the wave frequency vector and $F_{A_{\mathrm{m}}}$ and $F_{\tilde{A}}$ denote the discrete Fourier transforms of the time series $A_{\mathrm{m}}(x = x_{\mathrm{g}}, t)$ or $\tilde{A}(x = x_{\mathrm{g}}, t)$. The SSP is a normalized error metric, with $\mathrm{SSP} = 0$ indicating perfect agreement and $\mathrm{SSP} = 1$ implying a comparison against zero or phase-inverted surfaces. To visually illustrate the impact of the SSP metric, we systematically compared an exemplary wave envelope against various signals across a range of SSP and MSE values in the Appendix (Figure A1). Due to its straightforward error assessment and the applicability for comparing signals with differing sampling rates and lengths, the SSP has found application in recent research related to ocean wave prediction and reconstruction by Klein et al. [1,7], Wedler et al. [9,69], Desmars et al. [71,72], Lünser et al. [43], Kim et al. [73], and Ehlers et al. [2].

3. Results and Discussion

After a successful validation of the NLSE-PINN against an analytic solution to the NLSE (see Appendix A.2), we assessed its applicability to more realistic scenarios that involved data that do not entirely conform to the equation in the loss function, such as elevation measurements from the wave tank shown in Figure 1. The NLSE-PINN framework underwent training for all 630 generated wave measurement samples, where each was characterized by a distinct wave parameter combination and random phases. Our aim was to reconstruct wave envelopes throughout the spatio-temporal domain $3\,\mathrm{m} \leq x \leq 6\,\mathrm{m}$ and $0\,\mathrm{s} \leq t \leq 60\,\mathrm{s}$ solely based on two time series measured at the gauge locations $x_{\mathrm{g,meas}} \in \{3, 6\}$ m spaced 3 m apart. Additional measurements at locations $x_{\mathrm{g,test}} \in \{4, 5\}$ m were reserved exclusively for evaluating the PINN solutions post-training. Therefore, the first subsection (Section 3.1) focuses on the pure data assimilation task, where the NLSE coefficients in the PINN's loss function remained constant during training and determined a priori based on spectral wave properties and linear dispersion. In contrast, the second subsection (Section 3.2) explores treating the peak frequency and wavenumber, which form the NLSE coefficients, as identifiable variables that are adapted by the optimizer to fit the measurements best.

3.1. Data Assimilation with Constant NLSE Coefficients

In the following benchmark data assimilation task, the NLSE coefficients (Equation (6)) were kept constant during the PINN training and determined a priori based on the mean of the spectral peak frequencies $\omega_{\mathrm{p},3}$ and $\omega_{\mathrm{p},6}$ at the domain boundaries using Equation (8). This peak frequency $\overline{\omega}_{\mathrm{p}} = \frac{1}{2}(\omega_{\mathrm{p},3} + \omega_{\mathrm{p},6})$ was calculated for each of the 630 samples, where each individual sample was generated following the procedure outlined in Section 2.1 and followed the data structure shown in Figure 1b. The linear dispersion relation Equation (7) yielded the corresponding peak wavenumber $\overline{k}_{\mathrm{p}}$. Note that the calculated $\overline{\omega}_{\mathrm{p}}$ could slightly deviate from the peak frequency ω_{p} used to generate the JONSWAP spectrum for the HOSM. Each sample underwent individual PINN training for 15,000 epochs, which required approximately 1200 s of computational time on a NVIDIA GeForce RTX 3090 GPU.

Figure 5 depicts a representative training loss curve for the NLSE-PINN framework. During the initial epochs, a remarkable decrease in the PDE residual error components $\mathrm{MSE}_{U,\mathrm{res}}$ and $\mathrm{MSE}_{V,\mathrm{res}}$ was observed, while the data errors $\mathrm{MSE}_{U,\mathrm{data}}$ and $\mathrm{MSE}_{V,\mathrm{data}}$ remained high. As the data errors started to decrease and the PINN solution aligned more closely with the prescribed boundary data, the PDE errors experienced a transient increase before reaching a plateau. These observations could have arose from a zero envelope that caused high data errors in the early training, despite satisfying PDE residues that caused low PDE errors. However, as the envelope progressively matched the actual measurement data at the boundaries, fulfilling PDE residues in the remaining domain became more challenging, which led to a slight increase in the corresponding error.

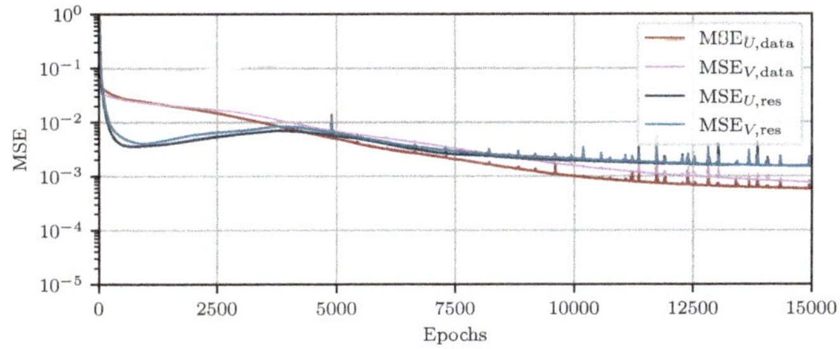

Figure 5. Exemplary NLSE-PINN training loss curve. The PDE residual error components $\mathrm{MSE}_{U,\mathrm{res}}$ and $\mathrm{MSE}_{V,\mathrm{res}}$ initially exhibited a strong decrease, but slightly increased as the data error components $\mathrm{MSE}_{U,\mathrm{data}}$ and $\mathrm{MSE}_{V,\mathrm{data}}$ were reduced. After around 10,000 epochs, the PDE residual errors reached a plateau, while the data errors continued to gradually improve.

Figure 6 presents a reconstruction of the real part $\tilde{U}(x,t)$ and imaginary part $\tilde{V}(x,t)$ of an envelope above the carrier wave sample no. 1, with $\epsilon = 0.075$, $\overline{\omega}_\mathrm{p} = 4.45\,\frac{\mathrm{rad}}{\mathrm{s}}$, and $\gamma = 6$. The PINN reconstructed a wave envelope structure across the entire computational domain, despite relying solely on measurement data from the boundaries marked in violet.

While some wave samples were successfully assimilated, challenges were evident for other samples. For instance, Figure 7 presents a PINN solution for sample no. 2, which was characterized by $\epsilon = 0.025$ and $\overline{\omega}_\mathrm{p} = 8.34\,\frac{\mathrm{rad}}{\mathrm{s}}$, with a broad-banded nature indicated by $\gamma = 1$. The average errors for the real and imaginary parts of the envelope were SSP $= 0.194$ and SSP $= 0.242$, respectively. Notably, increased errors were observed within the computational domain at $x_{\mathrm{g,test}} \in \{4, 5\}$ m, despite the satisfactory alignment of the boundary values with the ground truth at $x_{\mathrm{g,meas}} \in \{3, 6\}$ m. These errors seemed to originate from nonphysical envelope peaks that emerged from both boundaries toward the middle of the domain but met with an offset, which indicated a small error in the NLSE coefficient of the group velocity $c_\mathrm{g} = \frac{\overline{\omega}_\mathrm{p}}{2k_\mathrm{p}}$ for the underlying data. However, as the advanced method of Equation (8) was already utilized for $\overline{\omega}_\mathrm{p}$ determination, it became apparent that establishing a robust approach to determine this parameter consistently across all samples presented a challenge.

The trends observed in the aforementioned examples were repetitive across all 630 instances in the assimilation task, which is summarized in Figure 8, where the SSP value represents the average error across the real and imaginary parts of each sample. Each cell in the figure corresponds to the mean SSP achieved for all five samples of a specific ϵ–$\overline{\omega}_\mathrm{p}$–$\gamma$ combination, where the determined $\overline{\omega}_\mathrm{p}$ value was rounded to the nearest integer for this illustration. Notably, the errors tended to decrease as the $\overline{\omega}_\mathrm{p}$ value increased. The frequency dependence aligned with the inherent preference of PINNs for successive solving from lower to higher frequency components cf. [13] when considering that higher frequencies ω_p of the carrier wave $\eta(x,t)$ resulted in smoother, lower-frequency envelopes $A(x,t)$, while lower frequencies ω_p of the carrier wave $\eta(x,t)$ resulted in higher-frequency envelopes $A(x,t)$. Moreover, Figure 8 illustrates a general error increase as the peak enhancement factors γ decrease or the steepness values ϵ slightly increased. As proven by Appendix A.2, these observations were not attributable to the PINN method itself but rather aligned with the model misspecification. Specifically, the NLSE was limited for narrow-band and small-amplitude waves, while the ground truth HOSM data seem to exceed its validity range for increasing ϵ and a broader spectra.

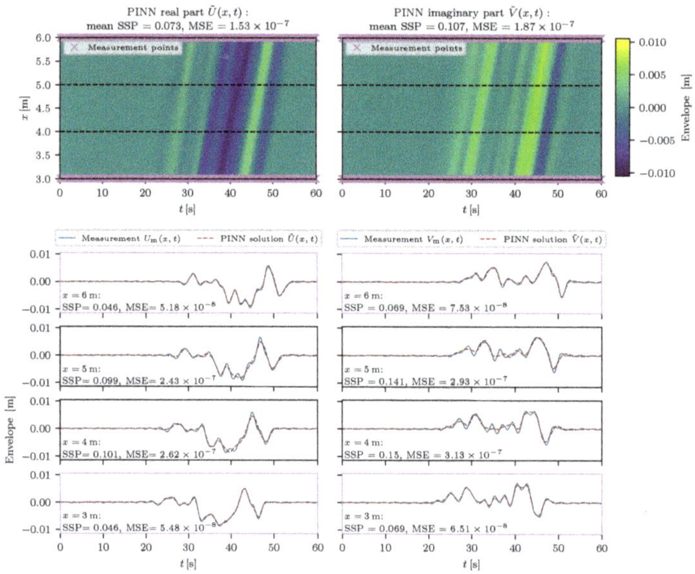

Figure 6. Real part and imaginary part of the envelope solution $\tilde{A}(x,t) = \tilde{U}(x,t) + i\tilde{V}(x,t)$ generated by the PINN using predetermined, constant NLSE coefficients for sample no. 1 (carrier wave with $\epsilon = 0.075$, $\overline{\omega}_\mathrm{p} = 9.45\,\frac{\mathrm{rad}}{\mathrm{s}}$, and $\gamma = 6$). The measurement envelope points U_m and V_m were provided on the domain boundaries only and are highlighted in violet. While the PINN solution at the boundaries aligned well with the measured data, it exhibited slight inaccuracies within the remaining computational domain.

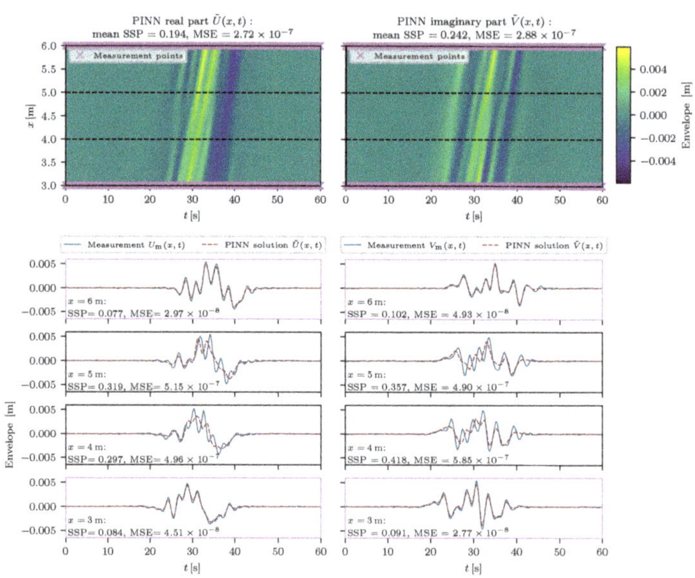

Figure 7. Real part and imaginary part of the envelope solution $\tilde{A}(x,t) = \tilde{U}(x,t) + i\tilde{V}(x,t)$ generated by the PINN using predetermined, constant NLSE coefficients for sample no. 2 (carrier wave with $\epsilon = 0.025$, $\overline{\omega}_\mathrm{p} = 8.34\,\frac{\mathrm{rad}}{\mathrm{s}}$, and $\gamma = 1$). Compared with sample no. 1 in Figure 6, the envelope maxima that developed from the boundaries was met with an offset inside the domain.

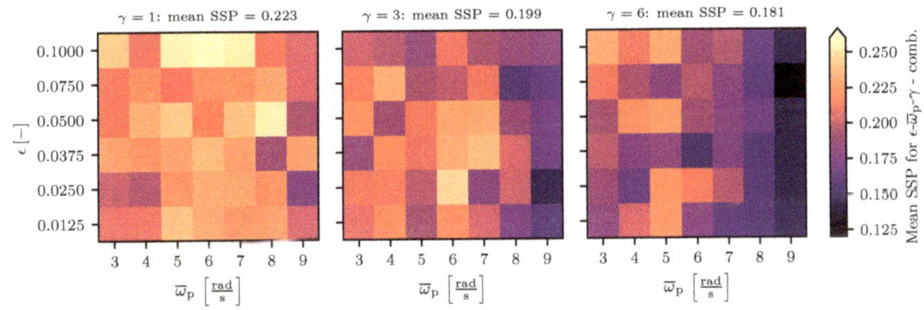

Figure 8. Reconstruction errors for all 630 samples from training the PINNs with predetermined, constant NLSE coefficients. In general, the envelopes above lower-frequency waves (lower $\overline{\omega}_{\mathrm{p}}$) were harder to reconstruct than for higher-frequency waves. Moreover, the errors increased for broader-banded samples (lower γ) and higher steepness samples (higher ϵ).

3.2. *Coefficient Fine Tuning Alongside Data Assimilation*

Despite employing an advanced formula (Equation (8)) to determine the peak frequency $\overline{\omega}_{\mathrm{p}}$ required for calculating the NLSE coefficients (Equation (6)), uncertainties in the $\overline{\omega}_{\mathrm{p}}$ determination persisted due to the short measurement durations and broader-banded signals, as discussed in the context of the envelope offset observed in Figure 7. Given the impracticality of exploring alternative possibilities for $\overline{\omega}_{\mathrm{p}}$ determination or iteratively adjusting it to recalculate the solution, we examined a data-driven opportunity to fine tune this parameter in the following subsection. Therefore, we treated the PDE coefficients not as constants but as additional trainable variables to address the inverse problems during the assimilation of the measurement data. Our objective was to enhance the reconstruction quality and eliminate the envelope offsets attributed to non-optimal constant coefficients by utilizing the results from the previous Section 3.1 as a benchmark. Since all the NLSE coefficients depended on $\overline{\omega}_{\mathrm{p}}$ and $\overline{k}_{\mathrm{p}}$, we refrained from directly setting c_g, δ, and ν as variables. Instead, we initially determined $\overline{\omega}_{\mathrm{p}}$ and $\overline{k}_{\mathrm{p}}$ using Equations (7) and (8), as demonstrated before, but subsequently treated them as additional PINN variables. Hence, the coefficients c_g, δ, and ν emerged from the current $\overline{\omega}_{\mathrm{p}}$ and $\overline{k}_{\mathrm{p}}$ in each epoch of training. Moreover, as water waves with increasing nonlinearity are known to deviate from the linear dispersion relation (Equation (7)), we decoupled the current $\overline{k}_{\mathrm{p}}$ and $\overline{\omega}_{\mathrm{p}}$ during training. As discussed in the previous subsection, each of the 630 samples, with a data structure shown in Figure 1b, underwent individual training for 15,000 epochs. However, now the training incorporated two additional trainable variables $\overline{\omega}_{\mathrm{p}}$ and $\overline{k}_{\mathrm{p}}$, as illustrated in the lower-left box of Figure 4.

The visual comparison of the new reconstruction for sample no. 2 that employed trainable variables $\overline{\omega}_{\mathrm{p}}$ and $\overline{k}_{\mathrm{p}}$ in Figure 9 with the previous solution that used constant coefficients in Figure 7 revealed a strong reduction in the envelope offset. Table 1 compares the initial peak frequency $\overline{\omega}_{\mathrm{p}}$ and wavenumber $\overline{k}_{\mathrm{p}}$ with their fine tuned values, along with the resulting NLSE coefficients. For this sample, the $\overline{\omega}_{\mathrm{p}}$ increased during the training, while the $\overline{k}_{\mathrm{p}}$ decreased, which resulted in a faster group velocity $c_g = \frac{\overline{\omega}_{\mathrm{p}}}{2\overline{k}_{\mathrm{p}}}$. Simultaneously, the nonlinear term received slightly less attention due to a decreased coefficient $\nu = -\overline{k}_{\mathrm{p}}^3$. These adjustments not only resulted in a visual reduction of the nonphysical envelope offsets but also improved the SSP error values. For the real part, the error decreased from SSP $= 0.194$ to SSP $= 0.150$, and for the imaginary part, from SSP $= 0.242$ to SSP $= 0.188$, both of which were an around 20% improvement for sample no. 2. This improvement is noteworthy considering that while adjustable coefficients can mitigate uncertainties in $\overline{\omega}_{\mathrm{p}}$ determination, they cannot fully overcome the NLSE's assumption of a single ideal carrier frequency, particularly for HOSM data with $\gamma = 1$.

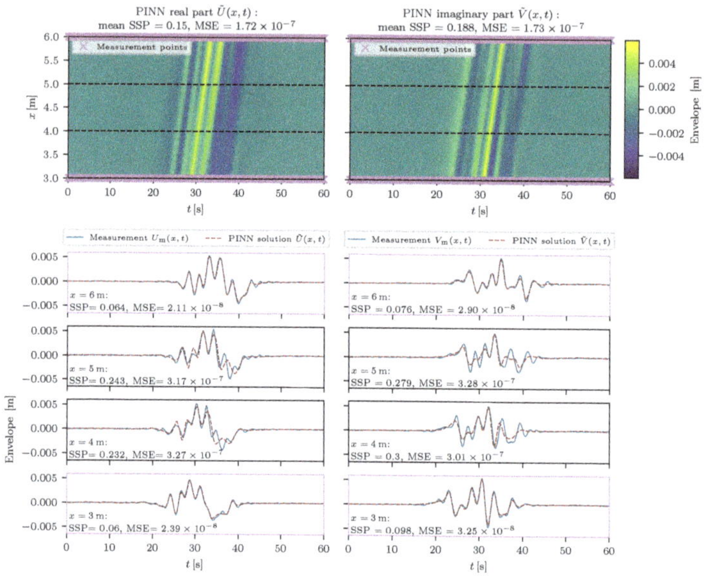

Figure 9. Real part and imaginary part of the envelope solution $\tilde{A}(x,t) = \tilde{U}(x,t) + i\tilde{V}(x,t)$ generated by the PINN using adaptable NLSE coefficients due to the learnable $\overline{\omega}_p$ and $\overline{k}_p$ for sample no. 2 (carrier wave with $\epsilon = 0.025$, $\overline{\omega}_p = 8.34\,\frac{\text{rad}}{\text{s}}$, and $\gamma = 1$). Compared with the reconstruction using constant NLSE coefficients in Figure 7, a reduction in the envelope offset was evident.

Examining the 630 samples collectively revealed that the solutions that used the trainable $\overline{\omega}_p$ and $\overline{k}_p$ values exhibited fewer envelope offsets. Figure 10 demonstrates an enhancement in the SSP values when comparing these results for fine tuned coefficients with results obtained with the constant and predetermined coefficients shown in Figure 8. While the trend persisted, where lower-frequency, higher-steepness, or broad-banded samples yielded comparatively high individual errors, the average SSP values experienced a reduction from SSP = 0.223 to SSP = 0.209 ($\gamma = 1$), from SSP = 0.199 to SSP = 0.185 ($\gamma = 3$), and from SSP = 0.181 to SSP = 0.165 ($\gamma = 6$). On average the reduction from SSP = 0.204 with constant coefficients to SSP = 0.186 with trainable coefficients represented an improvement of approximately 8.8%, while complicated instances, such as sample no. 2, also experienced considerably higher individual improvements (around 20%). Consequently, allowing the fine tuning of $\overline{\omega}_p$ and $\overline{k}_p$ values during PINN training confirmed the initial hypothsis, as it enhanced the measurable and visible reconstruction quality compared with using constant coefficients in pure data assimilation tasks.

Table 1. NLSE coefficients of sample no. 2 calculated from $\overline{\omega}_p$ and $\overline{k}_p$ for the pure data assimilation task (initial) compared with the optimized values learned during the parameter fine tuning. In addition, the corresponding SSP and MSE errors are given.

	NLSE Coefficients					Mean Error			
	$\overline{\omega}_p$	$\overline{k}_p$	$c_g = \frac{\overline{\omega}_p}{2\overline{k}_p}$	$\delta = -\frac{\overline{k}_p}{\overline{\omega}_p^2}$	$\nu = -\overline{k}_p^3$	SSP Re.	SSP Im.	MSE Re.	MSE Im.
Initial	8.344	7.097	0.588	−0.102	−357.4	0.194	0.242	2.72×10^{-7}	2.88×10^{-7}
Learned	9.355	6.202	0.754	−0.071	−256.7	0.150	0.188	1.72×10^{-7}	1.73×10^{-7}

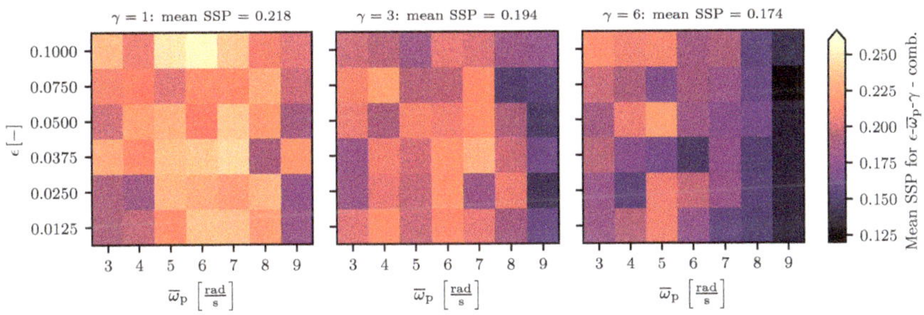

Figure 10. Reconstruction errors for all 630 samples from training the PINNs with tunable $\overline{\omega}_p$ and $\overline{k}_p$ values that caused adaptable NLSE coefficients. In general, the SSP errors were slightly reduced compared with the results obtained with constant coefficients in Figure 8. However, the general trend where lower $\overline{\omega}_p$ values, higher ϵ values, and lower γ values le d to increased errors remained.

During the coefficient fine tuning, the peak frequency increased and the wavenumber decreased while training of sample no. 2 shown in Figure 9 and Table 1. In contrast, other samples displayed varied trends, such as concurrent increasing or decreasing $\overline{\omega}_p$ and $\overline{k}_p$, or decreasing $\overline{\omega}_p$ with increasing $\overline{k}_p$. Figure 11 illustrates this variability across all 630 samples according to their respective ϵ–γ combination. The initial $\overline{\omega}_p$ and $\overline{k}_p$ values aligned with the blue graph of the linear dispersion relation, while the fine tuned values mostly deviated from this relation, as indicated by red crosses. The left and middle columns of this Figure reveal that medium- and broad-banded sea states ($\gamma = 3$ and $\gamma = 1$) tended to benefit from $\overline{\omega}_p$–$\overline{k}_p$ combinations above the linear dispersion relation. This suggests that increased group velocities $c_g = \frac{\overline{\omega}_p}{2\overline{k}_p}$ enhanced the reconstruction quality. In contrast, narrow-banded samples ($\gamma = 6$) showed learned combinations both above and below the linear dispersion curve, which reflected that the neural network effectively reduced the uncertainties in the initial determination of $\overline{\omega}_p$.

Figure 11 further illustrates a consistent trend: irrespective of the γ value, increasing steepness ϵ resulted in smaller changes in the $\overline{k}_p$ values during training. This was evident from the vertically oriented lines between the initial and final $\overline{\omega}_p$–$\overline{k}_p$ combinations in the upper subplots. This observation implies that as the steepness increases, the PINN prioritizes maintaining the nonlinear term of the NLSE, as represented by the coefficient $\nu = -\overline{k}_p^3$. This aligns with physical reasoning, as a higher ϵ corresponds to increased nonlinearities. Preserving this coefficient indicates that the PINN identifies nonlinearity as crucial. Consequently, alterations in the group velocity $c_g = \frac{\overline{\omega}_p}{2\overline{k}_p}$ primarily arise from alterations in $\overline{\omega}_p$ only.

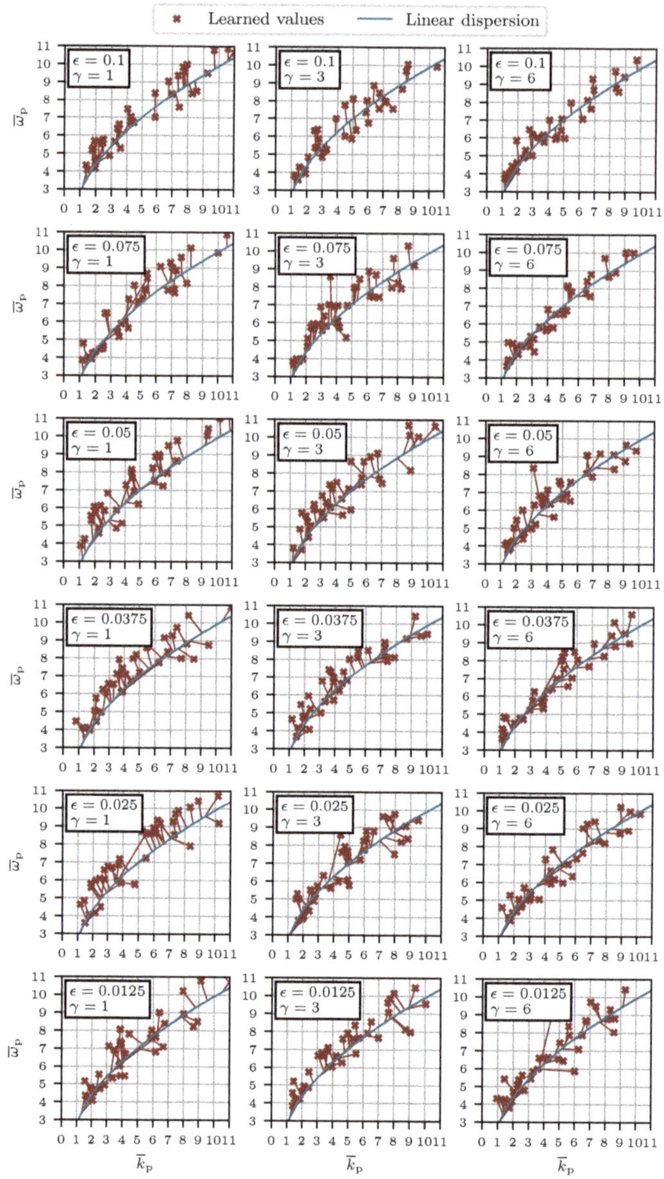

Figure 11. Initial and final $\overline{\omega}_{\mathrm{p}}$–$\overline{k}_{\mathrm{p}}$ combinations for all 630 samples from training the PINNs with tuneable $\overline{\omega}_{\mathrm{p}}$ and $\overline{k}_{\mathrm{p}}$ values that caused adaptable NLSE coefficients. While the initial values were determined following the linear dispersion relation, this constraint was removed during training. We observed that broader-banded samples (lower γ) tended to benefit from $\overline{\omega}_{\mathrm{p}}$–$\overline{k}_{\mathrm{p}}$ combinations above linear dispersion. Furthermore, with increasing steepness ϵ, the PINNs attempted to maintain the nonlinear term of the NLSE associated with $\nu = -\overline{k}_{\mathrm{p}}^{3}$ as far as possible by allowing minor changes to the $\overline{k}_{\mathrm{p}}$ value.

4. Conclusions

This study demonstrated realistic application scenarios of physics-informed neural networks (PINNs) in ocean engineering contexts. Specifically, the hydrodynamic nonlinear Schrödinger equation (NLSE) constrains the developed PINN framework to ensure the physical consistency of its solutions. First, the NLSE-PINN is employed for reconstructing synthetic wave elevation measurements between two gauges, a process known as data assimilation and a fundamental function of PINNs. However, the utilized wave data occasionally deviates from the small-steepness and narrow-band assumptions inherent to the NLSE, which poses a challenge known as model misspecification. This challenge must be anticipated whenever employing PINNs for real-world scenarios, where analytical solutions to the PDE utilized for a PINN's loss functions are typically not observable, and comprehensive knowledge of physical equations that cover all observed phenomena is rare. Moreover, this model misspecification and the short time intervals of elevation measurements complicate the determination of the required NLSE coefficients based on spectral properties. These uncertainties in the a priori determination of constant NLSE coefficients occasionally lead to non-physical envelope offsets in the reconstruction of certain instances.

To enhance the reconstruction quality of such cases, conventional numerical methods would typically require a systematic coefficient adjustment and numerical forward propagation to optimize the data fit between two measurement points. In contrast, our second research phase demonstrated that the NLSE-PINN method allowed for fine tuning the NLSE coefficients' base parameters $\overline{\omega}_{\mathrm{p}}$ and $\overline{k}_{\mathrm{p}}$ concurrently with the data assimilation in the same optimization process. This integrated parameter fine tuning effectively mitigated the previously observed non-physical offsets and reduced the mean error across all 630 wave samples from SSP $= 0.204$ with predetermined, constant coefficients during pure data assimilation to SSP $= 0.186$ with trainable coefficients. This represented an average improvement of 8.82%, while some samples experienced even greater improvements of up to 20%.

However, it is important to note that the reconstruction remains generally more challenging for broader wave spectra and samples with higher steepness. This challenge is not inherent to the PINN methodology itself, but rather associated with the characteristics of the utilized NLSE to constrain the loss function for data stemming from the more realistic HOSM. Therefore, future research can explore the integration of the modified nonlinear Schrödinger equation (MNLSE), which is also referred to as the Dysthe equation [74], into the PINN loss function. Due to its higher order of nonlinearity, the MNLSE may mitigate the steepness limitations inherent in the standard NLSE and provide new insights. Also, the broader-bandwidth variant of the MNLSE [75] might be particularly advantageous, as it additionally alleviates the limitations associated with narrow-band assumptions. Moreover, the development of a PINN capable of directly solving the fully nonlinear potential flow equations represents a promising direction, as it could offer improvements over relying on simplified models, such as the variants of the NLSE.

Moreover, concerning the successful fine tuning of $\overline{\omega}_{\mathrm{p}}$ and $\overline{k}_{\mathrm{p}}$ values in this study, further research can explore the feasibility of deriving a nonlinear dispersion relation as a function of wave parameters, such as the bandwidth, amplitude, or steepness. This research direction holds the potential to deepen our understanding of the interplay of NLSE coefficients and wave parameters and may contribute to the development of more flexible and accurate models for describing nonlinear water wave phenomena. Furthermore, this study employed synthetic wave data generated at two locations within a numerical wave tank with a fixed measurement distance. Hence, future research should investigate the impact of varying the distances between these two buoys. Moreover, applying the potential flow PINN or NLSE-PINN methods to real wave measurement data from wave tanks or open ocean scenarios, particularly for directional two-dimensional wave surfaces, would be valuable.

Author Contributions: Conceptualization, S.E., M.S. and N.H.; methodology, S.E., N.A.W. and A.S.; software, S.E.; validation, S.E. and N.H.; formal analysis, S.E.; investigation, S.E., N.A.W. and A.S.; resources, N.H.; data curation, M.K.; writing—original draft preparation, S.E.; writing—review and editing, N.A.W., A.S., M.K., N.H. and M.S.; visualization, S.E.; supervision, N.H. and M.S.; project administration, M.K. and N.H.; funding acquisition, N.H. All authors read and agreed to the published version of this manuscript.

Funding: This research was funded by the Deutsche Forschungsgemeinschaft (DFG—German Research Foundation) [project number 277972093: Excitability of Ocean Rogue Waves].

Data Availability Statement: The raw data and code supporting the conclusions of this article will be made available by the authors upon request.

Acknowledgments: The authors would like to express their gratitude to the lecturers George Em. Karniadakis and Khemraj Shukla, as well as to all organizers and funders for the Summer School on Physics-Informed Neural Networks and Applications at KTH, Stockholm, in 2023. Furthermore, appreciation is extended to the Studienstiftung des deutschen Volkes for the support provided through the Natural Science Collegium 2022/2023, during which time this work was initiated.

The investigations were conducted and the manuscript was written completely by the authors. Once the manuscript was completed, the authors used ChatGPT 3.5 (https://chat.openai.com/) in order to improve its grammar and readability. After using this tool, the authors reviewed and edited the content as needed and take full responsibility for the content of the publication.

Conflicts of Interest: The authors declare no conflicts of interest.

Abbreviations

The following abbreviations are used in this manuscript:

AD	Automatic differentiation
HOSM	High-order spectral method
MSE	Mean-squared error
NLSE	Nonlinear Schrödinger equation
PDE	Partial differential equation
PINN	Physics-informed neural network
SSP	Surface similarity parameter

Appendix A

Appendix A.1. Graphical Representation of SSP Error Metric

Although the surface similarity parameter (SSP, Equation (20)) was used in several studies to compare surface waves, it remains a relatively new metric. Therefore, Figure A1 presents the SSP for one-dimensional surfaces $\tilde{V}$ with increasing alignment with the reference surface V_{m}. To further illustrate the increasing surface alignment, the more established mean-squared error metric

$$\mathrm{MSE} = \sum_{i=1}^{n} \left(V_{\mathrm{m},i} - \tilde{V}_i\right)^2, \tag{A1}$$

with n being the number of grid points, is also provided for each signal. It is important to note that the MSE values are scale dependent, which poses a disadvantage for our comparative investigation, as the amplitudes of the 630 examined wave samples varied by several orders of magnitude due to the chosen wave parameter ranges for the data generation. Although the MSE could be normalized by the norm or the maximum of the reference signal V_{m} to mitigate the issue of differing magnitudes between wave samples, this would still treat phase shifts or frequency differences as amplitude errors. In contrast, the SSP distinguishes between amplitude, phase, and frequency errors cf. [69] by comparison in the frequency domain and combines the total error into a single value that always ranges between 0 and 1, regardless of the specific sample's amplitude.

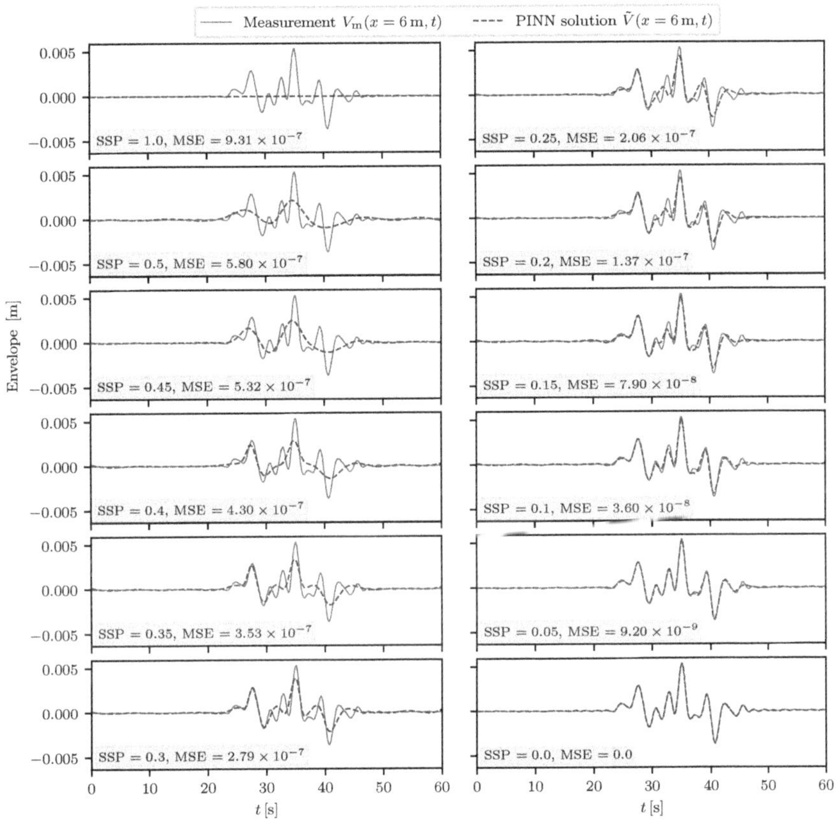

Figure A1. Graphical representation for assessing the meaning of different SSP values. The baseline signal V_m represents the imaginary part of sample no. 2 (also shown in Figures 7 and 9) at position $x = 6\,\text{m}$. It was compared with incrementally improving PINN solutions $\tilde{V}$, which resulted in decreasing SSP values. Additionally, the MSE values were calculated to provide further insights.

Appendix A.2. NLSE-PINN Validation Using Analytic Solution

To evaluate the assimilation and coefficient identification capabilities of the NLSE-PINN, distinguishing between errors originating from the neural network and those inherent to the NLSE is imperative. Given the NLSE's simplifying assumptions, it may not capture all the characteristics of the numerical wave tank data generated by the HOSM. Hence, we initially validated our method using the analytical NLSE solution of a Peregrine breather [76] in time-like form [51], which is defined as

$$A_{\text{per}} = a\left(-1 + \frac{4\left(1 - \frac{i\omega_p\epsilon^2 x}{c_g}\right)}{1 + 8k_p^2\epsilon^2(x - c_g t)^2 + \frac{\omega_p^2\epsilon^4 x^2}{c_g^2}}\right)\exp\left(-\frac{i\omega_p\epsilon^2 x}{2c_g}\right). \tag{A2}$$

As the NLSE entirely encompasses this analytic solution, the following analysis allowed for estimating the magnitude of errors attributable to the approximation by the neural network: Given the spatial range of 3 m covered by the numerical wave tank (Figure 1), we derived the analytical solution accordingly. The carrier wave frequency was set to $\omega_p = 9\,\frac{\text{rad}}{\text{s}}$, with amplitude $a = 0.02\,\text{m}$. The NLSE coefficients were calculated using Equation (6) and held constant during 15,000 epochs of PINN training. The reconstruction results in the spatio-temporal domain, which were solely based on the boundary data, are shown in Figure A2.

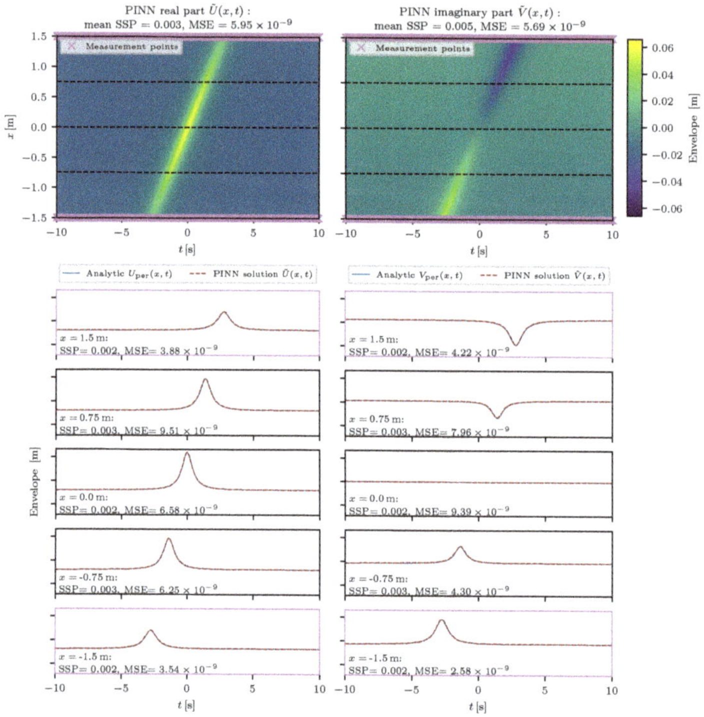

Figure A2. Real part and imaginary part of the envelope solution $\tilde{A}(x,t) = \tilde{U}(x,t) + i\tilde{V}(x,t)$ generated by the PINN using predetermined, constant NLSE coefficients for a Peregrine breather. The analytic envelope points U_{per} and V_{per} are provided only at the domain boundaries and are highlighted in violet. This analysis not only validated the efficiency of the developed NLSE-PINN framework but also allowed for accessing the small magnitude of approximation error attributable to the neural network.

Despite providing only two time series of the analytical solution at the outer boundaries $x = -1.5$ m and $x = 1.5$ m for computing the data loss (Equations (14) and(15)), the PINN effectively reconstructed physically consistent Peregrine envelopes within the remaining computational domain solely with the aid of the PDE residual loss components (Equations (16) and (17)). This is demonstrated in the lower subplots, where the time series of the PINN solution's real and imaginary parts, $\tilde{U}(x,t)$ and $\tilde{V}(x,t)$, are compared with the analytical solution, $U_{\text{per}}(x,t)$ and $V_{\text{per}}(x,t)$, at the cross-section of five points in space. Notably, despite the lower envelope amplitudes at the domain boundaries, the PINN accurately reconstructed the higher peak in the real part in the middle of the domain. This underscores the NLSE-PINN's efficacy to capture and calculate the underlying NLSE dynamics. Furthermore, this analysis revealed that the approximation errors attributable to the neural network were very small, where they averaged around MSE $= 5.8 \times 10^{-9}$ and SSP $= 0.004$ in this instance. Hence, any encountered errors were more likely attributable to model misspecification between the NLSE-PINN and the HOSM wave tank measurement data than to approximation errors.

References

1. Klein, M.; Dudek, M.; Clauss, G.F.; Ehlers, S.; Behrendt, J.; Hoffmann, N.; Onorato, M. On the deterministic prediction of water waves. *Fluids* **2020**, *5*, 9. [CrossRef]
2. Ehlers, S.; Klein, M.; Heinlein, A.; Wedler, M.; Desmars, N.; Hoffmann, N.; Stender, M. Machine learning for phase-resolved reconstruction of nonlinear ocean wave surface elevations from sparse remote sensing data. *Ocean. Eng.* **2023**, *288*, 116059. [CrossRef]
3. Jagtap, A.D.; Mitsotakis, D.; Karniadakis, G.E. Deep learning of inverse water waves problems using multi-fidelity data: Application to Serre–Green–Naghdi equations. *Ocean Eng.* **2022**, *248*, 110775. [CrossRef]
4. Karniadakis, G.E.; Kevrekidis, I.G.; Lu, L.; Perdikaris, P.; Wang, S.; Yang, L. Physics-informed machine learning. *Nat. Rev. Phy.* **2021**, *3*, 422–440. [CrossRef]
5. Willard, J.; Jia, X.; Xu, S.; Steinbach, M.; Kumar, V. Integrating physics-based modeling with machine learning: A survey. *arXiv* **2020**, arXiv:2003.04919.
6. Mohaghegh, F.; Murthy, J.; Alam, M.R. Rapid phase-resolved prediction of nonlinear dispersive waves using machine learning. *Appl. Ocean Res.* **2021**, *117*, 102920. [CrossRef]
7. Klein, M.; Stender, M.; Wedler, M.; Ehlers, S.; Hartmann, M.; Desmars, N.; Pick, M.-A.; Seifried, R.; Hoffmann, N. Application of machine learning for the generation of tailored wave sequences. In Proceedings of the International Conference on Offshore Mechanics and Arctic Engineering, Volume 5B: Ocean Engineering, Honoring Symposium for Professor Günther F. Clauss on Hydrodynamics and Ocean Engineering, Hamburg, Germany, 5–10 June 2022. [CrossRef]
8. Liu, Y.; Zhang, X.; Chen, G.; Dong, Q.; Guo, X.; Tian, X.; Lu, W.; Peng, T. Deterministic wave prediction model for irregular long-crested waves with Recurrent Neural Network. *J. Ocean Eng. Sci.* **2022**, *9*, 251–263. [CrossRef]
9. Wedler, M.; Stender, M.; Klein, M.; Hoffmann, N. Machine learning simulation of one-dimensional deterministic water wave propagation. *Ocean Eng.* **2023**, *284*, 115222. [CrossRef]
10. Zhao, M.; Zheng, Y.; Lin, Z. Sea surface reconstruction from marine radar images using deep convolutional neural networks. *J. Ocean Eng. Sci.* **2023**, *8*, 647–661. [CrossRef]
11. Raissi, M.; Perdikaris, P.; Karniadakis, G.E. Physics-informed neural networks: A deep learning framework for solving forward and inverse problems involving nonlinear partial differential equations. *J. Comput. Phys.* **2019**, *378*, 686–707. [CrossRef]
12. Baydin, A.G.; Pearlmutter, B.A.; Radul, A.A.; Siskind, J.M. Automatic differentiation in machine learning: A survey. *J. Mach. Learn. Res.* **2018**, *18*, 1–43. http://jmlr.org/papers/v18/17-468.html.
13. Cuomo, S.; Di Cola, V.S.; Giampaolo, F.; Rozza, G.; Raissi, M.; Piccialli, F. Scientific machine learning through physics-informed neural networks: Where we are and what's next. *J. Sci. Comput.* **2022**, *92*, 88. [CrossRef]
14. Kissas, G.; Yang, Y.; Hwuang, E.; Witschey, W.R.; Detre, J.A.; Perdikaris, P. Machine learning in cardiovascular flows modeling: Predicting arterial blood pressure from non-invasive 4D flow MRI data using physics-informed neural networks. *Comput. Methods Appl. Mech. Eng.* **2020**, *358*, 112623. [CrossRef]
15. Raissi, M.; Yazdani, A.; Karniadakis, G.E. Hidden fluid mechanics: Learning velocity and pressure fields from flow visualizations. *Science* **2020**, *367*, 1026–1030. [CrossRef]
16. Cai, S.; Mao, Z.; Wang, Z.; Yin, M.; Karniadakis, G.E. Physics-informed neural networks (PINNs) for fluid mechanics: A review. *Acta Mech. Sin.* **2021**, *37*, 1727–1738. [CrossRef]
17. Rasht-Behesht, M.; Huber, C.; Shukla, K.; Karniadakis, G.E. Physics-informed neural networks (PINNs) for wave propagation and full waveform inversions. *J. Geophys. Res. Solid Earth* **2022**, *127*, e2021JB023120. [CrossRef]
18. Zobeiry, N.; Humfeld, K.D. A physics-informed machine learning approach for solving heat transfer equation in advanced manufacturing and engineering applications. *Eng. Appl. Artif. Intell.* **2021**, *101*, 104232. [CrossRef]
19. Cai, S.; Wang, Z.; Wang, S.; Perdikaris, P.; Karniadakis, G.E. Physics-informed neural networks for heat transfer problems. *J. Heat Transf.* **2021**, *143*, 060801. [CrossRef]
20. Kashinath, K.; Mustafa, M.; Albert, A.; Wu, J.-L.; Jiang, C.; Esmaeilzadeh, S.; Azizzadenesheli, K.; Wang, R.; Chattopadhyay, A.; Singh, A.; et al. Physics-informed machine learning: Case studies for weather and climate modelling. *Philos. Trans. Ser. Math. Phys. Eng. Sci.* **2021**, *379*, 2194. [CrossRef]
21. Chen, Y.; Lu, L.; Karniadakis, G.E.; Dal Negro, L. Physics-informed neural networks for inverse problems in nano-optics and metamaterials. *Opt. Express* **2020**, *28*, 11618–11633. [CrossRef]
22. Nguyen-Thanh, V.M.; Zhuang, X.; Rabczuk, T. A deep energy method for finite deformation hyperelasticity. *Eur. J. Mech. A/Solids* **2020**, *80*, 103874. [CrossRef]
23. Wang, N.; Chen, Q.; Chen, Z. Reconstruction of nearshore wave fields based on physics-informed neural networks. *Coast. Eng.* **2022**, *176*, 104167. [CrossRef]
24. Chen, Q.; Wang, N.; Chen, Z. Simultaneous mapping of nearshore bathymetry and waves based on physics-informed deep learning. *Coast. Eng.* **2023**, *183*, 104337. [CrossRef]
25. Feng, D.; Tan, Z.; He, Q. Physics-informed neural networks of the Saint-Venant equations for downscaling a large-scale river model. *Water Resour. Res.* **2023**, *59*, e2022WR033168. [CrossRef]
26. Chiang, C.M.; Stiassnie, M.; Yue, D.K.-P. *Theory and Applications of Ocean Surface Waves*; World Scientific: Singapore, 2005. [CrossRef]
27. Zakharov, V.E. Stability of periodic waves of finite amplitude on the surface of a deep fluid. *J. Appl. Mech. Tech. Phys.* **1968**, *9*, 190–194. [CrossRef]

28. Shemer, L.; Kit, E.; Jiao, H.; Eitan, O. Experiments on nonlinear wave groups in intermediate water depth. *J. Waterw. Port. Coast. Ocean. Eng.* **1998**, *124*, 320–327. [CrossRef]
29. Trulsen, K.; Stansberg, C.T. Spatial evolution of water surface waves: Numerical simulation and experiment of bichromatic waves. In Proceedings of the ISOPE International Ocean and Polar Engineering Conference, ISOPE–I, Kitakyushu, Japan, 25–31 May 2001.
30. Dysthe, K.B.; Trulsen, K.; Krogstadd, H.E.; Socquet-Juglard, H. Evolution of a narrow-band spectrum of random surface gravity waves. *J. Fluid Mech.* **2003**, *478*, 1–10. [CrossRef]
31. Chabchoub, A.; Hoffmann, N.P. Rogue wave observation in a water wave tank. *Phys. Rev. Lett.* **2011**, *106*, 204502. [CrossRef]
32. Ruban, V.P. Gaussian variational ansatz in the problem of anomalous sea waves: Comparison with direct numerical simulation. *J. Exp. Theor. Phys.* **2015**, *120*, 925–932. [CrossRef]
33. Li, J.; Li, B. Solving forward and inverse problems of the nonlinear Schrödinger equation with the generalized PT-symmetric Scarf-II potential via PINN deep learning. *Commun. Theor. Phys.* **2021**, *73*, 125001. [CrossRef]
34. Pu, J.-C.; Li, J.; Chen, Y. Solving localized wave solutions of the derivative nonlinear Schrödinger equation using an improved PINN method. *Nonlinear Dyn.* **2021**, *105*, 1723–1739. [CrossRef]
35. Pu, J.-C.; Li, J.; Chen, Y. Soliton, breather, and rogue wave solutions for solving the nonlinear Schrödinger equation using a deep learning method with physical constraints. *Chin. Phys. B* **2021**, *30*, 060202. [CrossRef]
36. Wang, L.; Yan, Z. Data-driven rogue waves and parameter discovery in the defocusing nonlinear Schrödinger equation with a potential using the PINN deep learning. *Phys. Lett. A* **2021**, *404*, 127408. [CrossRef]
37. Jiang, X.; Wang, D.; Fan, Q.; Zhang, M.; Lu, C.; Lau, A.P.T. Physics-informed neural network for nonlinear dynamics in fiber optics. *Laser Photonics Rev.* **2022**, *16*, 2100483. [CrossRef]
38. Zhang, S.; Lan, P.; Su, J.-J. Wave-packet behaviors of the defocusing nonlinear Schrödinger equation based on the modified physics-informed neural networks. *Chaos Interdiscip. J. Nonlinear Sci.* **2021**, *31*, 113107. [CrossRef]
39. He, Q.Z.; Barajas-Solano, D.; Tartakovsky, G.; Tartakovsky, A.M. Physics-informed neural networks for multiphysics data assimilation with application to subsurface transport. *Adv. Water Resour.* **2020**, *141*, 103610. [CrossRef]
40. Von Saldern, J.G.R.; Reumschüssel, J.M.; Kaiser, T.L.; Sieber, M.; Oberleithner, K. Mean flow data assimilation based on physics-informed neural networks. *Phys. Fluids* **2022**, *34*, 115129. [CrossRef]
41. West, B.J.; Brueckner, K.A.; Janda, R.S.; Milder, D.M.; Milton, R.L. A new numerical method for surface hydrodynamics. *J. Geophys. Res. Ocean.* **1987**, *92*, 11803–11824. [CrossRef]
42. Ducrozet, G.; Bonnefoy, F.; Perignon, Y. Applicability and limitations of highly non-linear potential flow solvers in the context of water waves. *Ocean Eng.* **2017**, *142*, 233–244. [CrossRef]
43. Lünser, H.; Hartmann, M.; Desmars, N.; Behrendt, J.; Hoffmann, N.; Klein, M. The influence of characteristic sea state parameters on the accuracy of irregular wave field simulations of different complexity. *Fluids* **2022**, *7*, 243. [CrossRef]
44. Zou, Z.; Meng, X.; Karniadakis, G.E. Correcting model misspecification in physics-informed neural networks (PINNs). *arXiv* **2023**, arXiv:2310.10776. [CrossRef]
45. Hasselmann, K.; Barnett, T.P.; Bouws, E.; Carlson, H.; Cartwright, D.E.; Enke, K.; Ewing, J.A.; Gienapp, H.; Hasselmann, D.E.; Kruseman, P.; et al. Measurements of wind-wave growth and swell decay during the Joint North Sea Wave Project (JONSWAP). *ErgäNzung Zur. Dtsch. Hydrogr. Z. Reihe A* **1973**, *12*, 1–95.
46. Bouws, E.; Günther, H.; Rosenthal, W.; Vincent, C.L. Similarity of the wind wave spectrum in finite depth water: 1. Spectral form. *J. Geophys. Res.* **1985**, *90*, 975–986. [CrossRef]
47. Amir, M.A.U.; Wahid, A.S.A.; Appelbe, H.; Aziz, H.A.; Ahmad, M.A.; Azmir, H.; Jamil, R.A.; Radzi, Z.M. Optimal specification of wave flume in confined space. In Proceedings of the 2016 International Conference on Information and Communication Technology (ICICTM), Kuala Lumpur, Malaysia, 16–17 May 2016; pp. 136–140.
48. Khalilabadi, M.R.; Bidokhti, A.A. Design and Construction of an Optimum Wave Flume. *J. Appl. Fluid Mech.* **2013**, *5*, 99–103.
49. Hasimoto, H.; Ono, H. Nonlinear modulation of gravity waves. *J. Phys. Soc. Jpn.* **1972**, *33*, 805–811. [CrossRef]
50. Yuen, H.C.; Lake, B.M. Nonlinear Dynamics of Deep-Water Gravity Waves; *Adv. Appl. Mech.* **1982**, *22*, 67–229. [CrossRef]
51. Chabchoub, A.; Grimshaw, R.H.J. The hydrodynamic nonlinear Schrödinger equation: Space and time. *Fluids* **2016**, *1*, 23. [CrossRef]
52. Osborne, A.R. *Nonlinear Ocean Waves and the Inverse Scattering Transform*; Academic Press: Amsterdam, The Netherlands, 2010; Volume 97.
53. Sobey, R.J. Real Sea States—Wave Grouping. In *Real Sea States: Advanced Short Course Notes*; Leichtweiß-Institut für Wasserbau, Technische Universität Braunschweig: Braunschweig, Germany, 1999.
54. Mansard, E.P.D.; Funke, E.R. On the fitting of parametric models to measured wave spectra. In Proceedings of the 2nd International Symposium on Waves and Coastal Engineering, Hannover, Germany, 12–14 October 1988; pp. 12–14.
55. Huang, N.E.; Shen, Z.; Long, S.R. A new view of nonlinear water waves: The Hilbert spectrum. *Annu. Rev. Fluid Mech.* **1999**, *31*, 417–457. [CrossRef]
56. Thrane, N.; Wismer, J.; Konstantin-Hansen, H.; Gade, S. Application Note: Practical Use of the "Hilbert transform". Brüel & Kjær. 2011. Available online: https://www.bksv.com/media/doc/bo0437.pdf (accessed on 5 September 2024).

57. Glorot, X.; Bengio, Y. Understanding the difficulty of training deep feedforward neural networks. In *Proceedings of the Thirteenth International Conference on Artificial Intelligence and Statistics, Sardinia, Italy, 13–15 May 2010*; Teh, Y.W., Titterington, M., Eds.; PMLR: Sardinia, Italy, 2010; Volume 9, pp. 249–256. Available online: https://proceedings.mlr.press/v9/glorot10a.html (accessed on 10 September 2024) .
58. Jagtap, A.D.; Kawaguchi, K.; Karniadakis, G.E. Adaptive activation functions accelerate convergence in deep and physics-informed neural networks. *J. Comput. Phys.* **2020**, *404*, 109136. [CrossRef]
59. Jagtap, A.D.; Kawaguchi, K.; Karniadakis, G.E. Locally adaptive activation functions with slope recovery for deep and physics-informed neural networks. *Proc. R. Soc. A Math. Phys. Eng. Sci.* **2020**, *476*, 20200334. [CrossRef]
60. Jagtap, A.D.; Kharazmi, E.; Karniadakis, G.E. Conservative physics-informed neural networks on discrete domains for conservation laws: Applications to forward and inverse problems. *Comput. Methods Appl. Mech. Eng.* **2020**, *365*, 113028. [CrossRef]
61. Shukla, K.; Di Leoni, P.C.; Blackshire, J.; Sparkman, D.; Karniadakis, G.E. Physics-Informed Neural Network for Ultrasound Nondestructive Quantification of Surface Breaking Cracks. *J. Nondestruct. Eval.* **2020**, *39*, 1–20. [CrossRef]
62. Guo, Y.; Cao, X.; Peng, K. Solving nonlinear soliton equations using improved physics-informed neural networks with adaptive mechanisms. *Commun. Theor. Phys.* **2023**, *75*, 095003. [CrossRef]
63. McClenny, L.; Braga-Neto, U. Self-Adaptive Physics-Informed Neural Networks using a Soft Attention Mechanism. *arXiv* **2020**, arXiv:2009.04544.
64. Kingma, D.P.; Ba, J. Adam: A Method for Stochastic Optimization. In Proceedings of the 3rd International Conference on Learning Representations (ICLR 2015), San Diego, CA, USA, 12–15 May 2015; pp. 1–15 https://arxiv.org/abs/1412.6980.
65. Liu, D.C.; Nocedal, J. On the limited memory BFGS method for large scale optimization. *Math. Program.* **1989**, *45*, 503–528. [CrossRef]
66. Markidis, S. The Old and the New: Can Physics-Informed Deep-Learning Replace Traditional Linear Solvers? *Front. Big Data* **2021**, *4*, 669097. [CrossRef]
67. Reddi, S.J.; Kale, S.; Kumar, S. On the Convergence of Adam and Beyond. In Proceedings of the International Conference on Learning Representations, Vancouver, BC, Canada, 30 April–3 May 2018. Available online: https://openreview.net/pdf?id=ryQu7f-RZ (accessed on 5 September 2024).
68. Paszke, A.; Gross, S.; Massa, F.; Lerer, A.; Bradbury, J.; Chanan, G.; Killeen, T.; Lin, Z.; Gimelshein, N.; Antiga, L.; et al. PyTorch: An Imperative Style, High-Performance Deep Learning Library. In *Proceedings of the Advances in Neural Information Processing Systems, Vancouver, BC, Canada, 8–14 December 2019*; Curran Associates, Inc.: Red Hook, NY, USA, 2019; Volume 32, pp. 8024–8035. Available online: https://proceedings.neurips.cc/paper_files/paper/2019/file/bdbca288fee7f92f2bfa9f7012727740-Paper.pdf (accessed on 5 September 2024).
69. Wedler, M.; Stender, M.; Klein, M.; Ehlers, S.; Hoffmann, N. Surface Similarity Parameter: A New Machine Learning Loss Metric for Oscillatory Spatio-Temporal Data. *Neural Netw.* **2022**, *156*, 123–134. . [CrossRef]
70. Perlin, M.; Bustamante, M. A robust quantitative comparison criterion of two signals based on the Sobolev norm of their difference. *J. Eng. Math.* **2016**, *101*, 115–124. [CrossRef]
71. Desmars, N.; Hartmann, M.; Behrendt, J.; Klein, M.; Hoffmann, N. Reconstruction of Ocean Surfaces from Randomly Distributed Measurements Using a Grid-Based Method. In Proceedings of the International Conference on Offshore Mechanics and Arctic Engineering, Virtual Conference, 21–30 June 2021; Volume 6: Ocean Engineering. [CrossRef]
72. Desmars, N.; Hartmann, M.; Behrendt, J.; Klein, M.; Hoffmann, N. Nonlinear Reconstruction and Prediction of Regular Waves. In Proceedings of the International Conference on Offshore Mechanics and Arctic Engineering, Hamburg, Germany, 5–10 June 2022; . [CrossRef]
73. Kim, I.-C.; Ducrozet, G.; Bonnefoy, F.; Leroy, V.; Perignon, Y. Real-time phase-resolved ocean wave prediction in directional wave fields: Enhanced algorithm and experimental validation. *Ocean Eng.* **2023**, *276*, 114212. [CrossRef]
74. Dysthe, K.B. Note on a modification to the nonlinear Schrödinger equation for application to deep water waves. *Proc. R. Soc. Lond. A* **1979**, *369*, 105–114. [CrossRef]
75. Trulsen, K.; Dysthe, K.B. A Modified Nonlinear Schrödinger Equation for Broader Bandwidth Gravity Waves on Deep Water. *Wave Motion* **1996**, *24*, 281–289. [CrossRef]
76. Peregrine, D.H. Water waves, nonlinear Schrödinger equations and their solutions. *J. Aust. Math. Soc. Ser. B Appl. Math.* **1983**, *25*, 16. [CrossRef]

fluids MDPI

Article

Investigating the Effects of Labeled Data on Parameterized Physics-Informed Neural Networks for Surrogate Modeling: Design Optimization for Drag Reduction over a Forward-Facing Step

Erik Gustafsson * and Magnus Andersson

Department of Management and Engineering, Linköping University, SE-581 83 Linköping, Sweden; magnus.andersson@liu.se

* Correspondence: erik.gustafsson@liu.se

Abstract: Physics-informed neural networks (PINNs) are gaining traction as surrogate models for fluid dynamics problems, combining machine learning with physics-based constraints. This study investigates the impact of labeled data on the performance of parameterized physics-informed neural networks (PINNs) for surrogate modeling and design optimization. Different training approaches, including physics-only, data-only, and several combinations of both, are evaluated using fully connected (FCNN) and Fourier neural network (FNN) architectures. The test case focuses on reducing drag over a forward-facing step through optimal placement and sizing of an upstream obstacle. Results demonstrate that the inclusion of labeled data significantly enhances the accuracy and convergence rates of FCNNs, particularly in predicting flow separation and recirculation regions, and improves the stability of design optimization outcomes. Conversely, FNNs exhibit inconsistent responses to parameter changes when trained with labeled data, suggesting limitations in their applicability for certain design optimization tasks. The findings reveal that FCNNs trained with a balanced integration of data and physics constraints outperform both data-only and physics-only models, highlighting the importance of optimizing the training approach based on the specific requirements of fluid mechanics applications.

Keywords: machine learning; PINN; computational fluid dynamics; Navier–Stokes; laminar flow; data-driven fluid mechanics; design optimization; surrogate modeling; NVIDIA Modulus

1. Introduction

Physics-informed neural networks (PINNs) have emerged as a powerful tool in computational modeling, particularly for solving differential equations that describe physical phenomena. They integrate traditional neural networks (NNs) with the principles of physics, leveraging physical laws such as conservation equations, constitutive relationships, and boundary conditions (BCs) to guide the learning process. This integration allows PINNs to excel in scenarios where conventional methods might struggle, especially in inverse and ill-posed problems where BCs are not well defined or where data are sparse [1]. PINNs have been applied in various fields, including fluid mechanics [2,3], solid mechanics [2], and heat transfer [4].

Parametric PINNs extend the capabilities of standard PINNs by incorporating parameters that describe variations in the physical system, such as material properties [5,6], geometric dimensions [7,8], and operating conditions [7,9,10]. These parameters are treated as additional inputs to the NN, allowing the model to generate solutions across a range of conditions. This makes parametric PINNs particularly useful for surrogate modeling and design optimization, where the goal is to explore the design space efficiently and identify optimal configurations. Another notable feature of parametric PINNs for design optimization is their ability to provide gradients for the output variables concerning the design parameters, enabling gradient-based optimization. This approach is demonstrated by [11] with the design optimization of a one-dimensional heat transfer problem. In this instance, gradients are provided to the optimization from post-processing; however, it is also feasible to perform the design optimization concurrently with training the PINN by incorporating the design objective as a term in the loss function, as shown in, e.g., the study by [8] to optimize an airfoil geometry.

Despite their advantages, PINNs face several challenges that can complicate their implementation and reduce their effectiveness. These challenges include difficulties in training, sensitivity to hyperparameters, and the need for substantial computational resources. Training PINNs can be particularly challenging due to the complex interaction between the data-driven and physics-driven components of the loss function. Achieving a balance between these components is crucial for ensuring that the model learns both the data patterns and the underlying physical laws accurately. This can lead to challenges in optimization, as the loss landscape may contain multiple local minima corresponding to non-physical solutions [12–15]. Factors such as characteristic length and time scales, computational domain, and coefficients of differential equations influence the absolute scale of data and physics residuals. Adjusting loss weights through trial and error has become a standard procedure to achieve sufficiently accurate predictions [12].

The use of labeled data (beyond boundary and initial conditions) can significantly aid the training process of PINNs, improving their accuracy and convergence rates [9,16,17] or allowing data assimilation where some variables and closure terms are identified from sparse data [18]. For example, in a study comparing different turbulence models for PINNs on the backward-facing step, it was observed that none of the models performed well without labeled data. When adding a small amount of training data, all the models showed improved flow field accuracy and were able to predict the flow separation region behind the step [16]. In the aforementioned study, training data were added in a set of lines perpendicular to the flow direction, essentially providing additional BCs for the training process. Despite the demonstrated benefits in non-parametric cases, there is a notable lack of studies focusing on the use of labeled data for parametric PINNs. In parametric PINNs, the labeled data are spread over the parameter axis, representing different parameter configurations rather than spatial locations. This presents a unique challenge, as the parameter axis does not have derivatives defined in the partial differential equations (PDEs) governing the system. Additionally, when used for design optimization, there is a risk that the addition of data will lead to inconsistent accuracy between designs in the dataset and other designs. This might create unrealistic local minima in the objective function, thus limiting its usefulness for design optimization. Among the reviewed studies showing training of parametric PINNs with labeled data [5,9,10], only the work by [10] provides a comparison between physics-only and physics-plus-data training. The aforementioned study showed that a model trained with data and physics constraints, two discrete parameter values initially and then with only physics for three other discrete values, gives better accuracy compared to a model trained using only physics for the same parameter values. Furthermore, all of the reviewed studies use mean squared error (MSE) as the only performance metric, which might not pick up on problems with uneven objective functions when the model was used for design optimization.

This study makes a targeted effort to investigate the effects of labeled training data on parametric PINNs for a flow-related problem by comparing various PINN models against computational fluid dynamics (CFD) data obtained using the finite volume method. The performance of fully connected neural network (FCNN) and Fourier neural network (FNN) architectures (both of which are present in the reviewed studies) were evaluated using different training approaches, including physics-only, data-only, and combined physics–data constraints. The test case was set up to find the optimal placement and sizing of an upstream obstacle in order to reduce drag over a forward-facing step using design optimization. The comparison includes not only the mean error of the primitive flow variables (U, V, P) but also optimal designs found in design optimization and the objective function/response surface. The models were trained using NVIDIA Modulus 24.01 (NVIDIA, Santa Clara, CA, USA) [19,20], an open-source, engineering-focused platform for developing and deploying PINNs. Key contributions include (i) demonstrating that FCNNs significantly benefit from the inclusion of labeled data, achieving lower errors in flow variable predictions and more consistent responses across the parameter space; (ii) highlighting that FNNs exhibit increased variability and less-reliable optimization results when labeled data are used; (iii) emphasizing the importance of using metrics other than mean squared error (MSE) to validate PINN models; and (iv) providing insights into network architecture selection and the potential of FCNNs for robust design optimization in fluid dynamics applications.

2. Optimization Problem—Drag Reduction over Forward-Facing Step

In order to investigate the effects of labeled training data on the training of parameterized PINNs for design optimization, an optimization problem inspired by [21] was used. The aforementioned study describes how the flow through an abrupt contraction in a pipe or over a forward-facing step features two prominent vortex regions located just before and after the change in area and that the downstream vortex reduces the effective cross-sectional area, increasing flow resistance. To address this, an obstacle is placed upstream of the forward-facing step in order to decrease or eliminate the vortex region downstream of the step.

The specific setup used in this study is shown in Figure 1, including the dimensions and design parameters H_o and L_o, representing the height of the obstacle and the distance between the obstacle and step, respectively. The design objective is to find the optimal placement and height of the obstacle to minimize flow resistance in the laminar regime. The optimization problem is defined in Equation (1), where $\Delta P = P_1 - P_2$, P_1 is the static pressure at $x = -4$, P_2 is the static pressure at $x = 4$, ρ is the fluid density, and U_m^2 is the mean velocity upstream. The locations $x = \pm 4$ were chosen by analyzing the pressure profile along the x axis for designs with the obstacle placed near the variable limits. These locations fall within the linear region of the pressure profile with a sufficient margin.

$$\min \ \Delta C_p(H_o, L_o) = \frac{2\Delta P(Re, H_o, L_o)}{\rho U_m^2} \tag{1}$$

subject to $0.2 \leq L_o \leq 1, \quad 0.1 \leq H_o \leq 0.5, \quad Re = 100, 200, 300, \ldots, 1000.$

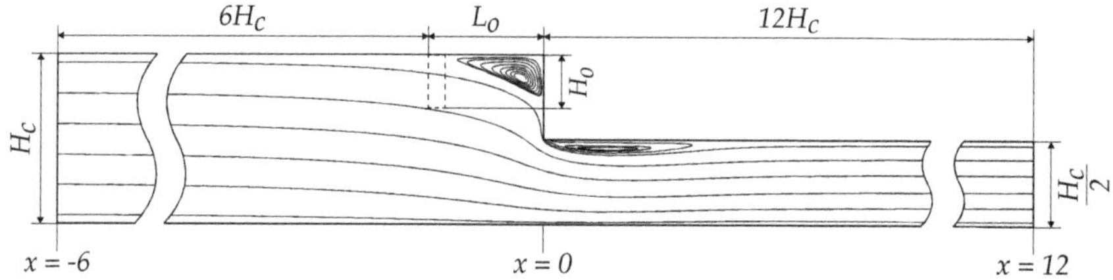

Figure 1. Test case setup for flow resistance reduction over a forward-facing step through the placement of an obstacle (dashed line) upstream of the step. By adding an obstacle, the downstream recirculation region can be minimized, reducing the flow resistance. The obstacle is placed with the design parameters H_o (height of the obstacle) and L_o (distance between the obstacle and step). The design objective is to find the optimal placement and height of the obstacle to minimize flow resistance.

Governing Equations, Boundary Conditions, and Case Data

The data for the test case, shown in Table 1, were chosen so that most quantities are equal to 1, which helps improve the PINN training process.

Table 1. Data for the test case, including channel height, inlet mean velocity, fluid density, and range for Reynolds number.

Property	Symbol	Value
Channel height	H_C	1 [m]
Inlet mean velocity	U_m	1 [m/s]
Fluid density	ρ	1 [kg/m^3]
Reynolds number	Re	100–1000

The flow is governed by the 2D, steady-state Navier–Stokes equations, which consist of the continuity equation (Equation (2)) and the momentum equations in the x direction (Equation (3)) and y direction (Equation (4)).

$$\frac{\partial U}{\partial x} + \frac{\partial V}{\partial y} = 0 \tag{2}$$

$$\rho\left(-\nu\frac{\partial^2 U}{\partial x^2} - \nu\frac{\partial^2 U}{\partial y^2} + U\frac{\partial U}{\partial x} + V\frac{\partial U}{\partial y}\right) + \frac{\partial P}{\partial x} = 0 \tag{3}$$

$$\rho\left(-\nu\frac{\partial^2 V}{\partial x^2} - \nu\frac{\partial^2 V}{\partial y^2} + U\frac{\partial V}{\partial x} + V\frac{\partial V}{\partial y}\right) + \frac{\partial P}{\partial y} = 0 \tag{4}$$

To vary Re for the optimization problem, the kinematic viscosity ν was adjusted according to Equation (5):

$$\nu = \frac{U_m H_C}{Re} \tag{5}$$

A fully developed laminar velocity profile was applied at the inlet boundary, defined by Equation (6):

$$U = 2U_m\left(1 - \left(\frac{|y|}{H_C/2}\right)^2\right) \tag{6}$$

where $|y|$ is the orthogonal distance from the channel centerline ($y = 0$). The outlet boundary uses a pressure condition with $P = 0$, and the remaining walls are treated as no-slip boundaries ($U, V = 0$).

3. Training and Validation Data

Training and validation data were generated using CFD simulations in a commercial FVM solver, Ansys CFX 2022 R2 (Ansys, Inc., Canonsburg, PA, USA, ansys.com). Meshing was conducted in Ansys Workbench 2022 R2, using triangular prism elements in the bulk region combined with hexahedral elements in the near-wall region. To mimic two-dimensional (2D) flow, only one element was used in the z direction. The mesh was refined locally in regions where velocity and pressure gradients were expected to be substantial. A total mesh size of approximately 2.8×10^4 elements was found to be sufficiently resolved with regard to the mesh-related errors on the flow resistance parameters and the purpose of the study (see Appendix A). The final mesh is shown in Figure 2. The flow was assumed to be laminar. A second-order high-resolution discretization scheme was adopted, ensuring boundedness and minimal false diffusion. Simulation convergence was ensured by a root-mean-square residual target of 10^{-6} as well as by monitoring the pressure drop.

Both training and validation data were generated in Ansys Workbench using Design of Experiments (DoEs) with Latin hypercube sampling for the parameters Re, H_o, and L_o. The training data consisted of 100 design points, and the validation data comprised 20 validation points, for all of which full simulations were run in the commercial solver to obtain the corresponding flow fields as labeled data.

Figure 2. Mesh used in the computational fluid dynamics (CFD) simulations to generate the training and validation data for the PINNs.

To also validate optimization results from the PINN models, optimization was performed in the commercial solver using response surface optimization. Using a DoE size of 100 (the same as for the training data) did not yield reliable results, and the identified optima varied greatly with small changes in parameter ranges or sampling techniques. DoE sizes up to 600 were tried without reaching consistency regarding the optima found during optimization. To obtain reliable results, separate DoEs and response surfaces were generated for discrete values of $Re = 100, 200, \ldots, 1000$, and the number of design points in each DoE was increased until further increases had negligible effects. To obtain reliable results, separate DoEs and response surfaces were generated for discrete values of $Re = 100, 200, \ldots, 1000$, and the number of design points in each DoE was increased until further increases had negligible effects. This resulted in a DoE size of 150 for each case and a total of 1500 design points. For completeness, optimization results from both models are used for comparison, where the discrete model (CFD Discrete) is considered the ground truth, and the continuous (CFD Continuous) model provides a comparison where the same dataset size is used for both response surface and PINN optimization.

4. Experiments—Training Process Analysis

The loss function used for training the PINNs, shown in Equation (7), is composed of three parts: L_{BC} corresponds to the BCs, L_{Data} to the labeled training data from CFD simulations, and L_{PDE} to the governing PDE (Equations (2)–(4)). ϕ represents the trainable parameters of the NN, with respect to which the loss is minimized; W_d is the weight for the data part of the loss, and W_p is the weight for the physics part of the loss.

$$L = W_p \times [L_{BC}(\phi) + L_{PDE}(\phi)] + W_d \times L_{Data}(\phi) \quad (7)$$

The NNs were trained using several different approaches in order to investigate the effects of labeled data on the training process. The tests can be divided into three groups: (I) training with physics only ($W_d = 0$), (II) training with labeled data only ($W_p = 0$), and (III–IV) training with both physics and labeled data ($W_p \neq 0, W_d \neq 0$). The physics-plus-labeled-data training was conducted using two approaches: one with full data (U, V, P) and one with only data for pressure P. The latter uses pressure data from six locations along the bottom wall ($y = -0.5$) to mimic data that could be obtained from experiments. Additionally, tests with labeled data were performed using two different weights, W_d, for the data component of the loss, L_{Data}. Each of these tests was also performed using two different NN architectures, FCNN and FNN, both of which have appeared in previous studies on parametric PINNs with labeled data. FNNs differ from FCNNs by incorporating an input encoding layer that transforms spatial and parameter inputs into a different feature space. In Modulus, this is achieved using a version of the method proposed by Tancik et al. [22], which uses sinusoidal functions to encode the inputs. This encoding mitigates the spectral bias commonly observed in neural networks, which tends to favor low-frequency solutions during training. In Modulus, FNNs are recommended for problems with sharp gradients and/or complicated geometries [23]. This led to a total of 12 approaches initially. During the analysis, it was noticed that training the data plus physics FNNs gave an uneven response for $\Delta C_p(L_o)$, so two additional tests were added, where W_d was set to 0 after 10^5 training steps, indicated by the parameter S_d. The approaches are

- Group I—Physics:PDE residuals and BCs
- Group II—Data: Only CFD data
- Group III—Data + Physics: Physics and data constraints with two different weights for the data, $W_d = \{0.1, 1\}$. Additionally, for FNN only, the same two approaches but with $S_d = 10^5$.
- Group IV—Pressure Data + Physics: Physics and pressure data at six locations along the bottom wall.

The trained models were evaluated using the validation data by comparing the normalized L_2 loss for the primitive flow variables and the absolute error of ΔC_p. Additionally, design optimization was performed using the different models, and the optimization results were compared to those obtained using response surface optimization in the commercial solver.

5. Parameterized Physics-Informed Neural Network

A NN was set up with coordinates (x, y) and parameters (Re, L_o, and H_o) as input and the primitive flow variables as output. For variables with large values, normalization was applied according to Table 2 in order to keep their ranges closer to $[-1, 1]$, facilitating the training process. The NN takes the following form:

$$NN_\phi \begin{pmatrix} x_{norm} \\ y \\ Re_{norm} \\ L_o \\ H_o \end{pmatrix} = \begin{pmatrix} U_{norm} \\ V \\ P_{norm} \end{pmatrix}$$

where ϕ are the trainable parameters of the NN.

Table 2. Normalization used to keep the variable ranges close to $[-1,1]$, facilitating the training process. Variables include Reynolds number (Re), x coordinate (x), x velocity (U), and pressure (P).

Variable	Normalization
Re	$Re_{norm} = Re/1000$
x	$x_{norm} = x/12$
U	$U_{norm} = U/3$
P	$P_{norm} = P/10$

Both NNs use mainly default settings in Modulus, with the exception of using adaptive activation functions and increasing the number of frequencies sampled for the FNN to $\frac{i}{2}$, for $i = 1, 2, 3, \ldots, 32$ from the default i, for $i = 1, 2, 3, \ldots, 10$. The default NN settings in Modulus consist of six hidden layers (between input and output layers), with each layer containing 512 neurons.

5.1. Dataset, Loss, and Training Procedure

For the BCs components of the loss (Equation (7)), i.e., L_{BC} and data L_{Data}, the training dataset contains input/output pairs for the NN in the form $[x, y, Re, H_o, L_o]$, $[U, V, P]$, where the output represents the target or desired output for the specific input. The input vector must be complete, but the output vector can contain one or more of the variables. For example, the no-slip BC includes points (x, y) located on the no-slip walls, along with the corresponding parameter values (Re, H_o, and L_o) in the input and target values for U and V (both of which are zero) in the output. The loss was calculated using the L_2 loss function, which is the default in Modulus. For the L_{PDE} component of the loss, the relevant derivatives of the NN are calculated using automatic differentiation to compute the residuals of the governing PDEs, which include the continuity equation (Equation (2)) and momentum equations (Equations (3) and (4)). These residuals are then added to the loss and minimized during training. In addition to minimizing PDE residuals, integral continuity was enforced by ensuring that the velocity integrated over an orthogonal line equals the volumetric flow rate, which is known from the inlet profile. In Modulus, this is achieved using Monte Carlo integration. The area parameter, which is constant for each point on a given line, makes the integrated quantity sensitive to the distribution of points. Since Modulus samples points randomly, this distribution is not necessarily even. Therefore, the number of points was increased in order to stabilize the corresponding loss component. For both the BCs and PDE parts of the loss, Modulus provides tools to handle this automatically given the geometry, parameter ranges, number of points to sample, and target output values.

Some initial tests were conducted with partial training of the FNN using only physics constraints and varying the number of frequencies sampled, the number of batches sampled, and different weights for the loss components. The setup that showed the most progress was selected for further tests. Here, the most important factor was the spatial weighting of the L_{PDE} loss components. Without spatial weighting, high loss was observed when the flow interacted with the obstacle and step which forced the flow downward and away from the sharp corners, hindering further improvement. Using wall distance to weight interior losses is a recommended practice in Modulus and is employed in several examples provided by NVIDIA (e.g., [23]) using a linear relationship, $W_{PDE} = C \times D_{Wall}$, where C is a constant, and D_{Wall} represents the wall distance. For the forward-facing step, this approach led to an imbalance between the loss upstream and downstream due to the difference in maximum wall distance. Therefore, two alternative functions were formulated heuristically (Figure 3), both of which use (tanh), as it provides a 'flat top' which enables better balance loss weighting between regions upstream and downstream of the step.

$$W_{PDE} = 0.5 + 0.5 \times \tanh(20 \times (D_{Wall} - 0.1)).$$

A batch of training data is shown in Figure 4, where inlet (red), no-slip (green), and outlet (blue) are used for L_{BC}, while interior (purple) and integral continuity (cyan) are used for L_{PDE}. Ten lines were sampled (spread over the x axis) for integral continuity in each batch. The figure also shows the relative densities of points sampled for the different loss components. A higher density was used for interior and no-slip close to the step. A total of 1000 batches were sampled for training, so the number of points in total is 1000 times that shown in the figure.

For the labeled data, the discretized solutions were read from CSV files, and the corresponding parameters (Re, H_o, L_o) were appended to all points $(x, y), (U, V, P)$ in each solution, resulting in a dataset with input (x, y, Re, H_o, L_o) and output (U, V, P). Each discretized solution contains roughly 2.5×10^4 points, giving a total of 2.5×10^6 points for the 100 solutions used for training. The total number of points was divided into 1800 batches to avoid running out of GPU memory during training, resulting in a batch size for the labeled training data of approximately 1400.

For the tests with labeled data only for pressure, a single point from the bottom wall boundary ($y = -0.5$) was selected as close as possible to the x locations $[-4, -2, 0, 3, 6, 9]$. This gives six points per solution and a total of 600 points of labeled training data. Given the small size of this dataset, it was not divided into batches, meaning that the same labeled data were included in every training batch. However, the batch sizes for the other components of the dataset, such as the PDE and boundary condition losses, remained unchanged.

Table 3 shows each component of the training dataset, including batch size, total number of points, and their individual weight, with which the corresponding loss is multiplied during training.

To minimize the number of trials, all models were trained using the same setup: the Adam optimizer with an initial learning rate (step size in the gradient descent) of 5×10^{-4} and a learning rate decay of 0.95 every 4000 steps. Furthermore, to avoid variation from the initialization of the NN trainable parameters ϕ, each test was initialized using the same parameters. This was achieved with two initialized but untrained NNs, one FNN and one FCNN, which were then trained using the different methods. All training was performed on a PC with an NVIDIA GeForce RTX 4090 GPU (NVIDIA, Santa Clara, CA, USA) with 24 GB of memory, an Intel(R) Xeon(R) CPU E5-2623 v3 @ 3.00GHz (Intel, Santa Clara, CA, USA), and 128 GB of system memory (8 × 16 GB @ 2133 MHz).

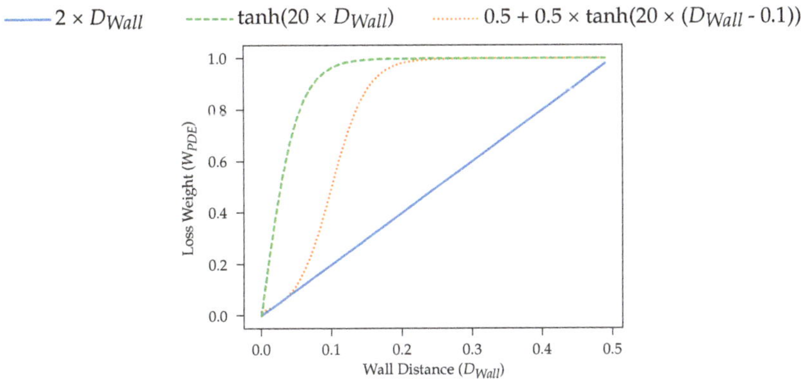

Figure 3. Comparison of different spatial loss weighting schemes for the PDE component (L_{PDE}) used in this study, visualized as functions of wall distance. The default approach utilized in many of the Modulus examples weights losses based on a linear relation to wall distance (solid line), while alternative schemes adjust weights using non-linear tanh-based functions to improve balance between upstream and downstream regions of the step given the difference in maximum wall distance. This study employs the alternative tanh-based weighting scheme (dotted line), as it demonstrated improved performance in preliminary tests.

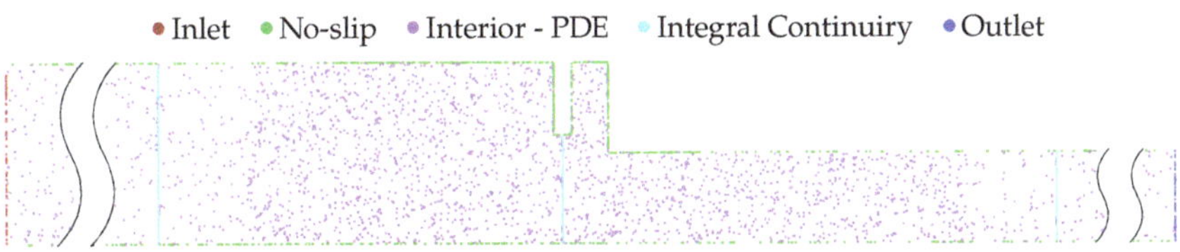

Figure 4. One batch out of 1000 from the physics dataset sampled using NVIDIA Modulus' geometry module. Points were sampled for inlet (red), no-slip (green), and outlet (blue) for L_{BC} and interior (purple) and integral continuity (cyan) for L_{PDE}. The figure shows the relative densities of points, with higher densities near the step.

Table 3. Components of the training dataset, including batch size, number of points, weight, and criteria for sampling.

Component	Batch Size	No. Points	Weight	Criteria
Inlet	64	6.4×10^4	10	-
Outlet	64	6.4×10^4	10	-
No-slip	1024	1.024×10^6	10	$y \leq 0$ or $x \notin (-1.5H_C, H_C/2)$
No-slip Hr	512	1.024×10^6	10	$y > 0$ and $x \in (-1.5H_C, H_C/2)$
Interior	2048	2.048×10^6	W_{PDE}	$x \notin (-2H_C, 2H_C)$
Interior Hr	2048	2.048×10^6	W_{PDE}	$x \in (-2H_C, 2H_C)$
Integral Continuity	512 × 10	5.12×10^6	0.1	-

5.2. Design Optimization

In order to evaluate ΔC_p for a design in the optimization process using the parameterized NN models, a set of 1024 points (x, y) was sampled along two lines (parallel to the y axis) in the domain at $x = -4$ and $x = 4$, totaling 2048 points. The corresponding parameter values (Re, H_o, L_o) for the design were appended to each point in order to infer pressure at each location and calculate the average pressures. These average pressures were then used to calculate ΔP and ΔC_p. To manage memory consumption, this evaluation was conducted in batches of 200 designs each. The optimization process was conducted using the Python multi-objective optimization package, pymoo 0.6.1.1 (Julian Blank, Michigan State University, MI, USA, pymoo.org) . Differential evolution was chosen as the optimization algorithm, with a population size of 200. The default stopping criteria provided by pymoo were applied.

6. Results

The performance of the PINNs was evaluated using various metrics and training approaches. The results are presented in terms of error metrics for the primitive flow variables and the objective function ΔC_p. Additionally, the response of ΔC_p to changes in the parameter L_o and optimization results are presented.

6.1. Flow Fields

Flow fields for the validation point (DP 171, on git) are presented. This is the validation point occurring most frequently when selecting the validation point with the highest normalized L_2 error of the primitive flow variables for each model.

Looking at the velocity magnitude predicted by the different models, as shown in Figure 5, it is evident that neither the FCNN nor the FNN was able to accurately predict the recirculation region in the cavity between the step and the obstacle without any labeled data. For the FCNN, the flow also separated from the top wall much further upstream than observed in the ground truth solution. However, with the addition of labeled data, both architectures successfully predicted both the flow separation and the recirculation region, and there were few noticeable differences between the data-only and the physics-plus-data training approaches. One notable difference was that the FNN trained with data only appeared to slightly mispredict the position of the obstacle, as the flow seemed to correspond to a configuration where the obstacle was placed further upstream. When trained with only pressure data and physics, both architectures provided predictions more similar to the FCNN trained with physics only, except for the FCNN with $W_d = 1$, where the flow bent upward upstream of the obstacle. For the FNNs trained initially with data only ($S_d = 10^5$), the flow closely aligned with the results from physics-only training. The pressure fields, as illustrated in Figure 6, exhibited a similar pattern in terms of which training methods yielded predictions closer to the ground truth.

The streamline plots in Figure 7 more clearly show that the data-only FNN struggled to predict the flow around the obstacle, appearing as if the flow corresponded to a configuration where the obstacle was placed further upstream (higher L_o). Adding physics improved the prediction in this area, especially for $W_d = 0.1$, but also led to an increase in the size of the upstream recirculation region, which did not agree with the CFD solution. With physics only, this recirculation region grew even larger, and the model failed to predict the recirculation in the cavity between the step and the obstacle.

Compared to the FNN, the streamline plots for the FCNN (Figure 8) show less variability between the data-only and data-plus-physics approaches, with all methods predicting both recirculation regions in a manner close to the CFD solution. However, for the physics-only solution, the increase in the size of the upstream recirculation region is much more noticeable, and, similar to the FNN, the recirculation region in the cavity was not predicted.

Figure 5. Velocity magnitude for the configuration $Re = 977.5$, $L_o = 0.58$, and $H_o = 0.45$ (DP171 on git) as predicted by the different models. W_d is the weight for the data part of the loss, L_{Data}, and S_d is the number of training steps before W_d is set to 0. Where no value for S_d is given, the default is 5×10^5, which is the full length of training.

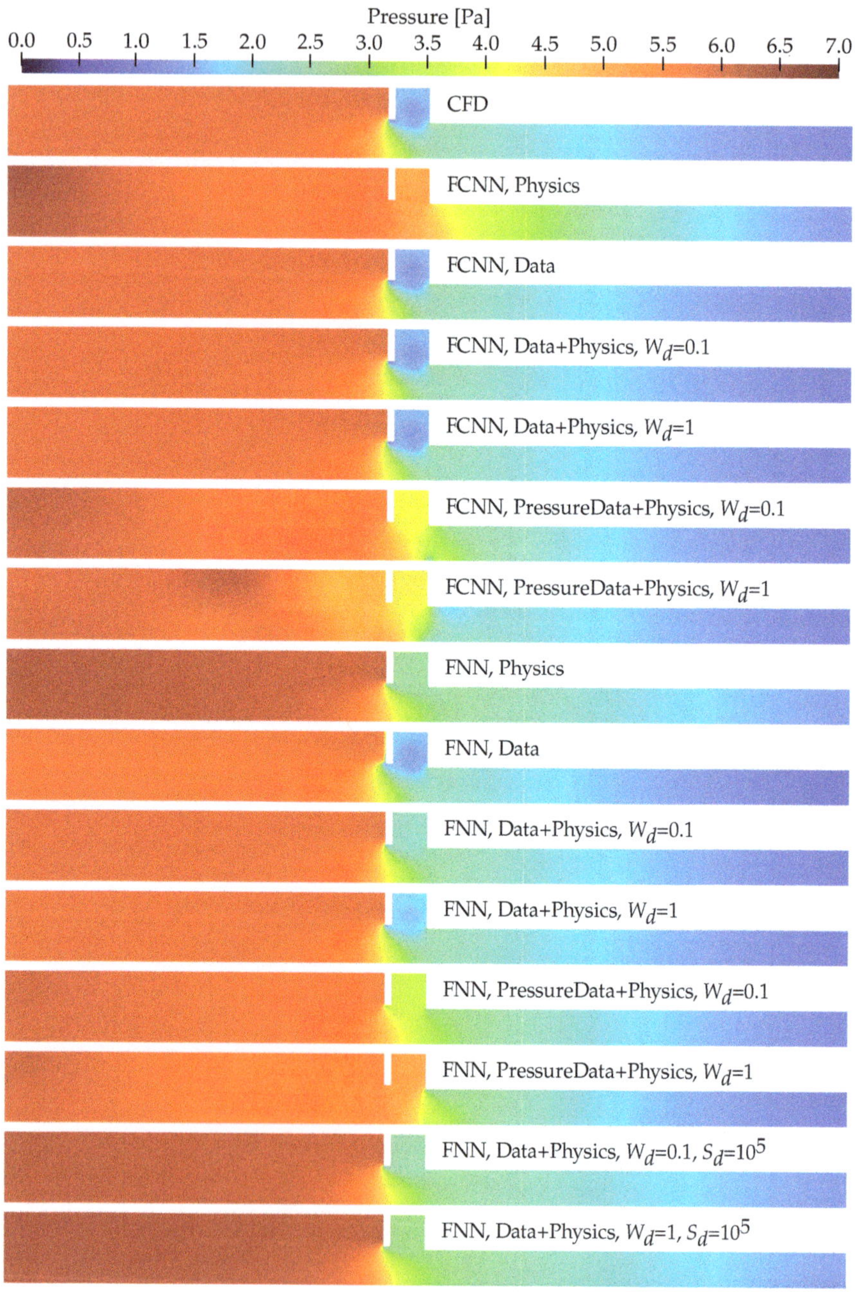

Figure 6. Pressure distribution for the configuration $Re = 977.5$, $L_o = 0.58$, and $H_o = 0.45$ (DP171 on git) as predicted by the different models. W_d is the weight for the data part of the loss, L_{Data} and S_d is the number of training steps before W_d is set to 0. Where no value for S_d is given, the default is 5×10^5, which is the full length of training.

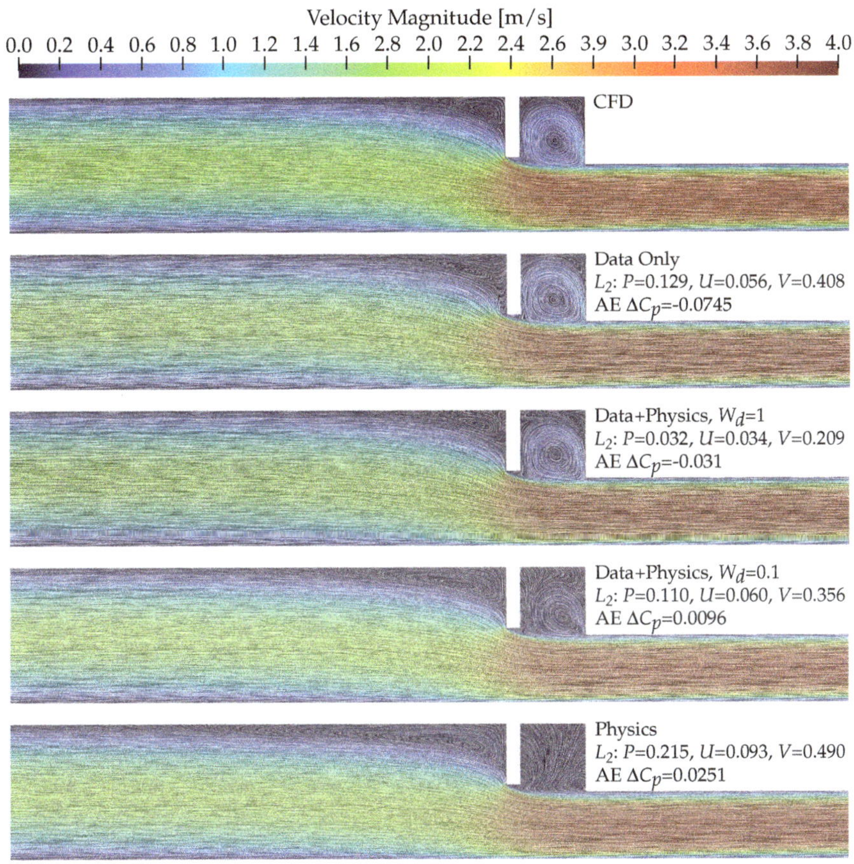

Figure 7. Streamlines showing solutions for the configuration $Re = 977.5$, $L_o = 0.58$, and $H_o = 0.45$ (DP171 on git) from the Fourier neural networks (FNN) with different training approaches and the corresponding validation errors.

6.2. Error Metrics for Primitive Flow Variables

The normalized L_2 error of the primitive flow variables was analyzed across the different training approaches, defined as

$$L_2(y, \hat{y}) = \sqrt{\frac{1}{n}\sum_{i=1}^{n}(y_i - \hat{y}_i)^2 \Big/ \mathrm{Var}(y)}, \quad (8)$$

where

- y_i is the true value of the i-th element in the discretized solution.
- $\hat{y}_i$ is the predicted value of the i-th element in the discretized solution.
- n is the total number of elements in the discretized solution.
- $\mathrm{Var}(y)$ is the variance of the true values in the discretized solution, defined as $\mathrm{Var}(y) = \frac{1}{n}\sum_{i=1}^{n}(y_i - \bar{y})^2$, where $\bar{y}$ is the mean of the true values.

For the FCNN (Figure 9), the addition of labeled data significantly reduced the error of all variables compared to using only physics constraints, and there was no noticeable difference in convergence rates between the different training approaches with data. The best-performing model for the velocity components U and V was the one trained with data only, whereas for the pressure variable P, the model that combined both data and physics constraints (with $W_d = 1$) achieved the best performance. Training with pressure data plus physics did not provide any performance improvements for U and V and led to only a slight improvement for P.

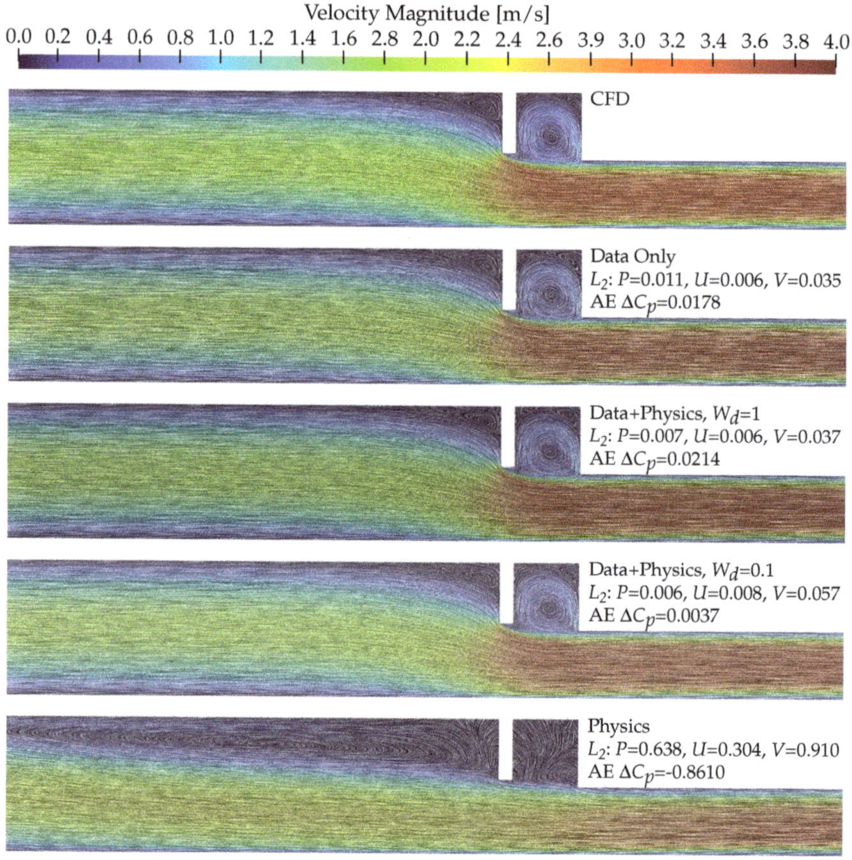

Figure 8. Streamlines showing solutions for the configuration $Re = 977.5$, $L_o = 0.58$, and $H_o = 0.45$ (DP171 on git) from the fully connected neural networks (FCNN) with different training approaches and the corresponding validation errors.

For the FNN (Figure 10), similar trends were observed, where the addition of labeled data improved performance. However, the improvement was less pronounced compared to the FCNN. This was due to the lower error of the FNN compared to the FCNN when trained with physics only, and the higher error of the FNN compared to the FCNN when trained with both data and physics. Additionally, for U and V, the difference between physics and data plus physics with $W_d = 0.1$ was mostly the convergence rate, where the addition of data significantly reduced the error early during training. Training with pressure data plus physics increased the error of all primitive flow variables.

Table 4 summarizes the normalized L_2 errors, showing that the FCNN consistently outperformed the FNN in terms of mean error and standard deviation, indicating higher and more stable performance across the design space. The exceptions to this trend were in the cases of training with no labeled data, where the FNN performed better, and training with pressure data plus physics, where both architectures showed low performance.

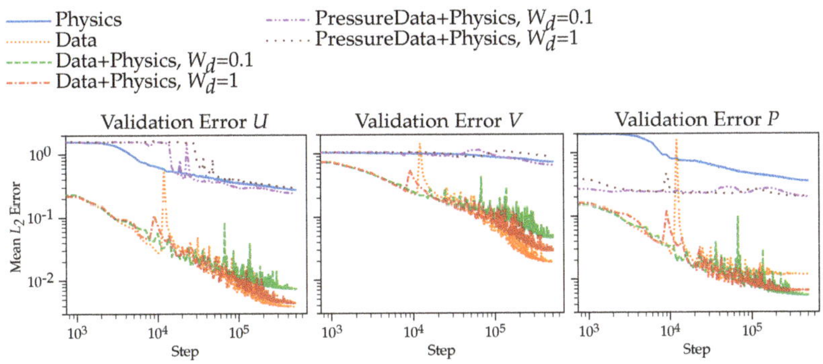

Figure 9. Mean normalized L_2 error of primitive flow variables plotted every 10^3 training steps for FCNNs. W_d is the weight for the data part of the loss, L_{Data}, and S_d is the number of training steps before W_d is set to 0.

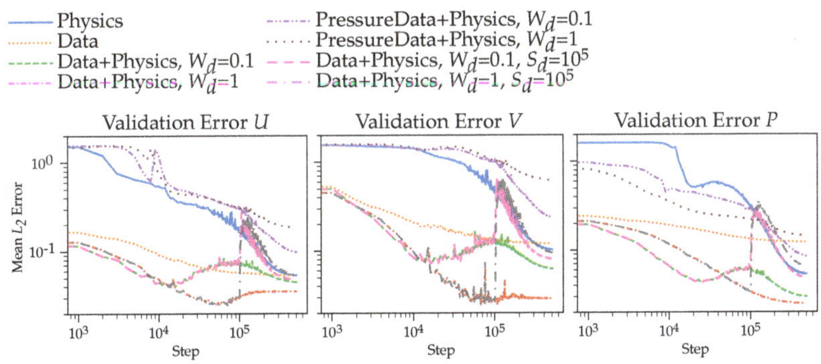

Figure 10. Mean normalized L_2 error of primitive flow variables plotted every 10^3 training steps training steps for FNNs. W_d is the weight for the data part of the loss, L_{Data}, and S_d is the number of training steps before W_d is set to 0. Where no value for S_d is given, the default is 5×10^5, which is the full length of training.

Table 4. Normalized L_2 error of primitive flow variables, given as mean (μ) and standard deviation (σ) over the validation points; the best value is in bold. W_d is the weight for the data part of the loss, L_{Data}, and S_d is the number of training steps before W_d is set to 0.

Training	W_d	S_d	Arch.	L_2 U		L_2 V		L_2 P	
				μ	σ	μ	σ	μ	σ
Physics	-	-	FCNN	0.273	0.031	0.730	0.107	0.373	0.121
Data	-	5×10^5	FCNN	**0.004**	**0.002**	**0.019**	0.011	0.012	0.018
Data + Physics	0.1	5×10^5	FCNN	0.007	**0.002**	0.047	0.011	**0.006**	**0.004**
Data + Physics	1	5×10^5	FCNN	**0.004**	**0.002**	0.030	**0.010**	0.007	0.008
Pressure Data + Physics	0.1	5×10^5	FCNN	0.239	0.035	0.646	0.133	0.209	0.096
Pressure Data + Physics	1	5×10^5	FCNN	0.293	0.023	0.885	0.100	0.216	0.102
Physics	-	-	FNN	0.056	0.018	0.240	0.140	0.084	0.046
Data	-	5×10^5	FNN	0.056	0.036	0.260	0.187	0.188	0.115
Data + Physics	0.1	5×10^5	FNN	0.047	0.009	0.183	0.087	0.050	0.026
Data + Physics	1	5×10^5	FNN	0.037	0.007	0.121	0.047	0.044	0.024
Pressure Data + Physics	0.1	5×10^5	FNN	0.101	0.035	0.377	0.181	0.129	0.092
Pressure Data + Physics	1	5×10^5	FNN	0.189	0.063	0.632	0.304	0.208	0.169
Data + Physics	0.1	1×10^5	FNN	0.050	0.014	0.210	0.128	0.076	0.042
Data + Physics	1	1×10^5	FNN	0.057	0.015	0.229	0.126	0.100	0.058

6.3. Error in Design Objective ΔC_p

The error of ΔC_p was evaluated in order to understand the accuracy of the PINNs in predicting the design objective. Figure 11 illustrates the errors for both FCNN and FNN models over the training process.

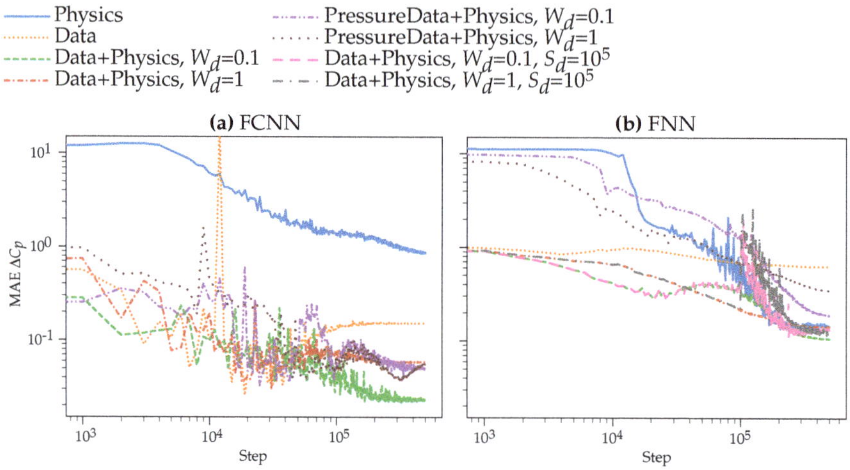

Figure 11. Mean absolute error (MAE) ΔC_p for FCNN (**a**) and FNN (**b**). W_d is the weight for the data part of the loss, L_{Data}, and S_d is the number of training steps before W_d is set to 0. Where no value for S_d is given, the default is 5×10^5, which is the full length of training.

For the FCNN, the addition of labeled data significantly reduced the error compared to using only physics constraints, and the model trained with both data and physics constraints ($W_d = 0.1$) showed the lowest mean absolute error (MAE) for ΔC_p. Examining the error of ΔC_p also shows a significant improvement when training with both data and physics constraints rather than data alone, unlike the normalized L_2 errors for the primitive flow variables. Similarly, training with pressure data plus physics now shows a reduction in error compared to both physics-and-data-only training, and the error is much closer to training with physics plus data. While the addition of data reduced the MAE for the FNN as well, the improvement was much smaller and the errors higher compared to the FCNN. Table 5 provides a detailed comparison of the errors for ΔC_p, highlighting the superior performance of the FCNN in terms of both mean and standard deviation.

6.4. Response of ΔC_p to Changes in L_o

The response of ΔC_p to changes in L_o was analyzed in order to evaluate the models' robustness for design optimization. Figure 12 shows the response curves for both FNNs and FCNNs.

For the FNN, the response of ΔC_p was highly variable, particularly for models trained with labeled data. This variability suggests that the FNN struggled to generalize across the parameter space when data were included. In contrast, the response for the FCNNs trained with data plus physics was more stable and consistent, demonstrating better generalization across the design space and notably also a response curve closer to CFD Discrete than CFD Continuous, which uses the same DoE size as the PINN models. Another thing worth noting is the response curves for FCNNs trained with pressure data plus physics which, although offset, show the correct general shape and a minimum located at roughly the same value of L_o.

Table 5. Prediction error of the design objective, ΔC_p, given as MAE, mean absolute percentage error (MAPE) as well as mean (μ) and standard deviation (σ) of the non-absolute error (E). The best values are in bold. W_d is the weight for the data part of the loss, L_{Data}, and S_d is the number of training steps before W_d is set to 0.

Training	W_d	S_d	Arch.	MAE ΔC_p	MAPE ΔC_p	E ΔC_p μ	E ΔC_p σ
Physics	-	-	FCNN	0.89	7.96	−0.89	0.26
Data	-	500	FCNN	0.15	0.95	−0.04	0.42
Data + Physics	0.1	500	FCNN	**0.02**	**0.19**	−0.01	0.03
Data + Physics	1	500	FCNN	0.06	0.39	−0.02	0.15
Pressure Data + Physics	0.1	500	FCNN	0.05	0.49	0.05	**0.02**
Pressure Data + Physics	1	500	FCNN	0.06	0.55	0.05	0.04
Physics	-	-	FNN	0.15	1.37	−0.10	0.18
Data	-	500	FNN	0.63	5.24	0.39	0.87
Data + Physics	0.1	500	FNN	0.11	0.89	−0.07	0.12
Data + Physics	1	500	FNN	0.14	1.19	−0.10	0.19
Pressure Data + Physics	0.1	500	FNN	0.19	1.80	−0.18	0.14
Pressure Data + Physics	1	500	FNN	0.35	2.71	−0.34	0.48
Data + Physics	0.1	100	FNN	0.13	1.28	−0.11	0.15
Data + Physics	1	100	FNN	0.12	1.18	−0.10	0.14

6.5. Optimization Results

The optimization results from all of the models were compared with those obtained in the commercial CFD solver in order to investigate the impact of labeled data on the optimization performance of the models. Figure 13 shows streamline plots for the optimal designs found in the commercial solver and the FCNN compared to the reference design without an obstacle for $Re = 1000$, illustrating how a well-placed obstacle reduces the size of the recirculation region downstream of the step and the ΔC_p value.

Optimal designs for other Re values are presented in Table 6 for both the discrete and continuous response surface models. The discrete model is considered the ground truth to which the PINN optimization results are compared. For Re values below 300, the optimal value for L_o is 0.2, which is the minimum allowed, indicating that the true optimum may be at a lower value; therefore, these results are excluded. Both methods give similar results in terms of the reduction in ΔC_p ($\delta\Delta C_p$) from the addition of an obstacle. This is the case even for the Res, where the optimal parameter values are quite different between the two, most notably for $Re = 300$, where the configurations $[L_o = 0.2,\ H_o = 0.41]$ and $[L_o = 0.35,\ H_o = 0.36]$ both lead to the same reduction (0.26).

Table 6. Optimal designs from response surface optimization in the CFD solver, both using discrete (for each Re) and continuous response surfaces (denoted: Discrete/Continuous in the table). The solutions from the discrete method are used as ground truth when comparing the different NN training approaches. L_o and H_o are the design parameters, and ΔC_p is the design objective. $\delta\Delta C_p$ shows the reduction in ΔC_p compared to ΔC_p^{ref} with no obstacle ($\Delta C_p^{ref} - \Delta C_p$). Both ΔC_p and ΔC_p^{ref} are from full simulations rather than response surface predictions.

Re	L_o [m]	H_o [m]	ΔC_p	ΔC_p^{ref}	$\delta\Delta C_p$
300	0.20/0.35	0.41/0.36	12.54/12.55	12.81	0.26/0.26
400	0.26/0.35	0.40/0.38	11.30/11.30	11.61	0.31/0.31
500	0.29/0.34	0.40/0.38	10.51/10.51	10.86	0.35/0.35
600	0.32/0.36	0.39/0.38	9.94/9.94	10.33	0.39/0.39
700	0.33/0.60	0.39/0.36	9.52/9.54	9.94	0.42/0.40
800	0.34/0.65	0.39/0.35	9.19/9.21	9.64	0.45/0.45
900	0.36/0.77	0.39/0.34	8.92/8.95	9.39	0.48/0.44
1000	0.38/0.67	0.38/0.35	8.69/8.71	9.20	0.50/0.49

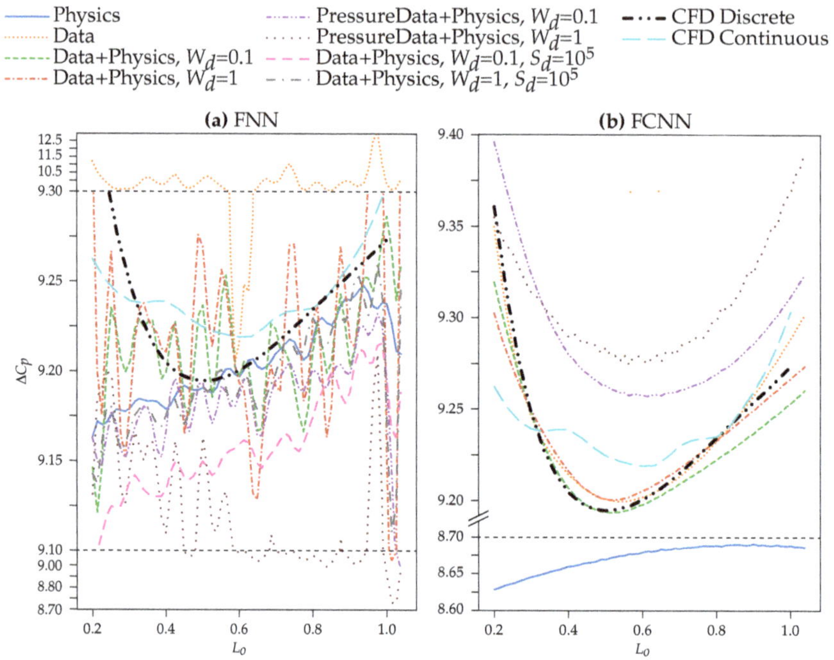

Figure 12. Response comparison for the different FNNs (**a**) and FCNNs (**b**). $Re = 800$; $H_o = 0.35$. W_d is the weight for the data part of the loss, L_{Data}, and S_d is the number of training steps before W_d is set to 0. Where no value for S_d is given, the default is 5×10^5, which is the full length of training.

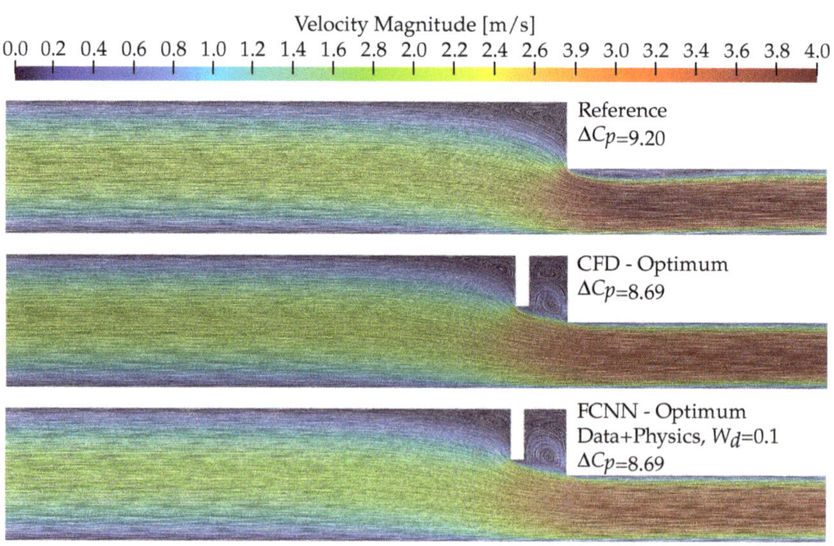

Figure 13. Streamlines visualizing the reduction in size of the recirculation region downstream of the step from the reference solution with no obstacle to the optimized configurations with an obstacle placed upstream of the step for $Re = 1000$. All solutions are from full simulations in the CFD solver.

In Table 7, which presents the optimal designs found using the FCNN models, it can be observed that for all training methods except for physics-only, the reduction in ΔC_p ($\delta \Delta C_p$)

was within a few percent of the ground truth optima found in the CFD solver ($\delta\hat{\Delta C_p}$). For some values of Re, the reduction was even larger than the ground truth; however, these differences were within the margin of error, and the solutions should be considered equivalent. The position of the obstacle varies between both the different training methods and the CFD solver, a similar trend to what was noticed when comparing the two CFD optimizations, CFD Discrete and CFD Continuous (Table 6). A closer look at the optimums found hints at several minima where the obstacle can be placed further upstream without a significant increase in ΔC_p if its height is also reduced. The data plus physics training method with $W_d = 0.1$ yielded configurations closest to the ground truth. Generally, the differences in configurations were larger for lower Re, although the design objective ΔC_p remained close to the ground truth solution, indicating that the sensitivity to both L_o and H_o was lower at low Re.

Table 7. Optimal designs from optimization using the FCNNs. L_o and H_o are the design parameters, and ΔC_p is the design objective. $\delta\Delta C_p$ shows the reduction in ΔC_p compared to the configuration with no obstacle, ΔC_p^{ref} in Table 6 . $\hat{L_o}$, $\hat{H_o}$, and $\delta\hat{\Delta C_p}$ show the difference compared to the ground truth solution (CFD Discrete in Table 6). ΔC_p was evaluated in the CFD solver to compare the solutions without including prediction errors from the NNs. W_d denotes the weight for the data part of the loss, L_{Data}. The best value(s) for each Re is highlighted in bold.

Model	Re	L_o [m]	$\hat{L_o}$ [m]	H_o [m]	$\hat{H_o}$ [m]	ΔC_p	$\delta\Delta C_p$	$\delta\hat{\Delta C_p}$[%]
Physics	300	0.20	−0.00	0.50	0.09	12.99	−0.18	−169.32
	400	0.20	−0.06	0.50	0.10	11.76	−0.15	−148.27
	500	0.20	−0.09	0.50	0.10	10.99	−0.13	−137.27
	600	0.20	−0.12	0.50	0.11	10.46	−0.12	−132.19
	700	0.20	−0.13	0.50	0.11	10.06	−0.12	−128.38
	800	0.20	−0.14	0.50	0.11	9.76	−0.12	−126.46
	900	0.20	−0.16	0.50	0.11	9.51	−0.12	−125.26
	1000	0.20	−0.18	0.50	0.12	9.32	−0.12	−123.96
Data	300	0.22	0.02	0.41	−0.00	12.54	0.27	**2.33**
	400	0.50	0.24	0.35	−0.05	11.32	0.29	−4.84
	500	0.20	−0.09	0.41	0.01	10.52	0.34	−1.95
	600	0.20	−0.12	0.41	0.02	9.95	0.38	−2.80
	700	0.24	−0.09	0.40	0.02	9.52	0.42	−1.46
	800	0.28	−0.07	0.40	0.01	9.19	0.44	−1.03
	900	0.29	−0.06	0.40	0.01	8.92	0.47	−0.93
	1000	0.31	−0.07	0.40	0.01	8.70	0.50	−0.28
Data + Physics $W_d = 0.1$	300	0.26	0.06	0.39	−0.02	12.54	0.27	2.06
	400	0.33	0.07	0.39	−0.01	11.30	0.31	**0.97**
	500	0.30	0.01	0.40	0.00	10.51	0.35	**0.98**
	600	0.29	−0.02	0.40	0.01	9.95	0.38	−1.30
	700	0.34	0.01	0.39	0.01	9.52	0.42	−0.57
	800	0.39	0.05	0.38	−0.00	9.19	0.45	**0.31**
	900	0.40	0.05	0.38	−0.00	8.92	0.48	**0.50**
	1000	0.41	0.03	0.38	−0.00	8.69	0.50	**0.35**
Data + Physics $W_d = 1$	300	0.49	0.29	0.35	−0.06	12.57	0.24	−7.95
	400	0.49	0.24	0.36	−0.04	11.32	0.30	−4.49
	500	0.20	−0.09	0.41	0.02	10.52	0.34	−1.52
	600	0.20	−0.12	0.42	0.03	9.95	0.38	−2.65
	700	0.20	−0.13	0.41	0.03	9.53	0.41	−4.05
	800	0.24	−0.11	0.40	0.02	9.19	0.44	−1.30
	900	0.28	−0.08	0.40	0.01	8.92	0.47	−0.54
	1000	0.30	−0.08	0.39	0.01	8.69	0.50	0.21

Table 7. *Cont.*

Model	Re	L_o [m]	$\hat{L_o}$ [m]	H_o [m]	$\hat{H_o}$ [m]	ΔC_p	$\delta \Delta C_p$	$\delta \Delta \hat{C}_p$[%]
Pressure Data + Physics	300	0.54	0.34	0.34	−0.07	12.57	0.24	−9.25
$W_d = 0.1$	400	0.51	0.25	0.35	−0.05	11.32	0.30	−4.78
	500	0.48	0.18	0.36	−0.03	10.51	0.34	−1.35
	600	0.47	0.16	0.37	−0.03	9.95	0.38	−1.72
	700	0.47	0.14	0.37	−0.02	9.52	0.42	−1.29
	800	0.46	0.12	0.37	−0.02	9.19	0.45	−0.51
	900	0.45	0.09	0.38	−0.01	8.92	0.48	−0.14
	1000	0.43	0.04	0.38	−0.00	8.69	0.50	**0.35**
Pressure Data + Physics	300	0.24	0.04	0.39	−0.02	12.54	0.26	−0.11
$W_d = 1$	400	0.47	0.21	0.36	−0.04	11.31	0.30	−3.32
	500	0.45	0.16	0.37	−0.03	10.51	0.35	−0.95
	600	0.33	0.01	0.40	0.00	9.94	0.39	**0.31**
	700	0.31	−0.03	0.40	0.01	9.52	0.42	**−0.47**
	800	0.31	−0.03	0.40	0.01	9.19	0.45	−0.22
	900	0.34	−0.02	0.39	0.01	8.92	0.48	0.25
	1000	0.41	0.02	0.38	−0.00	8.69	0.50	0.28

In contrast, the optimization performance with the FNN models (TableA1) was more uneven, and none of the training methods provided optima close to the ground truth for all *Re*.

7. Discussion

The findings of this study demonstrate that incorporating labeled data significantly impacts the performance of parametric PINNs, particularly in surrogate modeling and design optimization tasks. The results show that the impact of labeled data varies depending on the network architecture and the metrics used to evaluate performance.

For FCNN, the inclusion of labeled data improves both model accuracy and convergence rates. This is consistent with earlier studies on non-parametric PINNs, where labeled data were shown to enhance model performance [16]. Specifically, for the FCNN, training with only labeled data resulted in the best performance in terms of normalized L_2 validation error of the primitive flow variables U and V, which is also consistent with earlier studies [24]. Introducing the physics component into the training increased the error of U and V while reducing the error of P. The reduction in error of P was most notable when the weight for the data component of the loss, W_d, was set to 0.1 compared to 1, meaning less emphasis on data than on physics. This also translated to a decrease in the error of the design objective ΔC_p and improved optimization performance. When analyzing the model's response to changes in parameter values, the same trends were observed, where a mix of data and physics was the preferred training approach. Interestingly, this also highlighted that FCNNs trained with data plus physics provided a better response than the response surface optimization with the same DoE size in the commercial solver. All training approaches for FCNN models that included data yielded optimal designs with a reduction in ΔC_p ($\delta \Delta C_p$ in Table 7) within a few percent of the ground truth obtained from response surface optimization in the commercial solver. The best performance was achieved with data plus physics and W_d set to 0.1, indicating that a balanced integration of data and physics constraints yields optimal performance for design optimization tasks.

Adding only pressure data at six points on the boundary, with points both upstream and downstream of the step, led to improvements in the predictions of the design objective, ΔC_p, as well as in the response to parameter changes and optimization performance, even though there was little to no improvement in the normalized L_2 error of the primitive flow variables. The predicted flow fields from these NNs showed unphysical solutions, and the increased optimization performance is likely very case specific since the design objective is derived from the pressures upstream and downstream of the step. The same may hold true for the other training approaches combining data and physics compared to data only, as the addition of physics only led to a decreased normalized L_2 error of pressure, though the differences were much smaller.

In contrast, the performance of FNNs showed inconsistencies when trained with labeled data and led to uneven responses to parameter changes, which limited their effectiveness for design optimization. Despite these issues, the addition of labeled data still reduced the mean and standard deviation of both the normalized L_2 validation error of the primitive flow variables and the absolute error of the design objective ΔC_p. This highlights that these metrics alone may not fully capture the nuances of model performance across varying conditions. None of the FNNs produced valid optimization results, and plotting the design objective ΔC_p for a parameter sweep over L_o clearly showed an uneven response (Figure 12). FNNs rely on Fourier feature encoding, transforming inputs into a combination of sinusoidal functions in order to mitigate spectral bias. This encoding is evident in the response, where FNN predictions exhibit a periodic structure, reflecting the sinusoidal basis used for input encoding. While this encoding can enhance the network's ability to capture steep gradients and sharp corners, it might also contribute to inconsistencies when modeling non-periodic behaviors or smooth transitions. This inconsistency was hypothesized to be caused by the finite number of frequencies sampled, which may not accurately represent all the discretized solutions used for training. Consequently, this leads to conflicts between representing one data point and another, causing unreliable results. To test this theory, the FNN data plus physics with $W_d = 0.1$ was trained again with the number of frequencies increased, but this resulted in worse predictions for both primitive flow variables and the design objective (ΔC_p).

The choice of network architecture plays a crucial role in the model's performance. FCNNs demonstrated better generalization and robustness compared to FNNs when labeled data were included. However, without labeled training data, the FCNN failed to converge, while the FNN performed better. This suggests that the preferred neural network architecture may depend on the specific training approach. The initial tests in this study to tune the training process were conducted using the FNN and physics-only training. It is possible that the FCNN could also perform better with physics-only training if the training process were further optimized. While parametric Fourier PINNs have reportedly been trained with labeled data in order to improve performance in other studies, such as [9], the only metric reported was MSE, which this study revealed may not detect issues with uneven responses to design parameter changes. Furthermore, the aforementioned study utilized a split network approach, where each flow variable was predicted by a separate neural network, potentially mitigating some of the issues highlighted here.

The dataset size and distribution are also influential factors. The dataset used in this study included 100 design points, significantly more than the six design points used in the work by [9]. The larger dataset size may have contributed to the observed differences in model performance. It is plausible that there exists a critical dataset size where the performance of FCNNs begins to surpass that of FNNs. This hypothesis is supported by the observation that FCNNs perform better with 100 data points, while FNNs outperform FCNNs when no labeled data are used. The number of spatial points per design point also differs between studies, potentially impacting convergence speed and final model accuracy. In the aforementioned study, not all spatial points were used to speed up convergence, although no comparison was provided.

Based on the fact that the FCNNs were able to predict the flow fields when labeled data were included, the problems observed with physics-only training stem from the training process rather than the expressivity of the neural networks. This is in line with conclusions drawn by [13] for non-parametric PINNs, where training failure is attributed to the complexity of the loss landscape. Several studies (e.g., [13,17]) have reported that the PINN training process can be made more robust by employing different initialization strategies and using less-complex loss functions in the early stages of training. Utilizing an alternative weighting scheme, adjusting the point density for the various physics loss components, or modifying other training parameters such as the optimizer or learning rate might lead to improved results. However, as reported by [14], there are no current methods that works universally for all cases, making this tuning process time consuming and error prone. Reference [13] also notes that for more challenging problems, PINNs may fail to train even with extensive parameter tuning. If including labeled data can minimize or even eliminate the need for such tuning, it may offer a more practical solution, provided that the cost of generating data is reasonably low. Moreover, this also depends on both the size of the DoE and the fidelity required for the data. In this work, a DoE of size 100 was used, which is significantly larger than in other studies (e.g., [5,9,10]) based on the assumption that if a method does not work with a larger dataset, it likely will not work with a smaller one. Hence, further studies are required to address questions related to dataset size and data fidelity. Some studies (e.g., [5,16]) have shown that for non-parametric PINNs, where data are spread over principal axes (x, y, z, and/or t), relatively sparse data can suffice. Similarly, utilizing low-fidelity or noisy data has been shown to work well in combination with a small amount of high-fidelity data [25]. With a high-fidelity dataset of reasonable size, methods developed specifically for data-driven training, such as deep neural operators [26,27] and neural implicit flow [28], may also be considered for comparison. The potential benefit of using parameterized PINNs with some labeled data, compared to purely data-driven methods, is the possibility of using a smaller or lower-fidelity dataset, as the inclusion of physics could help prevent overfitting.

Limitations

This study included a single test case, which limits the generality of the findings herein. The flow was also considered laminar, providing a relatively small pressure drop and, hence, improvement capability when adding a upstream object. That said, it would be interesting to perform this analysis within a turbulent flow regime. Furthermore, learning rates, point density for the different dataset components, and other hyperparameters were not individually adjusted for each training approach, which could further improve the results. Future suggested research directions include

- Tuning the training process individually for different cases.
- Investigating the optimal size of the DoE. Since the FCNN trained with only data performed well, it is likely that the DoE size could be decreased.
- Using a subset of spatial points for each design point in order to reduce computational cost.
- Investigating the effects of noisy or low-fidelity data.
- Exploring the robustness of the models to different hyperparameters.

8. Conclusions

This study demonstrates the considerable impact of incorporating labeled data into parametric PINNs for surrogate modeling and design optimization tasks. The findings show that, for fully connected neural networks (FCNN), the addition of labeled data consistently improves model accuracy and convergence rates, with the best performance being achieved when using a lower weighting factor for the data part of the loss. FCNNs trained under this balanced approach produced optimized designs within a few percent of the ground-truth data obtained from simulations in a commercial CFD solver, while also showing a stable response to parameter variations.

Contrastingly, the Fourier neural networks (FNN) displayed inconsistent performance when trained with labeled data, with uneven responses to parameter changes affecting the reliability of the design optimization outcomes. Importantly, FNNs trained with data showed reduced validation errors for the primitive flow variables, highlighting that such validation metrics alone may not fully capture the differences in model performance across architectures and training approaches.

Furthermore, the study reveals that, while adding labeled data for pressure at six boundary points improved predictive accuracy for the design objective ΔC_p, the improvements in flow field accuracy are limited, indicating that performance may vary with case-specific objectives.

Author Contributions: Conceptualization, E.G. and M.A.; methodology, E.G. and M.A.; software, E.G.; validation, E.G. and M.A.; formal analysis, E.G.; investigation, E.G.; resources, E.G.; data curation, E.G.; writing—original draft preparation, E.G.; writing—review and editing, E.G. and M.A.; visualization, E.G.; supervision, M.A; project administration, E.G. and M.A. All authors have read and agreed to the published version of the manuscript.

Funding: This research received no external funding.

Data Availability Statement: All data including the dataset, trained models, and code are available at https://github.com/battlejacket/fwdFacingStep (accessed on 5 December 2024).

Acknowledgments: Andreas Hägg, Tove Hagstedt Palagyi (Siemens Energy): Support during study design and implementation. Chat GPT (OpenAI): Language support and proofreading. The computations were enabled by resources provided by the National Academic Infrastructure for Supercomputing in Sweden (NAISS), partially funded by the Swedish Research Council through grant agreement no. 2023/4-2, as well as the National Supercomputer Centre (NSC), funded by Linköping University.

Conflicts of Interest: The authors declare no conflicts of interest. The funders had no role in the design of the study; in the collection, analyses, or interpretation of data; in the writing of the manuscript; or in the decision to publish the results.

Abbreviations

The following abbreviations are used in this manuscript:

CFD	Computational Fluid Dynamics
NN	Neural Network
FCNN	Fully Connected Neural Network
FNN	Fourier Neural Network
PINN	Physics-Informed Neural Network
AE	Absolute Error
MAE	Mean Absolute Error
MAPE	Mean Absolute Percentage Error
DoE	Design of Experiments
L_2	L_2 Norm (Euclidean Distance)
PDE	Partial Differential Equation

Appendix A. Mesh Verification

The grid refinement factor was set to 1.4 between three sets of meshes (total number of elements: course = 1.3×10^4, medium = 2.8×10^4, and fine = 5.8×10^4) to ensure significantly different mesh densities [29]. The results from 100 design points were assessed using DoE with a continuous Latin hyperccube sampling to cover the design space of the parameters Re, L_o, and H_o. The mean absolute percentage error (MAPE) on ΔC_p between the medium and fine mesh was 0.5%, with all design points below 1% error. Further grid refinement was not considered relevant for the purposes of the study.

Appendix B. Optimal Designs from Optimization Using the FNNs

Table A1. Optimal designs from optimization using the FNNs. L_o and H_o are the design parameters, and ΔC_p is the design objective. $\delta\Delta C_p$ shows the reduction in ΔC_p compared to the configuration with no obstacles, ΔC_p^{ref} in Table 6 . $\hat{L_o}$, $\hat{H_o}$, and $\delta\hat{\Delta C_p}$ show the difference compared to the ground truth solution (CFD Discrete in Table 6). ΔC_p was evaluated in the CFD solver to compare the solutions without including prediction errors from the NNs. W_d denotes the weight for the data part of the loss, L_{Data}. The best value(s) for each Re is highlighted in bold.

Model	Re	L_o [m]	$\hat{L_o}$ [m]	H_o [m]	$\hat{H_o}$ [m]	ΔC_p	$\delta\Delta C_p$	$\delta\hat{\Delta C_p}$ [%]
Physics	300	0.20	−0.00	0.10	−0.31	12.83	−0.02	−107.45
	400	0.20	−0.06	0.34	−0.07	11.44	0.17	−43.76
	500	0.20	−0.09	0.33	−0.06	10.67	0.19	−46.11
	600	0.20	−0.12	0.33	−0.06	10.14	0.19	−50.22
	700	0.20	−0.13	0.33	−0.06	9.74	0.20	−53.25
	800	0.20	−0.14	0.33	−0.06	9.43	0.21	−54.12
	900	0.20	−0.16	0.33	−0.05	9.18	0.21	−55.27
	1000	0.20	−0.18	0.33	−0.05	8.99	0.21	−58.00
Data	300	0.60	0.40	0.43	0.02	12.81	0.00	−100.00
	400	0.60	0.34	0.43	0.03	11.53	0.09	−71.51
	500	0.59	0.30	0.43	0.03	10.70	0.16	−53.77
	600	0.85	0.53	0.31	−0.08	10.00	0.33	−14.35
	700	0.85	0.52	0.30	−0.08	9.57	0.37	−12.56
	800	0.85	0.51	0.31	−0.08	9.23	0.40	−9.89
	900	0.59	0.24	0.31	−0.08	8.96	0.43	**−9.12**
	1000	0.85	0.46	0.31	−0.07	8.73	0.46	−7.77
Data + Physics $W_d = 0.1$	300	0.28	0.08	0.39	−0.02	12.54	0.27	**1.95**
	400	0.22	−0.04	0.34	−0.07	11.42	0.19	−37.50
	500	0.22	−0.08	0.34	−0.05	10.63	0.23	−34.46
	600	0.22	−0.10	0.40	0.01	9.96	0.37	**−3.92**
	700	0.22	−0.11	0.40	0.01	9.53	0.41	**−3.92**
	800	0.22	−0.13	0.40	0.01	9.21	0.42	**−5.64**
	900	0.21	−0.15	0.35	−0.03	9.07	0.32	−32.23
	1000	0.22	−0.17	0.40	0.01	8.72	0.47	**−6.06**
Data + Physics $W_d = 1$	300	0.46	0.26	0.44	0.03	12.76	0.05	−82.50
	400	0.64	0.38	0.39	−0.01	11.41	0.21	−32.66
	500	0.64	0.35	0.39	−0.01	10.59	0.27	**−22.09**
	600	0.64	0.33	0.39	0.00	10.02	0.31	−19.48
	700	0.64	0.31	0.39	0.00	9.59	0.35	−17.46
	800	0.64	0.30	0.39	0.00	9.24	0.39	−12.57
	900	0.65	0.29	0.39	0.01	8.98	0.42	−12.80
	1000	0.64	0.26	0.31	−0.07	8.73	0.46	−7.68
Pressure Data + Physics $W_d = 0.1$	300	0.22	0.02	0.39	−0.02	12.55	0.26	−1.15
	400	0.22	−0.04	0.40	−0.00	11.31	0.31	−1.03
	500	0.22	−0.08	0.50	0.10	11.00	−0.14	−140.75
	600	0.21	−0.10	0.50	0.11	10.46	−0.13	−133.61
	700	0.21	−0.12	0.50	0.11	10.07	−0.13	−130.39
	800	0.21	−0.13	0.50	0.11	9.76	−0.13	−127.94
	900	0.21	−0.15	0.50	0.11	9.52	−0.13	−126.22
	1000	0.21	−0.17	0.50	0.12	9.32	−0.13	−125.34
Pressure Data + Physics $W_d = 1$	300	1.00	0.80	0.50	0.09	13.78	−0.97	−471.95
	400	1.00	0.74	0.50	0.10	12.42	−0.81	−361.57
	500	1.00	0.71	0.50	0.10	11.56	−0.70	−302.15
	600	1.00	0.68	0.50	0.11	10.96	−0.62	−260.95
	700	1.00	0.67	0.50	0.11	10.51	−0.57	−234.13
	800	1.00	0.66	0.50	0.11	10.15	−0.51	−214.17
	900	1.00	0.64	0.50	0.11	9.86	−0.47	−198.63
	1000	1.00	0.62	0.50	0.12	9.63	−0.43	−186.22

Table A1. *Cont.*

Model	Re	L_o [m]	$\hat{L_o}$ [m]	H_o [m]	$\hat{H_o}$ [m]	ΔC_p	$\delta \Delta C_p$	$\delta \hat{\Delta C_p}$ [%]
Data + Physics	300	0.22	0.02	0.33	−0.08	12.65	0.16	−38.90
$W_d = 0.1$	400	0.22	−0.04	0.34	−0.07	11.42	0.19	−37.95
$S_d = 10^5$	500	0.21	−0.08	0.34	−0.06	10.64	0.22	−37.96
	600	0.21	−0.11	0.34	−0.05	10.09	0.24	−39.10
	700	0.21	0.12	0.35	−0.04	9.67	0.27	−37.02
	800	0.20	−0.14	0.35	−0.04	9.36	0.28	−37.97
	900	0.20	−0.16	0.35	−0.04	9.11	0.28	−40.72
	1000	0.20	−0.18	0.35	−0.03	8.89	0.31	−38.66
Data + Physics	300	0.22	0.02	0.36	−0.05	12.59	0.22	−16.62
$W_d = 1$	400	0.21	−0.04	0.36	−0.04	11.38	0.24	−23.88
$S_d = 10^5$	500	0.21	−0.08	0.36	−0.04	10.60	0.26	−26.37
	600	0.21	−0.11	0.36	−0.04	10.05	0.28	−28.75
	700	0.21	−0.12	0.36	−0.03	9.64	0.30	−29.83
	800	0.21	−0.14	0.36	−0.03	9.32	0.31	−30.38
	900	0.21	−0.15	0.36	−0.03	9.06	0.33	−29.75
	1000	0.21	−0.18	0.36	−0.02	8.85	0.34	−31.61

References

1. Karniadakis, G.E.; Kevrekidis, I.G.; Lu, L.; Perdikaris, P.; Wang, S.; Yang, L. Physics-informed machine learning. *Nat. Rev. Phys.* **2021** , *3*, 422–440. [CrossRef]
2. Faroughi, S.A.; Pawar, N.M.; Fernandes, C.; Raissi, M.; Das, S.; Kalantari, N.K.; Kourosh Mahjour, S. Physics-Guided, Physics-Informed, and Physics-Encoded Neural Networks and Operators in Scientific Computing: Fluid and Solid Mechanics. *J. Comput. Inf. Sci. Eng.* **2024**, *24*, 040802. [CrossRef]
3. Cai, S.; Mao, Z.; Wang, Z.; Yin, M.; Karniadakis, G.E. Physics-informed neural networks (PINNs) for fluid mechanics: a review. *Acta Mech. Sin.* **2021**, *37*, 1727–1738. [CrossRef]
4. Cai, S.; Wang, Z.; Wang, S.; Perdikaris, P.; Karniadakis, G.E. Physics-Informed Neural Networks for Heat Transfer Problems. *J. Heat Transf.* **2021**, *143*, 060801. [CrossRef]
5. Haghighat, E.; Raissi, M.; Moure, A.; Gomez, H.; Juanes, R. A physics-informed deep learning framework for inversion and surrogate modeling in solid mechanics. *Comput. Methods Appl. Mech. Eng.* **2021**, *379*, 113741. [CrossRef]
6. Zhang, S.; Guo, X.W.; Li, C.; Zhao, R.; Yang, C.; Wang, W.; Zhong, Y. Multi-Viscosity Physics-Informed Neural Networks for Generating Ultra High Resolution Flow Field Data. *Int. J. Comput. Fluid Dyn.* **2023**, *37*, 279–297. [CrossRef]
7. Lu, Z.; Li, Y.; He, C.; Ren, J.; Yu, H.; Zhang, B.; Chen, Q. Multi-objective inverse design of finned heat sink system with physics-informed neural networks. *Comput. Chem. Eng.* **2024**, *180*, 108500. [CrossRef]
8. Sun, Y.; Sengupta, U.; Juniper, M. Physics-informed deep learning for simultaneous surrogate modeling and PDE-constrained optimization of an airfoil geometry. *Comput. Methods Appl. Mech. Eng.* **2023**, *411*, 116042. [CrossRef]
9. Ghosh, S.; Chakraborty, A.; Brikis, G.O.; Dey, B. RANS-PINN based Simulation Surrogates for Predicting Turbulent Flows. In Proceedings of the 1st Workshop on the Synergy of Scientific and Machine Learning Modeling @ ICML, Honolulu, HI, USA, 23–29 July 2023.
10. Wong, J.C.; Ooi, C.; Chiu, P.H.; Dao, M.H. Improved surrogate modeling of fluid dynamics with physics-informed neural networks. *arXiv* **2021**, arXiv:2105.01838.
11. Fuchi, K.W.; Wolf, E.M.; Makhija, D.S.; Wukie, N.A.; Schrock, C.R.; Beran, P.S. Investigation of Analysis and Gradient-Based Design Optimization Using Neural Networks. In Proceedings of the SMASIS, V001T08A002, Virtual, 15 September 2020. [CrossRef]
12. Rohrhofer, F.M.; Posch, S.; Gößnitzer, C.; Geiger, B.C. Data vs. Physics: The Apparent Pareto Front of Physics-Informed Neural Networks. *IEEE Access* **2023**, *11*, 86252–86261. [CrossRef]
13. Krishnapriyan, A.; Gholami, A.; Zhe, S.; Kirby, R.; Mahoney, M.W. Characterizing possible failure modes in physics-informed neural networks. In Proceedings of the Advances in Neural Information Processing Systems, Online, 6–14 December 2021; Ranzato, M., Beygelzimer, A., Dauphin, Y., Liang, P.S., Vaughan, J.W., Eds.; Curran Associates, Inc.: New York, NY, USA, 2021; Volume 34, pp. 26548–26560.
14. Monaco, S.; Apiletti, D. Training physics-informed neural networks: One learning to rule them all? *Results Eng.* **2023**, *18*, 101023. [CrossRef]
15. Chuang, P.Y.; Barba, L.A. Experience report of physics-informed neural networks in fluid simulations: Pitfalls and frustration. *arXiv* **2022**, arXiv:2205.14249.
16. Pioch, F.; Harmening, J.H.; Müller, A.M.; Peitzmann, F.J.; Schramm, D.; El Moctar, O. Turbulence Modeling for Physics-Informed Neural Networks: Comparison of Different RANS Models for the Backward-Facing Step Flow. *Fluids* **2023**, *8*, 43. [CrossRef]

17. Liu, X.; Zhang, X.; Peng, W.; Zhou, W.; Yao, W. A novel meta-learning initialization method for physics-informed neural networks. *Neural Comput. Appl.* **2022**, *34*, 14511–14534. [CrossRef]
18. Von Saldern, J.G.R.; Reumschüssel, J.M.; Kaiser, T.L.; Sieber, M.; Oberleithner, K. Mean flow data assimilation based on physics-informed neural networks. *Phys. Fluids* **2022**, *34*, 115129. [CrossRef]
19. NVIDIA Modulus Team. NVIDIA Modulus Symbolic (Container). Available online: https://github.com/NVIDIA/modulus-sym(accessed on 13 December 2024).
20. Hennigh, O.; Narasimhan, S.; Nabian, M.A.; Subramaniam, A.; Tangsali, K.; Fang, Z.; Rietmann, M.; Byeon, W.; Choudhry, S. NVIDIA SimNet™: An AI-Accelerated Multi-Physics Simulation Framework. In Proceedings of the Computational Science—ICCS 2021, Krakow, Poland, 16–18 June 2021; Paszynski, M., Kranzlmüller, D., Krzhizhanovskaya, V.V., Dongarra, J.J., Sloot, P.M., Eds.; Springer Cham: Cham, Switzerland, 2021; pp. 447–461. [CrossRef]
21. Ando, T.; Shakouchi, T. Flow characteristics over forward facing step and through abrupt contraction pipe and drag reduction. *Res. Rep. Fac. Eng. Mie Univ.* **2004**, *29*, 1–8.
22. Tancik, M.; Srinivasan, P.; Mildenhall, B.; Fridovich-Keil, S.; Raghavan, N.; Singhal, U.; Ramamoorthi, R.; Barron, J.; Ng, R. Fourier Features Let Networks Learn High Frequency Functions in Low Dimensional Domains. In Proceedings of the Advances in Neural Information Processing Systems, Virtual, 6–12 December 2020; Larochelle, H., Ranzato, M., Hadsell, R., Balcan, M., Lin, H., Eds. Curran Associates, Inc.: New York, NY, USA, 2020; Volume 33, pp. 7537–7547.
23. NVIDIA Modulus Team. NVIDIA Modulus Symbolic Documentation: FPGA Heat Sink with Laminar Flow. Available online: https://docs.nvidia.com/deeplearning/modulus/modulus-sym/user_guide/advanced/fpga.html (accessed on 13 December 2024).
24. Sun, L.; Gao, H.; Pan, S.; Wang, J.X. Surrogate modeling for fluid flows based on physics-constrained deep learning without simulation data. *Comput. Methods Appl. Mech. Eng.* **2020**, *361*, 112732. [CrossRef]
25. Meng, X.; Karniadakis, G.E. A composite neural network that learns from multi-fidelity data: Application to function approximation and inverse PDE problems. *J. Comput. Phys.* **2020**, *401*, 109020. [CrossRef]
26. Lu, L.; Jin, P.; Pang, G.; Zhang, Z.; Karniadakis, G.E. Learning nonlinear operators via DeepONet based on the universal approximation theorem of operators. *Nat. Mach. Intell.* **2021**, *3*, 218–229. [CrossRef]
27. Shukla, K.; Oommen, V.; Peyvan, A.; Penwarden, M.; Plewacki, N.; Bravo, L.; Ghoshal, A.; Kirby, R.M.; Karniadakis, G.E. Deep neural operators as accurate surrogates for shape optimization. *Eng. Appl. Artif. Intell.* **2024**, *129*, 107615. [CrossRef]
28. Pan, S.; Brunton, S.L.; Kutz, J.N. Neural Implicit Flow: A mesh-agnostic dimensionality reduction paradigm of spatio-temporal data. *J. Mach. Learn. Res.* **2023**, *24*, 1–60.
29. Celik, I.B.; Ghia, U.; Roache, P.J.; Freitas, C.J. Procedure for estimation and reporting of uncertainty due to discretization in CFD applications. *J. Fluids-Eng.-Trans. ASME* **2008**, *130*, 078001.

fluids

MDPI

Article

The Electrical Conductivity of Ionic Liquids: Numerical and Analytical Machine Learning Approaches

Theodoros E. Karakasidis, Filippos Sofos * and Christos Tsonos

Condensed Matter Physics Laboratory, Department of Physics, University of Thessaly, 35100 Lamia, Greece
* Correspondence: fsofos@uth.gr

Abstract: In this paper, we incorporate experimental measurements from high-quality databases to construct a machine learning model that is capable of reproducing and predicting the properties of ionic liquids, such as electrical conductivity. Empirical relations traditionally determine the electrical conductivity with the temperature as the main component, and investigations only focus on specific ionic liquids every time. In addition to this, our proposed method takes into account environmental conditions, such as temperature and pressure, and supports generalization by further considering the liquid atomic weight in the prediction procedure. The electrical conductivity parameter is extracted through both numerical machine learning methods and symbolic regression, which provides an analytical equation with the aid of genetic programming techniques. The suggested platform is capable of providing either a fast, numerical prediction mechanism or an analytical expression, both purely data-driven, that can be generalized and exploited in similar property prediction projects, overcoming expensive experimental procedures and computationally intensive molecular simulations.

Keywords: ionic liquids; electrical conductivity; machine learning; symbolic regression

1. Introduction

The investigation of complex materials has raised ever-growing interest among researchers in the area of fluid mechanics. Following an in-depth understanding of the internal atomic/molecular structure and the physics behind the imposed interaction mechanisms, advanced simulation techniques and experimental procedures are incorporated in order to extract the fluid properties and open the road to advances in the manufacturing and controlling of novel devices. The numerical modeling of such processes has always been an efficient, fast, and accurate choice for addressing these objectives, posing as the alternative to complex, time-consuming, and costly experiments. Among the proposed computational methods, machine learning (ML) techniques have now become a standard, showing remarkable efficiency, reduced processing time, and accuracy [1,2].

The existence of a certain number of data in a reliable database is a prerequisite for the adoption of ML. Data-driven approaches have been exploited to deal with complex physical processes, which are not described by analytical expressions and are mostly difficult to measure [3]. In most studies, the research data are obtained via limited experimental conditions. For fluid and material research, experimental results may not be sufficient to meet the ML demands, limiting its further development. Even when the research data are enriched with simulation results, and therefore sufficient, there may also be inherent processing difficulties because of the large number of input features to extract the desired prediction [4]. Therefore, high-quality training data production [5], along with ML adoption complimentary to simulation and experiments [6], can progress material discovery and investigation.

In material science and engineering, the field of application is enormous. Imaging data from microscopic studies and advanced informatic tools have been exploited for material characterization [7], and images from molecular dynamic (MD) simulations have been used to predict ice nucleation from ambient water [8]. The construction of potential energy surfaces (PESs), which had traditionally been a demanding ab initio simulation task, has been boosted by the Gaussian process regression (GPR) methods [9,10]. ML has also been successfully incorporated for the prediction of behavior from data in the fields of biological, biomedical, and behavioral sciences [11]. Fluid research has much to profit from reduced order models, turbulence modeling, fluid property extraction, and potential map creation with ab initio accuracy [12,13].

All these applications are only a small percentage of the true potential data science and ML have to offer. The new research directions focus on integrating physics-oriented parameters and domain knowledge with the proposed ML models. Physics-informed techniques have been suggested, integrating the knowledge of fundamental physics inside an algorithmic procedure [14]. Moreover, as a step toward explainable and generalizable ML, the method of symbolic regression (SR) has evolved, providing not only accurate predictions but also, more significantly, mathematical expressions that describe the phenomena under investigation, beyond classical regression methods [15,16].

In this paper, ML is approached from the perspective of ionic liquids (IL), a class of solvents that have lately attracted increasing attention due to their unique properties. Their important feature is that the melting point is so low that they remain liquid at ambient temperature, while common salts are usually solid at ambient temperature and melt at several hundred degrees Celsius [17]. Their other characteristic properties include negligible vapor pressure, high thermal and chemical stability, high ion conductivity, and nonflammability [18]. IL properties might as well be tuned for a specific application by the proper manipulation of anions and cations, from catalysis and electrochemistry to liquid crystal development, fuel production, and as electrolytes in lithium batteries, supercapacitors, and fuel cells [19]. ILs may serve as unique solvents in electrochemical processes where the use of water is forbidden [20], such as in electroplating and the electrodeposition of metals. Moreover, they are capable of dissolving organic compounds of great biological and ecological importance, such as enzymes, proteins, and cellulose [21].

The experimental measurements of ILs' physical quantities, such as conductivity, viscosity, and density, as a function of temperature or pressure, are usually performed with optofluidic and microscopic techniques [22,23], and empirical relations have been drawn to guide the experiments [24]. On the other hand, research efforts on property calculation have been mainly based on trial-and-error methods to bind anions and cations to constitute an IL of desired properties. Computational property estimation can bring research to the next level through the incorporation of novel ML methods [25]. Recent studies refer to ML techniques for ILS property prediction, such as viscosity and electrical conductivity [26], CO_2 capture capability [27], density, heat capacity, and thermal conductivity [28], among others, trying to depict the relationship between property and molecular structure, and environmental conditions.

Next, we present ML data-driven methods to extract the electrical conductivity, σ, of ionic liquids, both in numerical and analytical form. The incorporated data and preprocessing methods are presented, the ML techniques are described, and the validity of the predictions is discussed. We conclude that the proposed ML-based method is able to reproduce the electrical conductivity values for complex ILs, taking into account the environmental conditions (temperature and pressure) and the molecular weight of the IL of interest. To our knowledge, there has not been another numerical or analytical method able to extract IL properties from these three input parameters, and it can provide a fast and efficient choice to replace/complement timely and costly experiments, especially when the experimental conditions are extreme.

2. Materials and Methods

2.1. The Electrical Conductivity of Ionic Liquids

The physical properties of ILs, such as viscosity, conductivity, and density, are vital for the characterization of salt as being appropriate for a given application or not [29].

The electrical conductivity of ILs is primarily important for the understanding of their behavior and the applications that may profit from tuning their value. ILs can remain at a liquid state for a wide range of temperatures, and many electrochemical applications would incorporate them as solvents. Thus, it becomes clear that there is a need to define the possible parameters that affect electrical conductivity. In most of the studies in the literature, the temperature is the only parameter taken into account and is usually analyzed through the Vogel–Fulcher–Tammann (VFT) curves on the measured data [30,31]. The empirical VFT equation is given by

$$\sigma = \sigma_{\infty} e^{\frac{-B}{T-T_0}} \tag{1}$$

which is also examined in its linear form as

$$ln(\sigma) = ln(\sigma_{\infty}) - \frac{B}{T - T_0} \tag{2}$$

where the maximum conductivity is σ_{∞}, and the activation energy for conduction is $E_a = B \cdot k_B$, where k_B is the Boltzmann constant, both of which are derived from fitting the experimental measurements [32], and T_0 is the Vogel temperature.

Another empirical relation connects σ with the molecular volume V_m, i.e., the sum of ionic volumes of the constituent ions as [33].

$$\sigma = ce^{-dV_m} \tag{3}$$

where c and d are the empirical constants of the best fit on the experimental data, while approaches that replace experimental measurements with computational models have been also proposed [34].

2.2. Electrical Conductivity Data

For the computational model adopted in this paper, we followed the steps shown in the flowchart in Figure 1. The modeling started with database creation. High-quality experimental data (2274 points) from the NIST IL-Thermo database [35,36] were gathered for pure ionic liquids, with electrical conductivity being the property of interest. The parameters that affect electrical conductivity, as shown from the experiments, are the temperature, T, the pressure, P, and the liquid molecular weight, M_w. Table S1 (see Supplementary Material File S1) presents all the details for data origin and characteristics, while the IL database is provided in Supplementary Material File S2. The details on the incorporated experimental methods can be found in the respective references [32,37–60].

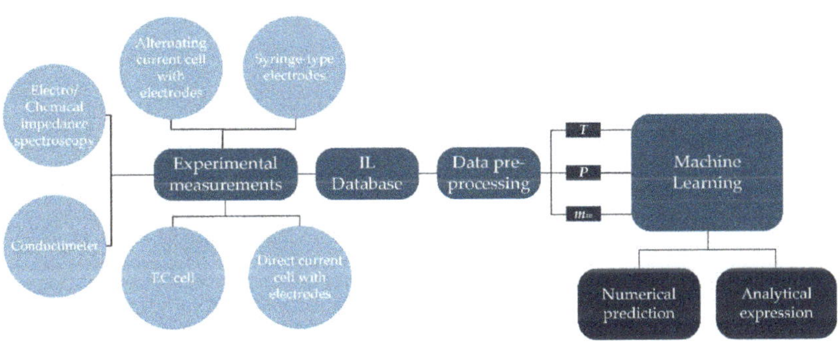

Figure 1. Machine learning model for electrical conductivity prediction, providing both numerical and analytical output.

2.3. Pre-Processing

It is common practice before entering the ML procedure, that data are normalized to restrict the input value range.

$$\bar{x} = \frac{x - x_{mean}}{x_{std}} \tag{4}$$

A correlation check was also performed in order to ensure that the input variables are not correlated to each other, and Figure 2 presents the correlation matrix. It is shown that no kind of correlation existed between the inputs, while the output was mostly correlated to temperature, *T*.

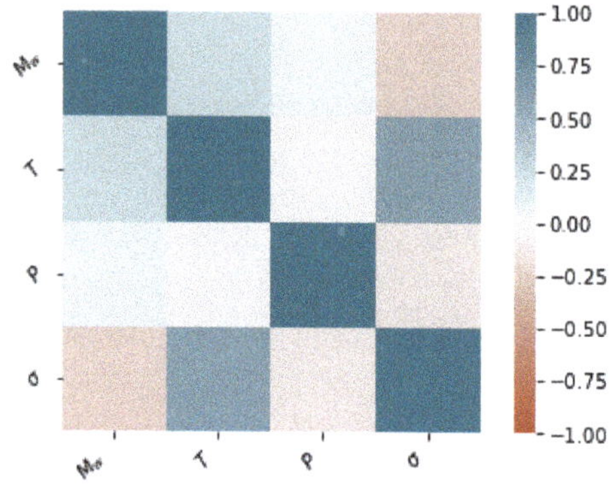

Figure 2. Correlation matrix for the three inputs and the output, σ.

The statistical information for the input data can be obtained from the pair plot diagram in Figure 3. The distribution of the three input parameters, *T*, *P*, and M_w, and the output, σ, is shown. The investigated ionic liquid is distinguished by the value of the molecular weight, which, in this paper, ranged from $108.1 \leq M_w \leq 556.18$. The temperature and pressure conditions were $203.4\ K \leq T \leq 528.55\ K$, and $0.1\ MPa \leq P \leq 250.9\ MPa$, respectively. The output was in the range of $3 \times 10^{-7}\ S/m \leq \sigma \leq 19.3\ S/m$.

2.4. Machine Learning

A supervised machine learning algorithm accepts a number of input data, is trained by a percentage of the data, and enters a computational process to extract the predicted values for the model's output(s) [61]. Data quality and quantity are important factors here. When representative data (uniformly distributed) existed, and their number was adequate to train the algorithm, the predicted output was obtained, as long as the incorporated algorithm was able to capture their behavior. The verification of the result was made by the remaining part of the input dataset (testing set). The training/testing set percentage on the total data points was taken here as 80/20.

Here, we incorporated six different numerical ML algorithms, namely the multiple linear regression (MLR), k-nearest neighbor (KNN), decision tree (DT), random forest (RF), gradient boosting regressor (GBR), and multi-layer perceptron (MLP) models, to propose the one that provided the best fit to our experimental data. These were implemented with aid of the respective functions from the sci-kit learn Python package [62]. Moreover, the symbolic regression (SR) algorithm was constructed and adjusted from a Julia package [63], in order to provide an analytical expression exclusively extracted from the data and generalizable to electrical conductivity predictions for all input cases, even those outside the data range.

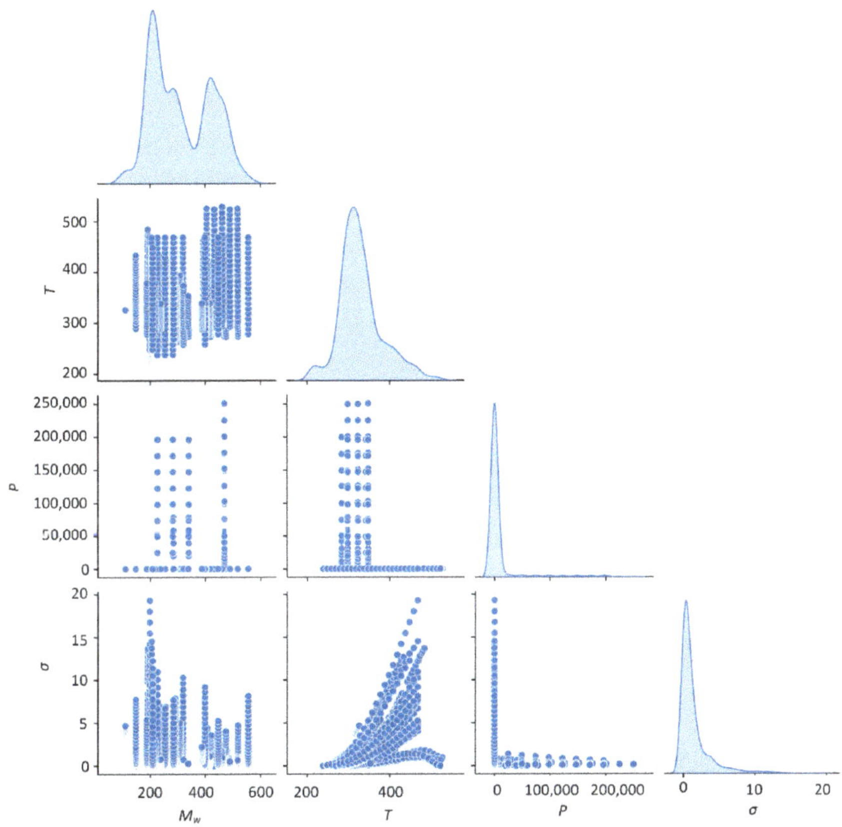

Figure 3. A pair plot diagram, showing input and output parameter distribution. The diagonal bar plots display the distribution of each parameter, while the remaining figures are scatter plots between all input pairs.

2.4.1. Multiple Linear Regression

Regression analysis refers to either a univariate method to analyze the relationship between a dependent variable and one independent variable or a model with one dependent variable and more than one independent variable, in which case it is called multiple linear regression (MLR) [64]. In MLR (Figure 4a), we consider n independent input variables, linearly combined to extract the dependent variable Y as

$$Y = \sum_{i=1}^{3} w_i X_i + b \tag{5}$$

where $w_1, w_2,$ and w_3 are the weights imposed on the three respective inputs $X_1 = T$, $X_2 = P$, $X_3 = M_w$, and b is a bias term.

2.4.2. k-Nearest Neighbors

The k-nearest neighbor (k-NN) algorithm selects k training points over a local region of a data point x and labels neighboring points on the basis of the Euclidean distance (Figure 4b). Each sample is a pair including an input vector and the desired output. After grouping the calculated distances from the lowest to the highest, the most prevalent outcome from the first k rows is the predicted result [65]. This algorithm is oftentimes accurate; however, there are cases where it may result in slow execution speed and large memory requirements [66].

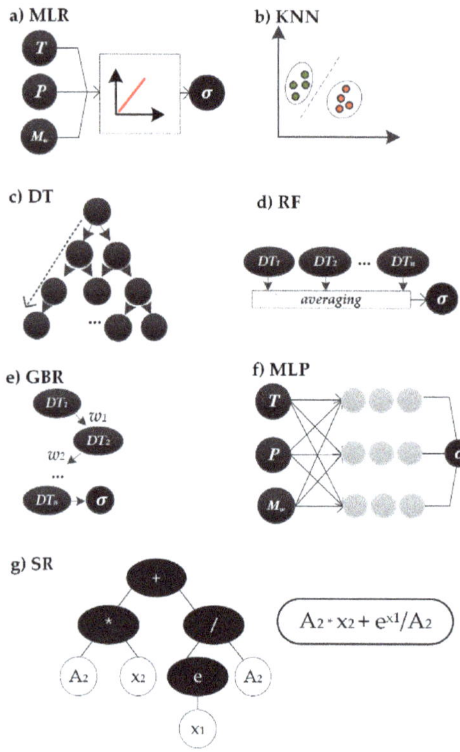

Figure 4. Graphical implementation of ML algorithms incorporated in this paper: (**a**) multiple linear regression; (**b**) k-nearest neighbors; (**c**) decision tree, (**d**) random forest; (**e**) gradient boosting regressor; (**f**) multi-layer perceptron; (**g**) symbolic regression.

2.4.3. Decision Trees

The decision tree (DT) algorithm functions in the sense of a tree flowchart, with nodes, branches, and leaves. Each node represents a test on a feature, and each branch represents the result of that test [67]. The DT mo'el's response is predicted by following the decisions from the start to the end node (the leaf), as shown by the dotted line in Figure 4c. The feature space is recursively partitioned based on the splitting attribute. Each final region is assigned a value to estimate the target output. The DT algorithm is considered easy to apply, although it might need contribution from other statistical methods to prevent overfitting [68].

2.4.4. Random Forest

A random forest (RF) algorithm consists of various DTs working in parallel (Figure 4d). Each tree outputs a different prediction, and their average is taken as the final prediction. Higher accuracy is usually obtained when the number of trees in the forest increase. In the literature, it has been shown that the random forest (RF) algorithm is much simpler to implement, than complex neural network structures, and has been the most accurate choice for fluid applications, such as slip length estimation [69] and the extraction of fluid transport properties [12,70]. All the trees' outputs are averaged (b is the trees' number) by

$$Y = \frac{1}{b}\sum_{j=1}^{b} Y_b(X') \tag{6}$$

providing an even more accurate result than the single-tree structure, hence less prone to overfitting.

2.4.5. Gradient Boosting Regressor

The gradient boosting regressor (GBR) algorithm is another implementation of a decision tree algorithm that combines various simple functions (learners) that constitute an ensemble function. Initial learners may be weak, but when combined, they may form strong learners. GBR follows three main steps sequentially: It optimizes the loss function, spots the weaker learner, and improves it by adding more trees to increase accuracy [71]. As shown in Figure 4e, the sequential DTs were incorporated, and the output of each one was weighted to enter the next DT. The weights were selected in a way to minimize the induced errors [72].

2.4.6. Multi-Layer Perceptron

The traditional perceptron, when presented in multiple layers, i.e., the input, the output, and a number of internal hidden layers, constitute the multi-layer perceptron (MLP) algorithm. The number of hidden layers is usually determined by trial and error, although there have also been various methods proposed, such as genetic programming [73]. Here, we considered three hidden layers, each one with 20 nodes, with Adam stochastic solver [74] and a learning rate equal to 0.5 (Figure 4f). The data flow between neurons depends on the activation functions applied in every internal node and a weight function imposed on every input. These weights are adjusted so that the predicted output resembles the expected output with minimum error. The training of the MLP was performed iteratively, with backward computation capability.

2.4.7. Symbolic Regression

SR can also be represented by tree structures (Figure 4g); however, here, the tree nodes are mathematical operators, and leaf nodes correspond to input variables/constants [75]. The algorithm begins by considering a random parent tree structure, calculates the mean squared error (MSE) of the specific implementation, and follows an iterative procedure, in which a node or a branch of nodes is substituted until the minimum MSE is achieved, with low complexity. Complexity refers to the number of leaves and nodes used in the proposed SR tree. The Julia-based SR algorithm by Cranmer et al. [63], which we have widely incorporated in similar works [16,76], accepts a set of mathematical operators $\{+, -, *, /, \hat{}, e, log\}$ and the input variables $\{T, P, M_w\}$ and creates an equation pool, from which it selects the best candidates that adhere to the Pareto front, i.e., those that present the minimum MSE values and small complexity, along with physical correspondence to the problem.

Although more computationally intensive and demanding, SR is capable of providing an analytical expression at hand, which, if it fits the dataset under investigation, is superior to other ML techniques, since it can be easily applied for a wide range of inputs. However, care has to be taken so that this expression remains simple and is bound to physical laws [76].

2.4.8. Metrics of Accuracy

A number of popular metrics were applied to every algorithmic result to determine which one best satisfies the accuracy criteria. These were the coefficient of determination,

R^2, the mean absolute error (MAE), the mean squared error (MSE), and the average absolute deviation (AAD) [77], as shown in Equations (7)–(10):

$$R^2 = 1 - \frac{\sum_{i=1}^{n} \left(y^*_{exp.,i} - \overline{y}^*_{\exp.}\right)^2}{\sum_{i=1}^{n} \left(y^*_{exp.,i} - y^*_{pred,i}\right)^2} \quad (7)$$

with $\overline{y}^*_{exp.}$ the mean value of the expected output:

$$MAE = \frac{1}{n}\sum_{i=1}^{n} \left|Y_i - \overline{Y}\right| \quad (8)$$

where $Y_i = y^*_{exp,i} - y^*_{pred,i}$ and $\overline{Y} = \frac{1}{n}\sum_{i=1}^{n} Y_i$.

$$MSE = \frac{1}{n}\sum_{i=1}^{n} \left(Y_i - \overline{Y}\right)^2 \quad (9)$$

$$AAD(\%) = \frac{100}{n}\sum_{i=1}^{n} \frac{\left|y^*_{exp,i} - y^*_{pred,i}\right|}{y^*_{exp,i}} \quad (10)$$

3. Results and Discussion

3.1. Partial Dependence

To analyze the effect of each input parameter on the acquired electrical conductivity value, σ, a partial dependence plot was constructed (Figure 5). The partial dependence plot calculates the average marginal effect on the σ prediction when only one input variable changes its value, and, in parallel, the remaining inputs remain constant. The estimation of the partial dependence (normalized value) is shown on the vertical axis and the respective input on the horizontal axis. In Figure 5a, it is observed that the molecular weight significantly affected σ, especially on smaller values around 200–230. The effect of temperature was prominent, especially for the values above 270 K (Figure 5b). On the other hand, pressure had only a slight, inversely linear effect on σ for small pressure values, since partial dependence decreased as the pressure increased (Figure 5c). Furthermore, it seems that σ was practically unaffected by P for the values above 100 kPa.

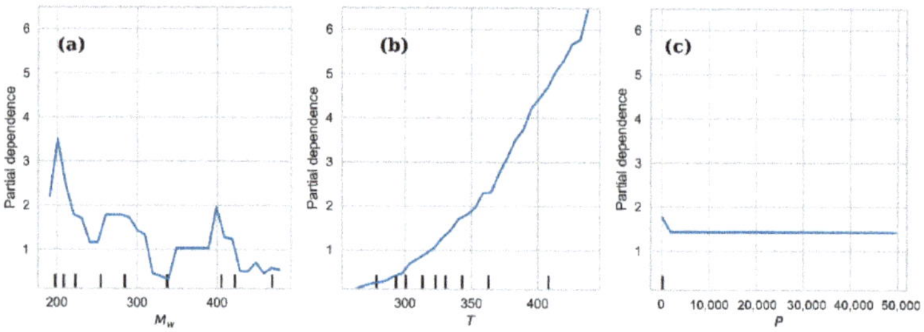

Figure 5. Partial dependence plot, where each input is investigated on the effect on electrical conductivity (σ). (**a**) molecular weight; (**b**) temperature; (**c**) pressure.

3.2. Machine Learning Results

The results from the application of the numerical ML algorithms on the electrical conductivity dataset are gathered in Figure 6a–f, in identity plots that estimate the model's accuracy by fitting the experimental and predicted data on the 45° diagonal line. The prediction is more accurate when the data points are set close to the line [78].

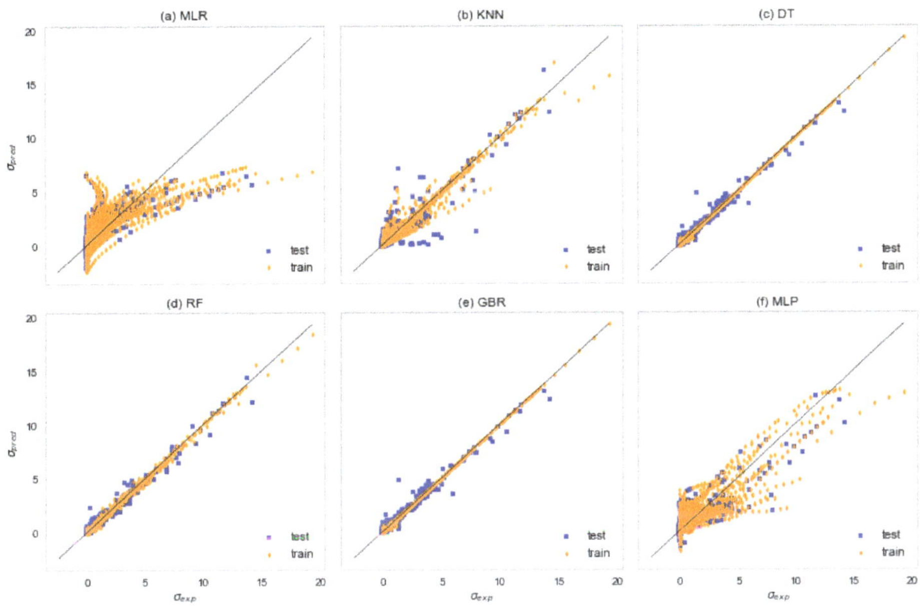

Figure 6. Experimental vs. predicted values for electrical conductivity with six different ML algorithms: (**a**) MLR; (**b**) KNN; (**c**) DT; (**d**) RI (**e**) GBR; (**f**) MLP. The 45° straight line denotes perfect match.

The linear regression method (MLR) in Figure 6a presented a rather poor fit for the ionic liquid data. This is somehow expected if we take into account the empirical relations from Equations (1)–(3), where the electrical conductivity value seems to have logarithmic dependence on the temperature or molecular volume. Thus, we expected that nonlinear ML methods would achieve better results. The KNN algorithm in Figure 6b showed better performance than MLR. Nevertheless, the three tree-based algorithms that follow in Figure 6c–e, i.e., DT, RF, and GBR, respectively, fit well on the experimental data, as it seems that their tree structure was better suited to the problem. The neural network (NN) architecture in Figure 6f did not achieve adequate prediction capability for the specific implementation (three hidden layers of 20 nodes each). We also tested different architectures with trial-and-error procedures but did not manage to obtain better results. However, NNs are a distinct field of investigation, and further investigation is needed to find the optimal architecture, which is beyond the scope of this paper. Conclusively, it was shown that most of the algorithms investigated here (except for MLR) achieved acceptable prediction performance on the available dataset.

The accuracy metrics for the fittings shown in Figure 6, such as R^2, MSE, MAE, and AAD, are shown in Table 1. The table values confirmed our observations that the three-based algorithms achieved the best fit on the data, as the coefficient of the determination reached values close to unity (R^2 = 0.99), while a minimum number of errors were expressed by MAE and MSE, compared with the remaining algorithms. However, the AAD values differed significantly. The AAD value expresses the average sum of the errors derived from the distance of the predicted data around the experimental base data. From Table 1, it is evident that the GBR method is superior to RF, followed by DT.

Table 1. Accuracy metrics and comparison of 6 ML algorithms for the ionic liquids' dataset.

Algorithm	R^2	*MAE*	*MSE*	*AAD*
MLR	0.69701	1.02	2.287	663720.1
KNN	0.91344	0.381	0.755	1265.019
DT	0.98916	0.138	0.098	536.7186
RF	0.98919	0.16	0.097	1635.613
GBR	0.98886	0.137	0.1	271.6886
MLP	0.86707	0.706	1.107	35444.57

Let us now turn our attention to finding the most important input feature that controls the internal mechanism of the algorithmic decisions for the GBR. The feature importance plot in Figure 7a presents an estimation of the importance of each input variable on the prediction of the electrical conductivity value. Temperature, T, was found to be the most important parameter that guided the decisions between the DTs and the branches of the GBR architecture. This is in agreement with the widely used empirical Equations (1)–(2), where T is the only parameter that affects electrical conductivity. The next important feature was the molecular weight, M_w, as it was the main parameter in the proposed model that differentiated between the various types of incorporated ILs. Pressure, P, had only a small effect on the final result. As is also shown in the partial dependence plot of Figure 5, P affected σ only for small values (around $P = 100$ kPa), and no effect was observed above this limit.

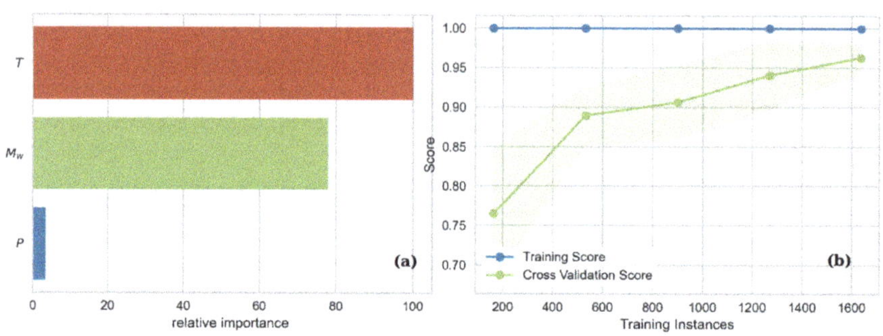

Figure 7. Interpretation output diagrams from the application of GBR algorithm on the prediction of electrical conductivity: (**a**) feature importance plot; (**b**) learning curve diagram.

Another important outcome to aid in the interpretation of the ML model is the learning curve diagram in Figure 7b, which reveals if the proposed algorithm was efficiently trained on the dataset. This is connected to the ability of the algorithm to make new predictions. Here, we observe that the cross-validation score increased as the number of training data points increased, reaching the highest value after about 1500 data points. This is evidence that the dataset used in this model (2274 data points) is capable of providing accurate predictions that could be generalized in research cases inside and outside the range of the parameters that constitute the dataset.

3.3. Obtaining an Analytical Expression

Symbolic regression is capable of providing analytical expressions to fit the dataset under investigation, without a priori knowledge of the system. This means that the SR algorithm starts with the random construction of expressions and iteratively searches for the best candidate equation. The proposed equations are of various complexity levels, and the choice of a simple or a more complicated one depends on the application and the desired accuracy. Here, we present three possible expressions that describe the electrical conductivity, σ, of ionic liquids, with input variables T, M_w, and P. Table 2 presents the mathematical expressions, along with calculated metrics.

Table 2. The three SR-extracted equations for electrical conductivity and their respective accuracy metrics. *Comp.* is the equation complexity.

Equation	*Comp.*	R^2	*MAE*	*MSE*	*AAD*
$\sigma_1 = e^{(T-M_w)/P}$	6	0.760	1.234	2.846	2,660,488.8
$\sigma_2 = \frac{17.335(T-259.461)(T-210.814)}{(M_w+0.113 \cdot T)(M_w+P)}$	19	0.857	0.727	1.392	149,183.5
$\sigma_3 = \frac{7.558}{\frac{0.955\sqrt{\lvert T(2P-M_w)\rvert+M_w}}{T-262.014}}$	20	0.883	0.728	1.160	3,551,375.7

The SR algorithm proposed three different classes of solutions, namely an exponential form (σ_1), a nonlinear fractional form (σ_2), and a combined fractional/square root form (σ_3). These forms appeared at most in the output expression pool. We have to note here that the SR output included a total of 20 equations, with increasing complexity *Comp.* = 1–20, per iteration run, for 40 parallel runs (for more details refer to [76]), i.e., 800 candidate expressions.

We observe that Equation σ_1 captured the exponential behavior shown in empirical Equation 1; however, one could not directly compare the two equations since Equation σ_1 considered M_w and P apart from T. Nonetheless, this is a simple equation that captures the ionic liquid physical behavior, with satisfying error metrics, but its disadvantage is the high *AAD* value, denoting increased distance from the real experimental values. The increased complexity of Equations σ_2 and σ_3 yielded better error metrics, with Equation σ_3 reaching R^2 = 0.883 and Equation σ_2 having the smallest *AAD* = 149,183.5.

4. Conclusions

Ionic liquid research is a field of investigation mainly based on experimental measurements, and fundamental information is hard to obtain. The need for incorporating novel computational techniques has opened the road to ML techniques that can assist in this direction.

We incorporated various ML algorithms in this paper that showed a good fit on the employed ionic liquid dataset for the electrical conductivity prediction. The best fit was obtained for the GBR algorithm, for which its tree-based procedure and the ensemble approach to processing the data successfully captured the electrical conductivity behavior. Notwithstanding the fact that numerical ML algorithms performed well on their predictions, the SR-based investigation also presented in this paper approached the problem analytically, providing mathematical expressions that can be used without further implications, thus overcoming the black-box nature of numerical ML algorithms.

We believe that by further enriching the dataset with the values deriving from either experiments or carefully established molecular simulations, ML data-driven techniques can become part of the property calculations of ionic liquids. It is of importance to suggest novel evolutional processes, reliable pre- and post-processing techniques, and physics-oriented justification to establish an integrated computational platform that can be used by scientists and engineers who wish to harness the vast volume of data involved in their field.

Supplementary Materials: The following supporting information can be downloaded at: https://www.mdpi.com/10.3390/fluids7100321/s1, Table S1: The database of ILs incorporated for our model, with 2274 data points; Table S2: Ionic Liquids Database.

Author Contributions: Conceptualization, C.T.; funding acquisition, F.S.; methodology, F.S.; software, F.S.; supervision, T.E.K.; visualization, F.S.; writing—original draft preparation, F.S. and T.E.K.; writing—review and editing, T.E.K. and C.T. All authors have read and agreed to the published version of the manuscript.

Funding: F.S. acknowledges support from the project CAMINOS (No. 5600.03.08.03), which is implemented in the context of a grant by the Center of Research Innovation and Excellence of U. Th., funded by the Special Account for Research Grants of U. Th.

Institutional Review Board Statement: Not applicable.

Informed Consent Statement: Not applicable.

Data Availability Statement: All data are extracted from NIST Ionic Liquid Database (ILThermo) available online: https://ilthermo.boulder.nist.gov/ (accessed on 4 August 2022) and provided in Supplementary Material File S2.

Acknowledgments: We acknowledge computational time granted from the National Infrastructures for Research and Technology S.A. (GRNET S.A.) in the National HPC facility—ARIS.

Conflicts of Interest: The authors declare no conflict of interest.

References

1. Ward, L.; Wolverton, C. Atomistic calculations and materials informatics: A review. *Curr. Opin. Solid State Mater. Sci.* **2017**, *21*, 167–176. [CrossRef]
2. Frank, M.; Drikakis, D.; Charissis, V. Machine-Learning Methods for Computational Science and Engineering. *Computation* **2020**, *8*, 15. [CrossRef]
3. Ramprasad, R.; Batra, R.; Pilania, G.; Mannodi-Kanakkithodi, A.; Kim, C. Machine learning in materials informatics: Recent applications and prospects. *npj Comput. Mater.* **2017**, *3*, 54. [CrossRef]
4. Gao, C.; Min, X.; Fang, M.; Tao, T.; Zheng, X.; Liu, Y.; Wu, X.; Huang, Z. Innovative Materials Science via Machine Learning. *Adv. Funct. Mater.* **2022**, *32*, 2108044. [CrossRef]
5. Craven, G.T.; Lubbers, N.; Barros, K.; Tretiak, S. Machine learning approaches for structural and thermodynamic properties of a Lennard-Jones fluid. *J. Chem. Phys.* **2020**, *153*, 104502. [CrossRef]
6. Sun, W.; Zheng, Y.; Yang, K.; Zhang, Q.; Shah, A.A.; Wu, Z.; Sun, Y.; Feng, L.; Chen, D.; Xiao, Z.; et al. Machine learning-assisted molecular design and efficiency prediction for high-performance organic photovoltaic materials. *Sci. Adv.* **2019**, *5*, eaay4275. [CrossRef]
7. Voyles, P.M. Informatics and data science in materials microscopy. *Curr. Opin. Solid State Mater. Sci.* **2017**, *21*, 141–158. [CrossRef]
8. Davies, M.B.; Fitzner, M.; Michaelides, A. Accurate prediction of ice nucleation from room temperature water. *Proc. Natl. Acad. Sci. USA* **2022**, *119*, e2205347119. [CrossRef]
9. Deringer, V.L.; Bartók, A.P.; Bernstein, N.; Wilkins, D.M.; Ceriotti, M.; Csányi, G. Gaussian Process Regression for Materials and Molecules. *Chem. Rev.* **2021**, *121*, 10073–10141. [CrossRef]
10. Boussaidi, M.A.; Ren, O.; Voytsekhovsky, D.; Manzhos, S. Random Sampling High Dimensional Model Representation Gaussian Process Regression (RS-HDMR-GPR) for Multivariate Function Representation: Application to Molecular Potential Energy Surfaces. *J. Phys. Chem. A* **2020**, *124*, 7598–7607. [CrossRef]
11. Alber, M.; Buganza Tepole, A.; Cannon, W.R.; De, S.; Dura-Bernal, S.; Garikipati, K.; Karniadakis, G.; Lytton, W.W.; Perdikaris, P.; Petzold, L.; et al. Integrating machine learning and multiscale modeling—Perspectives, challenges, and opportunities in the biological, biomedical, and behavioral sciences. *npj Digit. Med.* **2019**, *2*, 115. [CrossRef] [PubMed]
12. Sofos, F.; Stavrogiannis, C.; Exarchou-Kouveli, K.K.; Akabua, D.; Charilas, G.; Karakasidis, T.E. Current Trends in Fluid Research in the Era of Artificial Intelligence: A Review. *Fluids* **2022**, *7*, 116. [CrossRef]
13. Schran, C.; Thiemann, F.L.; Rowe, P.; Müller, E.A.; Marsalek, O.; Michaelides, A. Machine learning potentials for complex aqueous systems made simple. *Proc. Natl. Acad. Sci. USA* **2021**, *118*, e2110077118. [CrossRef]
14. Karniadakis, G.E.; Kevrekidis, I.G.; Lu, L.; Perdikaris, P.; Wang, S.; Yang, L. Physics-informed machine learning. *Nat. Rev. Phys.* **2021**, *3*, 422–440. [CrossRef]
15. Alam, T.M.; Allers, J.P.; Leverant, C.J.; Harvey, J.A. Symbolic regression development of empirical equations for diffusion in Lennard-Jones fluids. *J. Chem. Phys.* **2022**, *157*, 014503. [CrossRef] [PubMed]
16. Papastamatiou, K.; Sofos, F.; Karakasidis, T.E. Machine learning symbolic equations for diffusion with physics-based descriptions. *AIP Adv.* **2022**, *12*, 025004. [CrossRef]
17. Marsh, K.N.; Boxall, J.A.; Lichtenthaler, R. Room temperature ionic liquids and their mixtures—A review. *Fluid Phase Equilibria* **2004**, *219*, 93–98. [CrossRef]
18. Gao, T.; Itliong, J.; Kumar, S.P.; Hjorth, Z.; Nakamura, I. Polarization of ionic liquid and polymer and its implications for polymerized ionic liquids: An overview towards a new theory and simulation. *J. Polym. Sci.* **2021**, *59*, 2434–2457. [CrossRef]
19. Mousavi, S.P.; Atashrouz, S.; Nait Amar, M.; Hemmati-Sarapardeh, A.; Mohaddespour, A.; Mosavi, A. Viscosity of Ionic Liquids: Application of the Eyring's Theory and a Committee Machine Intelligent System. *Molecules* **2021**, *26*, 156. [CrossRef]
20. Earle, M.J.; Esperança, J.M.S.S.; Gilea, M.A.; Canongia Lopes, J.N.; Rebelo, L.P.N.; Magee, J.W.; Seddon, K.R.; Widegren, J.A. The distillation and volatility of ionic liquids. *Nature* **2006**, *439*, 831–834. [CrossRef]

21. Armand, M.; Endres, F.; MacFarlane, D.R.; Ohno, H.; Scrosati, B. Ionic-liquid materials for the electrochemical challenges of the future. *Nat. Mater.* **2009**, *8*, 621–629. [CrossRef] [PubMed]
22. Koutsoumpos, S.; Giannios, P.; Stavrakas, I.; Moutzouris, K. The derivative method of critical-angle refractometry for attenuating media. *J. Opt.* **2020**, *22*, 075601. [CrossRef]
23. Tsuda, T.; Kawakami, K.; Mochizuki, E.; Kuwabata, S. Ionic liquid-based transmission electron microscopy for herpes simplex virus type 1. *Biophys. Rev.* **2018**, *10*, 927–929. [CrossRef] [PubMed]
24. Xu, L.; Cui, X.; Zhang, Y.; Feng, T.; Lin, R.; Li, X.; Jie, H. Measurement and correlation of electrical conductivity of ionic liquid [EMIM][DCA] in propylene carbonate and γ-butyrolactone. *Electrochim. Acta* **2015**, *174*, 900–907. [CrossRef]
25. Koutsoukos, S.; Philippi, F.; Malaret, F.; Welton, T. A review on machine learning algorithms for the ionic liquid chemical space. *Chem. Sci.* **2021**, *12*, 6820–6843. [CrossRef]
26. Duong, D.V.; Tran, H.-V.; Pathirannahalage, S.K.; Brown, S.J.; Hassett, M.; Yalcin, D.; Meftahi, N.; Christofferson, A.J.; Greaves, T.L.; Le, T.C. Machine learning investigation of viscosity and ionic conductivity of protic ionic liquids in water mixtures. *J. Chem. Phys.* **2022**, *156*, 154503. [CrossRef] [PubMed]
27. Daryayehsalameh, B.; Nabavi, M.; Vaferi, B. Modeling of CO_2 capture ability of [Bmim][BF4] ionic liquid using connectionist smart paradigms. *Environ. Technol. Innov.* **2021**, *22*, 101484. [CrossRef]
28. Beckner, W.; Ashraf, C.; Lee, J.; Beck, D.A.C.; Pfaendtner, J. Continuous Molecular Representations of Ionic Liquids. *J. Phys. Chem. B* **2020**, *124*, 8347–8357. [CrossRef]
29. Bulut, S.; Eiden, P.; Beichel, W.; Slattery, J.M.; Beyersdorff, T.F.; Schubert, T.J.S.; Krossing, I. Temperature Dependence of the Viscosity and Conductivity of Mildly Functionalized and Non-Functionalized [Tf2N]−Ionic Liquids. *ChemPhysChem* **2011**, *12*, 2296–2310. [CrossRef]
30. Leys, J.; Wübbenhorst, M.; Preethy Menon, C.; Rajesh, R.; Thoen, J.; Glorieux, C.; Nockemann, P.; Thijs, B.; Binnemans, K.; Longuemart, S. Temperature dependence of the electrical conductivity of imidazolium ionic liquids. *J. Chem. Phys.* **2008**, *128*, 064509. [CrossRef]
31. Rodil, E.; Arce, A.; Arce, A.; Soto, A. Measurements of the density, refractive index, electrical conductivity, thermal conductivity and dynamic viscosity for tributylmethylphosphonium and methylsulfate based ionic liquids. *Thermochim. Acta* **2018**, *664*, 81–90. [CrossRef]
32. Bandrés, I.; Montaño, D.F.; Gascón, I.; Cea, P.; Lafuente, C. Study of the conductivity behavior of pyridinium-based ionic liquids. *Electrochim. Acta* **2010**, *55*, 2252–2257. [CrossRef]
33. Slattery, J.M.; Daguenet, C.; Dyson, P.J.; Schubert, T.J.S.; Krossing, I. How to Predict the Physical Properties of Ionic Liquids: A Volume-Based Approach. *Angew. Chem. Int. Ed.* **2007**, *46*, 5384–5388. [CrossRef] [PubMed]
34. Beichel, W.; Preiss, U.P.; Verevkin, S.P.; Koslowski, T.; Krossing, I. Empirical description and prediction of ionic liquids' properties with augmented volume-based thermodynamics. *J. Mol. Liq.* **2014**, *192*, 3–8. [CrossRef]
35. Ionic Liquids Database-ILThermo. NIST Standard Reference Database #147. Available online: https://ilthermo.boulder.nist.gov/ (accessed on 4 August 2022).
36. Dong, Q.; Muzny, C.D.; Kazakov, A.; Diky, V.; Magee, J.W.; Widegren, J.A.; Chirico, R.D.; Marsh, K.N.; Frenkel, M. ILThermo: A Free-Access Web Database for Thermodynamic Properties of Ionic Liquids. *J. Chem. Eng. Data* **2007**, *52*, 1151–1159. [CrossRef]
37. Zec, N.; Bešter-Rogač, M.; Vraneš, M.; Gadžurić, S. Physicochemical properties of (1-butyl-1-methylpyrrolydinium dicyanamide+γ-butyrolactone) binary mixtures. *J. Chem. Thermodyn.* **2015**, *91*, 327–335. [CrossRef]
38. Vila, J.; Fernández-Castro, B.; Rilo, E.; Carrete, J.; Domínguez-Pérez, M.; Rodríguez, J.R.; García, M.; Varela, L.M.; Cabeza, O. Liquid–solid–liquid phase transition hysteresis loops in the ionic conductivity of ten imidazolium-based ionic liquids. *Fluid Phase Equilibria* **2012**, *320*, 1–10. [CrossRef]
39. Harris, K.R.; Kanakubo, M.; Kodama, D.; Makino, T.; Mizuguchi, Y.; Watanabe, M.; Watanabe, T. Temperature and Density Dependence of the Transport Properties of the Ionic Liquid Triethylpentylphosphonium Bis(trifluoromethanesulfonyl)amide, [P222,5][Tf2N]. *J. Chem. Eng. Data* **2018**, *63*, 2015–2027. [CrossRef]
40. Harris, K.R.; Kanakubo, M.; Tsuchihashi, N.; Ibuki, K.; Ueno, M. Effect of Pressure on the Transport Properties of Ionic Liquids: 1-Alkyl-3-methylimidazolium Salts. *J. Phys. Chem. B* **2008**, *112*, 9830–9840. [CrossRef]
41. Vranes, M.; Dozic, S.; Djeric, V.; Gadzuric, S. Physicochemical Characterization of 1-Butyl-3-methylimidazolium and 1-Butyl-1-methylpyrrolidinium Bis(trifluoromethylsulfonyl)imide. *J. Chem. Eng. Data* **2012**, *57*, 1072–1077. [CrossRef]
42. Kanakubo, M.; Harris, K.R.; Tsuchihashi, N.; Ibuki, K.; Ueno, M. Temperature and pressure dependence of the electrical conductivity of the ionic liquids 1-methyl-3-octylimidazolium hexafluorophosphate and 1-methyl-3-octylimidazolium tetrafluoroborate. *Fluid Phase Equilibria* **2007**, *261*, 414–420. [CrossRef]
43. Kanakubo, M.; Harris, K.R.; Tsuchihashi, N.; Ibuki, K.; Ueno, M. Effect of Pressure on Transport Properties of the Ionic Liquid 1-Butyl-3-methylimidazolium Hexafluorophosphate. *J. Phys. Chem. B* **2007**, *111*, 2062–2069. [CrossRef] [PubMed]
44. Oleinikova, A.; Bonetti, M. Critical Behavior of the Electrical Conductivity of Concentrated Electrolytes: Ethylammonium Nitrate in n-Octanol Binary Mixture. *J. Solut. Chem.* **2002**, *31*, 397–413. [CrossRef]
45. Kanakubo, M.; Harris, K.R. Density of 1-Butyl-3-methylimidazolium Bis(trifluoromethanesulfonyl)amide and 1-Hexyl-3-methylimidazolium Bis(trifluoromethanesulfonyl)amide over an Extended Pressure Range up to 250 MPa. *J. Chem. Eng. Data* **2015**, *60*, 1408–1418. [CrossRef]

46. Vila, J.; Ginés, P.; Pico, J.M.; Franjo, C.; Jiménez, E.; Varela, L.M.; Cabeza, O. Temperature dependence of the electrical conductivity in EMIM-based ionic liquids: Evidence of Vogel–Tamman–Fulcher behavior. *Fluid Phase Equilibria* **2006**, *242*, 141–146. [CrossRef]
47. Machanová, K.; Boisset, A.; Sedláková, Z.; Anouti, M.; Bendová, M.; Jacquemin, J. Thermophysical Properties of Ammonium-Based Bis{(trifluoromethyl)sulfonyl}imide Ionic Liquids: Volumetric and Transport Properties. *Chem. Eng. Data* **2012**, *57*, 2227–2235. [CrossRef]
48. Rad-Moghadam, K.; Hassani, S.A.R.M.; Roudsari, S.T. N-methyl-2-pyrrolidonium chlorosulfonate: An efficient ionic-liquid catalyst and mild sulfonating agent for one-pot synthesis of δ-sultones. *J. Mol. Liq.* **2016**, *218*, 275–280. [CrossRef]
49. Fleshman, A.M.; Mauro, N.A. Temperature-dependent structure and transport of ionic liquids with short-and intermediate-chain length pyrrolidinium cations. *J. Mol. Liq.* **2019**, *279*, 23–31. [CrossRef]
50. Nazet, A.; Sokolov, S.; Sonnleitner, T.; Makino, T.; Kanakubo, M.; Buchner, R. Densities, Viscosities, and Conductivities of the Imidazolium Ionic Liquids [Emim][Ac], [Emim][FAP], [Bmim][BETI], [Bmim][FSI], [Hmim][TFSI], and [Omim][TFSI]. *J. Chem. Eng. Data* **2015**, *60*, 2400–2411. [CrossRef]
51. Abdurrokhman, I.; Elamin, K.; Danyliv, O.; Hasani, M.; Swenson, J.; Martinelli, A. Protic Ionic Liquids Based on the Alkyl-Imidazolium Cation: Effect of the Alkyl Chain Length on Structure and Dynamics. *J. Phys. Chem. B* **2019**, *123*, 4044–4054. [CrossRef]
52. Bandrés, I.; Carmen López, M.; Castro, M.; Barberá, J.; Lafuente, C. Thermophysical properties of 1-propylpyridinium tetrafluoroborate. *J. Chem. Thermodyn.* **2012**, *44*, 148–153. [CrossRef]
53. García-Mardones, M.; Bandrés, I.; López, M.C.; Gascón, I.; Lafuente, C. Experimental and Theoretical Study of Two Pyridinium-Based Ionic Liquids. *J Solut. Chem.* **2012**, *41*, 1836–1852. [CrossRef]
54. Yamamoto, T.; Matsumoto, K.; Hagiwara, R.; Nohira, T. Physicochemical and Electrochemical Properties of K[N(SO2F)2]–[N-Methyl-N-propylpyrrolidinium][N(SO2F)2] Ionic Liquids for Potassium-Ion Batteries. *J. Phys. Chem. C* **2017**, *121*, 18450–18458. [CrossRef]
55. Cabeza, O.; Vila, J.; Rilo, E.; Domínguez-Pérez, M.; Otero-Cernadas, L.; López-Lago, E.; Méndez-Morales, T.; Varela, L.M. Physical properties of aqueous mixtures of the ionic 1-ethl-3-methyl imidazolium octyl sulfate: A new ionic rigid gel. *J. Chem. Thermodyn.* **2014**, *75*, 52–57. [CrossRef]
56. Stoppa, A.; Zech, O.; Kunz, W.; Buchner, R. The Conductivity of Imidazolium-Based Ionic Liquids from (−35 to 195) °C. A. Variation of Cation's Alkyl Chain. *J. Chem. Eng. Data* **2010**, *55*, 1768–1773. [CrossRef]
57. Zech, O.; Stoppa, A.; Buchner, R.; Kunz, W. The Conductivity of Imidazolium-Based Ionic Liquids from (248 to 468) K. B. Variation of the Anion. *J. Chem. Eng. Data* **2010**, *55*, 1774–1778. [CrossRef]
58. Nazet, A.; Sokolov, S.; Sonnleitner, T.; Friesen, S.; Buchner, R. Densities, Refractive Indices, Viscosities, and Conductivities of Non-Imidazolium Ionic Liquids [Et3S][TFSI], [Et2MeS][TFSI], [BuPy][TFSI], [N8881][TFA], and [P14][DCA]. *J. Chem. Eng. Data* **2017**, *62*, 2549–2561. [CrossRef]
59. Benito, J.; García-Mardones, M.; Pérez-Gregorio, V.; Gascón, I.; Lafuente, C. Physicochemical Study of n-Ethylpyridinium bis(trifluoromethylsulfonyl)imide Ionic Liquid. *J Solut. Chem.* **2014**, *43*, 696–710. [CrossRef]
60. Kasprzak, D.; Stępniak, I.; Galiński, M. Electrodes and hydrogel electrolytes based on cellulose: Fabrication and characterization as EDLC components. *J. Solid State Electrochem.* **2018**, *22*, 3035–3047. [CrossRef]
61. Brunton, S.L.; Noack, B.R.; Koumoutsakos, P. Machine Learning for Fluid Mechanics. *Annu. Rev. Fluid Mech.* **2020**, *52*, 477–508. [CrossRef]
62. Pedregosa, F.; Varoquaux, G.; Gramfort, A.; Michel, V.; Thirion, B.; Grisel, O.; Blondel, M.; Prettenhofer, P.; Weiss, R.; Dubourg, V.; et al. Scikit-learn: Machine Learning in Python. *J. Mach. Learn. Res.* **2011**, *12*, 2825–2830.
63. Cranmer, M.; Sanchez Gonzalez, A.; Battaglia, P.; Xu, R.; Cranmer, K.; Spergel, D.; Ho, S. Discovering symbolic models from deep learning with inductive biases. In Proceedings of the 34th Conference on Neural Information Processing Systems (NeurIPS 2020), Vancouver, BC, Canada, 6–12 December 2020; Larochelle, H., Ranzato, M., Hadsell, R., Balcan, M.F., Lin, H., Eds.; Curran Associates, Inc.: New York, NY, USA, 2020; Volume 33, pp. 17429–17442.
64. Uyanık, G.K.; Güler, N. A Study on Multiple Linear Regression Analysis. *Procedia-Soc. Behav. Sci.* **2013**, *106*, 234–240. [CrossRef]
65. Rahman, J.; Ahmed, K.S.; Khan, N.I.; Islam, K.; Mangalathu, S. Data-driven shear strength prediction of steel fiber reinforced concrete beams using machine learning approach. *Eng. Struct.* **2021**, *233*, 111743. [CrossRef]
66. Song, Y.; Liang, J.; Lu, J.; Zhao, X. An efficient instance selection algorithm for k nearest neighbor regression. *Neurocomputing* **2017**, *251*, 26–34. [CrossRef]
67. Loh, W.-Y. Classification and regression trees. *WIREs Data Min. Knowl. Discov.* **2011**, *1*, 14–23. [CrossRef]
68. Schmidt, J.; Marques, M.R.G.; Botti, S.; Marques, M.A.L. Recent advances and applications of machine learning in solid-state materials science. *npj Comput. Mater.* **2019**, *5*, 83. [CrossRef]
69. Sofos, F.; Karakasidis, T.E. Nanoscale slip length prediction with machine learning tools. *Sci. Rep.* **2021**, *11*, 12520. [CrossRef]
70. Allers, J.P.; Harvey, J.A.; Garzon, F.H.; Alam, T.M. Machine learning prediction of self-diffusion in Lennard-Jones fluids. *J. Chem. Phys.* **2020**, *153*, 034102. [CrossRef]
71. Sandhu, A.K.; Batth, R.S. Software reuse analytics using integrated random forest and gradient boosting machine learning algorithm. *Softw Pract. Exp.* **2021**, *51*, 735–747. [CrossRef]
72. Jerome, H. Friedman Greedy function approximation: A gradient boosting machine. *Ann. Stat.* **2001**, *29*, 1189–1232. [CrossRef]
73. Stathakis, D. How Many Hidden Layers and Nodes? *Int. J. Remote Sens.* **2009**, *30*, 2133–2147. [CrossRef]

74. Kingma, D.P.; Ba, J. Adam: A Method for Stochastic Optimization. *arXiv* **2015**, 6908. [CrossRef]
75. Wang, Y.; Wagner, N.; Rondinelli, J.M. Symbolic regression in materials science. *MRS Commun.* **2019**, *9*, 793–805. [CrossRef]
76. Sofos, F.; Charakopoulos, A.; Papastamatiou, K.; Karakasidis, T.E. A combined clustering/symbolic regression framework for fluid property prediction. *Phys. Fluids* **2022**, *34*, 062004. [CrossRef]
77. Agrawal, A.; Deshpande, P.D.; Cecen, A.; Basavarsu, G.P.; Choudhary, A.N.; Kalidindi, S.R. Exploration of data science techniques to predict fatigue strength of steel from composition and processing parameters. *Integr. Mater. Manuf. Innov.* **2014**, *3*, 90–108. [CrossRef]
78. Bengfort, B.; Bilbro, R. Yellowbrick: Visualizing the Scikit-Learn Model Selection Process. *J. Open Source Softw.* **2019**, *4*, 1075. [CrossRef]

fluids

Article

DropletAI: Deep Learning-Based Classification of Fluids with Different Ohnesorge Numbers during Non-Contact Dispensing

Pranshul Sardana [1,2], Mohammadreza Zolfaghari [3], Guilherme Miotto [2], Roland Zengerle [1,2], Thomas Brox [3], Peter Koltay [1] and Sabrina Kartmann [1,2,*]

1 Laboratory for MEMS Applications, Department of Microsystems Engineering—IMTEK, University of Freiburg, Georges-Koehler-Allee 103, 79110 Freiburg, Germany
2 Hahn-Schickard, Georges-Koehler-Allee 103, 79110 Freiburg, Germany
3 Department of Computer Science, University of Freiburg, Georges-Köhler-Allee 106, 79110 Freiburg, Germanybrox@informatik.uni-freiburg.de (T.B.)
* Correspondence: sabrina.kartmann@hahn-schickard.de

Abstract: The reliable non-contact dispensing of droplets in the pico- to microliter range is a challenging task. The dispensed drop volume depends on various factors such as the rheological properties of the liquids, the actuation parameters, the geometry of the dispenser, and the ambient conditions. Conventionally, the rheological properties are characterized via a rheometer, but this adds a large liquid overhead. Fluids with different Ohnesorge number values produce different spatiotemporal motion patterns during dispensing. Once the Ohnesorge number is known, the ratio of viscosity and surface tension of the liquid can be known. However, there exists no mathematical formulation to extract the Ohnesorge number values from these motion patterns. Convolutional neural networks (CNNs) are great tools for extracting information from spatial and spatiotemporal data. The current study compares seven different CNN architectures to classify five liquids with different Ohnesorge numbers. Next, this work compares the results of various data cleaning conditions, sampling strategies, and the amount of data used for training. The best-performing model was based on the ECOmini-18 architecture. It reached a test accuracy of 94.2% after training on two acquisition batches (a total of 12,000 data points).

Keywords: open microfluidics; non-contact dispensing; drop-on-demand; deep learning; video classification

1. Introduction

Droplet dispensing is the process of ejecting discrete fluidic volumes with two immiscible fluids [1]. The ejection process can happen in various droplet breakup regimes, one of them being an on-demand breakup. The dispense is performed by providing the fluid with enough energy as a short actuation pulse in the form of thermal fluctuations or mechanical vibrations to overcome energies holding the fluid back, such as capillary forces.

When this actuation pulse is applied, the droplet takes a teat shape due to the positive pressure and the opposing surface tension where the fluid touches the orifice [2]. Next, a tail is formed, caused by the inertia of the liquid, and this tail is broken by the upward pull of the negative pressure created. After leaving the nozzle, the droplet continuously changes its shape from the initial cylindrical-like shape to wobbling between tear-like and spherical shapes, until it becomes stably spherical due to the surface tension of the fluid. Hence, during the initial phase of the ejection, the droplet undergoes a shape evolution before reaching a stable spherical state, i.e., it follows a specific spatiotemporal trajectory. For non-contact dispensers (such as PipeJet P9, BioFluidix GmbH, Freiburg im Breisgau, Germany [3]), this spatiotemporal motion pattern depends on various factors: the geometries involved (such as the characteristic length (D) of the dispenser), the actuation dynamics (such as the stroke length (SL) and downstroke velocity (DV)), the rheological (such as viscosity) and physical properties (such as density) of the liquid being dispensed, the surface effects, and the environmental variables (such as temperature and relative humidity). This primary droplet can also be followed by secondary droplets known as satellites.

To simplify the modeling of the ejection events, characteristic numbers can be used. These numbers are dimensionless quantities and hence systems with the same characteristic numbers behave similarly regardless of factors such as their scale. The dimensionless numbers which are interesting for the droplet ejection process are the Weber number (*We*) and the Ohnesorge number (*Oh*) [1]. The droplets will follow different spatiotemporal trajectories when ejected either with a different *We* and/or with a different *Oh*. Figure 1 shows the output of a simulation conducted wherein fluid is ejected at various *Oh* [4].

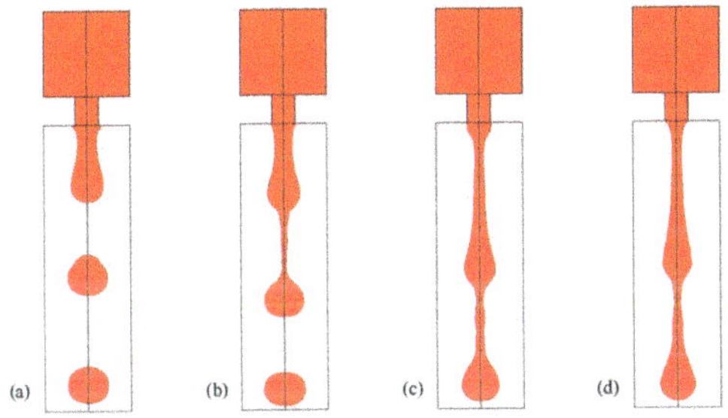

Figure 1. Droplet ejection simulation for various *Oh*: (**a**) 0.066, (**b**) 0.656, (**c**)1.312, (**d**) 1.967 [4]. The figure is reproduced with permission from the author.

We is the ratio between the kinetic energy of a fluid and its surface energy [1]. Therefore, it depends on the actuation parameters and the liquid properties, whereas *Oh* relates the following: internal viscous dissipation to inertia and surface tension. Therefore, it depends on the liquid properties and the characteristic length. The relation between *We*, *Oh*, and the Reynolds number (*Re*) is given in Equation (1)

$$Oh = \sqrt{We}/Re = \eta/\sqrt{\rho \cdot D \cdot \sigma} \tag{1}$$

where η is the dynamic viscosity of the liquid, ρ is the density, D is the characteristic length, and σ is the surface tension. The spatiotemporal trajectories formed by the ejected droplets can be imaged using a high-speed camera.

Even though mathematical models exist to create these droplet trajectories from the initial conditions [5], solving the inverse problem (i.e., estimating the initial conditions from the trajectories) is not trivial. Nonetheless, such a solver is useful in practical scenarios. It could be used, for instance, to adjust dispenser parameters on the fly as the liquid properties change due to temperature fluctuations. As neural networks are universal approximators [6], these can be applied to solve the aforementioned problem. Convolutional neural networks (CNNs) are great tools for capturing the spatial and spatiotemporal content [7]. Hence, these can be used to identify liquids with different *Oh* from the captured images and image sequences (videos) of the dispensed droplets.

Convolutional neural networks (CNNs), which belong to the class of machine learning algorithms known as deep learning (DL) [8], build complex representations by taking simpler representations as input in an end-to-end manner. This ability to automatically build new constructs on top of existing representations gives DL the ability to extract the non-obvious complex representations from the underlying data. This can then be used to perform tasks such as distinguishing between different classes present in the data, i.e., classification. The training for this classification process can be conducted in a supervised fashion if the type of liquid in the image, i.e., the label, is known.

As droplets with different *Oh* follow different spatiotemporal trajectories, they may form unique shapes. This indicates that neural networks with only 2D extractors may be enough and 3D information processing may not be needed. To test this, different CNN architectures with pure 2D convolutions, pure 3D convolutions, or 3D–2D mixed convolutions were compared.

2. Data Acquisition

Data acquisition is one of the major tasks in the DL pipeline. To fulfill the need of acquiring a large amount of labeled data, an automated data acquisition setup, named TestRig, was used. For each droplet dispense, a sequence of 250 frames at 6806 frames per second was recorded. Figure A2 shows an example sequence. During the data acquisition process, various sources for shortcut learning [9] such as poor lighting, poor focus, tilting of the dispenser nozzle, etc., were avoided. The following topics are discussed in the following subchapters: hardware and software of the TestRig (Section 2.1), the test liquids used (Section 2.2), as well as hardware configuration and acquisition protocols (Section 2.3).

2.1. Hardware and Software of the TestRig

Figure 2 shows the TestRig. It was built on top of an optical breadboard. The droplets were dispensed using a nanoliter non-contact droplet dispenser (BioFluidix PipeJet®, Freiburg im Breisgau, Germany [3]) mounted on a three-axis precision stage. The nozzles had an inner diameter of 500 µm. The falling droplets were imaged as shadows by positioning the dispenser nozzle between a light-emitting diode and a high-speed camera (MotionBLITZ EoSens® mini2, Mikrotron GmbH, Unterschleissheim, Germany). To reduce the influence of convection on the trajectory of the droplets, the setup was covered with walls and a lid.

The liquid pressure was regulated during the acquisition using a pressure control module (ActivePCR, BioFluidix GmbH, Freiburg im Breisgau, Germany). Additionally, the weight of the falling droplets was measured using a micro-weighing balance (SC2 Ultra-Microbalance, Sartorius AG, Goettingen, Germany) placed under the orifice, and the ambient temperature, pressure, and humidity were logged with a thermo-hygrometer.

The TestRig was controlled via an in-house developed software called ♯Drop [10]. It was used to trigger the camera, configure the parameters of the PipeJet® actuation, and log mass, ambient temperature, relative humidity, and pressure readings. Next, MotionBLITZ® Director2 was used to configure and operate the camera. Furthermore, Biofluidix Pipejet® Control software was used to configure and operate the pressure control unit.

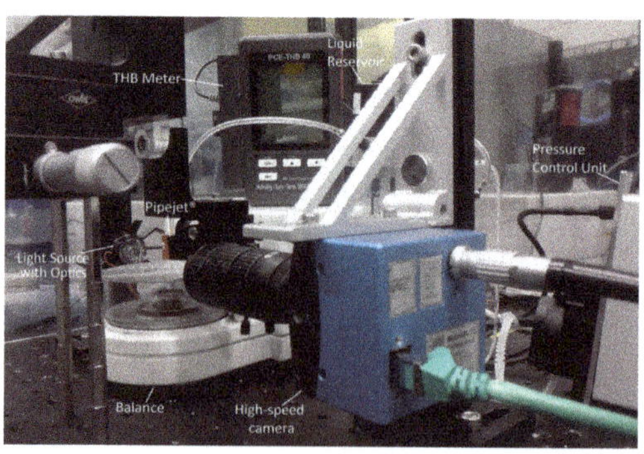

Figure 2. TestRig. It consists of a dispenser (BioFluidix PipeJet®, Freiburg im Breisgau, Germany), a high-speed camera (MotionBLITZ EoSens® mini2, Mikrotron GmbH, Unterschleissheim, Germany), a pressure control module (ActivePCR, BioFluidix GmbH, Freiburg im Breisgau, Germany), a balance (SC2 Ultra-Microbalance, Sartorius AG, Goettingen, Germany), and a thermo-hygrometer.

2.2. Used Liquids

This study uses five of the eight model liquids developed in [11], covering the whole range of viscosity and surface tension of 646 of the most common in vitro diagnostics reagents. Figure 3 presents the surface tension vs. viscosity for the model liquids. Table 1 summarizes fluid properties, the *Oh* for a 500 µm nozzle, and the components shortlisted from [11]. There were little fluctuations in the ambient temperatures (23.2 °C to 28.6 °C) during acquisition, but the resulting error in the *Oh* calculation (for water) was not significant (0.54×10^{-3}) as all the other liquids used in this study have significantly larger *Oh*.

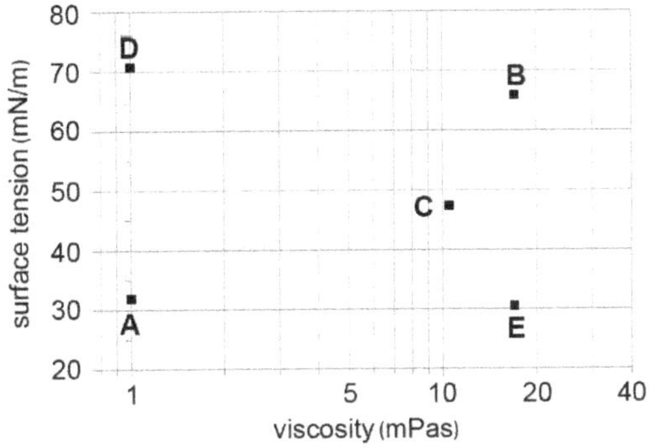

Figure 3. Model liquids with different surface tension and viscosity [11]. The figure is reproduced with permission from the author. The properties of these fluids are provided in Table 1.

Table 1. Test liquids used, viscosity (η), density (ρ), surface tension (σ), Ohnesorge number produced with a 500 µm nozzle at 25 °C, and components of the liquids.

Fluid	η (mPas)	ρ (kg/m^3)	σ (mN/m)	Ohnesorge Number (*Oh*)	Components
A	1.0	998	31.9	7.93×10^{-3}	Deionized water (DI), Sympatens®
B	16.9	1169	65.9	86.11×10^{-3}	DI water, Glycerin
C	10.5	1139	47.3	63.98×10^{-3}	DI water, Glycerin, Myrj® S100®
D	1.0	998	70.8	5.32×10^{-3}	DI water
E	16.9	1169	30.5	126.57×10^{-3}	DI water, Glycerin, Sympatens®

2.3. Hardware Parameters and Acquisition Protocol

A set of 64 configurations was chosen for driving the Pipejet®. Each of these configurations was repeated 25 times consecutively, leading to a total of 1600 dispenses. Each configuration was a combination of one of the eight SLs (10, 12, 14, 16, 18, 21, 23, or 25 µm) and one of the eight DVs (60, 70, 80, 90, 100, 110, 120, or 130 µm/ms) with all the DVs sequenced in ascending order with a single SL before moving to a higher SL. An acquisition batch is a set containing one acquisition cycle from all five test liquids. The data acquisition was conducted in three acquisition batches.

Depending on the viscosity of the individual liquids, some actuation configurations did not provide enough energy to eject the liquid. These invalid configurations were filtered out of the dataset. This resulted in 31 to 64 valid configurations based on the liquid, nozzle condition, and environmental conditions.

The camera was configured by defining a frame size with a height of 750 pixels and a width of 144 pixels. The height was experimentally determined to ensure that, even for the drop with the longest tail (produced with the liquid with the highest *Oh*, liquid E), there should exist at least one frame that contains the entire droplet, from head to tail. Next, the width was obtained by observing the number of pixels covered by the nozzle and multiplying it by 3 to account for the tilting of the nozzle. The configurations resulted in a maximum frame rate of 6806 fps for the current hardware. For each dispense, 250 images were acquired at this frame rate.

The tip of the nozzle was kept in the frame. This was primarily because the droplets have a high degree of spatiotemporal changes when leaving the nozzle, which can be interesting for the classification task. Moreover, the nozzle of known diameter can also be used as a reference object for camera zoom and focus. A shadowgraph of the field of view in the imaging plane is given in Figure 4.

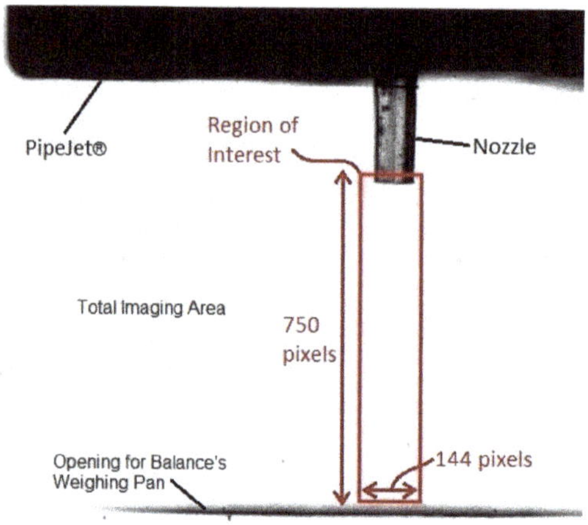

Figure 4. A shadowgraph of the field of view in the imaging plane.

The time difference between two consecutive dispenses was 31 s for the first two acquisition batches to save all the acquired images. It was increased to 33 s in the third acquisition batch to give the dataset a bit of diversity and improve the generalization of the trained model on new, unseen datasets. Gravimetric weight analysis was performed according to the GRM method [12,13] with settling and measuring times of eight seconds each in the first two acquisition batches and which were increased to 15 s in the third to measure the slowest falling droplets accurately.

2.4. First Appearance of Droplet

Each image sequence had 250 images, in which some of the initial frames were empty. The number of empty frames was dependent on the PipeJet® configuration used and was independent of the liquids used in this study. The first frame where the droplet appears was identified by running a data pre-processing script. The first appearance ranged from frame 71 (SL = 25 µm, DV = 60 µm/ms)–186 (SL = 10 µm, DV = 130 µm/ms) (Appendix A).

3. Deep Learning

3.1. Architectures

In the current study, seven different neural network architectures were compared. These neural networks can be classified based on the type of data stream (pure 2D, pure 3D, or both 2D and 3D) or based on the network depth (shallower or deeper).

There were two pure 2D architectures: ResNet-18 [14] (which contains 18 2D-convolution layers) and ResNet-C (a 14-layer variant of ResNet-18). Both of these were relatively shallow networks. The ResNet (or Residual Network) architecture introduced skip connections that are effective in reducing the vanishing gradients problem of deep networks. There was one purely 3D architecture (3D-ResNet18 [15]). This is a variant of Resnet-18 where the pure 2D convolutions are replaced with pure 3D convolutions to capture the spatiotemporal content of the data. The remaining four architectures contained both the 2D and 3D convolutions. These are ECO Lite, ECO Full [7], ECOmini-18, and ECOmini-C. ECO Lite and ECO Full are the deeper networks used in this study. These were originally made for action recognition on videos, where the relevant information is also spread out over multiple frames. These networks were created on an assumption that, even though the information relevant for classification spans multiple frames, single frames also contain significant information regarding the task.

In both these architectures, the image frames are initially fed to 2D convolution layers. This is achieved using the BNInception architecture until the inception-3c layer [16]. In the ECO Lite architecture, the representations from these individual frames are stacked together and fed to the 3D-ResNet18 architecture. The ECO Full architecture further augments the ECO lite by adding a 2D convolution stream parallel to the 3D-ResNet18 architecture. This 2D stream is made using the BNInception architecture from the inception-4a layer to the last pooling layer. The two steams are fused later in the network by appending the outputs in a single fully connected layer.

The last two architectures, ECOmini-18 and ECOmini-C, were made on a blueprint similar to the ECO Full architecture. In ECOmini-18, the layers up to inception-3c from ECO Full were replaced by the first nine of ResNet-18, and the rest of the layers were used in parallel to 3D-ResNet18, whereas in ECOmini-C, the first seven layers of ResNet-18 replaced the layers up to inception-3c and used the rest parallel to 3D-ResNet18. As these architectures use a ResNet-18 backbone, they were relatively shallower. Table 2 summarizes the seven architectures and the type of convolutions utilized.

Table 2. Seven deep learning architectures used in this study and the convolution types in the architectures.

Architecture	Convolutions	Description
ResNet-18 [14]	2D only	An 18 layers shallow network containing convolutions and skip connections
ResNet-C	2D only	A 14 layer variant of ResNet-18
3D-ResNet18 [15]	3D only	A variant of ResNet-18 containing purely 3D convolutions
ECO Lite [7]	2D + 3D	Contains 2D convolution layers of BNInception for 2D feature extraction which are fed to the 3D-ResNet18 architecture
ECO Full [7]	2D + 3D	ECO Lite augmented by adding a 2D convolution stream parallel to the 3D layers
ECOmini-18 [14]	2D + 3D	A shallower variant of ECO Full where 2D convolutions are replaced by layers of ResNet-18
ECOmini-C [14]	2D + 3D	A shallower variant of ECO Full where 2D convolutions are replaced by layers of ResNet-C

3.2. *Hyperparameters and Hyperparameter Optimization*

While training the neural network models, different hyperparameters such as learning rate (LR), drop out, batch size, etc., were tuned. The choice of optimal hyperparameters was made using a sequential model-based algorithm configuration (SMAC) [17] run for a maximum of 20 epochs. For each training run, SMAC was run for a maximum of 24 h either on an NVIDIA® Tesla® V100 or on an NVIDIA® GeForce® RTX 2080 Ti. The set's accuracy was obtained from the epoch with the highest validation accuracy. The hyperparameters (and their SMAC limits) were as follows: LR (10^{-6}–10^{-1}), drop out (0.01–0.99), and batch size (8–32).

In addition to the above, a decaying LR strategy was used within each SMAC instance. In this strategy, the LR was reduced to one-tenth of its value if the accuracy did not improve for the "n" number of epochs.The choice of this "n" count was also a hyperparameter (3–10). Next, the networks were provided with an option to retain self-normalization after the dropout layer. This was achieved using a flag that switches between dropout and alpha dropout as a hyperparameter. Lastly, a hyperparameter was used to choose the data augmentation level. Control over the amount of augmentation is required to ensure that the augmented data still represent the actual process of droplet ejection.

3.3. Data Augmentation

Three techniques were used to perform data augmentation: horizontal flipping with a flipping probability of 50%, contrast and brightness jitter with a random value up to 5%, and random cropping with different augmentation levels. There were six augmentation levels (0–5) that represent a maximum reduction in side length due to random cropping and were used as a hyperparameter which was optimized via SMAC. An augmentation level of 0 represented no cropping, whereas an augmentation level of 5 represented a maximum cropping of 10% of the side length, preserving the aspect ratio. The crops were upscaled to the original image size before feeding to the neural network.

3.4. Experimental Setup

The current work was a five-class classification task, trained with stochastic gradient descent, using cross-entropy as a loss function. Acquisition Batch 1 was used for training and Batch 2 for testing. To analyze the effect of adding more data, acquisition batch 3 was also included in the training data. To ensure reproducibility, the GPU and CPU pseudorandom generators were seeded with the same value of 0 for deterministic runs.

The dataset was used either uncleaned (referred to as CV ∞) or cleaned with CV values of 5% or 2% calculated from the droplet mass values obtained during data acquisition. As a pre-processing step, the grayscale images were shrunk by a factor of 6, resulting in an image size of 125×24 pixels. The last pooling layer was made with rectangular kernels with a 4:1 height-to-width ratio.

The input images were obtained by sampling down the acquired videos to n-images (4, 8, 12, or 16) with either the random sampling (RS) [7] or sequential sampling (SS) technique. In random sampling, the image sequence is temporally separated into n-segments of equal size, and one image is sampled from each segment. This strategy helps the network to see the images from the entire temporal domain. In sequential sampling, n-images are sampled sequentially from a predefined starting point. By this strategy, the network is fed with sequential localized information from the region of interest. The starting point for this strategy was selected at a point after the first appearance of the droplet. Therefore, the images can be sampled from the initial region of dispense where there are a significant number of changes in the shape of droplets or these can be from a later part in the sequence where the droplet is mostly spherical. Figure 5 illustrates both sequential and random sampling techniques.

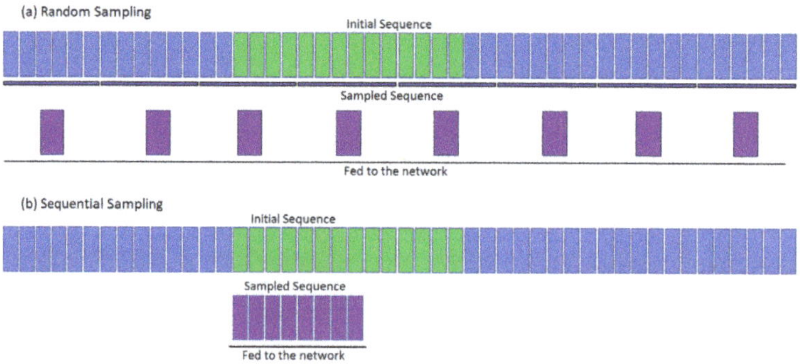

Figure 5. Illustration of (**a**) random sampling and (**b**) sequential sampling techniques.

4. Results and Discussions

The first tests were performed on deeper architectures. These networks were trained for 70 epochs. 3D-Resnet18, ECO Lite, and ECO Full had 87.40%, 87.62%, and 89.68% respective accuracies. Extensive testing was not performed on these architectures because they were overfitting. The rest of this section compares the shallower architectures, compares the sampling strategies, analyzes the effect of data cleaning, and analyzes the effect of adding the third acquisition batch. All these tests use eight segments.

4.1. Comparing Shallower Architectures

As there were various reasons (sampling strategy, data cleaning, and the number of batches used for training) which may lead to different accuracies for the same neural network architecture, this section compares the performance of the networks over these variables.

Figure 6 compares the pure 2D architectures on various filtering intensities (expressed as CV thresholds) and sampling conditions. ResNet-C has a lower performance as compared to ResNet-18. This shows that, for the current task, the higher number of layers in ResNet-18 makes it better at extracting the underlying features for every sampling and cleaning condition.

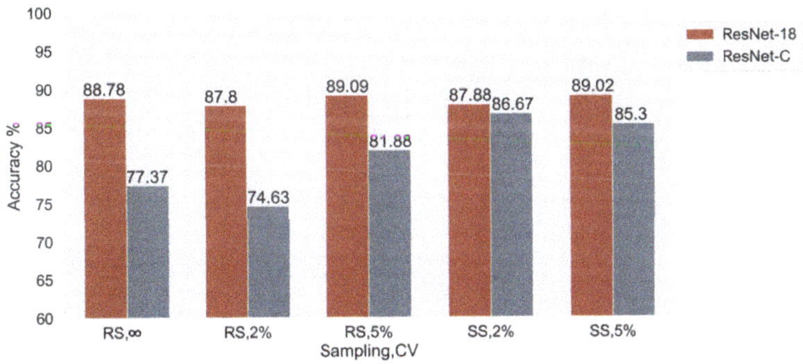

Figure 6. A comparison of pure 2D architectures on various filtering intensities and sampling conditions.

Next, Figure 7 compares ECOmini-18 and ECOmini-C on various CV and sampling conditions. ECOmini-18 and ECOmini-C showed similar performance on the given dataset. This shows that adding the 3D layers with the ResNet-C helps it to extract more useful information from the current dataset.

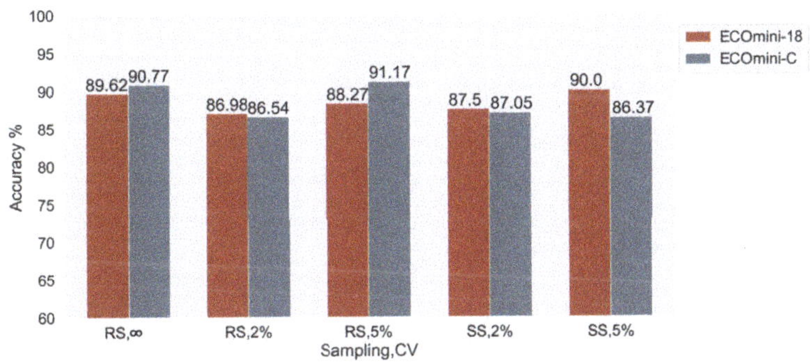

Figure 7. ECOmini-18 vs. ECOmini-C on various coefficients of variation (CV) and sampling conditions.

To analyze the effect of adding a 3D component with a good 2D extractor, a comparison between ECOmini-18 and ResNet-18 is shown in Figure 8. ECOmini-18 and ResNet-18 showed similar performance. This shows that the spatiotemporal trajectories produced during dispensing at the current *Oh* produce distinct enough underlying frames that the 2D components extracted by ResNet-18 alone are enough for the classification task. A high difference between the *Oh* leads to very different underlying images [4]; hence, the current 2D extractors can differentiate between the *Oh* based on the single frames. Hence, for datasets similar to the one used in this study where the difference in the *Oh* is large, a 2D architecture can be used to make inferences with a single frame.

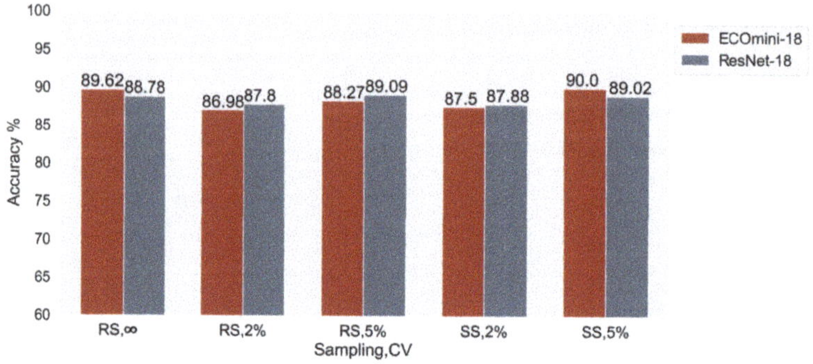

Figure 8. ECOmini-18 vs. ResNet-18 on various coefficients of variation (CV) and sampling conditions.

4.2. Comparing Sampling Strategies

Depending on the non-contact dispensing parameters, the droplets had a different first frame of appearance and were dispensed at different velocities. These lead to a different number of empty frames before the dispense and after the droplet had left the region of interest. Therefore, a random sampling strategy can have a significant number of empty frames. Figure 9 shows a performance comparison between ResNet-18 and ECOmini-18 while using the two sampling strategies.

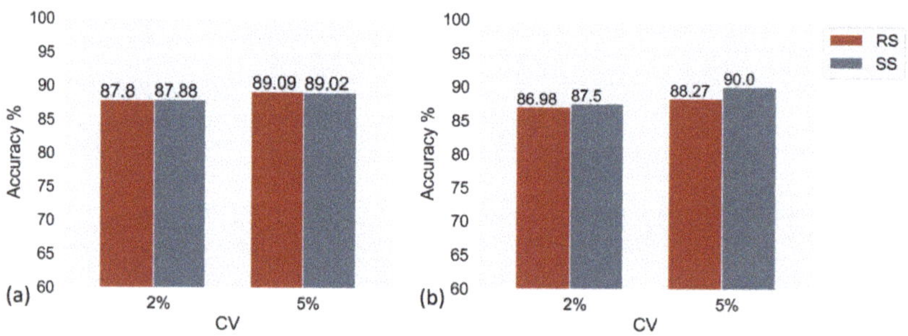

Figure 9. Comparison between random sampling (RS) and sequential sampling (SS) while using (**a**) ResNet-18 and (**b**) ECOmini-18.

In the case of ResNet-18, both sampling strategies had very similar performances. As it is a 2D architecture, ResNet-18 does not extract any patterns that exist between the frames. Hence, it is not affected by the type of sampling, whereas in the case of ECOmini-18, sequential sampling outperforms random sampling for the current dataset. This can be because, for the current dataset, the useful information on shape evolution lies in the initial part of the dispenses. To verify this, sequential sampling with different offsets was performed. Figure 10 shows a sequential sampling of eight frames with no offset, a five-frame offset, and a ten-frame offset for ECOmini-18 and ResNet-18.

The neural networks perform the best when fed with images from the initial part of the sequence. Hence, the relevant information is relatively more localized in the initial part of the sequences.

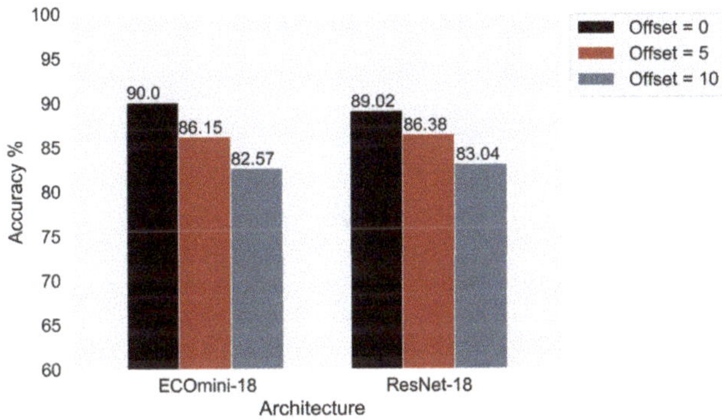

Figure 10. Sequential sampling of eight frames with no offset, a five-frame offset, and a ten-frame offset.

4.3. Effect of Data Cleaning

The data were cleaned based on the CV of the droplet mass obtained within the 25 repeats of the same dispenser configuration. The lower the CV threshold, the lower the amount of remaining data. It is worth noting that CV ∞ represents the uncleaned data. Figure 11 compares the performance of the ECOmini-18 and ResNet-18 over different CV thresholds and sampling strategies.

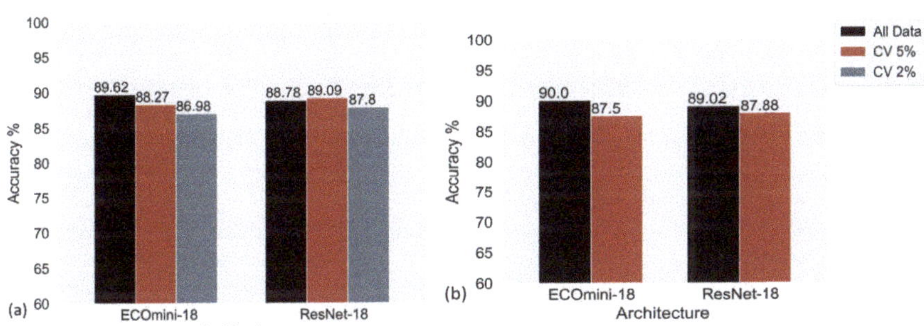

Figure 11. Performance of ECOmini-18 and ResNet-18 over different coefficient of variation (CV) thresholds with (**a**) random sampling and (**b**) sequential sampling strategy.

The performance of the neural networks decreases with the decrease in the amount of data as the amount of filtering is increased.

4.4. Effect of Adding New Acquisition Batch

Training on more than one acquisition batch helps the network reduce the shortcut learning that can arise by training on a single batch. The increase in the amount of diverse data helps the neural network to better generalize. Figure 12 shows a comparison between the networks trained on a single acquisition batch and the networks trained on two acquisition batches. The data here are cleaned with a CV threshold of 5%.

The performance of the neural networks increases when the training is performed on acquisition batch 1 and 3 jointly. The ECOmini-18 model obtained by training on acquisition batch 1 and 3 jointly was the best model in the entire study.

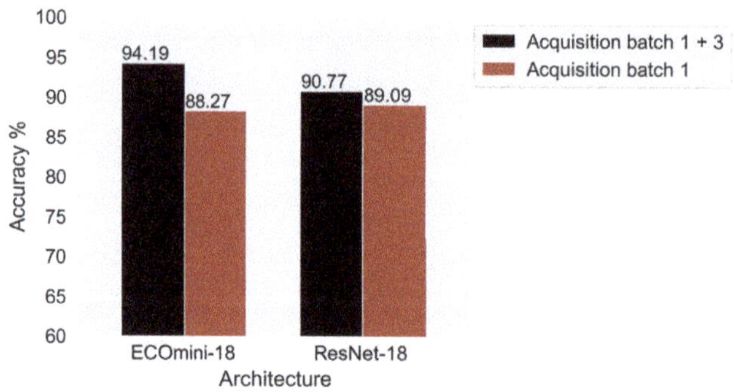

Figure 12. Performance of networks when trained on a single or two acquisition batches.

5. Conclusions

The current work compares seven different neural network architectures over different sampling techniques, data cleaning conditions, and number of acquisition batches. In conclusion, deeper architectures (such as ECO Full) tend to overfit while finding the Ohnesorge number from dispensed droplet images. A shallower architecture with both 2D and 3D convolutions (ECOmini-18) performs slightly better than a shallower architecture with only 2D convolutions (ResNet-18). Though it has a slightly lower accuracy, inference can be made using ResNet-18 with as little as one image. The results also show that the CNNs with both 2D and 3D convolutions perform better when the frames are sampled sequentially, starting from the frame of the first appearance of the droplet. Next, it can be seen that data cleaning causes a decrease in classification accuracy due to a reduction in the amount of training data. Lastly, it can be seen that training on more than one acquisition batch improves the accuracy of the trained model. ECOmini-18 showed the maximum accuracy of 94.2%, trained on data from acquisition batches 1 and 3 jointly. This study is a proof-of-concept for neural networks to predict Ohnesorge numbers from droplet images. Moving forward, the technique can be further developed into a regression task, where the Ohnesorge number is a continuous scalar, instead of restricted to discrete classes. The CNNs can further be applied to other applications such as predicting the pinch-off time. Furthermore, the current CNNs can be trained in a physics-informed setup, reducing the size of datasets and biases from the experimental data.

Author Contributions: Conceptualization, R.Z., P.K., S.K., P.S., M.Z., and T.B.; Methodology, R.Z., P.K., S.K., P.S., M.Z., and T.B.; software, R.Z., P.K., S.K., P.S., M.Z., and T.B.; validation, R.Z., P.K., and T.B.; formal analysis, R.Z., P.K., S.K., P.S., M.Z., and T.B.; investigation, P.S.; resources, R.Z., P.K., S.K., M.Z., and T.B.; data curation, P.S.; writing—original draft preparation, P.S.; writing—review and editing, P.S., S.K., M.Z., G.M., P.K., R.Z., and T.B.; visualization, P.S.; supervision, S.K., M.Z., P.K., R.Z., and T.B.; project administration, S.K., M.Z., P.K., R.Z., and T.B.; funding acquisition, S.K., P.K., and R.Z. All authors have read and agreed to the published version of the manuscript.

Funding: This research received no external funding.

Data Availability Statement: The code for the work can be found at https://github.itap.purdue.edu/psardana/DropletAI. A sample dataset containing the first 200 dispenses from every acquisition batch for all five fluid classes can be found at https://osf.io/yxqjn/. The authors can provide the full dataset upon request.

Acknowledgments: The authors would like to thank Baden-Württemberg's HPC for providing resources (BwForCluster NEMO) for a big part of the work. We acknowledge support by the Open Access Publication Fund of the University of Freiburg.

Conflicts of Interest: The authors declare no conflict of interest.

Abbreviations

The following abbreviations and variables are used in this manuscript:

CNNs	Convolutional Neural Networks
D	Characteristic Length
SL	Stroke Length
DV	Downstroke Velocity
η	Dynamic Viscosity
ρ	Density
σ	Surface Tension
We	Weber Number
Oh	Ohnesorge Number
Re	Reynolds Number
DL	Deep Learning
LR	Learning Rate
SMAC	Sequential Model-Based Algorithm Configuration
RS	Random Sampling
SS	Sequential Sampling

Appendix A. Detailed Dataset Analysis

The complete dataset had 18,550 videos distributed over three acquisition batches and five classes. Figure A1 gives an overview of data distribution over different acquisition batches and classes.

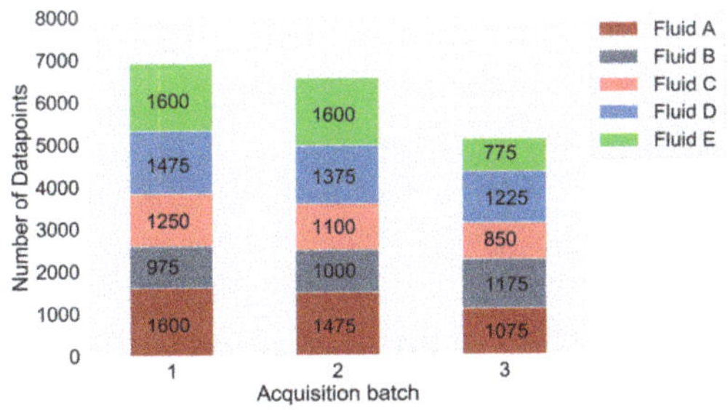

Figure A1. An overview of data distribution over different acquisition batches and classes.

Each video was a sequence of 250 frames with some of the initial frames empty because the fluid takes some time to respond to the actuation pulse applied. Figure A2 shows the entire sequence of 250 frames obtained for Fluid A when dispensed at SL = 21 µm and DV = 90 µm/ms.

To perform a more detailed analysis, the distribution of the dispenser configurations used in each class was also observed. A heatmap for the coefficient of variation (CV) distribution is shown in Figure A3.

In the figure, an upper CV threshold (of 50%) was used to make the visualizations clear. The current gravimetric weight analysis is sensitive to the used measurement time. Hence, it produced a high CV for droplets falling with low velocity or dispensing with satellites. The "-" in the image are the cases for which no droplet was ejected when the dispenser was driven at those configurations. Next, a similar analysis for the frame of the first appearance of the droplet was made. The results show that the frame of the first appearance of the droplet remained almost the same (the maximum difference was 1 frame and there was no clear pattern for its origin) irrespective of the liquid used. Figure A4 summarizes the relationship between the dispenser configuration used and the frame with the first appearance of the droplet. This analysis was only conducted for a set of dispenses with a CV of 5% or less.

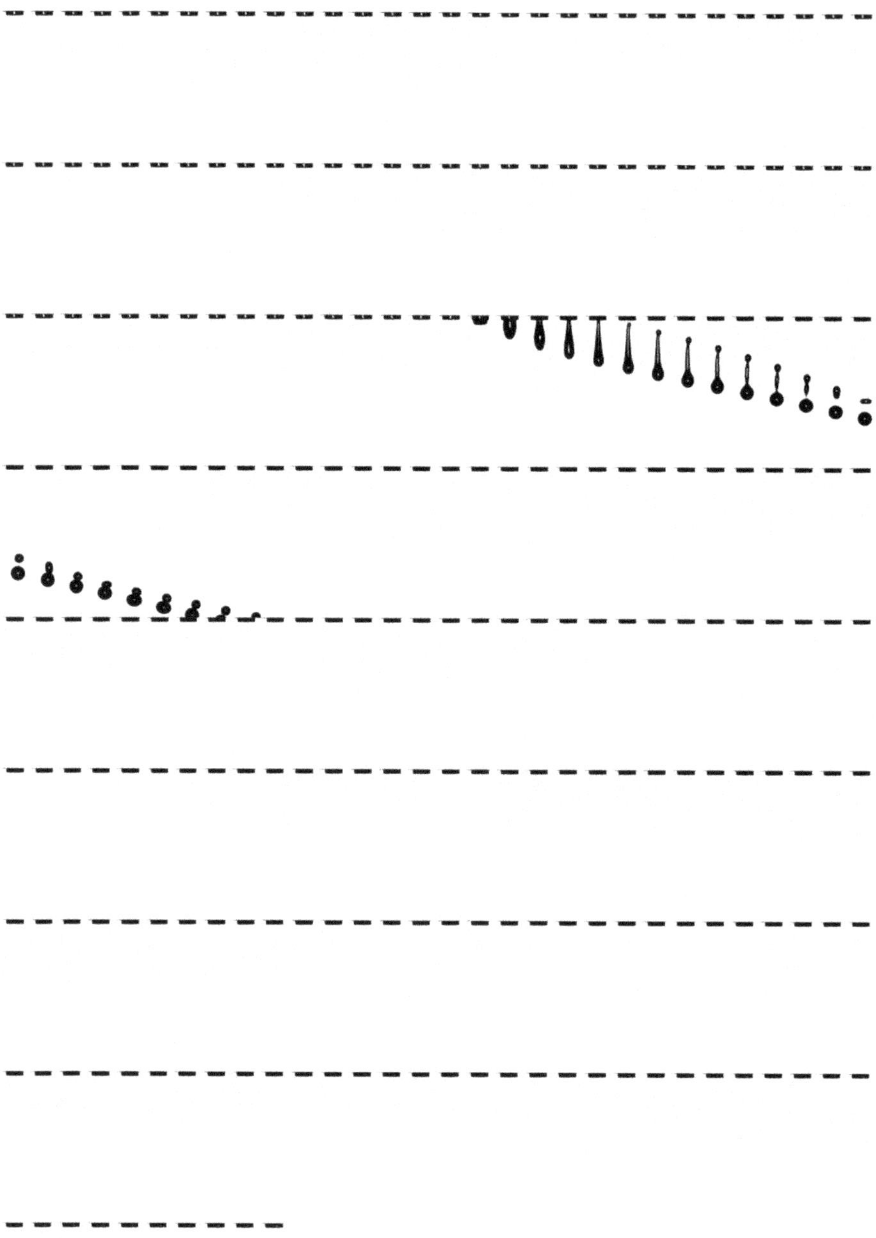

Figure A2. The sequence of 250 frames obtained for Fluid A when dispensed at SL = 21 µm and DV = 90 µm/ms.

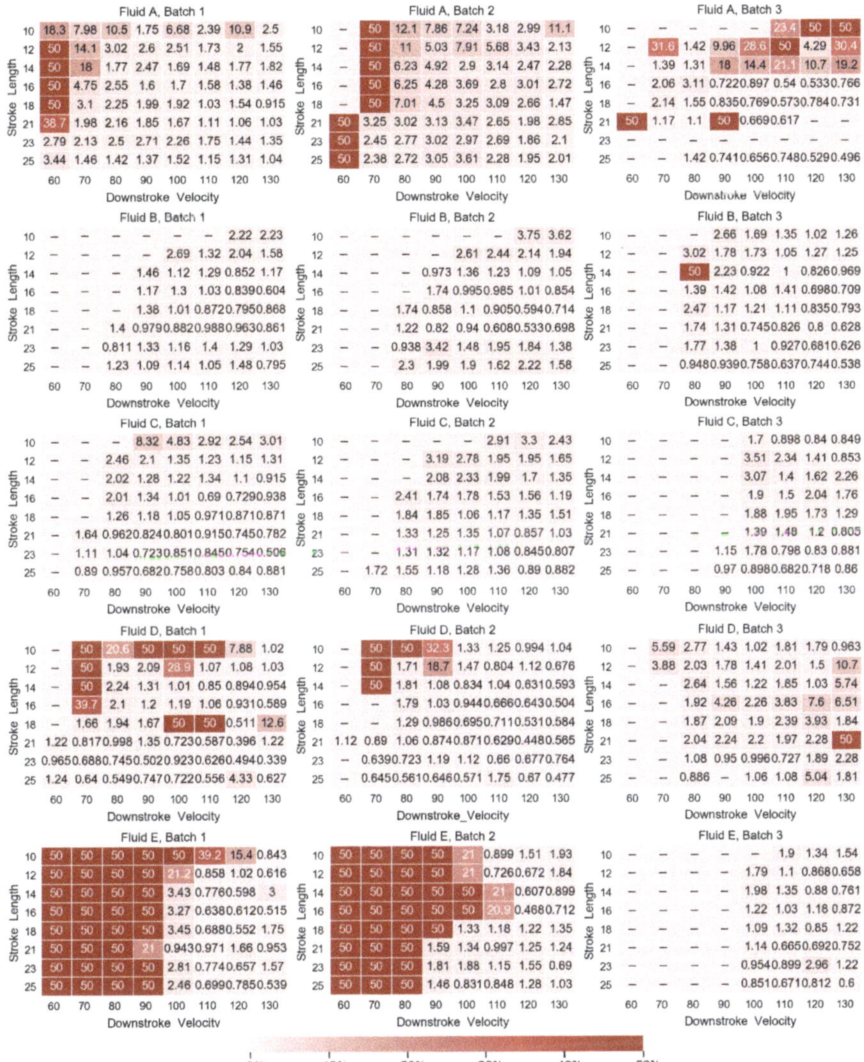

Figure A3. A heatmap for the coefficient of variation (CV) distribution. The heatmap was trimmed at an upper threshold for better visualization.

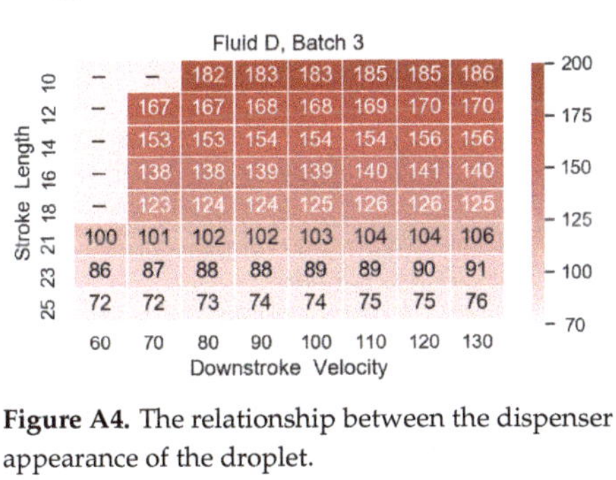

Figure A4. The relationship between the dispenser configuration used and the frame with the first appearance of the droplet.

References

1. Lindemann, T.; Zengerle, R. Droplet Dispensing. *Encycl. Microfluid. Nanofluidics* **2013**, 1–14. [CrossRef]
2. Wu, H.C.; Lin, H.J. Effects of Actuating Pressure Waveforms on the Droplet Behavior in a Piezoelectric Inkjet. *Mater. Trans.* **2010**, *51*, 2269–2276. [CrossRef]
3. BioFluidix GmbH. *PipeJet Nanodispenser Kit*; BioFluidix GmbH: Freiburg, Germany, 2020.
4. Lindeman, T. Droplet Generation-from the Nanoliter to the Femtoliter Range. Ph.D. Thesis, Albert-Ludwigs-Universität Freiburg, Freiburg im Breisgau, Germany 2006.
5. Fuster, D.; Agbaglah, G.; Josserand, C.; Popinet, S.; Zaleski, S. Numerical simulation of droplets, bubbles and waves: State of the art. *Fluid Dyn. Res.* **2009**, *41*, 065001. [CrossRef]
6. Hornik, K.; Stinchcombe, M.; White, H. Multilayer feedforward networks are universal approximators. *Neural Netw.* **1989**, *2*, 359–366. [CrossRef]
7. Zolfaghari, M.; Singh, K.; Brox, T. *ECO: Efficient Convolutional Network for Online Video Understanding*; Springer: Berlin/Heidelberg, Germany, 2018; Volume 11206, pp. 713–730. [CrossRef]
8. Krizhevsky, A.; Sutskever, I.; Hinton, G.E. ImageNet Classification with Deep Convolutional Neural Networks. *Commun. ACM* **2017**, *60*, 84–90. [CrossRef]
9. Geirhos, R.; Jacobsen, J.H.; Michaelis, C.; Zemel, R.; Brendel, W.; Bethge, M.; Wichmann, F.A. Shortcut learning in deep neural networks. *Nat. Mach. Intell.* **2020**, *2*, 665–673. [CrossRef]
10. Liang, D.; Zhang, J.; Muniyogeshbabu, T.G.; Bammesberger, S.; Tanguy, L.; Ernst, A.; Koltay, P.; Zengerle, R. Multi-principle droplet calibration technology. *microTEC Suedwest* **2012**. Available online: https://www.imtek.de/data/lehrstuehle/app/dokumente/conferences-pdf/conferences-2012/liang-multi-principle-droplet-calibration (accessed on 8 June 2023).
11. Losleben, N. Contact-Free Dispensing for In-Vitro Diagnostics: Challenges of the Reagent Diversity on the Performance of Appropriate Dispensing Technologies. Ph.D. Thesis, Albert-Ludwigs-Universität Freiburg, Freiburg im Breisgau, Germany, 2015.
12. Liang, D.; Steinert, C.; Bammesberger, S.; Tanguy, L.; Ernst, A.; Zengerle, R.; Koltay, P. Novel gravimetric measurement technique for quantitative volume calibration in the sub-microliter range. *Meas. Sci. Technol.* **2013**, *24*, 025301. [CrossRef]
13. *ISO 23783-2:2022*; Automated Liquid Handling Systems—Part 2: Measurement Procedures for the Determination of Volumetric Performance. International Organization for Standardization: Geneva, Switzerland 2022.
14. He, K.; Zhang, X.; Ren, S.; Sun, J. Deep Residual Learning for Image Recognition. In Proceedings of the 2016 IEEE Conference on Computer Vision and Pattern Recognition (CVPR), Las Vegas, NV, USA, 27–30 June 2016; pp. 770–778. [CrossRef]
15. Hara, K.; Kataoka, H.; Satoh, Y. Can Spatiotemporal 3D CNNs Retrace the History of 2D CNNs and ImageNet? In Proceedings of the IEEE Conference on Computer Vision and Pattern Recognition (CVPR), Salt Lake City, UT, USA, 18–23 June 2018; pp. 6546–6555.
16. Szegedy, C.; Ioffe, S.; Vanhoucke, V.; Alemi, A. Inception-v4, Inception-ResNet and the Impact of Residual Connections on Learning. In Proceedings of the AAAI Conference on Artificial Intelligence, San Francisco, CA, USA, 4–9 February 2017; Volume 31.
17. Hutter, F.; Hoos, H.H.; Leyton-Brown, K. *Sequential Model-Based Optimization for General Algorithm Configuration*; Springer: Berlin/Heidelberg, Germany, 2011; Volume 6683, pp. 507–523. [CrossRef]

fluids MDPI

Article

Application of Machine Learning Algorithms in Predicting Rheological Behavior of BN-diamond/Thermal Oil Hybrid Nanofluids

Abulhassan Ali [1,*], Nawal Noshad [2], Abhishek Kumar [3], Suhaib Umer Ilyas [1,*], Patrick E. Phelan [4], Mustafa Alsaady [1], Rizwan Nasir [1] and Yuying Yan [5]

1 Department of Chemical Engineering, University of Jeddah, Jeddah 23890, Saudi Arabia; malsaady@uj.edu.sa (M.A.); rnasir@uj.edu.sa (R.N.)
2 Department of Chemical Engineering, University of Gujrat, Gujrat 50700, Pakistan; nawal.noshad.ict@gmail.com
3 Petroleum Engineering Department, Universiti Teknologi PETRONAS, Seri Iskandar 32610, Perak Darul Ridzuan, Malaysia; abhi.upes01@gmail.com
4 School for Engineering of Matter, Transport & Energy, Arizona State University, Tempe, AZ 85281, USA; phelan@asu.edu
5 Fluids & Thermal Engineering Research Group, Faculty of Engineering, University of Nottingham, Nottingham NG7 2RD, UK; yuying.yan@nottingham.ac.uk
* Correspondence: aquddusi@uj.edu.sa (A.A.); suilyas@uj.edu.sa (S.U.I.)

Abstract: The use of nanofluids in heat transfer applications has significantly increased in recent times due to their enhanced thermal properties. It is therefore important to investigate the flow behavior and, thus, the rheology of different nanosuspensions to improve heat transfer performance. In this study, the viscosity of a BN-diamond/thermal oil hybrid nanofluid is predicted using four machine learning (ML) algorithms, i.e., random forest (RF), gradient boosting regression (GBR), Gaussian regression (GR) and artificial neural network (ANN), as a function of temperature (25–65 °C), particle concentration (0.2–0.6 wt.%), and shear rate (1–2000 s^{-1}). Six different error matrices were employed to evaluate the performance of these models by providing a comparative analysis. The data were randomly divided into training and testing data. The algorithms were optimized for better prediction of 700 experimental data points. While all ML algorithms produced R^2 values greater than 0.99, the most accurate predictions, with minimum error, were obtained by GBR. This study indicates that ML algorithms are highly accurate and reliable for the rheological predictions of nanofluids.

Keywords: artificial neural network; gradient boost regression; Gaussian regression; machine learning; nanofluids; random forest; rheology

1. Introduction

The increase in the significance and demand of thermal management for heat transfer applications led to the introduction of nanofluids, owing to their exceptional thermal properties. Nanofluids are colloidal suspensions formed by dispersing nanoparticles of diameters less than 100 nm into base fluids such as water, ethylene glycol, water, etc. Hybrid nanofluids can be prepared through the dispersion of two or more dissimilar nanoparticles in the base fluid in the form of a composite or a mixture, and they are found to have greater thermal, hydrodynamic, and rheological properties than unitary nanofluids [1,2]. Suspensions of metals, metal oxides, and other compound nanoparticles allow for the enhancement of thermal conductivity. However, other thermophysical properties, such as the nanofluid's density, specific heat, and dynamic viscosity, also play an important role in improving its heat transfer performance in several applications [3,4]. Nanofluids have been utilized in heating and cooling systems, electronics, transportation, and power generation, amongst other applications [5]. The flow behavior of several different nanofluids and the effect of other parameters such as magnetic field, solar energy, and entropy are now being studied for more advanced applications [6–8].

Viscosity is an essential thermophysical property because of its direct influence on the heat transfer characteristics of a nanofluid [9] and its effect on the pressure drop and pumping power. There is a higher pumping work demand for nanofluids because of their increased viscosity, in comparison to base fluids that do not contain nanoparticles. Consequently, the operating cost of the system increases [10,11]. The viscosity of nanofluids depends on several factors such as the type of base fluid and temperature, size and shape of particles, pH value, shear rate, etc. [12,13].

Several conventional models have been investigated to predict the viscosity of nanofluids. These include the Einstein [14], Brinkman [15], Wang [16], Krieger and Dougherty (K-D) [17], and Batchelor [18] models, among others. However, due to certain limitations, these models do not accurately predict the viscosity of nanofluids [19]. Researchers have also developed mathematical or empirical correlations for viscosity predictions with consideration of more parameters [20]. Recently, there has been a surge in the use of artificial intelligence (AI) and machine learning (ML) algorithms, particularly supervised learning techniques for the prediction of different thermophysical properties [21–23]. Despite being black-box models, the use of AI and ML is found to be fairly useful due to the challenges of processing data for the characterization of nanofluids [24]. Moreover, these predictions are also found to be highly accurate, granted that enough data points are available in the form of experimental data [25].

Amongst the many models available, artificial neural network (ANN) has been the most popular among researchers so far. Studies show that ANN is more accurate than conventional models and empirical correlations for viscosity predictions of nanofluids [26]. Li et al. [27] investigated the thermal performance of Al_2O_3/ethylene glycol nanofluids via a proposed correlation and ANN. The values for the coefficient of determination (R^2) for ANN predictions were found to be 0.9984 and 0.9997 for viscosity and thermal conductivity, respectively, showing more precise results than the correlation used. Davood et al. [28] used ANN to predict the viscosity of (WO_3)-MWCNTs/engine oil hybrid nanofluids as a function of the volume fraction of nanoparticles, shear rate, and temperature. The ANN results showed acceptable accuracy. In another study [29], the ANN prediction for the viscosity of CuO-MWCNT/engine oil hybrid nanofluids with inputs of temperature and volume fraction was found to be accurate and in good agreement with the experimental data. Similar results in favor of the accuracy of ANN compared to data fitting in an empirical correlation for MCWNT/water nanofluids were obtained in another study; ANN produced a maximum deviation of 0.28%, where solid volume fraction and temperature were used as inputs [30]. Thermophysical properties besides viscosity have also been predicted via ANN. Vafaei et al. [31] determined the thermal conductivity ratio of MgO-MWCNT/ethylene glycol hybrid nanofluids with a maximum deviation of 0.8% from the best ANN model. The ANN-predicted results in another study also provided good accuracy in comparison to experimental results for thermal conductivity ratios, density, viscosity ratio, and specific heat [32].

Researchers have also employed ML algorithms such as random forest (RF), multilayer perceptron (MLP), support vector regression (SVR), least-squared support vector machine (LSSVM), gradient boosting regression (GBR), extreme gradient boost (XGBoost), boosting regression trees (BRT), Gaussian regression (GR), decision tree regression (DTR), etc., and have drawn a comparison to determine which of these models exhibits a better performance for that particular nanosuspension. Ilyas et al. [33] employed ANN, RF, and GBR to predict the viscosity of diamond–graphene/mineral oil hybrid nanofluids based on input parameters of temperature, shear rate, and concentration. RF produced more accurate predictions than ANN and GBR. Gholizadeh et al. [34] assessed the predictive performance of RF against that of MLP and SVR, using five input parameters i.e., solid volume fraction, viscosity of the base fluid, temperature, density, and size of nanoparticle. The RF model provided the best prediction for the viscosity of nanofluids (R = 0.989). Bhanuteja et al. [35]

developed viscosity prediction models via four ML algorithms: GBR, RF, Adaboost, and voting regression. RF and GBR models demonstrated the best fit, with an R^2 greater than 0.99 and a maximum deviation of ±4.2%.

A recent study [30] estimated the specific heat capacity of water-based hybrid nanofluids considering the influence of several parameters using the GR, XGBoost, and SVR models. While all three algorithms showed excellent results, XGBoost outperformed the other two models, suggesting a much more feasible alternative to experimental approaches to determine the specific heat capacity of nanofluids. Ning et al. [36] used ANN and LSSVM for the prediction of filtration loss and shear stress of SiO_2/water drilling fluid. Both the models achieved R^2 values greater than 0.99, which demonstrated excellent accuracy. Similar results were obtained through ANN modeling to predict thermal conductivity (with the input parameters being temperature and concentration) and zeta potential (with the inputs being pH and concentration) of Fe_3O_4/water [37]. In another study [38], a GBR predictive model was developed to estimate the density of Al_2N_3, Si_3N_4, and TiN, suspended in ethylene glycol based on mass fraction, size and molecular weights of nanoparticles, and temperature. The proposed model exhibited high precision and was in good agreement with the experimental results. The results of some other studies that utilized ML algorithms for the prediction of thermophysical properties are given in Table 1.

Table 1. Use of ML techniques for high-accuracy predictions.

Nanofluid	Parameters	Inputs	Ranges	ML Techniques	Remarks	Ref.
TiO_2/water	Thermal Conductivity	Size Volume Fraction Temperature Thermal Conductivity	10–51.87 nm 0.002–4% 10–90 °C 0.6–1.455 W/mK	ANN, GBR, SVR, DTR, RF	GBR was found to be the most accurate, with an R^2 value of 0.99 for both testing and training data.	[39]
rGO-Fe_3O_4-TiO_2/ethylene glycol	Density Viscosity	Temperature Nanoparticle Concentration Shear Rate	25–50 °C 0.01–0.25% 1–1000 s^{-1}	BRT, SVR, ANN	R value of BRT for both density (0.9989) and viscosity (0.9979) was higher than that of SVR and ANN.	[40]
20 different Nanofluids	Thermal Conductivity	Temperature Volume Concentration Particle Size Mixture Ratio Acentric Factor of Base Fluid Thermal Conductivity Nanoparticle Density	20–70 °C 0–3.5 1.5–70 nm 0.15–0.85 0.343–0.659 0.16–1.44 W/mK 1000–10,500 kg/m^3	MLP-ANN, SVR	R^2 values of 0.99997 and 0.99788 were obtained by ANN and SVR, respectively.	[41]
ZrO_2/water	Viscosity	Temperature Concentration	10–65 °C 0.0125–0.2%	MLP-ANN	R^2 value of 0.99858 was obtained by ANN, demonstrating high-accuracy predictions.	[42]
ZnO-MWCNT (30:70)/W30 engine oil	Viscosity	Volume Fraction Temperature Shear Rate	0.05–1% 5–55 °C 50–1000 rpm	ANN	ANN produced high-accuracy predictions (compared to correlations), with an R^2 value of 0.9973.	[43]

ML algorithms have also been utilized for other applications besides evaluating the thermal performance of nanofluids, e.g., in the energy and environment sectors. Nie et al. [44] proposed a prediction model for electrical energy consumption in buildings using GBR. In another study [45], GR performed well compared to BRT in evaluating the performance of a small-scale solar organic Rankine cycle system. Cai et al. [46] used GBR for the prediction of net ecosystem carbon exchange (NEE), and RF for identifying the significant parameters influencing NEE. Comparison with three other models concluded that GBR performed high-accuracy predictions.

There is scarcely any literature available using some of the more recent ML algorithms, let alone on the predictions of thermophysical properties, particularly viscosity, despite being significantly influential to thermal transfer, which is the value added by this research.

In this study, the viscosity of a boron nitride (BN)–diamond (1:1)/thermal oil hybrid nanofluid was predicted via four models, i.e., random forest (RF), gradient boosting regression (GBR), Gaussian regression (GR), and artificial neural network (ANN), using 700 experimental data points. The rheological properties of BN-diamond nanofluids have not been subjected to any prior predictive investigation via machine learning techniques or otherwise. A comparative analysis of the four algorithms is presented to evaluate the performance and the accuracy of each model.

2. Methodology

2.1. Experimental Dataset

A BN-diamond/thermal oil hybrid nanofluid was employed as the experimental dataset for the prediction of viscosity in this study [47]. The nanomaterials included spherical diamond nanoparticles with a diameter of 3–10 nm and purity > 98.3%, as well as hexagonal BN with a size of 70–80 nm and 99.8% purity. The nanomaterials were mixed in equal-mass proportions (1:1) to form a hybrid system. The base fluid used was commercial thermal oil, used to disperse the hybrid nanomaterials. The details are included in reference [47].

The nanofluid was prepared using the two-step method, which involved the formation of a homogeneous suspension by adding the nanoparticles in thermal oil, i.e., the base fluid, in a 1:1 ratio, as it was stirred continuously. An inorganic surfactant (Span-85) was added to stabilize the nanofluids [47]. In reference [47], the rheology of BN-diamond/thermal nanosuspensions was evaluated as a function of temperature, particle concentration, and shear rate ranging from 25 to 65 °C, 0.2 to 0.6 wt.%, and 1 to 2000 s^{-1}, respectively.

In this study, BN-diamond/thermal oil was subjected to rheological predictions via four machine learning techniques i.e., random forest (RF), gradient boosting regression (GBR), Gaussian regression (GR), and artificial neural network (ANN), using 700 experimental data points, with inputs of temperature, particle concentration, and shear rate. The distribution of the 700 data points used, with respect to the three input parameters over their experimental ranges (25–65 °C, 0.2–0.6 wt.%, and 1–2000 s^{-1}), is illustrated in a violin plot (Figure 1). The distribution of data points was even for temperature and particle concentration, while shear rate and viscosity (the output of the models) showed more variation, with medians in the lower region. Experimental results showed that there was a significant change in viscosity for shear rates under 100 s^{-1} (where the data points are largely concentrated), and a Newtonian behavior was observed [47].

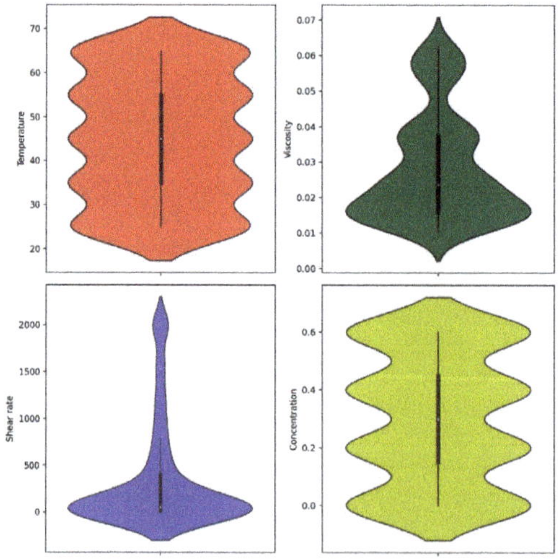

Figure 1. Violin plot of the experimental dataset used for modeling.

The accuracy of these predictions was evaluated by comparison with the measured viscosity from the aforementioned study. (An overview of the methodology is presented in Figure 2.) Six error matrices i.e., coefficient of determination (R^2), mean absolute error (MAE), root mean square error (RMSE), bias, Legate and McCabe Index ($E_{L,M}$) and Willmot Index of Agreement (I_A), were used for this purpose, and they are given in Equations (1)–(6), respectively [33].

$$R^2 = 1 - \frac{\sum_1^n (y_i - \hat{y}_i)^2}{\sum_1^n (y_i - \bar{y}_i)^2}. \quad (1)$$

$$MAE = \frac{1}{N}\sum_1^n |y_i - \hat{y}_i| \quad (2)$$

$$RMSE = \sqrt{\frac{\sum_{i=1}^n (y_i - \hat{y}_i)^2}{N}} \quad (3)$$

$$bias(\hat{\theta}) = E_\mu(\hat{\theta}) - \theta \quad (4)$$

$$E_{L,M} = 1 - \frac{\sum_{i=1}^{i=N} |y_i - \hat{y}_i|}{\sum_{i=1}^{i=N} |y_i - \bar{y}_i|} \quad (5)$$

$$I_A = 1 - \frac{\sum_{i=1}^{i=N} (y_i - \hat{y}_i)^2}{\sum_{i=0}^{i=N} (|y_i - \bar{y}_i| + (\hat{y}_i - \bar{y}_i))^2} \quad (6)$$

where y_i is the given data point, $\bar{y}_i$ is the mean of all the data points, $\hat{y}_i$ denotes the predicted values, and N is the number of observations.

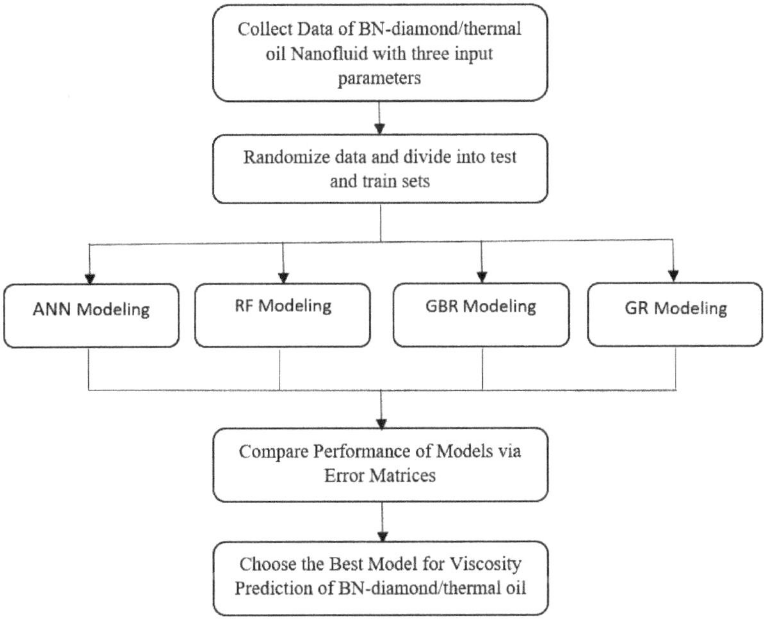

Figure 2. An overview of the methodology applied for predicting rheology via machine learning algorithms.

2.2. Predictions via Random Forest (RF)

Random Forest (RF) is a fast and reliable model that makes decisions based on the characteristics of several decision trees. The more diverse these trees are, the more consistent the RF model is [39]. It utilizes a statistical technique called bagging.

2.3. Predictions via Gradient Boosting Regression (GBR)

In gradient boosting regression, the learning procedure sequentially and consecutively adds new models to the ensemble, which produces an accurate prediction of the response variable [48]. Gradient boosting regression utilizes an algorithm that minimizes the amount of loss that occurs [49] by fitting a regression tree on the negative gradient of a specified loss function [39]. The loss function applied can be arbitrary, which allows GBR to be flexible for a particular dataset.

2.4. Predictions via Gaussian Regression (GR)

GR is a probabilistic regression-based model, i.e., it is associated with learning a continuous function from training data. Several covariance functions can be integrated with GR to obtain optimized results from testing data. The relationship between inputs and outputs in GR can be described as shown in Equation (7) [50].

$$y = f(x) + \xi\xi \sim N\left(0, {v_n}^2\right) \tag{7}$$

where N is the number of pairs of feature vectors and ξ is distributed with zero mean noise.

2.5. Predictions via Artificial Neural Network (ANN)

ANN models contain a hidden layer which includes the number of neurons, the activation function, the number of layers, and a training algorithm. ANN finds the connection between the data provided to the model; the output depends on the back propagation of the randomly initialized weights of the input data [39]. The neural network in this study was modeled using the Levenberg–Marquardtt training algorithm with three hidden layers. It was modeled using a number of neurons ranging from 2 to 12. For proper optimization, the errors were all generated for different numbers of neurons. As demonstrated in Figure 3, the least error (in accordance with all error matrices) was detected at 12 neurons.

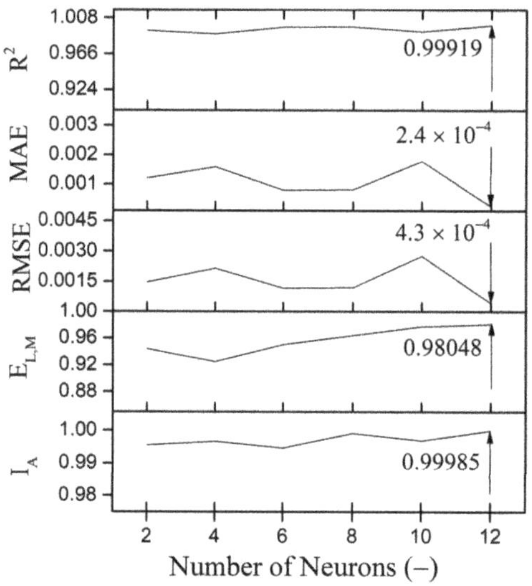

Figure 3. Error analysis to find the optimum number of neurons for ANN.

3. Results and Discussion

The rheology of BN-diamond/thermal oil nanosuspensions was predicted via random forest (RF), gradient boosting regression (GBR), Gaussian regression (GR), and artificial neural network (ANN) machine learning algorithms as a function of three input parameters: temperature, particle concentration, and shear rate.

The data points were randomized and divided into testing and training data. It is important to note that all data points were used for modeling the algorithms. The original experimental data (taken from [47]) were not refined or normalized and no data points were excluded for rheological predictions. This signifies that the errors generated also include data points which may show a certain degree of deviation.

3.1. Random Forest (RF)

RF showed the most deviation in data and the least accurate predictions amongst the four ML algorithms. The distribution of the data points for RF against absolute difference, which is the relative error between the testing and training data, is shown in Figure 4. Data far from the origin represent more deviation. It is noticeable from the figure that data points are significantly scattered for the both training and testing sets, and they are less concentrated around nil absolute difference.

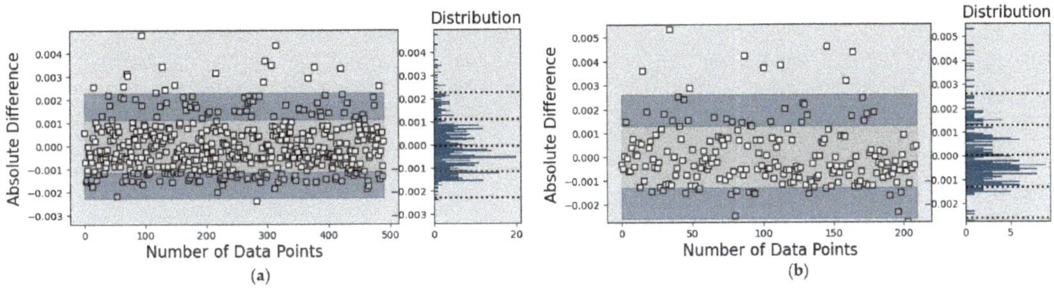

Figure 4. Distribution of data points for (**a**) RF training data and (**b**) RF testing data.

One of the key matrices used to validate the performance of the algorithms is the coefficient of determination (R^2), which represents the relationship between the predicted and the experimentally measured viscosity; an R^2 value close to 1 indicates accuracy in the predictions. Although all four ML algorithms produced good predictions, with an R^2 greater than 0.99 for the testing and training data, the least effective of them is RF, i.e., R^2_{train} = 0.994368 and R^2_{test} = 0.99258 (Figure 5).

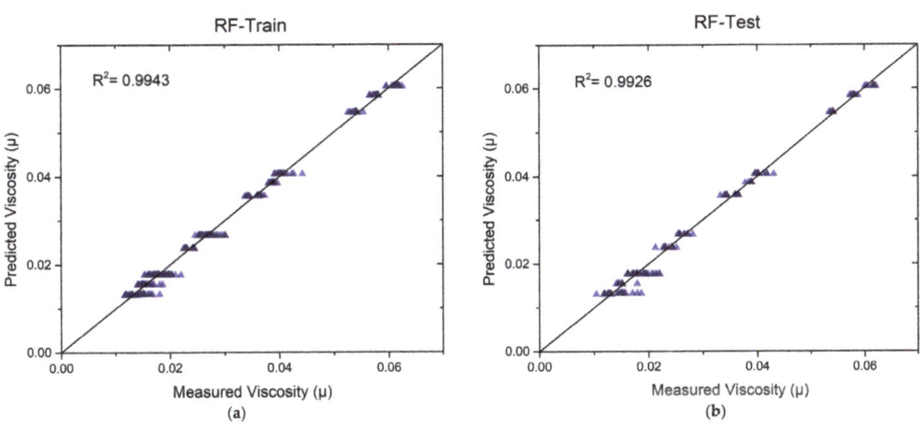

Figure 5. Measured vs predicted viscosity for (**a**) RF training data and (**b**) RF testing data.

3.2. Gradient Boosting Regression (GBR)

GBR training and testing data, illustrated in Figure 6, both show high accumulation of data at absolute difference of zero. Very few data points lie outside the ranges of −0.00025 up to 0.00025 and −0.0005 up to 0.0005 for the training and testing data, respectively. The most accurate predictions for the viscosity of the BN-diamond/thermal oil nanofluid were obtained by GBR, with R^2 values of R^2_{train} = 0.999939 and R^2_{test} = 0.99953. The plots in Figure 7 are well accumulated to the 45° reference, exhibiting an excellent fit to the model. Predicted data scarcely show any deviation from the experimentally measured viscosity, demonstrating the high precision and accuracy of the rheological behavior predictions and the reliability of the GBR model. This was enabled by the iterative boosting of a weaker learner into a stronger learner.

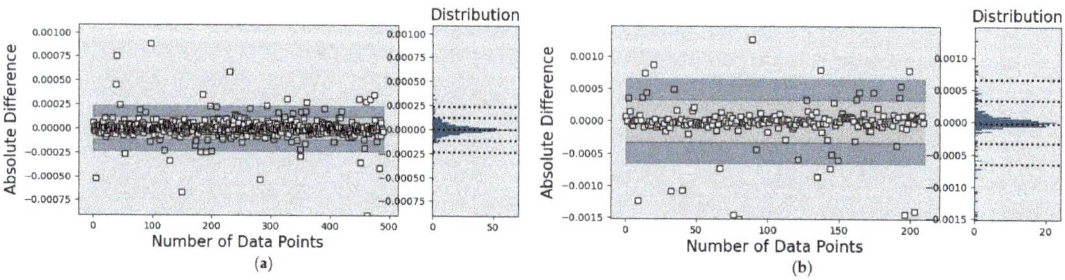

Figure 6. Distribution of data points for (**a**) GBR training data and (**b**) GBR testing data.

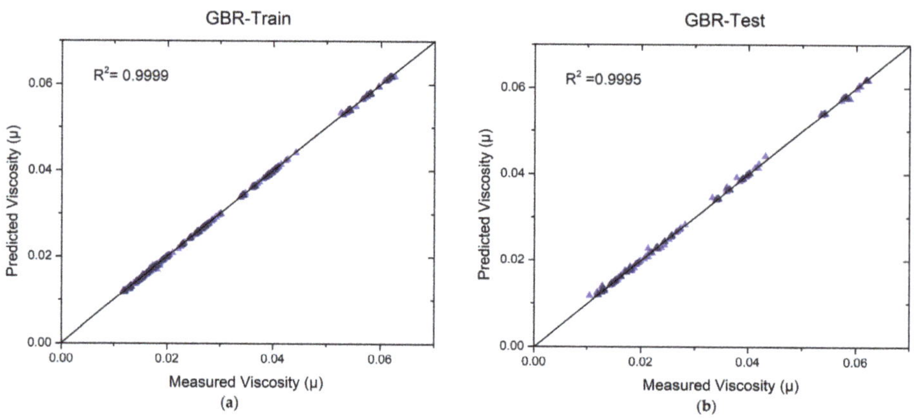

Figure 7. Measured vs. predicted viscosity for (**a**) GBR training data and (**b**) GBR testing data.

3.3. Gaussian Regression (GR)

Rheological predictions via GR were good, but not the most accurate out of the algorithms used. More convergence of data was observed in the training set (plots accumulated between −0.001 and 0.001) than in the testing set, as seen in Figure 8. This reflects on the greater deviation shown in Figure 9, where the R^2 value of the testing set, i.e., 0.994911 was lower than that of the training set, i.e., 0.998587. This result of GR may be attributed to its potential to not scale very well with an increased number of data points.

3.4. Artificial Neural Network (ANN)

ANN produced the second most accurate predictions, after GBR. The absolute difference ranged from −0.001 to 0.003, with plots more concentrated up to 0.001, as illustrated in Figure 10; a higher level of convergence of data was observed in comparison to GR and RF. From the regression plot depicted in Figure 11, a little more deviation from the reference line in the testing set is observed. However, the R^2 values for both the training and testing data (R^2_{train} = 0.999193 and R^2_{test} = 0.998561) were found to be satisfactory. ANN is a highly reliable model owing to its ability to optimize data by selecting the number of neurons by trial and error. Optimization techniques reduce error, thus increasing the accuracy of the predictions.

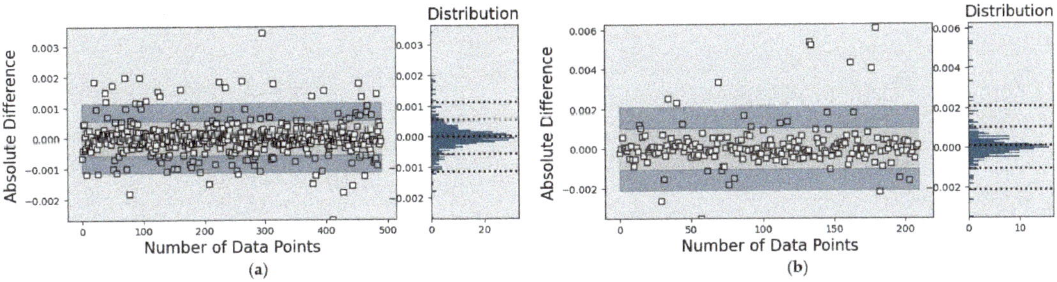

Figure 8. Distribution of data points for (**a**) GR training data and (**b**) GR testing data.

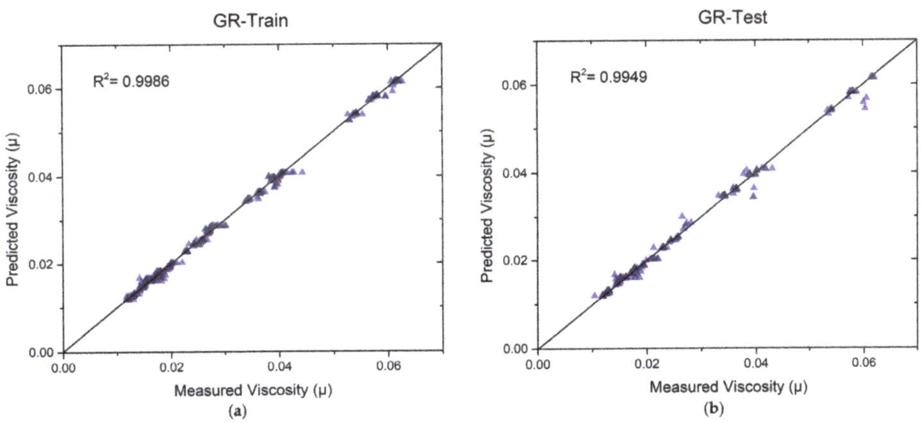

Figure 9. Measured vs predicted viscosity for (**a**) GR training data and (**b**) GR testing data.

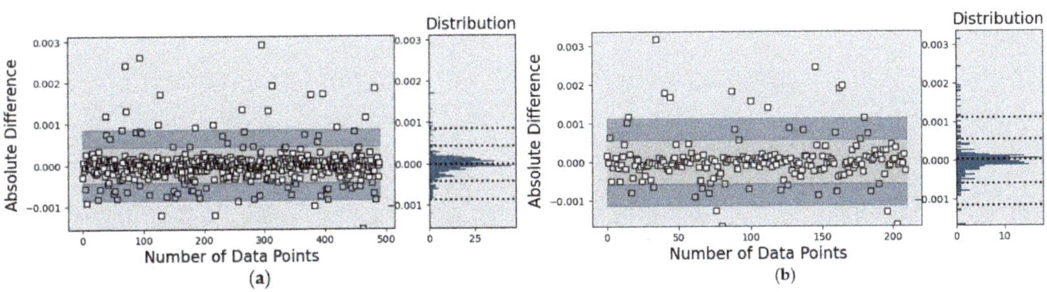

Figure 10. Distribution of data points for (**a**) ANN training data and (**b**) ANN testing data.

These results were further verified by using five other error matrices, i.e., mean absolute error (MAE), root mean square error (RMSE), bias, the Legate and McCabe Index ($E_{L,M}$), and the Willmot Index of Agreement (I_A). The comparison of these error matrices (and R^2) with RF, GBR, GR, and ANN for testing and training data is presented in Figure 12. MAE is the average variance between the significant and projected values of the same dataset, RMSE is the average difference between the experimentally measured and the predicted values, and bias refers to the error between the average predictions and the measured values. $E_{L,M}$ and I_A are measures of how well the model fits. Higher values of

MAE, RMSE, and bias errors and lower values of $E_{L,M}$ and I_A indicate less accuracy in the predictions obtained by the ML algorithms. It can be observed from Figure 12 that errors in RF were much greater in number than those in GR, ANN, and GBR (in the same order), and GBR had the least errors and the highest values of $E_{L,M}$ and I_A, i.e., 0.9952 and 0.99998 for training data and 0.98755 and 0.99988 for testing data, respectively.

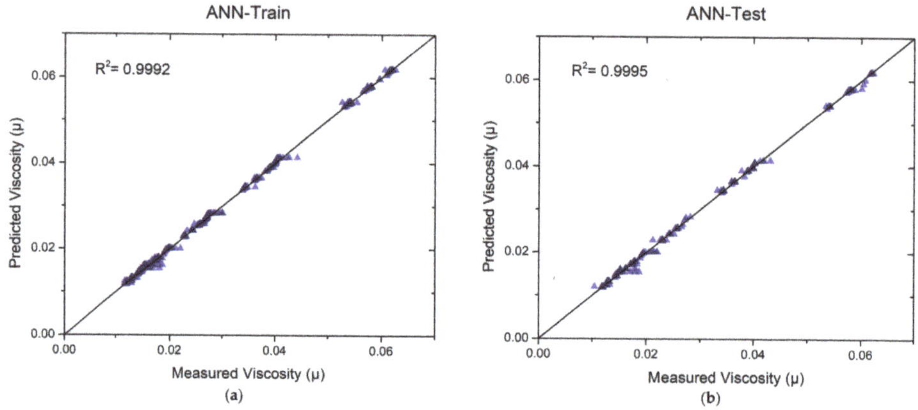

Figure 11. Measured vs predicted viscosity for (**a**) ANN training data and (**b**) ANN testing data.

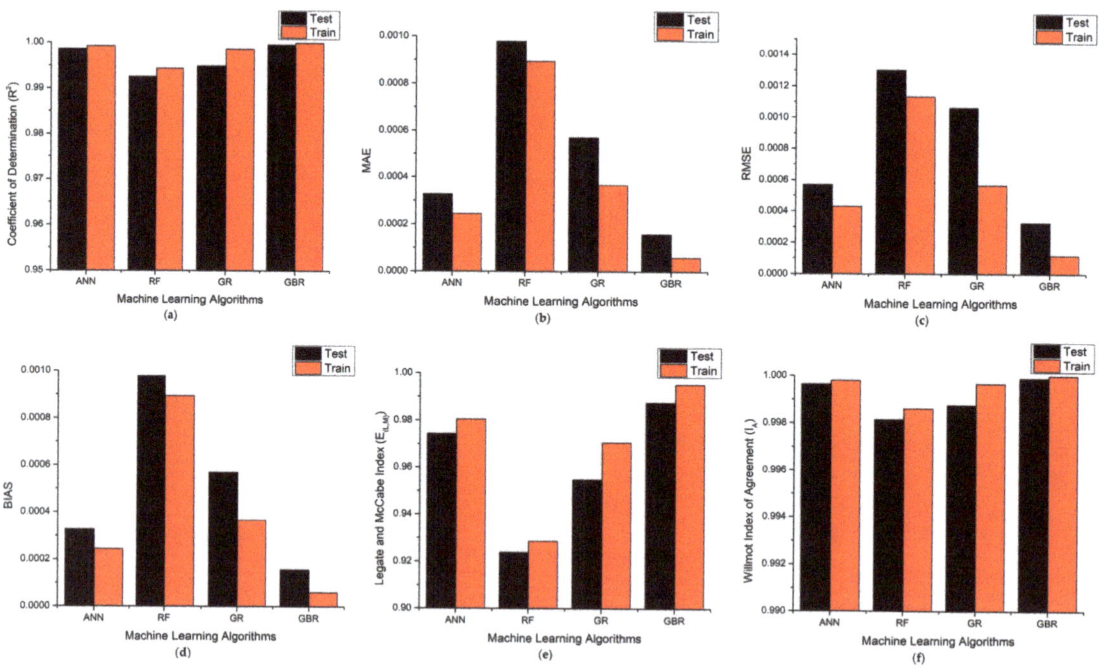

Figure 12. Comparison of error matrices between testing and training data of RF, GBR, GR, and ANN. (**a**) Coefficient of determination (R^2); (**b**) mean absolute error (MAE); (**c**) root mean square error (RMSE); (**d**) bias; (**e**) Legate and McCabe Index ($E_{L,M}$); (**f**) Willmot Index of Agreement (I_A).

4. Conclusions

In light of the limited studies available on the rheological behavior predictions of hybrid nanofluids, especially via machine learning (ML) algorithms, this study aimed to predict the viscosity of BN-diamond/thermal oil using four different ML algorithms, i.e., random forest (RF), gradient boosting regression (GBR), Gaussian regression (GR), and artificial neural network (ANN). Temperature, particle concentration, and shear rate were the three input parameters of the models, with the output being viscosity. The performance of these models was evaluated using six error matrices, namely, coefficient of determination (R^2), mean absolute error (MAE), root mean square error (RMSE), bias, the Legate and McCabe Index ($E_{L,M}$), and the Willmot Index of Agreement (I_A). Results showed that all R^2 values were greater than 0.99; the highest R^2 was obtained using GBR, followed by ANN, GR, and RF (which showed the most deviation from the experimentally measured viscosity). Overall, ML algorithms were found to be highly accurate and reliable for rheological predictions, specifically in comparison to conventional and empirical models.

Author Contributions: Conceptualization, S.U.I., P.E.P. and Y.Y.; Data curation, S.U.I.; Formal analysis, A.A., N.N., A.K. and M.A.; Funding acquisition, A.A.; Investigation, M.A. and R.N.; Methodology, A.A., A.K., S.U.I., P.E.P., M.A. and Y.Y.; Resources, S.U.I. and M.A.; Software, A.K.; Supervision, P.E.P. and Y.Y.; Validation, A.K. and S.U.I.; Visualization, N.N. and R.N.; Writing—original draft, A.A. and N.N.; Writing—review and editing, P.E.P. and Y.Y. All authors have read and agreed to the published version of the manuscript.

Funding: This work was funded by the University of Jeddah, Jeddah, Kingdom of Saudi Arabia, under grant No. UJ-21-ICI-13.

Data Availability Statement: The data can be available on request.

Acknowledgments: The authors acknowledge the University of Jeddah, KSA, for technical and financial support.

Conflicts of Interest: The authors declare no conflicts of interest.

Nomenclature

Symbols	
E_{LM}	Legate and McCabe Index, LM Index (-)
I_A	Willmot Index of Agreement, W Index (-)
N	Number of observations in Equations (1)–(7)
R^2	Coefficient of determination (-)
T	Temperature (°C)
y_i	Given data point in Equations (1)–(6)
$\hat{y}_i$	Predicted values in Equations (1)–(6)
$\overline{y}_i$	Mean of the given values in Equations (1)–(6)
Greek Symbols	
φ	Nanomaterial concentration (%)
γ	Shear rate (1/s)
μ	Viscosity (Pa·s)
Abbreviations	
AI	Artificial intelligence
ANN	Artificial neural network
BRT	Boosting regression trees
DTR	Decision tree regression
GBM	Gradient boosting machine
GR	Gaussian regression
LSSVM	Least squared support vector machine
MAE	Mean absolute error
ML	Machine learning
MLP	Multilayer perceptron
RF	Random forest

RMSE	Root mean square error
R^2_{test}	Coefficient of determination for testing set
R^2_{train}	Coefficient of determination for training set
SVR	Support vector regression
XGBoost	Extreme gradient boost

References

1. Babar, H.; Ali, H.M. Towards Hybrid Nanofluids: Preparation, Thermophysical Properties, Applications, and Challenges. *J. Mol. Liq.* **2019**, *281*, 598–633. [CrossRef]
2. Nabil, M.F.; Azmi, W.H.; Hamid, K.A.; Zawawi, N.N.M.; Priyandoko, G.; Mamat, R. Thermo-Physical Properties of Hybrid Nanofluids and Hybrid Nanolubricants: A Comprehensive Review on Performance. *Int. Commun. Heat Mass Transf.* **2017**, *83*, 30–39. [CrossRef]
3. Ali, A.R.I.; Salam, B. A Review on Nanofluid: Preparation, Stability, Thermophysical Properties, Heat Transfer Characteristics and Application. *SN Appl. Sci.* **2020**, *2*, 1636. [CrossRef]
4. Xuan, Y.; Li, Q. Heat Transfer Enhancement of Nanofluids. *Int. J. Heat Fluid Flow* **2000**, *21*, 58–64. [CrossRef]
5. Bashirnezhad, K.; Bazri, S.; Safaei, M.R.; Goodarzi, M.; Dahari, M.; Mahian, O.; Dalkiliça, A.S.; Wongwises, S. Viscosity of Nanofluids: A Review of Recent Experimental Studies. *Int. Commun. Heat Mass Transf.* **2016**, *73*, 114–123. [CrossRef]
6. Dawar, A.; Wakif, A.; Thumma, T.; Shah, N.A. Towards a New MHD Non-Homogeneous Convective Nanofluid Flow Model for Simulating a Rotating Inclined Thin Layer of Sodium Alginate-Based Iron Oxide Exposed to Incident Solar Energy. *Int. Commun. Heat Mass Transf.* **2022**, *130*, 105800. [CrossRef]
7. Salahuddin, T.; Ali, Z.; Awais, M.; Khan, M.; Altanji, M. A Flow Behavior of Sutterby Nanofluid near the Catalytic Parabolic Surface. *Int. Commun. Heat Mass Transf.* **2022**, *131*, 105821. [CrossRef]
8. Upadhya, S.M.; Raju, S.V.S.R.; Raju, C.S.K.; Shah, N.A.; Chung, J.D. Importance of Entropy Generation on Casson, Micropolar and Hybrid Magneto-Nanofluids in a Suspension of Cross Diffusion. *Chin. J. Phys.* **2022**, *77*, 1080–1101. [CrossRef]
9. Minakov, A.V.; Rudyak, V.Y.; Pryazhnikov, M.I. About Rheology of Nanofluids. *AIP Conf. Proc.* **2018**, *2027*, 030141. [CrossRef]
10. Said, Z.; Sundar, L.S.; Tiwari, A.K.; Ali, H.M.; Sheikholeslami, M.; Bellos, E.; Babar, H. Recent Advances on the Fundamental Physical Phenomena behind Stability, Dynamic Motion, Thermophysical Properties, Heat Transport, Applications, and Challenges of Nanofluids. *Phys. Rep.* **2022**, *946*, 1–94. [CrossRef]
11. Rashidi, M.M.; Nazari, M.A.; Mahariq, I.; Assad, M.E.H.; Ali, M.E.; Almuzaiqer, R.; Nuhait, A.; Murshid, N. Thermophysical Properties of Hybrid Nanofluids and the Proposed Models: An Updated Comprehensive Study. *Nanomaterials* **2021**, *11*, 3084. [CrossRef] [PubMed]
12. Mahbubul, I.M.; Saidur, R.; Amalina, M.A. Latest Developments on the Viscosity of Nanofluids. *Int. J. Heat Mass Transf.* **2012**, *55*, 874–885. [CrossRef]
13. Mishra, P.C.; Mukherjee, S.; Nayak, S.K.; Panda, A. A Brief Review on Viscosity of Nanofluids. *Int. Nano Lett.* **2014**, *4*, 109–120. [CrossRef]
14. Einstein, A. Eine Neue Bestimmung der Moleküldimensionen. Ph.D. Thesis, ETH Zurich, Zurich, Switzerland, 1905. [CrossRef]
15. Brinkman, H.C. The Viscosity of Concentrated Suspensions and Solutions. *J. Chem. Phys.* **1952**, *20*, 571. [CrossRef]
16. Wang, X.; Xu, X.; Choi, S.U.S. Thermal Conductivity of Nanoparticle—Fluid Mixture. *J. Thermophys. Heat Transf.* **2012**, *13*, 474–480. [CrossRef]
17. Krieger, I.M.; Dougherty, T.J. A Mechanism for Non-Newtonian Flow in Suspensions of Rigid Spheres. *Trans. Soc. Rheol.* **1959**, *3*, 137. [CrossRef]
18. Batchelor, G.K. The Effect of Brownian Motion on the Bulk Stress in a Suspension of Spherical Particles. *J. Fluid Mech.* **1977**, *83*, 97–117. [CrossRef]
19. Ilyas, S.U.; Narahari, M.; Theng, J.T.Y.; Pendyala, R. Experimental Evaluation of Dispersion Behavior, Rheology and Thermal Analysis of Functionalized Zinc Oxide-Paraffin Oil Nanofluids. *J. Mol. Liq.* **2019**, *294*, 111613. [CrossRef]
20. Sundar, L.S.; Sharma, K.V.; Naik, M.T.; Singh, M.K. Empirical and Theoretical Correlations on Viscosity of Nanofluids: A Review. *Renew. Sustain. Energy Rev.* **2013**, *25*, 670–686. [CrossRef]
21. Tan, K.X.; Ilyas, S.U.; Pendyala, R.; Shamsuddin, M.R. Assessment of Thermal Conductivity and Viscosity of Alumina-Based Engine Coolant Nanofluids Using Random Forest Approach. *AIP Conf. Proc.* **2022**, *2610*, 030002. [CrossRef]
22. Krishna, S.; Ridha, S.; Ilyas, S.U.; Campbell, S.; Bhan, U.; Bataee, M. Application of Deep Learning Technique to Predict Downhole Pressure Differential in Eccentric Annulus of Ultra-Deep Well. In Proceedings of the ASME 2021 40th International Conference on Ocean, Offshore and Arctic Engineering, Virtual, 21–30 June 2021. [CrossRef]
23. Kumar, A.; Ridha, S.; Ganet, T.; Vasant, P.; Ilyas, S.U. Machine Learning Methods for Herschel-Bulkley Fluids in Annulus: Pressure Drop Predictions and Algorithm Performance Evaluation. *Appl. Sci.* **2020**, *10*, 2588. [CrossRef]
24. Sharma, P.; Said, Z.; Kumar, A.; Nižetić, S.; Pandey, A.; Hoang, A.T.; Huang, Z.; Afzal, A.; Li, C.; Le, A.T.; et al. Recent Advances in Machine Learning Research for Nanofluid-Based Heat Transfer in Renewable Energy System. *Energy Fuels* **2022**, *36*, 6626–6658. [CrossRef]
25. Jirasek, F.; Hasse, H. Perspective: Machine Learning of Thermophysical Properties. *Fluid Phase Equilibria* **2021**, *549*, 113206. [CrossRef]

26. Ramezanizadeh, M.; Ahmadi, M.H.; Nazari, M.A.; Sadeghzadeh, M.; Chen, L. A Review on the Utilized Machine Learning Approaches for Modeling the Dynamic Viscosity of Nanofluids. *Renew. Sustain. Energy Rev.* **2019**, *114*, 109345. [CrossRef]
27. Li, L.; Zhai, Y.; Jin, Y.; Wang, J.; Wang, H.; Ma, M. Stability, Thermal Performance and Artificial Neural Network Modeling of Viscosity and Thermal Conductivity of Al2O3-Ethylene Glycol Nanofluids. *Powder Technol.* **2020**, *363*, 360–368. [CrossRef]
28. Toghraie, D.; Aghahadi, M.H.; Sina, N.; Soltani, F. Application of Artificial Neural Networks (ANNs) for Predicting the Viscosity of Tungsten Oxide (WO3)-MWCNTs/Engine Oil Hybrid Nanofluid. *Int. J. Thermophys.* **2020**, *41*, 163. [CrossRef]
29. Aghaei, A.; Khorasanizadeh, H.; Sheikhzadeh, G.A. Measurement of the Dynamic Viscosity of Hybrid Engine Oil -Cuo-MWCNT Nanofluid, Development of a Practical Viscosity Correlation and Utilizing the Artificial Neural Network. *Heat Mass Transf.* **2018**, *54*, 151–161. [CrossRef]
30. Afrand, M.; Nadooshan, A.A.; Hassani, M.; Yarmand, H.; Dahari, M. Predicting the Viscosity of Multi-Walled Carbon Nanotubes/Water Nanofluid by Developing an Optimal Artificial Neural Network Based on Experimental Data. *Int. Commun. Heat Mass Transf.* **2016**, *77*, 49–53. [CrossRef]
31. Vafaei, M.; Afrand, M.; Sina, N.; Kalbasi, R.; Sourani, F.; Teimouri, H. Evaluation of Thermal Conductivity of MgO-MWCNTs/EG Hybrid Nanofluids Based on Experimental Data by Selecting Optimal Artificial Neural Networks. *Phys. E Low-Dimens. Syst. Nanostructures* **2017**, *85*, 90–96. [CrossRef]
32. Kumar, V.; Pare, A.; Tiwari, A.K.; Ghosh, S.K. Efficacy Evaluation of Oxide-MWCNT Water Hybrid Nanofluids: An Experimental and Artificial Neural Network Approach. *Colloids Surfaces A Physicochem. Eng. Asp.* **2021**, *620*, 126562. [CrossRef]
33. Ilyas, S.U.; Ridha, S.; Sardar, S.; Estellé, P.; Kumar, A.; Pendyala, R. Rheological Behavior of Stabilized Diamond-Graphene Nanoplatelets Hybrid Nanosuspensions in Mineral Oil. *J. Mol. Liq.* **2021**, *328*, 115509. [CrossRef]
34. Gholizadeh, M.; Jamei, M.; Ahmadianfar, I.; Pourrajab, R. Prediction of Nanofluids Viscosity Using Random Forest (RF) Approach. *Chemom. Intell. Lab. Syst.* **2020**, *201*, 104010. [CrossRef]
35. Bhanuteja, S.; Srinivas, V.; Moorthy, C.V.K.N.S.N.; Kumar, S.J.; Raju, B.L.L. Prediction of Thermophysical Properties of Hybrid Nanofluids Using Machine Learning Algorithms. *Int. J. Interact. Des. Manuf.* **2023**, *2023*, 1–14. [CrossRef]
36. Ning, Y.C.; Ridha, S.; Ilyas, S.U.; Krishna, S.; Dzulkarnain, I.; Abdurrahman, M. Application of Machine Learning to Determine the Shear Stress and Filtration Loss Properties of Nano-Based Drilling Fluid. *J. Pet. Explor. Prod. Technol.* **2022**, *13*, 1031–1052. [CrossRef]
37. Sahin, F.; Genc, O.; Gökcek, M.; Çolak, A.B. An Experimental and New Study on Thermal Conductivity and Zeta Potential of Fe3O4/Water Nanofluid: Machine Learning Modeling and Proposing a New Correlation. *Powder Technol.* **2023**, *420*, 118388. [CrossRef]
38. Singh, H.M.; Sharma, D.P.; Alade, I.O. GBR-GSO Based Machine Learning Predictive Model for Estimating Density of Al2N3, Si3N4, and TiN Nanoparticles Suspended in Ethylene Glycol Nanofluids. *Eur. Phys. J. Plus* **2022**, *137*, 587. [CrossRef]
39. Sharma, P.; Ramesh, K.; Parameshwaran, R.; Deshmukh, S.S. Thermal Conductivity Prediction of Titania-Water Nanofluid: A Case Study Using Different Machine Learning Algorithms. *Case Stud. Therm. Eng.* **2022**, *30*, 101658. [CrossRef]
40. Said, Z.; Cakmak, N.K.; Sharma, P.; Sundar, L.S.; Inayat, A.; Keklikcioglu, O.; Li, C. Synthesis, Stability, Density, Viscosity of Ethylene Glycol-Based Ternary Hybrid Nanofluids: Experimental Investigations and Model -Prediction Using Modern Machine Learning Techniques. *Powder Technol.* **2022**, *400*, 117190. [CrossRef]
41. Adun, H.; Wole-Osho, I.; Okonkwo, E.C.; Bamisile, O.; Dagbasi, M.; Abbasoglu, S. A Neural Network-Based Predictive Model for the Thermal Conductivity of Hybrid Nanofluids. *Int. Commun. Heat Mass Transf.* **2020**, *119*, 104930. [CrossRef]
42. Çolak, A.B. A Novel Comparative Analysis between the Experimental and Numeric Methods on Viscosity of Zirconium Oxide Nanofluid: Developing Optimal Artificial Neural Network and New Mathematical Model. *Powder Technol.* **2021**, *381*, 338–351. [CrossRef]
43. Sepehrnia, M.; Mohammadzadeh, K.; Veyseh, M.M.; Agah, E.; Amani, M. Rheological Behavior of Engine Oil Based Hybrid Nanofluid Containing MWCNTs and ZnO Nanopowders: Experimental Analysis, Developing a Novel Correlation, and Neural Network Modeling. *Powder Technol.* **2022**, *404*, 117492. [CrossRef]
44. Nie, P.; Roccotelli, M.; Fanti, M.P.; Ming, Z.; Li, Z. Prediction of Home Energy Consumption Based on Gradient Boosting Regression Tree. *Energy Rep.* **2021**, *7*, 1246–1255. [CrossRef]
45. Said, Z.; Sharma, P.; Tiwari, A.K.; Le, V.V.; Huang, Z.; Bui, V.G.; Hoang, A.T. Application of Novel Framework Based on Ensemble Boosted Regression Trees and Gaussian Process Regression in Modelling Thermal Performance of Small-Scale Organic Rankine Cycle (ORC) Using Hybrid Nanofluid. *J. Clean. Prod.* **2022**, *360*, 132194. [CrossRef]
46. Cai, J.; Xu, K.; Zhu, Y.; Hu, F.; Li, L. Prediction and Analysis of Net Ecosystem Carbon Exchange Based on Gradient Boosting Regression and Random Forest. *Appl. Energy* **2020**, *262*, 114566. [CrossRef]
47. Ali, A.; Naseer, H.; Ilyas, S.U.; Phelan, P.E.; Nasir, R.; Alsaady, M.; Yan, Y. Thermal and Rheological Behavior of Hybrid Nanofluids Containing Diamond and Boron Nitride in Thermal Oil for Cooling Applications. *Arab. J. Sci. Eng.* **2023**. [CrossRef]
48. Natekin, A.; Knoll, A. Gradient Boosting Machines, a Tutorial. *Front. Neurorobotics* **2013**, *7*, 21. [CrossRef]

49. Singh, H.M.; Sharma, D.P. Machine Learning Model to Predict the Efficiency of Thermal Conductivity of Oxide Nanofluids. *Waves Random Complex Media* **2023**, *2023*, 1–20. [CrossRef]
50. Dai, X.; Andani, H.T.; Alizadeh, A.; Abed, A.M.; Smaisim, G.F.; Hadrawi, S.K.; Karimi, M.; Shamsborhan, M.; Toghraie, D. Using Gaussian Process Regression (GPR) Models with the Matérn Covariance Function to Predict the Dynamic Viscosity and Torque of SiO_2/Ethylene Glycol Nanofluid: A Machine Learning Approach. *Eng. Appl. Artif. Intell.* **2023**, *122*, 106107. [CrossRef]

fluids

MDPI

Article

AI-Driven Optimization of Breakwater Design: Predicting Wave Reflection and Structural Dimensions

Mohammed Loukili [1], Soufiane El Moumni [2] and Kamila Kotrasova [3,*]

[1] Institut de Recherche de l'Ecole Navale (EA 3634, IRENav), 29160 Brest, France; mohammed.loukili@ecole-navale.fr
[2] Digital Engineering for Leading Technology and Automation Laboratory (DELTA Lab), Hassan II University of Casablanca, Casablanca 20670, Morocco; soufianeelmoumni@gmail.com
[3] Institute of Structural Engineering and Transportation Structures, Faculty of Civil Engineering, Technical University of Kosice, Vysokoskolska 4, 042 00 Kosice, Slovakia
* Correspondence: kamila.kotrasova@tuke.sk; Tel.: +421-55-602-4294

Abstract

Coastal defense structures play a crucial role in mitigating wave impacts; yet, existing breakwater designs often face challenges in balancing wave reflection, energy dissipation, and structural stability. This study leverages machine learning (ML) to predict the optimal 2D dimensions of rectangular breakwaters in two configurations: submerged at the bottom of a wave tank and positioned at the free surface. Further, the objective is to achieve controlled wave reflection allowing a specific wave run-up and optimized energy dissipation, while ensuring maritime stability. Thus, we used an analytical equation modeling the reflection coefficient versus relative water depth (KH), for different immersion ratios of obstacle (h/H), and relative length (l/H). Two datasets of 32,000 data points were generated for underwater and free-surface breakwaters, with an additional 10,000 data points for validation, totaling 42,000 data points per case. Five ML algorithms—Random Forest, Support Vector Regression, Artificial Neural Network, Decision Tree, and Gaussian Process—were applied and evaluated. Results demonstrated that Random Forest and Decision Tree balanced accuracy with computational efficiency, while the Gaussian Process closely matched analytical results but demanded higher computational resources. These findings support ML as a powerful tool to optimize breakwater design, complementing traditional methods and contributing to more sustainable and resilient coastal defense systems.

Keywords: machine learning (ML); breakwater design; wave reflection; maritime engineering; coastal defense

1. Introduction

Artificial intelligence (AI) has emerged as a transformative technology, revolutionizing multiple sectors by processing large datasets, recognizing patterns, and making data-driven decisions with minimal human intervention [1–4]. In the medical field, AI improves diagnostic accuracy and treatment plans through advancements such as medical image analysis [5], while deep learning techniques in computational hemodynamics enable non-invasive diagnostic methods and treatments for vascular conditions [6]. AI applications extend beyond healthcare and engineering, finding use in industrial optimization through computational fluid dynamics and numerical techniques to enhance efficiency and reduce environmental impact in glass manufacturing [7]. Moreover, it supports pipeline integrity management by employing transient test techniques to detect partial blockages in pressurized systems [8] and facilitates the prediction of ionic liquid properties using machine learning approaches, offering a data-driven alternative to experimental procedures and molecular simulations [9]. In education, intelligent tutoring systems offer personalized learning experiences [10]. The maritime industry has seen significant advancements through AI, enhancing safety, efficiency, and sustainability [11]. AI-powered Autonomous Surface Vehicles (ASVs) use machine learning algorithms for real-time route optimization, collision avoidance, and obstacle detection, reducing human error [12–14]. Predictive maintenance systems analyze sensor data to forecast equipment failures, minimizing downtime and maintenance costs [15]. For instance, reduced-order CFD models have been developed to support wind turbine maintenance by accurately assessing the impact of leading-edge erosion on blade performance, enabling precise maintenance planning [16]. Additionally, AI-driven weather routing systems optimize fuel usage and contribute to eco-friendly shipping practices [17]. As AI continues to evolve, its potential for transforming maritime operations and other industries remains substantial, positioning it as a critical tool for addressing complex challenges and driving innovation.

The focus of our scientific work is to leverage machine learning (ML) to predict the optimal 2D dimensions of rectangular breakwaters. Two cases are treated in this work: underwater breakwater and free-surface breakwater. For the sake of clarity, accurate prediction of breakwater dimensions is crucial for balancing and controlling wave reflection, energy dissipation, and structural stability. By modeling these interactions using machine learning, the researchers and engineers will be able to optimize the structure's performance in terms of energy dissipation, ensuring the desired outcomes of wave behavior. For more clarity, this is a continuation of our research work cited in the reference [18].

Extensive research has been conducted on various types of breakwaters. However, due to the complex operating conditions of breakwaters, further studies are needed to address existing knowledge gaps. A brief summary of past work is provided here. Dean [19] explored how wave amplitudes affect the reflection of surface waves by a submerged flat barrier. Takano [20] examined the effect of waves passing beneath a rectangular breakwater. Patarapanich [21] investigated wave reflection and transmission using the finite element method (FEM) on a submerged thin horizontal plate. Liu and Jiankng [22] used the matched asymptotic method to study wave transmission through a submerged slit on a vertical barrier. Stamos et al. [23] employed parametric experiments to compare reflection and transmission coefficients from waves interacting with different submerged water-filled breakwater models of hemi-cylindrical and rectangular shapes. Molin et al. [24] conducted lab experiments on wave interaction with a rigid vertical plate, followed by Lui et al. [25], who studied Bragg reflections caused by waves on multiple submerged semi-circular breakwaters.

The submerged rectangular step is commonly used as a breakwater to shield shorelines, mitigate wave damage, reduce coastal erosion, and protect coastal structures [26]. Recently, submerged rectangular breakwaters have gained more attention than traditional emerged structures due to their esthetic appeal, ability to promote water circulation, and allow fish passage. Numerous experimental, analytical, and numerical studies have investigated wave reflection and transmission in relation to these structures. Mei and Black [27] used the variational method to study the scattering properties of bottom and surface obstacles. Massel [28] analyzed wave interaction with rectangular submerged breakwaters of infinite and finite lengths. Andrew et al. [29] conducted experimental and numerical research using the boundary element method (BEM) to explore wave propagation over a submerged rectangular impermeable obstacle. More recently, Szmidt [30] used the finite difference method (FDM) to numerically assess wave interaction with a rectangular breakwater fixed to the bottom of a numerical wave tank, evaluating its effectiveness in protecting coastal shelf zones from open sea waves.

Nowadays, the application of artificial intelligence (AI) to predict optimal breakwater dimensions has gained significant attention. Machine learning (ML) models have been used to predict wave reflection and transmission based on varying breakwater geometries. For example, in [31,32], the authors applied Artificial Neural Network (ANN) to model wave-structure interactions, allowing the prediction of optimal breakwater dimensions with minimal computational effort. Similarly, the work by MAO et al. demonstrated the use of genetic algorithms to optimize the geometry of submerged breakwaters, effectively reducing wave energy [33]. These AI-driven approaches provide a more efficient and accurate way to determine the optimal breakwater dimensions for desired wave reflection and transmission coefficients, offering a powerful tool for engineers working on maritime structure design.

The cited articles, while innovative, exhibit several limitations that open the door for further research and improvement. For instance, the authors in [31,32] focused primarily on neural networks but did not extensively explore the generalization capabilities of their model to different wave conditions or more complex breakwater shapes. The model's predictive accuracy under non-regular or irregular waves remains under-explored, which could limit its application in real-world, unpredictable maritime environments. Moreover, MAO et al. used genetic algorithms to optimize submerged breakwater geometries, but their method faced challenges in terms of computational complexity and convergence speed [33]. The algorithm's efficiency in handling high-dimensional design spaces or incorporating multiple environmental factors simultaneously is still an area of concern. These limitations highlight the need for more robust, scalable, and flexible AI models capable of handling a broader range of scenarios, including varying wave conditions and more complex geometries, which will be the focus of the work presented in this paper.

In addition, while previous works primarily focused on single algorithms or optimization techniques, our research aims to overcome this by presenting and comparing five different machine learning algorithms to predict the optimal dimensions of the breakwater. This approach allows for a more comprehensive evaluation and deeper comparison, helping to identify the most suitable algorithm for this specific problem. By doing so, we address the need for a more flexible and accurate predictive model that can adapt to varying wave conditions and geometric complexities, ultimately improving the reliability and efficiency of breakwater design in real-world scenarios.

This paper is structured as follows: Section 2 outlines the problem formulation and the adopted methodology, providing the foundation for the analysis. Section 3 presents the results and discusses their implications, highlighting key findings. Finally, Section 4 concludes the study, summarizing the main contributions and offering directions for future research.

2. Problem Formulation and Methodology

This section presents the analytical formulation of reflection coefficients for the interactions of regular waves with a rectangular obstacle positioned at the bottom of the wave tank [18], as well as the interaction between water waves and a rectangular obstacle located at the free surface.

Based on this theoretical framework, we focus on the application of artificial intelligence (AI) to further analyze wave-structure interactions. Our strategy is to integrate these theoretical models with AI-driven methods to automate and optimize problem-solving processes (see Figure 1 below).

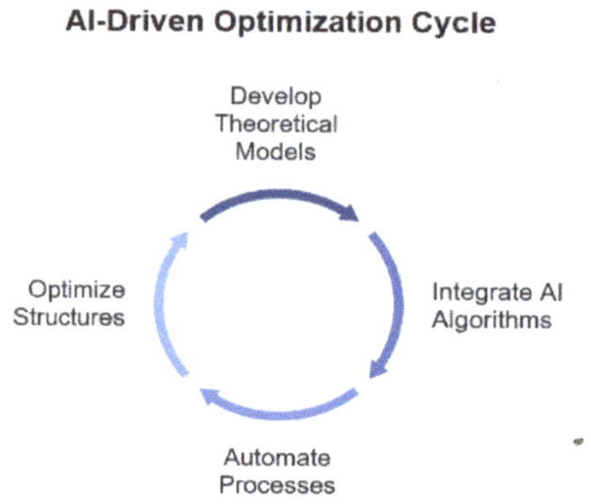

Figure 1. AI-driven optimization cycle.

For more clarity, we aim to use five machine-learning methods to investigate effective techniques for analyzing wave-structure interactions. Additionally, the datasets utilized in these machine-learning methods are derived from a theoretical approach that has already been validated in our latest research [18]. By training AI models on datasets generated from various wave conditions and breakwater configurations, we can identify patterns and relationships that may not be immediately apparent through traditional analytical methods. This approach not only enhances our understanding of wave dynamics but also enables us to predict performance outcomes for different design scenarios.

Furthermore, we will explore artificial intelligence (AI) techniques to refine breakwater design parameters, identifying optimal configurations that maximize wave attenuation while minimizing material costs and environmental impacts.

The theoretical formulation is divided into two parts. The first part addresses the expression of the reflection coefficient based on the conservation of flow rate and the continuity of velocity potential for wave-structure interactions cited at the bottom of the wave tank. The second part focuses on expressing the reflection coefficient for wave-structure interactions at the free surface.

2.1. Part 1: Wave-Structure Interactions Cited at the Bottom of Wave Tank

To theoretically analyze the interactions of waves with a rectangular structure located at the bottom of the wave tank (Figure 2 below), we establish an analytical formulation based on the conservation of flow rate and the continuity of velocity potential to express the reflection coefficient.

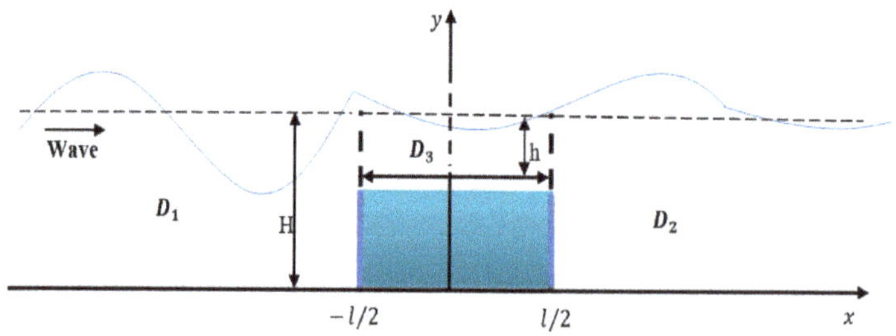

Figure 2. Configuration of bottom submerged breakwater (for analytical calculation).

The velocity potential for each subdomain is expressed as follows:

- At the subdomain D_1,

$$\phi_1 = a[exp(-jk(x+l/2)) + R_b exp(jk(x+l/2))]ch(k(y+H)), \tag{1}$$

- At the subdomain D_2,

$$\phi_2 = aT_b exp(-jk(x-l/2))ch(k(y+H)), \tag{2}$$

- At the subdomain D_3,

$$\phi_3 = a[Cexp(-j\sigma x) + Dexp(j\sigma x)]ch(\sigma(y+h)). \tag{3}$$

The continuity flow rate conservation is defined at $x = -l/2$ and $x = l/2$ as follows:

- At the position $x = -l/2$,

$$\int_{-h}^{0} \phi_1(-l/2,y)ch(\sigma(y+h))dy = \int_{-h}^{0} \phi_3(-l/2,y)ch(\sigma(y+h))dy \tag{4}$$

$$\int_{-h}^{0} \frac{\phi_1(-l/2,y)}{\partial x} ch(k(y+H))dy = \int_{-h}^{0} \frac{\partial\phi_3(-l/2,y)}{\partial x} ch(k(y+H))dy, \tag{5}$$

- At the position $x = l/2$,

$$\int_{-h}^{0} \phi_1(l/2,y)ch(\sigma(y+h))dy = \int_{-h}^{0} \phi_3(l/2,y)ch(\sigma(y+h))dy, \tag{6}$$

$$\int_{-h}^{0} \frac{\partial\phi_1(l/2,y)}{\partial x} ch(k(y+H))dy = \int_{-h}^{0} \frac{\partial\phi_3(l/2,y)}{\partial x} ch(k(y+H))dy, \tag{7}$$

Following the expression of the boundary conditions at the positions $x = l/2$ and $x = -l/2$, we obtain an algebraic system represented as follows:

$$\begin{cases} I_1(1+R) = I_2(Cz + D\overline{z}) \\ I_1 T_r = I_2(C\overline{z} + Dz) \\ kI_3(1-R) = \sigma I_1(Cz - D\overline{z}) \\ kI_3 T_r = \sigma I_1(C\overline{z} - Dz) \end{cases} \tag{8}$$

The system of Equation (8) is written in matrix form as follows:

$$\begin{bmatrix} A & B \\ B & A \end{bmatrix} \begin{bmatrix} T_b \\ 0 \end{bmatrix} = \begin{bmatrix} \overline{z^2} & 0 \\ 0 & z^2 \end{bmatrix} \begin{bmatrix} A & B \\ B & A \end{bmatrix} \begin{bmatrix} 1 \\ R_b \end{bmatrix} \tag{9}$$

Then, the reflection and coefficients are expressed as follows:

$$R_b = \frac{z^2 - \overline{z^2}}{\overline{z^2}B/A - z^2 A/B}, \tag{10}$$

where

$$A = \frac{I_1}{I_2} + \frac{kI_3}{\sigma I_1} \tag{11}$$

$$B = \frac{I_1}{I_2} - \frac{kI_3}{\sigma I_1}. \tag{12}$$

$$I_1 = \int_{-h}^{0} ch(k(y+d))ch(\sigma(y+d))dy, \tag{13}$$

$$I_2 = \int_{-h}^{0} ch^2(\sigma(y+h))dy, \tag{14}$$

$$I_3 = \int_{-h}^{0} ch^2(k(y+H))dy, \tag{15}$$

2.2. Part 2: Wave-Structure Interactions Cited at the Free Surface of Wave Tank

To conduct a theoretical analysis of the interactions between waves and a rectangular structure positioned at the free surface of the wave tank (Figure 3 below), we develop an analytical formulation based on the conservation of flow rate and the continuity of velocity potential to determine the reflection coefficient.

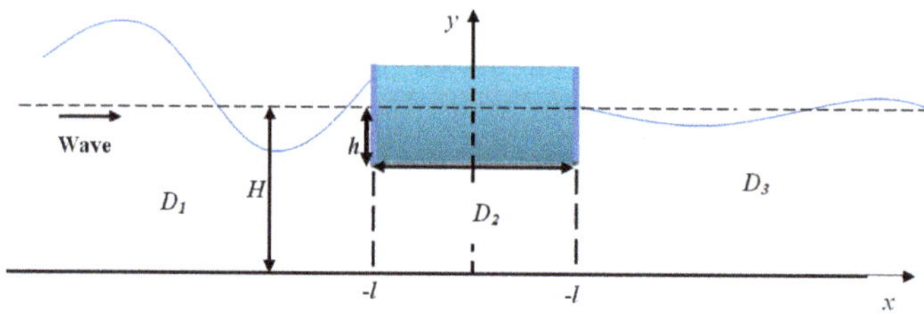

Figure 3. Configuration of free-surface submerged breakwater (for analytical calculation).

The velocity potential for each subdomain is expressed as follows:

- At the subdomain D1,

$$\varphi_1'(x,y,t) = a.(exp(-jk(x+l)) + R'.exp(jk(x+l))).ch(ky).exp(j\omega t) \tag{16}$$

$$\varphi'_2(x,y,t) = \left(\frac{E'}{2} + \frac{F'}{2l}.x\right).exp(j\omega t) \tag{17}$$

$$\varphi'_3(x,y,t) = a.T'.exp(-jk(x-l)).ch(ky).exp(j\omega t) \tag{18}$$

Next, we express the continuity of potentials at $x = -l$ as follows:

$$\varphi'_1(-l,y) = \varphi'_2(-l,y) \tag{19}$$

$$\int_0^{\mathrm{d}} \varphi'_1(-l,y)dy = \int_0^{\mathrm{d}} \varphi'_2(-l,y)dy \tag{20}$$

$$(1+R')\int_0^{\mathrm{dy}} ch(ky).dy = \frac{1}{2}(E'-F')\int_0^{\mathrm{d}} dy \tag{21}$$

$$E' - F' = 2(1+R')\frac{I'_1}{I''_2} \tag{22}$$

The flow rate conservation is expressed as follows:

$$\int_0^{\mathrm{H}} \frac{\partial\varphi'_1(-l,y)}{\partial x}dy = \int_0^{\mathrm{d}} \frac{\partial\varphi'_2(-l,y)}{\partial x}dy \tag{23}$$

$$jk(-1+R')\int_0^{\mathrm{H}} ch(ky).dy = \frac{F'}{2l}\int_0^{\mathrm{d}} dy \tag{24}$$

$$jk(-1+R')I'_3 = \frac{F}{2l}I'_2 \tag{25}$$

Further, we express the continuity of potentials at $x = l$ as follows:

$$\varphi'_3(l,y) = \varphi'_2(l,y) \tag{26}$$

$$\int_0^{\mathrm{d}} \varphi'_2(l,y)dy = \int_0^{\mathrm{d}} \varphi'_3(l,y)dy \tag{27}$$

$$\frac{1}{2}(E' + F')\int_0^{\mathrm{d}} dy = T'\int_0^{\mathrm{d}} ch(ky).dy \tag{28}$$

$$E' + F' = 2T'\frac{I'_1}{I'_2} \tag{29}$$

The flow rate conservation is expressed as follows:

$$\int_0^{\mathrm{H}} \frac{\partial \varphi'_3(l,y)}{\partial x} dy = \int_0^{\mathrm{d}} \frac{\partial \varphi'_2(l,y)}{\partial x} dy \tag{30}$$

$$\frac{F'}{2l}\int_0^{\mathrm{d}} dy = -jkT'\int_0^{\mathrm{H}} ch(ky).dy \tag{31}$$

$$\frac{F'}{2l}I'_2 = -jkT' I'_3 \tag{32}$$

Then, we obtain a system of equation as follows:

$$\begin{cases} E' - F' = 2(1 + R')\frac{I'_1}{I'_2} \\ E' + F' = 2T'\frac{I'_1}{I'_2} \\ jk(-1 + R')I'_3 = \frac{F'}{2l}I'_2 \\ \frac{F}{2l}I'_2 = -jkT' I'_3 \end{cases} \tag{33}$$

where

$$I'_1 = \int_0^{\mathrm{d}} ch(ky).dy = \frac{\sinh(kH(1 - h/H))}{k} \tag{34}$$

$$I'_2 = \int_0^{\mathrm{d}} dy \tag{35}$$

$$I'_3 = \int_0^{\mathrm{H}} ch(ky).dy = \frac{\sinh(kH)}{k} \tag{36}$$

Finally, by combining the system of Equations (33), we obtain the reflection coefficient as follows:

$$R' = \frac{1}{1 - i\frac{\sinh(k(H-h))}{k.l.\sinh(KH)}} = \frac{1}{1 - i\frac{\sinh(kK(1-h/H))}{kH.(l/h).\sinh(KH)}} \tag{37}$$

Our study is structured into three key phases: (1) predicting the wave reflection coefficient based on the dimensions of a breakwater; (2) identifying the optimal height–width pair to achieve a desired level of reflection; and (3) extending this approach to a breakwater located at the free surface.

a. Data and Training Dataset Generation

To achieve these objectives, we used an analytical equation that models the reflection coefficient based on several parameters:

- KH: the relative water depth;
- h/H: the immersion ratio;
- l/H: the relative length;
- R_b: the reflection coefficient in the case of underwater breakwater;
- R': the reflection coefficient in the case of free-surface breakwater.

Based on this equation, we generated two primary datasets, each consisting of 32,000 data points:

- Underwater Breakwater: data representing various configurations of an obstacle submerged at the bottom of the wave tank;

- Free-Surface Breakwater: data for an obstacle located at the water's free surface, using similar parameters (h/H and l/H).

Each dataset was divided into 80% for training and 20% for testing. Additionally, 10,000 new data points were generated for a rigorous validation test, providing a total of 42,000 annotated data points for each case (underwater and free surface).

b. Choice and Justification of AI Algorithms

To model wave-structure interactions for both underwater and free-surface scenarios, we selected five AI algorithms based on their complementary features and suitability to the problem:

- Random Forest (RF) [34]: It was chosen for its ability to handle complex, nonlinear relationships between design parameters and the reflection coefficient. RF also provides variable importance, offering valuable insights into the influence of geometric parameters;
- Support Vector Regression (SVR) [35]: It is well-suited for multidimensional regression tasks, providing high accuracy in capturing the nuances of wave-structure interactions, even in complex configurations;
- Artificial Neural Network (ANN) [36]: Neural networks are well-suited for modeling nonlinear relationships, enabling the capture of subtle interactions between variables, essential for predicting wave reflection behavior;
- Decision Tree [37]: Its speed and interpretability make it an excellent choice for initial exploratory analysis. This model allows direct visualization of the impact of each parameter on the reflection coefficient;
- Gaussian Process (GP) [38]: This probabilistic model not only provides predictions but also confidence intervals, allowing the evaluation of result reliability—a valuable advantage in a maritime environment with variable conditions.

This diversity of algorithms allows us to cover a wide range of techniques, ensuring a thorough and accurate performance evaluation for each configuration.

c. Modeling Process and Performance Evaluation

The modeling process was carried out in two distinct steps:

i. Prediction of the Reflection Coefficient: In this first phase, the algorithms were trained to predict the reflection coefficients based on the parameters of the breakwater, including the submersion-to-height ratio (h/H), width (l/H), and relative water depth (KH). This phase validated the algorithms' ability to replicate the results of the analytical equation and accurately model wave-obstacle interactions;
ii. Prediction of the Optimal Height–Width Pair for a Given Reflection: In the second phase, the selected algorithm was used to reverse the initial reflection calculation and identify the optimal height–width pair (h/H, l/H) that met target reflection coefficients, effectively creating a "reciprocal function" of the analytical equation.

Each model configuration was carefully selected to optimize predictive accuracy and efficiency:

- Random Forest Regressor was configured with 100 estimators (Decision Tree) and a fixed random state of 42, ensuring reproducibility and robustness by averaging multiple decision tree outputs;
- Support Vector Regressor (SVR) used its default configuration to model complex, nonlinear relationships between input parameters and the reflection coefficient;
- Artificial Neural Network (ANN) was structured with two hidden layers of 50 neurons each and a maximum of 1000 iterations. ReLU activation was applied, along with a

transformation to prevent negative outputs, ensuring the network accurately modeled the relationships while stabilizing training;

- Decision Tree Regressor was initialized with a random state of 42, allowing a single-tree model to capture relationships without ensembling, which makes the approach simpler and faster to train;
- Gaussian Process Regressor was employed to provide probabilistic predictions, modeling uncertainties in the reflection coefficient by assuming a Gaussian distribution of outputs.

Each model's performance was assessed based on Mean Squared Error (MSE), Root Mean Squared Error (RMSE), Mean Absolute Error (MAE), R^2, training duration, and prediction duration as mentioned in the Results Section. This rigorous approach, applied across both underwater and free-surface configurations, ensured robust and reliable optimization of breakwater dimensions under varying maritime conditions.

3. Results and Discussions

In this section, we present the performance of each artificial intelligence model used to predict the wave reflection coefficient and optimize the breakwater dimensions. The models were evaluated using several performance metrics to measure prediction accuracy, computational efficiency, and each model's ability to generalize under varying conditions. These metrics include Mean Squared Error (MSE), Root Mean Squared Error (RMSE), Mean Absolute Error (MAE), Coefficient of Determination (R^2), and the training and prediction durations. These parameters provide a comprehensive assessment of model quality, with each metric playing a specific role in judging the algorithms' suitability for accuracy and robustness in maritime environments.

3.1. Definitions and Utility of Performance Metrics

Each model's performance was assessed based on performance metrics below:

- Mean Squared Error (MSE): MSE calculates the average of the squared differences between predicted and actual values, penalizing larger errors more heavily. See Equation (38) below:

$$MSE = \frac{1}{n}\sum_{i=1}^{n}(y_i - \hat{y}_i)^2 \tag{38}$$

where

- ○ y_i: actual value of the i-th data point;
- ○ $\hat{y}_i$: predicted value of the i-th data point;
- ○ n: number of data points.

In this study, MSE is used to evaluate each model's overall accuracy, particularly to identify those that minimize prediction errors.

- Root Mean Squared Error (RMSE): It is the square root of MSE. Using RMSE alongside MSE provides complementary insights into model performance. See Equation (39) below:

$$RMSE = \sqrt{MSE} \tag{39}$$

- Mean Absolute Error (MAE): MAE measures the average absolute difference between predictions and actual values (see Equation (40) below).

$$MAE = \frac{1}{n}\sum_{i=1}^{n}|y_i - \hat{y}_i| \tag{40}$$

While MSE gives more weight to larger errors, making it useful when large deviations are particularly undesirable, MAE offers a straightforward and robust alternative for assessing the average error magnitude without overemphasizing outliers.

- Coefficient of Determination (R^2): This score measures how well the predictions match the actual values, relative to the variance in the data. It is defined as follows:

$$R^2 = 1 - \frac{\sum_{i=1}^{n}(y_i - \hat{y}_i)^2}{\sum_{i=1}^{n}(y_i - \overline{y})^2} \tag{41}$$

where $\overline{y}$ is the mean of the actual values.

While MSE and MAE measure the magnitude of errors, R^2 adds value by providing a relative understanding of model performance compared to a baseline, such as always predicting the mean. R^2 is easier to interpret as it quantifies how much of the variation in the target variable is explained by the model. Additionally, it allows for comparison across datasets and helps detect overfitting when evaluated on different data splits.

- Training Time (TT): It refers to the duration required to train the model. It is measured as follows:

$$TT = t_{end} - t_{start} \tag{42}$$

where t_{start} and t_{end} are the timestamps indicating when training begins and ends, respectively.

This metric is important in our study as it determines whether a model can be quickly recalibrated when new data become available, a valuable asset in dynamic maritime applications.

- Prediction Time (PT): It refers to the time required to make predictions on new data. For a single input, it can be expressed as follows:

$$PT = t_{end} - t_{start} \tag{43}$$

where t_{start} and t_{end} are the timestamps indicating when prediction begins and ends, respectively.

This measure is crucial for real-time applications, where rapid predictions are necessary to adjust breakwater parameters in response to changing wave conditions.

By combining all these metrics, we obtain an overview of each model's performance, considering both accuracy and operational efficiency. This approach allows us to identify the models best suited for precise and fast predictions in a maritime environment.

3.2. Reflection Prediction

In this phase, we present the performance results of our proposed AI models for wave reflection prediction in underwater and free-surface breakwater.

Table 1 below shows the performance comparison of AI models for wave reflection prediction in underwater breakwater.

Table 1. Performance comparison of AI models for wave reflection prediction in underwater breakwater.

Model	MSE	RMSE	MAE	R^2	TT (s)	PT (s)
RF	6.25×10^{-8}	0.00025002	0.00011301	0.99999531	3.2104	0.1852
SVR	4.43×10^{-3}	0.06658152	0.05894506	0.66763488	2.446	0.3195
ANN	8.11×10^{-5}	0.00900543	0.00536504	0.99391981	2.9037	0.0079
DT	1.02×10^{-7}	0.00032000	0.00018931	0.99999232	0.083	0.005
GP	1.49×10^{-5}	0.00386001	0.00235702	0.99888291	215.006	60.685

In predicting wave reflection for underwater breakwaters, Random Forest (RF) and Decision Tree (DT) models provided the best balance of accuracy and efficiency. Both models achieved near-perfect fit, with extremely low error metrics (MSE, RMSE, and MAE) and high R^2 scores, indicating excellent predictive capability. Decision Tree further stood out with the fastest training and prediction times, making it highly efficient and suitable for real-time applications.

The Artificial Neural Network (ANN) also showed high accuracy, with the fastest prediction times among all models, making it ideal for scenarios requiring rapid predictions. However, it exhibited a slight overestimation near peak values, which may benefit from further tuning.

Gaussian Process (GP) was highly accurate but suffered from long training and prediction times, limiting its practicality to small datasets or offline analysis where computational time is less critical.

Support Vector Regressor (SVR) showed the weakest accuracy, with relatively high errors and lower R^2, making it unsuitable without further optimization. In summary, Random Forest, Decision Tree, and ANN are the most effective models for wave reflection prediction in underwater breakwater settings, balancing high accuracy with practical computational efficiency.

Moving from underwater to free-surface conditions, we next evaluate the performance of these AI models for wave reflection prediction in a free-surface breakwater context.

Table 2 below shows the performance comparison of AI models for wave reflection prediction in free-surface breakwater.

Table 2. Performance comparison of AI models for wave reflection prediction in free-surface breakwater.

Model	MSE	RMSE	MAE	R^2	TT (s)	PT (s)
RF	9.07×10^{-8}	0.00030116	0.00010446	0.999996963	3.5724	0.1968
SVR	4.30×10^{-3}	0.06560362	0.05871522	0.855901961	0.7296	0.0282
ANN	6.11×10^{-5}	0.00781446	0.00471538	0.997955438	3.6293	0.0115
DT	1.97×10^{-7}	0.00044414	0.00019752	0.999993396	0.0949	0.0055
GP	1.48×10^{-7}	0.00038483	0.00019241	0.999995042	212.3107	60.3129

Based on these results, we can observe that Random Forest, Decision Tree, and ANN maintain consistent performance across both breakwater settings, making them reliable choices for wave reflection prediction. SVR performs slightly better and faster in the free-surface environment, while the Gaussian Process remains accurate but computationally expensive in both.

These results suggest that while model accuracy remains largely stable across environments, computational efficiency may vary slightly, particularly for SVR and ANN in the free-surface context.

To visually reinforce these performance results, we present comparison curves that highlight the predictive accuracy of each AI model across varying values of the dimensionless parameter KH. Each curve represents the predictions from one of the five AI models—Random Forest, Support Vector Regressor (SVR), Artificial Neural Network (ANN), Decision Tree, and Gaussian Process—alongside the theoretical model.

These comparisons were based on a test dataset of 10,000 data points reserved specifically for validation. This extensive test set allows for a robust assessment of each model's ability to approximate the theoretical reflection behavior, providing a clear visual representation of each model's strengths and limitations relative to the analytical solution.

The graphs below show a validation test for underwater reflection, comparing theoretical model values (from an analytical equation) with predictions from various AI algorithms—Random Forest, SVR (Support Vector Regression), ANN (Artificial Neural Network), Decision Tree, and Gaussian Process. This test uses fixed values which were not present during the training or performance testing. It emphasizes the importance of model selection, particularly when extrapolating previously unseen data.

Here are the plots comparing the analytical reflection values with the predictions from each AI algorithm for different obstacle dimensions (h/H and L). See Figure 4 below (plots from a to d).

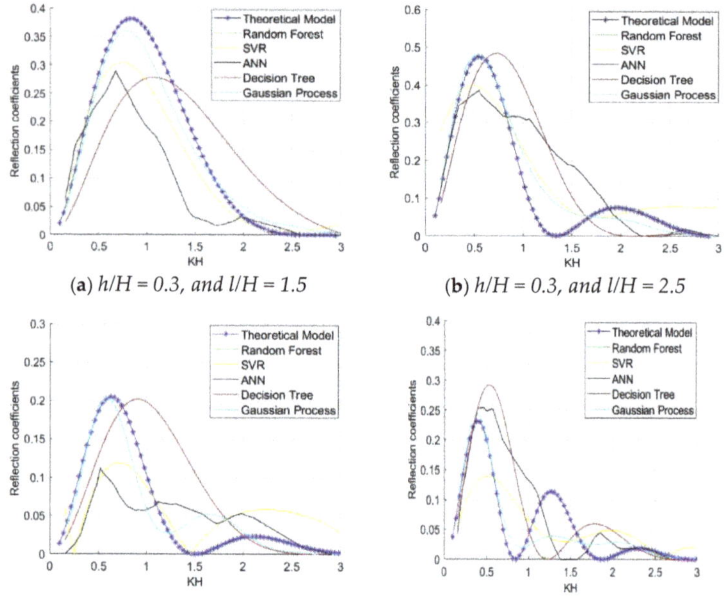

(**a**) *h/H = 0.3, and l/H = 1.5* (**b**) *h/H = 0.3, and l/H = 2.5*

(**c**) *h/H = 0.6, and l/H = 1.5* (**d**) *h/H = 0.6, and l/H = 2.5*

Figure 4. Plots comparing the analytical reflection values with the predictions from AI algorithms for different obstacle dimensions (h/H and l/H).

The discussion across all scenarios highlights consistent performance trends among the models evaluated for underwater reflection. Decision Tree (DT) and Random Forest (RF) capture the overall trend of reflection R effectively, particularly in the first harmonic (low KH), but struggle to model complex oscillations in intermediate and higher harmonics. Their rigid, "step-like" predictions show increasing deviations from the analytical equation as KH increases, with RF's lack of diversity in trees limiting its potential advantage over DT. Similarly, Support Vector Regression (SVR) captures the general trend but exhibits rigidity and produces non-physical negative values in some regions (h/H = 0.6, L1.5 and KH > 1.5, or h/H = 0.6 L = 2.5 and KH > 2.5) which was corrected using absolute values. These corrections ensured realistic outputs but introduced distortions in the oscillation structure. Artificial Neural Network (ANN) effectively models simple trends but fails to align peaks and troughs in more complex oscillations, sometimes producing artificial or misaligned variations due to suboptimal architecture or training.

Gaussian Process (GP) consistently outperforms all other models, offering high accuracy and fidelity across the scenarios tested in this study. It closely follows the analytical curve in all harmonics, capturing both global trends and fine oscillations, even in regions with rapid and dynamic variations. GP's flexibility allows it to model continuous transitions and complex interactions between waves and obstacles, with only minor deviations in extreme cases observed within the scope of the test conditions. This makes it the most reliable model for wave-obstacle interactions within the limits of this study, whereas the other models, although suitable for capturing general trends, require significant adjustments to handle complex and oscillatory behaviors effectively.

In the other case, Figure 5 shows a validation test for free-surface reflection, comparing theoretical model values (from an analytical equation) with predictions from various AI algorithms (RF, DT, SVR, GP, and ANN). See Figure 5 below (plots from a to d).

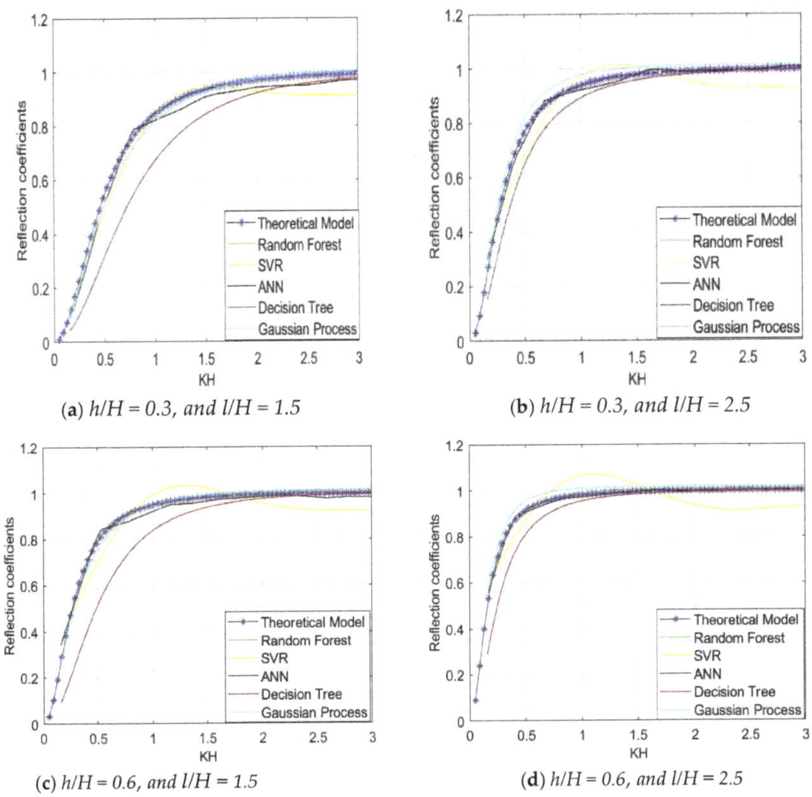

(**a**) $h/H = 0.3$, and $l/H = 1.5$ (**b**) $h/H = 0.3$, and $l/H = 2.5$

(**c**) $h/H = 0.6$, and $l/H = 1.5$ (**d**) $h/H = 0.6$, and $l/H = 2.5$

Figure 5. Plots comparing the analytical reflection values with the predictions from AI algorithms for different obstacle dimensions (h/H and l/H).

Across the four scenarios, the performance of the predictive models in replicating the analytical reference for wave reflection varies significantly. Decision Tree (DT) and Random Forest (RF) consistently capture general trends but exhibit notable deviations in regions with rapid variations in the reflection coefficient R, particularly for KH < 1.5 or in regions of complex relationships. Their comparable performances suggest limitations in the RF model's optimization or training data coverage. Gaussian Process (GP) demonstrates strong alignment in areas with gradual R variations and dense data coverage, but its predictions falter in regions with complex nonlinearities or sparse data. Notably, GP frequently exceeds the physical limit R = 1 in several scenarios, indicating the need for constraints to enforce physical consistency.

Support Vector Regression (SVR) exhibits acceptable predictions in some regions but shows significant deviations in others, often exceeding R = 1 or failing to generalize well for KH > 1.7. Its sensitivity to hyperparameters and lack of inherent physical constraints hinder its reliability. In contrast, the Artificial Neural Network (ANN) consistently emerges as the most robust and reliable model across all scenarios, maintaining minimal deviations and a strong alignment with the analytical reference throughout the entire range of KH. While some models benefit from improvements in hyperparameters, data enrichment, and kernel or architecture selection, ANN stands out as the most effective approach for accurately modeling wave reflection, combining flexibility with physical coherence.

3.3. Dimensions Prediction

In this phase, we present the performance results of our proposed AI models for the prediction of obstacle dimensions in underwater and free-surface cases.

Table 3 below shows the performance comparison of AI models for the prediction of obstacle dimensions in the underwater case.

Table 3. Performance comparison of AI models for the prediction of obstacle dimensions for underwater breakwater.

Model	Target	MSE	RMSE	MAE	R^2	TT (s)	PT (s)
RF	h/H	0.002955	0.054363	0.017301	0.928489	3.271199	0.077791
	l/H	0.092398	0.30397	0.110707	0.861997	4.188477	0.093852
DT	h/H	0.006158	0.078474	0.016343	0.850993	0.04668	0.002157
	l/H	0.165272	0.406537	0.109399	0.753153	0.062463	0.002334
SVR	h/H	0.025798	0.160617	0.129903	0.375774	26.310649	3.836463
	l/H	0.607661	0.779526	0.653736	0.092411	34.439204	5.889853
GP	h/H	0.022301	0.149336	0.117598	0.460379	244.665434	68.324839
	l/H	0.449268	0.670274	0.555867	0.328984	239.60938	68.330051
ANN	h/H	0.021895	0.147971	0.113052	0.470201	11.363988	0.013058
	l/H	0.414486	0.643806	0.51954	0.380934	195.429634	0.03317

The results show that Random Forest (RF) is the best-performing model, combining high accuracy (R^2 = 0.928 for h/H and 0.862 for L) with reasonable computational efficiency. Decision Tree (DT), while less accurate (R^2 = 0.851 for h/H), is highly efficient and serves as a viable alternative when resources are limited. In contrast, Support Vector Regression (SVR) and Gaussian Process (GP) deliver poor accuracy and are computationally expensive, making them unsuitable. Artificial Neural Network (ANN) shows moderate accuracy but higher computational demands, limiting its practicality.

Moving from underwater to free-surface conditions, we next evaluate the performance of these AI models for obstacle dimension prediction in a free-surface context.

Table 4 below shows the performance comparison of AI Models for the prediction of obstacle dimensions in a free-surface case.

Table 4. Performance comparison of AI models for the prediction of obstacle dimensions in a free-surface case.

Model	Target	MSE	RMSE	MAE	R^2	TT (s)	PT (s)
RF	h/H	0.001825	0.042724	0.012152	0.956379	3.781168	0.072074
	l/H	0.029419	0.171519	0.049657	0.95551	3.681144	0.073746
DT	h/H	0.003793	0.061584	0.013919	0.909367	0.058323	0.002116
	l/H	0.059431	0.243784	0.052862	0.910123	0.054024	0.002139
SVR	h/H	0.026049	0.161396	0.138118	0.377508	31.094358	4.719603
	l/H	0.422047	0.649651	0.526329	0.361739	36.449994	6.32239
GP	h/H	0.013696	0.117029	0.086317	0.672708	270.447472	68.376187
	l/H	0.371135	0.609208	0.48067	0.438733	247.67819	68.148208
ANN	h/H	0.017425	0.132003	0.103471	0.583595	20.12097	0.014777
	l/H	0.395293	0.628723	0.515053	0.402199	24.952677	0.015778

In free-surface conditions, Random Forest (RF) remains the best-performing model, achieving the highest accuracy ($R^2 = 0.956$ for both h/H and L) with reasonable computation times. Decision Tree (DT) offers a good trade-off between accuracy ($R^2 = 0.909$) and efficiency, making it ideal for resource-limited scenarios. In contrast, Support Vector Regression (SVR) and Gaussian Process (GP) perform poorly with low accuracy and high computational costs, while Artificial Neural Network (ANN) provides moderate accuracy but is less practical due to higher training times.

4. Conclusions

This paper investigates the interaction between waves and two types of wave attenuators: a free-surface breakwater and an underwater breakwater fixed to the bottom of a channel. The focus of the study is the variation in the reflection coefficient as a function of the dimensionless parameter KH. Our findings indicate that the free-surface breakwater acts as a low-pass filter, with its reflective power increasing as its length grows and its immersion decreases. In contrast, the underwater breakwater exhibits a reflection curve that peaks before decreasing for shorter wavelengths, functioning as a band-pass filter. The study further reveals that the reflective power of the underwater breakwater intensifies with its length, with its most effective wave attenuation occurring at shallow immersion depths.

Additionally, this research highlights the transformative potential of machine learning (ML) in optimizing the design of both submerged and free-surface rectangular breakwaters. Utilizing five AI algorithms—Random Forest, Support Vector Regression, Artificial Neural Network, Decision Tree, and Gaussian Process—we successfully predicted wave reflection coefficients and determined optimal dimension ratios (h/H and L) for achieving specific wave behavior targets. The results demonstrate that Random Forest and Decision Tree offer a favorable balance between accuracy and computational efficiency, while Gaussian Process, though highly accurate, requires more computational resources. These findings emphasize the potential of AI to complement traditional approaches, offering a robust framework for addressing complex maritime engineering challenges. Future research should aim to refine these models, integrate additional environmental factors, and explore hybrid methodologies that combine multiple algorithms, ultimately improving predictive accuracy and enhancing the design of sustainable, cost-effective coastal defense systems.

Author Contributions: Conceptualization, M.L., S.E.M. and K.K.; methodology, M.L., S.E.M. and K.K.; software, M.L., S.E.M. and K.K.; validation, M.L., S.E.M. and K.K.; formal analysis, M.L., S.E.M. and K.K.; investigation, M.L., S.E.M. and K.K.; resources, M.L., S.E.M. and K.K.; writing—original draft preparation, M.L., S.E.M. and K.K.; writing—review and editing, M.L., S.E.M. and K.K.;

visualization, M.L., S.E.M. and K.K.; supervision, M.L., S.E.M. and K.K.; project administration, M.L., S.E.M. and K.K.; funding acquisition, M.L., S.E.M. and K.K. All authors have read and agreed to the published version of the manuscript.

Funding: This research was founded by TUKE by the Ministry of Education, Science, Research and Sport of the Slovak Republic.

Data Availability Statement: Data is contained within the article.

Acknowledgments: The paper presented was supported by the project VEGA 1/0307/23 of the Scientific Grant Agency of the Ministry of Education, Science, Research and Sport of the Slovak Republic.

Conflicts of Interest: The authors declare no conflict of interest.

Nomenclature

a	The amplitude of the incident wave
R_b	The reflection coefficient in the case of bottom breakwater
R'	The reflection coefficient in the case of free-surface breakwater
T_b	The transmission coefficient in the case of bottom breakwater
ϕ	Velocity potential in the case of bottom breakwater study
φ'	Velocity potential in the case of free-surface breakwater study
$\mathrm{d} = \mathrm{H} - \mathrm{h}$	The height under the obstacle in the case of free-surface breakwater
T'	The transmission coefficient
H	The water depth at D_1 and D_2 locations
h/H	The immersion ratio
l/H	The relative length
KH	The relative water depth
k	The wave number that verifies the dispersion relation $\frac{\omega^2}{g} = \mathrm{kth}(\mathrm{kH})$
σ	The wave number above of the obstacle that verifies the dispersion relation $\frac{\omega^2}{g} = \sigma\mathrm{th}(\sigma\mathrm{h})$

References

1. Zhang, C.; Lu, Y. Study on artificial intelligence: The state of the art and future prospects. *J. Ind. Inf. Integr.* **2021**, *23*, 100224. [CrossRef]
2. Haleem, A.; Javaid, M.; Qadri, M.A.; Singh, R.P.; Suman, R. Artificial intelligence (AI) applications for marketing: A literature-based study. *Int. J. Intell. Netw.* **2022**, *3*, 119–132. [CrossRef]
3. Castiglioni, I.; Rundo, L.; Codari, M.; Di Leo, G.; Salvatore, C.; Interlenghi, M.; Gallivanone, F.; Cozzi, A.; D'Amico, N.C.; Sardanelli, F. AI applications to medical images: From machine learning to deep learning. *Phys. Medica* **2021**, *83*, 9–24. [CrossRef] [PubMed]
4. Spanaki, K.; Karafili, E.; Despoudi, S. AI applications of data sharing in agriculture 4.0: A framework for role-based data access control. *Int. J. Inf. Manag.* **2021**, *59*, 102350. [CrossRef]
5. Rajpurkar, P.; Chen, E.; Banerjee, O.; Topol, E.J. AI in health and medicine. *Nat. Med.* **2022**, *28*, 31–38. [CrossRef] [PubMed]
6. Taebi, A. Deep learning for computational hemodynamics: A brief review of recent advances. *Fluids* **2022**, *7*, 197. [CrossRef]
7. Cravero, C.; De Domenico, D.; Marsano, D. The use of uncertainty quantification and numerical optimization to support the design and operation management of air-staging gas recirculation strategies in glass furnaces. *Fluids* **2023**, *8*, 76. [CrossRef]
8. Brunone, B.; Maietta, F.; Capponi, C.; Duan, H.-F.; Meniconi, S. Detection of partial blockages in pressurized pipes by transient tests: A review of the physical experiments. *Fluids* **2023**, *8*, 19. [CrossRef]
9. Karakasidis, T.E.; Sofos, F.; Tsonos, C. The electrical conductivity of ionic liquids: Numerical and analytical machine learning approaches. *Fluids* **2022**, *7*, 321. [CrossRef]
10. Holmes, W.; Tuomi, I. State of the art and practice in AI in education. *Eur. J. Educ.* **2022**, *57*, 542–570. [CrossRef]
11. Wang, S.; Zhang, Y.; Zhang, X.; Gao, Z. A novel maritime autonomous navigation decision-making system: Modeling, integration, and real ship trial. *Expert Syst. Appl.* **2023**, *222*, 119825. [CrossRef]

12. Tsai, C.-M.; Lai, Y.-H.; Perng, J.-W.; Tsui, I.-F.; Chung, Y.-J. Design and application of an autonomous surface vehicle with an AI-based sensing capability. In Proceedings of the 2019 IEEE Underwater Technology (UT), Taiwan, China, 16–19 April 2019; IEEE: New York, NY, USA, 2019; pp. 1–4.
13. Chae, C.-J. The Evolution of Maritime Technology Development: A Dynamic Positioning System Perspective of Maritime Autonomous Surface Ship. *WMU J. Marit. Aff.* **2024**, 1–29. [CrossRef]
14. Qiao, Y.; Yin, J.; Wang, W.; Duarte, F.; Yang, J.; Ratti, C. Survey of deep learning for autonomous surface vehicles in marine environments. *IEEE Trans. Intell. Transp. Syst.* **2023**, *24*, 3678–3701. [CrossRef]
15. Jimenez, V.J.; Bouhmala, N.; Gausdal, A.H. Developing a predictive maintenance model for vessel machinery. *J. Ocean Eng. Sci.* **2020**, *5*, 358–386. [CrossRef]
16. Carraro, M.; De Vanna, F.; Zweiri, F.; Benini, E.; Heidari, A.; Hadavinia, H. CFD modeling of wind turbine blades with eroded leading edge. *Fluids* **2022**, *7*, 302. [CrossRef]
17. Tay, Z.Y.; Hadi, J.; Chow, F.; Loh, D.J.; Konovessis, D. Big data analytics and machine learning of harbour craft vessels to achieve fuel efficiency: A review. *J. Mar. Sci. Eng.* **2021**, *9*, 1351. [CrossRef]
18. Loukili, M.; Dutykh, D.; Nadjib, C.; Ning, D.; Kotrasova, K. Analytical and numerical investigations applied to study the reflections and transmissions of a rectangular breakwater placed at the bottom of a wave tank. *Geosciences* **2021**, *11*, 430. [CrossRef]
19. Dean, W.R. On the reflexion of surface waves by a submerged plane barrier. *Math. Proc. Camb. Philos. Soc.* **1945**, *41*, 231–238. [CrossRef]
20. Takano, K. Effets d'un obstacle parallélépipédique sur la propagation de la houle. *Houille Blanche* **1960**, *3*, 247–267. [CrossRef]
21. Patarapanich, M. Forces and moment on a horizontal plate due to wave scattering. *Coast. Eng.* **1984**, *83*, 279–301. [CrossRef]
22. Liu, P.L.F.; Wu, J. Wave transmission through submerged apertures. *J. Waterw. Port Coast. Ocean Eng.* **1987**, *113*, 660–671. [CrossRef]
23. Stamos, D.J.; Hajj, M.R.; Telionis, D.P. Performance of hemi-cylindrical and rectangular submerged breakwaters. *Ocean Eng.* **2003**, *30*, 813–828. [CrossRef]
24. Molin, B.; Kimmoun, O.; Liu, Y.; Remy, F.; Bingham, H.B. Experimental and numerical study of the wave run-up along a vertical plate. *J. Fluid Mech.* **2010**, *654*, 363–386. [CrossRef]
25. Liu, Y.; Li, H.J.; Zhu, L. Bragg reflection of water waves by multiple submerged semi-circular breakwaters. *Appl. Ocean Res.* **2016**, *11*, 67–78. [CrossRef]
26. Scyphers, S.B.; Powers, S.P.; Heck, K.L. Ecological Value of Submerged Breakwaters for Habitat Enhancement on a Residential Scale. *Environ. Manag.* **2015**, *55*, 383–391. [CrossRef]
27. Mei, C.; Black, J. Scattering of surface waves by rectangular obstacles in waters of finite depth. *J. Fluid Mech.* **1969**, *38*, 499–511. [CrossRef]
28. Massel, S. Harmonic generation by waves propagating over a submerged step. *Coast. Eng.* **1983**, *7*, 357–380. [CrossRef]
29. Driscoll, A.; Dalrymple, R.; Grilli, S. Harmonic generation and transmission past a submerged rectangular obstacle. *Coast. Eng.* **1992**, 1142–1152.
30. Szmidt, K. Finite difference analysis of surface wave scattering by underwater rectangular obstacles. *Arch. Hydro-Eng. Environ. Mech.* **2010**, *57*, 179–198.
31. Akoz, M.S.; Cobaner, M.; Kirkgoz, M.S.; Oner, A.A. Prediction of geometrical properties of perfect breaking waves on composite breakwaters. *Appl. Ocean Res.* **2011**, *33*, 178–185. [CrossRef]
32. Kuntoji, G.; Rao, M.; Rao, S. Prediction of wave transmission over submerged reef of tandem breakwater using PSO-SVM and PSO-ANN techniques. *ISH J. Hydraul. Eng.* **2020**, *26*, 283–290. [CrossRef]
33. Mao, P.; Chen, C.; Chen, X.; Zhang, Q.; Bao, Y.; Yang, Q. An innovative design for floating breakwater with Multi-objective genetic optimal method. *Ocean Eng.* **2024**, *312*, 119202. [CrossRef]
34. Breiman, L. Random forests. *Mach. Learn.* **2001**, *45*, 5–32. [CrossRef]
35. Drucker, H.B.; Christopher, J.; Kaufman, L.; Smola, A.; Vapnik, V. Support vector regression machines. In *Advances in Neural Information Processing Systems*; MIT Press: Cambridge, MA, USA, 1996; p. 9.
36. Zupan, J. Introduction to artificial neural network (ANN) methods: What they are and how to use them. *Acta Chim. Slov.* **1994**, *41*, 327.
37. Quinlan, J.R. Learning decision tree classifiers. *ACM Comput. Surv. (CSUR)* **1996**, *28*, 71–72. [CrossRef]
38. Mackay, D.J.C. Introduction to Gaussian processes. In *NATO ASI Series F Computer and Systems Sciences*; Springer Verlag: Berlin/Heidelberg, Germany, 1998; Volume 168, pp. 133–166.

fluids

MDPI

Article

AI-Driven Surrogate Model for Room Ventilation

Jaume Luis-Gómez [1], Francisco Martínez [2], Alejandro González-Barberá [1], Javier Mascarós [2], Guillem Monrós-Andreu [1], Sergio Chiva [1], Elisa Borrás [2] and Raúl Martínez-Cuenca [1,*]

[1] Department of Mechanical Engineering and Construction, Universitat Jaume I, 12071 Castelló de la Plana, Spain; jaluis@uji.es (J.L.-G.); gonzalal@uji.es (A.G.-B.); gmonros@uji.es (G.M.-A.); schiva@uji.es (S.C.)

[2] Ingeniería Global Omnium, Avanqua Oceanogràfic, S.L., 46024 València, Spain; fmartinez@globalomnium.com (F.M.); jamasvi@globalomnium.com (J.M.); eborras@oceanografic.org (E.B.)

* Correspondence: rcuenca@uji.es

Abstract

The control of ventilation systems is often performed by automatic algorithms which often do not consider the future evolution of the system in its control politics. Digital twins allow system forecasting for a more sophisticated control. This paper explores a novel methodology to create a Machine Learning (ML) model for the predictive control of a ventilation system combining Computational Fluid Dynamics (CFD) with Artificial Intelligence (AI). This predictive model was created to forecast the temperature and humidity evolution of a ventilated room to be implemented in a digital twin for better unsupervised control strategies. To replicate the full range of annual conditions, a series of CFD simulations were configured and executed based on seasonal data collected by sensors positioned inside and outside the room. These simulations generated a dataset used to develop the predictive model, which was based on a Deep Neural Network (DNN) with fully connected layers. The model's performance was evaluated, yielding final average absolute errors of 0.34 degrees Kelvin for temperature and 2.2 percentage points for relative humidity. The presented results highlight the potential of this methodology to create AI-driven digital twins for the control of room ventilation.

Keywords: computational fluid dynamics; artificial intelligence; surrogate models

1. Introduction

Nowadays, multiple industries rely on a wide variety of automatic control strategies to control different processes and facilities. Among all these strategies, some of them are based on forecasting the future behavior of the system based on sparse real-time information available through sensors. Digital twins can be found in the previous group, which employ real-time data from interest variables and system actuators gathered by sensors to replicate the dynamical behavior of the system under study. The core simulator of a digital twin can be a mathematical model ranging from analytical models to sophisticated systems of partial differential equations solved by proper numerical methods.

This virtual simulation serves as a base to analyze the dynamics of the system, optionally comparing it with historical data, to implement optimized control strategies. The representation of the physical system can also be used to establish a comprehensive analysis of the real system [1]. In an era where a massive amount of data is accessible and large computing clusters are readily available, creating in-house digital twins to control systems has become not only feasible but a growing trend across diverse industries. They can be applied to traditional sectors such as agriculture [2], as well as industrial processes [3] and manufacturing [4]. Indeed, they can be useful in complex workflow like urban planning [5], showcasing their versatility and value in driving operational efficiency and innovation [6].

To implement a digital twin in a system where one or more fluids are involved, the modeling core can be deployed with Computational Fluid Dynamic (CFD) simulations, since they implement the set of equations that model the behavior of such fluids and their interactions with external bodies and surfaces. The use of CFD simulations to model room ventilation is not new, and it is possible to find reviews analyzing papers based on CFD for the past decades [7,8]. Most of the works conducted in this area aim at reproducing in a virtual environment a ventilated room. This reproduction allows one to understand the effects of the different elements of the ventilation system to improve its design and performance as well as increasing room comfort [9].

Furthermore, it is possible not only to model the hydrodynamics and heat transfer of such systems, but also to model the evaporation and condensation of water in different scenarios. Different works can be cited in this field, and it is of interest to mention [10], where energy transfer and psychrometric variables are included to model the condensation of water vapor, validating the work with experimental data. In [11], a similar study is conducted, using a pilot reactor with controlled conditions to validate the simulations of the water condensation. The aforementioned works study reduced cases to test their different models, whereas in [12], an underground facility has been modeled with CFD, evaluating different air conditioning systems to avoid water condensation. Some works can be found regarding water evaporation using CFD simulations. In [13], an indoor swimming pool facility has been modeled, including the ventilation system, to study the evaporation and transport of water vapor. Similarly interesting works can be found in [14–16]. On a more detailed level, [17,18] study the effect of convective flow and air conditions on the evaporation rate of indoor swimming pools. The present paper implements an evaporation source similar to the one used by Elazm et al. [19]. They study the effect of air velocity at the water interphase on water evaporation, with a source term proposed by the American Society of Heating, Refrigerating, and Air-Conditioning Engineers (ASHRAE) [20].

The cited works focus mainly on essential phenomena related to water vapor generation or condensation independently. Several works can be found where only one of both effects is included, depending on the scope of the study or the characteristics of the system. One of the distinctive features of the present paper is that it includes both effects as they play an important role in the studied problem.

Despite the great advantages that CFD simulations may provide, the integration of such tools into ventilation control is usually suboptimal. As an illustrative example, in [21], CFD simulations are included in a control loop with a PID controller for room ventilation. For small systems where the time required to obtain a converged simulation result is minor compared with the inertia of the system, these kinds of proposals are realizable. Unfortunately, this may be rather the exception than the norm, typically requiring these simulations several hours to finish and being unfit to be implemented in real-time controls.

Especially for digital twins, the key requirement is to reproduce the dynamics of the system in real time; that is, the predicted state of the system (in terms of the spatial distribution of velocity, temperature, and relative humidity) has to be known prior to the occurrence of that state so that a control strategy can be studied and implemented. This requirement strongly limits the implementation of CFD simulations in digital twins, since the necessary time to complete such simulations can span from minutes to days depending on the dimensions and complexity of the system.

To solve this problem, a novel methodology is proposed and studied in the present work, creating and training an Artificial Intelligence (AI) model using physics-informed data from CFD simulations to learn the behavior of a ventilated room with presence of water vapor, approximating future states of the system in milliseconds. To the authors' best knowledge, the use of AI to predict the state variables of a ventilated room based on sensor data is new, a recent work from Quang et al. [22] being the closest one. Quang uses different ML models trained with CFD simulations to predict the velocity and temperature distribution in an indoor building caused by natural convection, although there is no prediction of the relative humidity. The approach proposed in the present work for ventilated rooms with presence of water vapor bridges the gap between detailed CFD analysis and practical, actionable insights, allowing the implementation of a digital twin for automated ventilation control.

The integration of CFD with AI is a state-of-the-art field with lots of research branches [23] that can encompass simple regression models that predict CFD results to more complex systems that control machines with the help of Deep Reinforcement Learning (DRL). The reason behind its growing spread in multiple areas of technology is due to its predictive speed and computational efficiency, along with the ability to scale large datasets and complex systems, making them suitable for various real-world applications. Some researches remark the use of Physics-Informed Neural Networks (PINNs) [24–26]. Although their implementation provides better inferences fulfilling physical governing equations, they are mostly restricted to general conservation equations. However, in the present case, the problem requires to predict transported species with complex sink and source terms. The inclusion of such specific equations is not yet clear for a PINN, and thus, this approach has been discarded for the present work.

It is important to remark that the deployment of digital twins implementing AI models is a research topic in development, and there exist a wide variety of control systems based on this solution, tackling problems for many different industries. A reader willing to explore previous works in this area may be referred to the following review [27]. Despite the considerable amount of work performed to implement digital twins with AI models, and the amount of research conducted in the previous years coupling AI with CFD simulations, to the knowledge of the present authors, almost no work has been carried out to use these CFD–AI models to create a digital twin. To put a number on this statement, in another review [28], 149 studies have been analyzed, finding only one work where the use of CFD-based AI models is integrated into a digital twin. In this mentioned study [29], the authors use AI models based on CFD simulations to predict simple variables representative of the system in order to build a digital twin. Moreover, there is an increasing interest not only in digital twins but also in the surrogate models they use with different architectures and approaches [30,31] or even in the promising field of Deep Learning explainability, where they try to explain the different behaviors of the fluid in CFD [32]. Furthermore, hybrid CFD solvers where ML helps to accelerate the final converged solution for steady-state problems or skips timesteps in transient ones is also an increasing interest in the research community [33,34].

To wrap up, the present study combines different research problems found in the bibliography into a joint study. Regarding the modeling of ventilated rooms, in this work, a CFD model of a complex ventilated room will be obtained, both incorporating the phenomenon of water evaporation and condensation. This poses an extra challenge compared with previous works, as now both opposing phenomena need to be included and the results will be influenced by its equilibrium. In addition, data from sensors needs to be gathered around a year to be able to create a suitable and realistic dataset to train the surrogate model. Furthermore, it is a considerable challenge to train a predictive generative model capable of inferring the complete state of a complex system using only few input data. Combined with that, another problem of the present research work is to select the best suitable sensor position inside the room to optimize the predictive performance of the model by maximizing the quality of the input data.

With the aforementioned research problems presented, the novelty of this work lies in exploring the methodology of using a CFD–AI surrogate model to approximate the complete state of a ventilated room based on real-time data from sensors. This predictive tool can be envisioned as a computing core to build a digital twin of the ventilated room to guide the control strategies in advanced ventilation systems. By taking into account the 3D state of the system, deeper spatial analysis will be possible, enabling the study of more complex and accurate actuation rules, ultimately leading to more efficient operational uses.

The following sections are devoted to describing the methodology followed in the present work and the results obtained. Materials and Methods introduces the ventilated room studied in this work, describing its ventilation systems and the main variables describing its operational state. Furthermore, the CFD models used in the simulations are listed, and the mesh generation process and configuration of boundary conditions are presented. Finally, the architecture of the CFD–AI surrogate model is described and the optimization process depicted. In Results, the main findings of the work are shown, illustrated with one example case simulation for the CFD results, and one predicted case for the accuracy of the predictive model. The overall assessment of the CFD–AI model is also included for completion. Last, sections Discussion and Conclusions gather the main interpretations of the findings and wrap up some final comments.

2. Materials and Methods

In this section, all details of the methodology used to carry out the present research work are described. First, the case study is described, presenting the facility where this work was developed. From this case, the set of variables that permits a complete characterization of the facility is identified and described. Representative sensor data is provided, and the generation of the dataset is given. To continue, the CFD model used to simulate the described domain is depicted, presenting all relevant equations and simplifications. Then, the generated mesh is shown, along with the description of the configuration for the relevant boundary conditions of the system. Finally, the architecture of the CFD–AI surrogate model is introduced, showing the point grid used to extract the data from the CFD simulations to train the model. The stages for the optimization of the CFD–AI model are depicted.

2.1. Context and Facility Description

The present manuscript leverages AI to create a surrogate model for the ventilation of a specific room within the Oceanogràfic (https://www.oceanografic.org/), a public aquarium and marine complex situated in the Jardí del Túria in the city of València. Oceanogràfic is a large facility composed of multiple underground galleries and connected rooms. To ensure the well-being of the marine fauna, a complex of underground chambers with technical equipment extends next to each aquarium, controlling the conditions of the water for each environment while purifying and cleaning it from impurities. These technical rooms contain technical equipment (pumps, mixers, control electronics, etc.) and are usually located behind the aquariums, in direct contact with them by windows and chambers above the water level. To develop the presented work, only a room with one connection to the main gallery was selected to work with.

To represent and describe the selected space for the study, Figures 1 and 2 present a 2D and 3D view of the studied room. In Figure 1, two regions can be appreciated: one at ground level, named technical room from now on, and one elevated from the ground that communicates with a water surface (light blue in the scheme), which will be named chamber. Both regions communicate with a set of apertures, named windows from now on, that must remain opened for maintenance purposes. In Figure 2, the purple surface represents the only connection of the room with the gallery and will be referred to as swap surface.

Without proper ventilation, the water vapor generated at the water surface would be transported from the chamber to the technical room by means of diffusion and natural air currents. When psychrometric conditions are unfavorable, the evaporation rate rises, resulting in subsequent condensation on the walls, roof, and technical equipment. This condensation is responsible for increased corrosion on devices causing apparatus malfunction and structural damages.

To palliate these problems and to ensure air renovation, ventilation circuits are installed in the rooms, consisting of a supply circuit and an exhaust circuit. Furthermore, in the room object of the present study, a drying circuit was implemented to provide extra dehumidification. The locations for the ventilation grilles are illustrated in Figure 1. The supply circuit grilles are painted dark blue, the exhaust circuit grilles red, the intake grilles of the drying circuit are black (corresponding to two dehumidifiers), and the discharge grilles are green. For the sake of simplicity, only the final section of the ventilation circuits has been modeled, since other circuits and pipes are already inside the room and act as obstacles for the air. Figure 2 shows a 3D view of the simulated domain, with the grilles acting as inlets (discharge) or outlets (intake) colored according to each circuit. Finally, the swap surface is colored in purple (note that the air flow can go in both directions across this surface depending on the flow conditions).

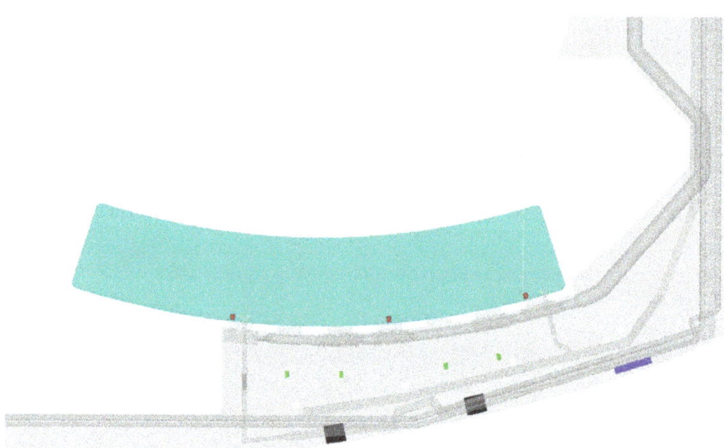

Figure 1. Two-dimensional view of the technical room, with the ventilation grilles (red—exhaust intake, green—drying discharge, black—drying intake, blue—supply discharge) and the water surface in the cavity marked with light blue.

In this configuration, the supply circuit introduces fresh air from the exterior of the facility into the technical room, causing the main air current in the domain of the study. Note that the temperature and humidity of the supplied air depends on external meteorological conditions (typically between 60% and 90%). The exhaust system is designed to remove stale and humid air from the chamber. By extracting air from this location, the air flow through the windows is directed towards the chamber. This prevents humidity diffusion into the technical room, thus maintaining optimal air quality and the buildup of excess moisture in its interior. When the net flow between exhaust and supply is positive, i.e., exhaust flow is greater than the supply one, the defect of air is covered by an inflow from the gallery via the swap surface. When the net flow is negative, the excess of air from the supply circuit exits the room via the swap surface to the gallery. Finally, the drying circuit reduces the moisture content of the air supplied from the exterior and the gallery. Each dehumidifier draws in humid air, cooling it to condense the moisture, and then reheats the air before releasing it through the discharge drying grilles. This process ensures that the air remains dry and comfortable in the technical room. The resulting dry air is injected towards the windows to further prevent moisture diffusion into the technical room. Therefore, the combination of the different ventilation systems leads to multiple behaviors and situations that need to be taken into account and have their representation on the generated data used to train the predictive model.

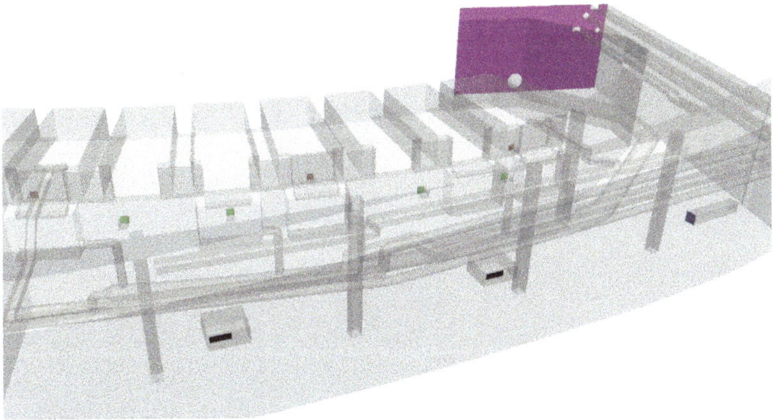

Figure 2. Three-dimensional view of the technical room with the ventilation grilles for each circuit and the swap surface (colored in purple).

2.2. CFD Model

The CFD simulations forming the training dataset were conducted using OpenFOAM v2306, an open-source software widely employed in academia and research for its flexibility and the ability to integrate custom code to enhance its functionality. The solver *buoyant-BoussinesqSimpleFoam* was used, a choice motivated by the expected low variation of air density due to the low variations of temperature in the domain. It was also assumed that the low pressure changes due to buoyant and velocity change effects would produce almost negligible changes on the density. In such cases, the Boussinesq approximation can be used, assuming a linear variation of air density with temperature as given in Equation (1):

$$\rho = \rho_0[1 - \beta(T - T_{ref})], \tag{1}$$

where ρ is the air density, ρ_0 is the reference density at T_{ref} temperature, β is the thermal expansion coefficient (3×10^{-3} K^{-1}), and T_{ref} is the reference temperature (300 K).

Under these assumptions, according to Ferziger and Peric [35], the density can be treated as constant in the unsteady and convection terms, but treated as variable in the gravitational term. As the density is considered constant for the majority of the terms, and the velocities are expected to be low, an incompressible approach for the air can be used. Thus, the Averaged Navier–Stokes equation is modified as shown below:

$$\nabla \cdot (\vec{u}\vec{u}) = -\frac{1}{\rho_0}(\nabla p - \rho \vec{g}) + \nabla \cdot \left(\nu_{Eff} \nabla \vec{u} \right). \tag{2}$$

With ν_{Eff}, the modeled effective viscosity is computed as $\nu_{Eff} = \nu + \nu_t$, with ν_t being the turbulent viscosity modeled with the corresponding turbulence model. The time derivative term has been removed, as the converged steady-state solution is desired to train the model.

Several works devoted to simulating the ventilation of rooms with large volumes as the one in the present work use the k–ε model or similar variants [9–13] due to limitations in the computational resources required to properly resolve the boundary layer. It is important to stress the fact that the present work is not devoted to investigating the effect of the turbulence modeling on the evaporation and condensation rates. Rather, its main focus lays on obtaining a surrogate model to predict the approximate equilibrium state of a ventilated room in pursuit of being implemented in a control algorithm. Expecting low velocity gradients in the majority of the domain, and being aware that the computational resources required to resolve the boundary layer with more sophisticated turbulence models, the k–ε turbulence model has been used. This model offers a good tradeoff between accuracy and resource efficiency.

To provide closure to the system of equations, the energy conservation equation has been modified and rearranged under the simplifications of incompressible flow and Boussinesq approach as

$$\nabla \cdot (T\vec{u}) = \nabla \cdot \left(\left(\frac{\nu}{Pr} + \frac{\nu_t}{Pr_t} \right) \nabla \vec{u} \right) + S_T. \tag{3}$$

where Pr is the Prandtl number (taken as 0.7), Pr_t the turbulent Prandtl number (taken as 0.85), and S a source term of energy coming from radiation, convection, or conduction from external sources or boundary surfaces.

To model the presence of relative humidity, the simulated fluid was chosen to be dry air, and the water vapor was transported as a scalar variable over the air. This simplification was made after studying the error committed in density estimation by not accounting for the mass of evaporated water present in the air. The extreme case would be at high temperature (40 °C in summer) with maximum saturation of water vapor. In this situation, the difference of density between dry air and moist air would be of a 2.75%. It is assumed that this error is acceptable to generate the results used to test the proposed methodology of this work. The transport of relative humidity is governed according to the following equation:

$$\nabla \cdot (w\vec{u}) = \nabla \cdot \left(\left(\frac{\nu + \nu_t}{Sct} \right) \nabla w \right) + S_w, \tag{4}$$

with w being the specific humidity (kg_{H2O}/m^3), while ν and ν_t stand for the kinematic and turbulent kinematic viscosity, respectively, and Sct is the Schmidt number. Finally, S_w represents a source/sink term for the specific humidity, being a source when evaporation from the water surface occurs, and a sink when condensation happens in the walls of the domain. The experimental correlation used to model the source/sink term of water vapor is a variant of Carrier's empirical correlation [36] included in the ASHRAE's Application Handbook [20]. As this correlation does not consider the effect of vapor density difference, it tends to over-predict the evaporation rate. To improve its accuracy, Smith et al. [37] proposed a correction for indoor and outdoor swimming pools, multiplying Carrier's correlation by 0.76. Therefore, the final empirical correlation used in Equation (5) was

$$\phi_w = A \cdot 0.76(0.08893 + 0.07835v)\frac{(P_{v,s} - RH \cdot P_{v,\text{inf}})}{h_{fg}}, \tag{5}$$

where A is the surface area of the patch where the condensation/evaporation occurs, v is the air velocity next to the water surface, and $P_{v,s}$ is the saturated vapor pressure at the surface temperature, whereas $P_{v,\text{inf}}$ is the saturated vapor pressure at air temperature at the cell center. RH is the relative humidity and h_{fg} is the latent heat of vaporization. The previous equation relates to the total amount of vapor exchange by a surface [kg/s]; to implement the source term S_w, it is necessary to divide by the cell volume, resulting in $S_w = \phi_w / V_{cell}$.

2.3. Studied Variables and Sensor Data

The prediction of the steady-state configuration of the room characteristics needs the knowledge of a limited set of variables. These variables can be separated into three subsets: distribution of psychrometric variables in the initial state (spatial distribution of temperature and relative humidity), psychrometric variables in the boundaries of the domain (swap surface and ventilation grilles), and the values for the actuators that activate the flow in the three circuits (supply, exhaust, and drying).

The knowledge of the distribution of the psychrometric variables in the initial state is a cumbersome task, as the values of temperature and relative humidity present changes across the domain. Consequently, one would need a considerable amount of data to accomplish this requirement. However, the structure of these fields is determined by the fluid dynamics inside the volume, so the resulting distribution might be properly described by a reduced number of sensors. In this approach, 4 sampling locations were selected to account for the initial conditions. This amount of sensors is not the result of a casual choice, but because of technical limitations. This is so because only four more sensors were available and installed in the studied room. It is understood by the authors that more sensors would increase the predictive accuracy of the NN; this is a limitation of this work. The exact positions of the rest of the sensors were not known a priori, but resulted from the analysis of the CFD results (the procedure is detailed in Section 2.5.3).

The knowledge of the psychrometric variables in the boundary conditions is also required. The temperature and relative humidity of the supply discharge play a central role, as they are related to the main air current in the domain. Also, the knowledge of these magnitudes in the discharge drying grilles is of primary importance, as they represent the main source of air treatment in the domain. The temperatures in the walls are also important as they have a strong influence in the condensation. Finally, the psychrometric conditions at the swap surface are relevant for those cases when there is a net negative flow into the room. Due to its importance on the psycrhometric conditions in the vicinity of the room, one sensor from the set destined to monitoring the internal state of the room was placed near the swap surface, and its location was fixed.

To sum up, the necessary variables regarding psychrometric conditions are as follows:

- Room temperature at 3 locations (K);
- Room humidity at the same 3 locations (%);
- Supply temperature (K);
- Supply relative humidity (%);
- Swap temperature (K);
- Swap relative humidity (%);
- Dehumidifier temperature (K);
- Dehumidifier relative humidity (%);

- Wall temperature (K).

Regarding the actuators of the ventilation system, three levels of functioning have been defined for each circuit. The supply and exhaust circuits have a frequency converter to control the air flow between 50% and 100% of its nominal flow (3000 m^3/h). To reduce the amount of dataset combinations, three states where considered for both the supply and exhaust fans: switched off, 50% of its maximum flow rate, and its maximum flow rate. The surrogate model should be able to properly predict the results for any flow rate between 50% and the maximum. The dehumidifiers have only two operational points: on and off, with a constant flow rate of 1500 m^3/h per dehumidifier. Combining both machines, the possible flow rates for the drying circuit are 0, 1500, and 3000 m^3/h. To sum up, the necessary variables regarding actuators are

- Supply flow rate (m^3/h);
- Exhaust flow rate (m^3/h);
- Drying flow rate (m^3/h).

From this analysis, the set of 15 variables forming the inputs of the predictive model was 4 temperatures and 4 relative humidities for the initial conditions and swap surface, the temperature and humidity of the supply and drying circuits, and the three flow rates for the actuators.

Note that the wall temperature has not been considered as an input for the predictive model, which is mostly motivated by the lack of proper experimental data. The studied domain presents a total area covered by walls of roughly 2300 m^2; this represents a great complexity in terms of accurately setting a suitable temperature for the different wall patches and regions. Several technical limitations prevented the implementation of permanent temperature sensors in contact with the walls, so no temporal study could be made. To overcome this issue, it was decided to simplify the complexity by reducing the different surfaces located in different positions of the room to a single data point (used in the CFD simulations), being fully aware of the limitations of this approximation. During one visit to the facility, a thermal gun was used to measure the temperature of the walls in a state where the ventilation was switched off and the room was in equilibrium (sharing the same temperature as the one in the gallery). The measurements yielded an average difference of 2 degrees between the temperature of the main walls and the air in the gallery. Lacking more accurate experimental data, we decided to extrapolate this to the weather conditions of the rest of the year. The idea behind it is to lean on the fact that the air in the gallery is not treated, and since the facilities are underground, they can be regarded as isolated from the exterior. In that case, the air inside the gallery will, after an infinite time, approach the wall's temperature, the dynamics of the wall being considerably slower than the evolution of the ventilated room. If experimental data for the wall temperature had been readily available, it would have been included in the CFD simulations and as input for the predictive model, as it provides valuable information. This is one of the limitations of the present study.

After identifying the input variables, a suitable dataset must be created to train the surrogate model. The goal is to conduct realistic CFD simulations, incorporating initial and boundary conditions that typically occur during operation. To accomplish this, sensor data for psychrometric variables was collected over one year, allowing for the identification of typical psychrometric values for each season. Table 1 shows the average values for the psychrometric variables for each season. Note that the conditions for the exit of the dehumidifiers are not present, since they depend on the state of the air inside the room.

Also, mean room temperature and mean relative humidity were calculated from a set of sensors distributed inside the room.

The data in Table 1 served to provide the first four initial conditions for the simulations (one typical day per season, assuming uniform temperature and relative humidity distributions in the room). Then, each condition was run with three different configurations of the actuators (switched off, half flow rate, and maximum flow rate), leading to a set of twelve simulations, namely "First Generation", which started with initial uniform distribution of temperature and relative humidity.

Table 1. Typical values for the different monitored variables at each season.

Variables	Spring	Summer	Autumn	Winter
Mean room temperature (K)	293	298	294	290
Mean room relative humidity (%)	95	78	85	90
Swap temperature (K)	290	302	292	286
Swap relative humidity (%)	60	75	60	70
Supply temperature (K)	290	297	294	288
Supply relative humidity (%)	85	80	82	87

Then, these twelve converged non-uniform states served as the initial conditions for another set of 120 CFD simulations, with a total of 142 cases. For each one of the First Generation simulations, 10 simulations were run by changing just one of the variables among the psychrometric conditions of the supply/drying circuits, the swap surface, or the flow rate of one of the ventilation circuits. To provide clearer information about the dataset creation, a repository has been created (https://zenodo.org/records/15025264, accessed on 14 March 2025), providing a table including the values used to set each boundary condition described in Table 2.

Table 2. Summary of boundary conditions for each patch of the simulation domain.

Patch	Type	Boundary Condition	Relative Humidity and Temperature BC
Supply	Inlet	Fixed velocity	Fixed value
Exhaust	Outlet	Fixed pressure	Zero gradient
Dehumidifier's supply	Inlet	Fixed velocity	Fixed value
Dehumidifier's exhaust	Outlet	Fixed pressure	Zero gradient
Water surface	Wall	Free slip	Humidity source term
Swap surface	Inlet–Outlet	Flux dependent	Fixed value for inflow
Walls	Wall	No slip	Humidity sink term

The total amount of available cases in the dataset may be considered small to train a model capable of delivering such an amount of outputs. Nonetheless, it is important to stress the fact that the different conditions chosen for each variable were not created utilizing any function to choose randomly between a predefined range. This is important because random combinations of psychrometric variables would lead to unreal weather conditions, e.g., it is not within the typical expected weather conditions in this location to find a day with 32% relative humidity at 7 °C. Thus, the different combinations were manually selected having in mind these limitations. This explains in part why the number of cases has that size. Furthermore, if the number of cases increases, it also increases the risk of repeating similar conditions for two different cases, thus leading the trained model to lose its ability to generalize (leading to overfitting).

2.4. Mesh Generation and CFD Configuration

The simulated domain is 53.5 m long and 25.6 m wide, with a maximum height of 6.6 m, yielding a total volume of 1607 m^3. These numbers will aid the reader in understanding the difficulty of controlling the ventilation system of such a technical room and the necessity of developing a control algorithm based on a 3D digitalization of the room. Due to the complexity of the geometry, a CAD model was created using SolidWorks® (https://www.solidworks.com/) using built-in drawings kindly provided by Avanqua, the company responsible for the maintenance of the facility, complemented with in situ measurements.

From the CAD model, different STL files were created, differentiating for each specific boundary patch and separating the walls into different types of geometries to increase the refinement control: small pipes, large pipes, ventilation ducts, and small details. The mesh was generated using OpenFOAM's meshing tool SnappyHexMesh, with an initial cell size of 10 cm. Then, different surface refinement levels were defined, along with specific volumetric refinement zones to reproduce all the details of the room. The smaller cell size corresponding to the region near the ventilation grilles and other geometric details is 6.25 mm. The total area represented by walls in the system is 2282 m^2; considering the number of cells required to properly resolve the wall boundary layer for the total wall surface by means of an appropriate inflation, it is clear that the computational resources required would make impractical the creation of the desired dataset of simulations. Having this limitation in mind, no inflation layers were included, expecting to encounter low velocity gradients near the walls. Thus, the wall boundary layer was modeled relying on the use of wall functions that are well known and widely used in many research works using a RANS approach ($k - \varepsilon$ model). This configuration yielded a mesh with 7.9 million cells. In Figure 3, some mesh details are shown, with a 3D image of the upper zone and a 2D slice at 4.55 m from the ground.

Additionally, a mesh test was performed to assess its effect on CFD convergence. To this end, two more meshes were evaluated, comprising 5 and 12 million cells. A case with maximum ventilation flow was used to compare the effect of the mesh. The converged velocity profiles were obtained at different distances from the supply grill. Figure 4 depicts the three profiles at different distances for each mesh, showing that the obtained velocities converge towards the 12 million cell mesh. A noticeable difference is present between meshes with 5 and 8 million cells, although velocity profiles converge towards the finest mesh. The results indicate that the intermediate mesh used in this study (7.9 million cells) can be regarded as sufficiently refined to yield good results to train a predictive model, optimizing the computational cost of creating the dataset.

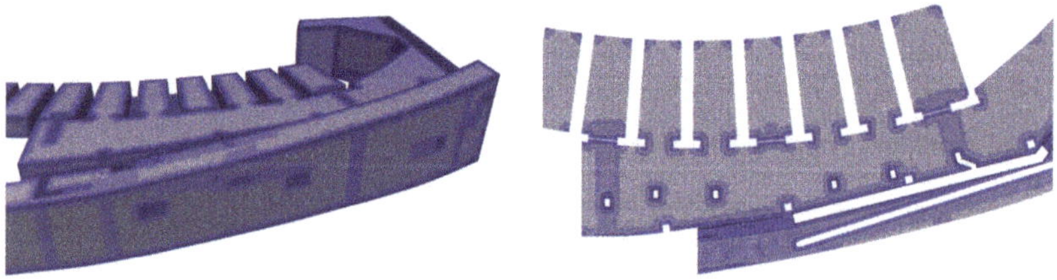

(a) (b)

Figure 3. Three-dimensional detail of final mesh used (**a**) and two-dimensional slice at 4.55 m from ground (**b**).

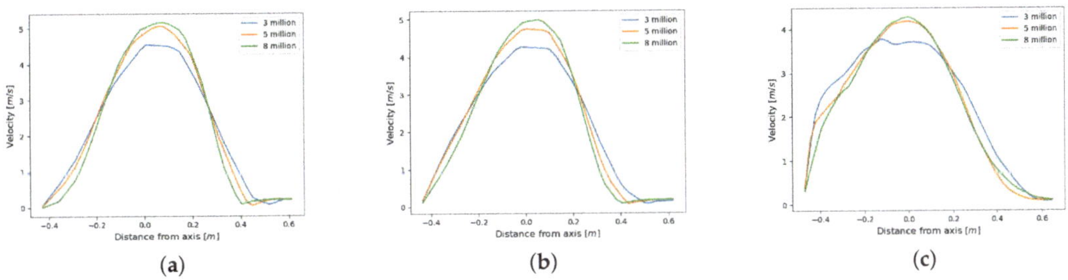

Figure 4. Converged velocity profiles of the supply circuit inlet at 1 m (**a**), 2 m (**b**), and 4 m (**c**).

After the mesh is generated, the boundary conditions need to be defined for the different patches of the domain. Each special patch has been introduced in Figure 1, and then variables with special treatment associated with these patches have been listed in Section 2.3. The summary of the complete configuration for the main types of boundaries is presented in Table 2. Relative humidity and temperature of the supply circuit were set accordingly with the typical psychrometric conditions of each corresponding season from data extracted from a weather station located near the Oceanogràfic. The ranges for both variables are temperature $= [280, 308]$ K and relative humidity $= [30, 85]\%$, considering days and nights within the whole year, excluding extreme conditions outside the typical ranges of each season. The relative humidity at the exhaust grilles of the drying circuit was set at the operating point of the machine (60%), although lower values below 30% were considered for a subset of the dataset. The temperature of the water surface was kept constant to 293 K for the complete dataset. This configuration is justified because the water from the aquarium was kept at constant conditions by the machinery present in the technical room due to the biological requirements of the marine fauna present in the aquarium. Finally, the psychrometric conditions of the swap surface were chosen among a defined range obtained from the study of the typical values of temperature and relative humidity of the gallery at different times of the year. The ranges for both variables are temperature $= [288, 297]$ K and relative humidity $= [55, 87]\%$.

Since the goal of the simulations is to obtain the final equilibrium state of the room for a given set of psychrometric variables and actuators settings, steady-state simulations were performed until reaching convergence. To this end, the temporal derivative was not considered in any equation—as shown in Equations (2), (3), and (5)—and thus, there was no need to adjust the time step to ensure a Courant number smaller than 1. Second-order advective schemes were selected for the majority of the variables, excluding the turbulent variables -k and ε-, where first order was used to ensure its boundedness and provide stability to the simulation. The convergence criteria used were 1×10^{-3} for the pressure and 1×10^{-4} for the velocity, temperature, and specific humidity.

Employing the previously described CFD models and boundary conditions configured accordingly with the defined set of cases (https://zenodo.org/records/15025264, accessed on 14 March 2025), the complete dataset was generated.

2.5. AI Model

2.5.1. Data Preprocess for Training

As explained in Section 2.3, the obtained dataset was composed of 142 CFD cases, and each case was defined by a set of 15 variables (related to initial conditions, boundary conditions, and actuators) that served as inputs to the CFD–AI surrogate model. Then, to train the predictive model to infer the equilibrium state of the whole domain, the resulting 3D fields of temperature and relative humidity were extracted from the CFD simulations.

Due to the immense amount of cell data and its non-uniform spatial distribution, a secondary grid was constructed, featuring points uniformly distributed. The points were arranged aligned with the main axis of the domain, separated from each other by 35 cm. The extracted temperature and relative humidity values were sampled by linearly interpolating the data of the surrounding cell centers of the primary mesh. This sampling grid resulted in roughly 62,000 points for each predicted variable, yielding a total of 124,000 output points. In Figure 5, the 3D spatial distribution of the sampling points is shown, along with a 2D horizontal slice, showing that the sampling grid adapts to the domain shape and internal cavities of the primary mesh.

Thus, the CFD data used to train the CFD–AI surrogate model can be summarized in the following: 15 input data comprising boundary conditions, actuator flow rates, and 3 virtual sensors inside the domain; and 124,000 output data from the 3D fields interpolated to the regular mesh shown in Figure 5, comprising temperature and relative humidity data.

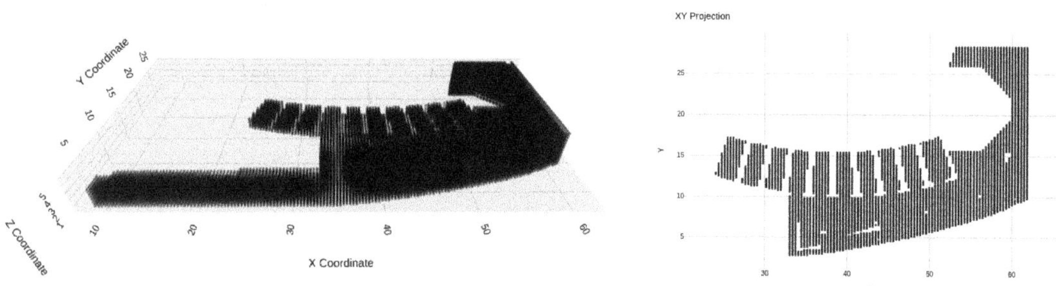

(a) (b)

Figure 5. Three-dimensional spatial distribution of the sampling points (**a**) and two-dimensional detail of a horizontal slice (**b**) of the secondary grid used to extract the data used to train the predictive model.

2.5.2. Model Architecture

To construct the surrogate model, a neural network (NN) based on the Multilayer Perceptron (MLP) architecture was employed. The MLP was characterized by its fully connected layers, meaning each neuron is connected to every neuron in the adjacent layers. This dense connectivity significantly increases the number of parameters in the neural networks, demanding substantial computational resources [38]. Despite this complexity, MLPs are widely used due to their versatility in various applications. Recent works can still be found where the use of MLP can deliver results up to the required standards. To this regard, the work of Mansour et al. [39] can be cited, where an MLP-based architecture is used for image reconstruction. Authors find that this architecture enables image reconstruction performance slightly better than a U-Net NN. Similarly, Albasiouny et al. [40] used an MLP-based architecture for a generative model, finding that this model proved to be more efficient and robust than a model solely based on convolutions. These works encouraged the use of the present architecture to tackle the problem given in this work.

The key complexity of this work is the high ratio between the number of inputs and outputs (15 to 124,000). MLPs have proven to be a suitable architecture when such a high generation of data is needed. Another complexity arises from the low amount of training data; as explained in Section 2.3, this comes from a limitation in the possible combination of psychrometric variables, which leads to low batch sizes and low cases available for training. As a precedent work, Liu et al. [41] employed deep NN based on MLP for high dimensionality data with low sample size with satisfactory results, which also supports the present selection of NN architecture. Furthermore, more sophisticated NN like convolution-based or graph-based were not tested, as the low amount of training data was expected to be insufficient to properly train such models.

It should be noted that, in this work, simpler ML models like Decision Trees, Random Forests, or Support Vector Machines were not tested. This decision was based on the large number of outputs, as these models are often used for tabular data and lower dimensional problems. However, some state-of-the-art works remark the capability of models like XGBoost to handle these types of problems and should be explored in the future [42].

To prevent overfitting —a common issue where the model performs well on training data but poorly on unseen data— dropout layers were included. Dropout is a regularization technique where a fraction of the neurons is randomly set to zero during training, preventing the network from becoming too reliant on specific neurons and promoting generalization. The total dataset was split into 80% of cases for training the predictive model and the remaining 20% was used to validate the accuracy of the model. All the accuracy results shown in the Results section are extracted using the validating subset.

When it comes to architecture definition and training hyperparameters, several degrees of freedom must be considered for this problem. These parameters include number of hidden layers, number of neurons per layer, dropout rates, and learning rate. In order to optimize the CFD–AI surrogate model to its maximum accuracy, a framework for hyperparameter optimization (Optuna) was used. Optuna [43] uses an efficient sampling algorithm to explore the parameter space, balancing exploration and exploitation to find the best possible solution. It supports various optimization strategies, including Bayesian optimization, grid search, and random search, making it highly versatile for a wide range of applications. A generalized scheme of the optimization algorithm is depicted in Algorithm 1, which serves to illustrate the optimization workflow. There, the studied parameters in line 3 can be number of layers, number of nodes, etc.

Algorithm 1 Optimization process with Optuna.

1: Initialize Optuna study with direction='minimize'
2: **for** each trial in study **do**
3:     Suggest value for the studied variables
4:     Set data and architecture based on suggested values
5:     Train model
6:     Predict temperature and humidity fields using the trained model
7:     Evaluate loss based on the predicted fields compared to CFD data
8:     Return loss as the objective metric
9: **end for**
10: **Optuna** $\rightarrow$ minimize objective metric
11: Retrieve best variable values

2.5.3. Optimization of Sensor Locations

After the architecture is refined, an additional source of variability remains, as the locations of the 3 sensors inside the technical room are not predefined and therefore can be located anywhere inside. These virtual sensors are responsible for conveying information about the present state of the room to the predictive model. As stated before, it is not possible to know a priori which locations are going to deliver more information to the predictive model and therefore yield better results. As a consequence, different sets of sensor locations need to be tested to obtain the best suitable combination.

Assuming independence between the different stages of the optimization process, the optimization was performed in a sequential manner. First, the optimal neural network architecture was determined, as previously described. Subsequently, a second optimization stage was carried out using the Optuna framework to identify the most suitable combination of sensor locations. In this stage, Optuna was employed following the procedure outlined in Algorithm 1, where the parameters under study were the indices corresponding to the potential input (sensor) locations. By treating the architecture selection and sensor placement as decoupled stages, this approach simplifies the overall optimization workflow while still enabling the identification of an effective configuration for the sensing system.

To carry out the described process, different sensor placements were defined from a preliminary study of the CFD simulations, identifying the most representative locations of the state of the room. From those locations, 50 different combinations of 3 sensors were defined, varying the X, Y, and Z coordinates. This process aimed to identify the optimal sensor configuration that would minimize the model's loss function and enhance its predictive accuracy. In Figure 6, a schematic of all possible sensor placements is illustrated, including variations along the Z axis.

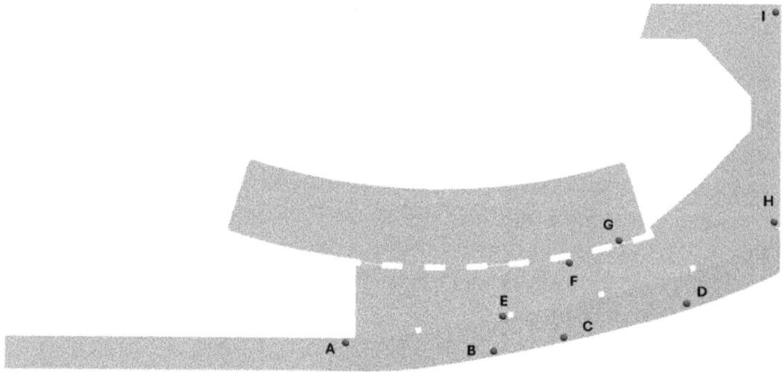

Figure 6. Possible sensor placements (A–I) shown in a XY plane.

3. Results

In this section, the results for the different stages of the present work are shown. First, the obtained simulation results are investigated for a specific case to study the coherence of the converged solutions with the CFD models. Then, the outcomes of the optimizing process are shown, indicating the final architecture and training configuration, as well as the selected points for the virtual sensors that will provide input data to the predictive model. Finally, the accuracy of the model is investigated, showing an example case from the validation subset and computing the overall predictive error for the temperature and relative humidity.

3.1. Simulation Results

In this subsection, a case from the training dataset will be analyzed to illustrate the interaction between the different elements and ventilation circuits in the room. To demonstrate the effect of the supply and dryer circuits, a case with maximum ventilation flow and typical temperature and humidity is studied, with the configuration detailed in Table 3.

The converged results of the simulations can be seen in Figure 7. The streamlines in Figure 7a represent the air currents from the discharge grilles of the drying circuit. The relative humidity ranges from 60% at the grills to almost 100% in the interior of the chamber, showing that the dried air entered the chamber and slightly reduced the relative humidity in the interior of the cavity. This validates the design idea of using the dehumidifier system to directly lower the moisture content inside the chamber, especially the ceilings, where the majority of the water condensation occurs. Following a similar philosophy, the grilles of the exhaust circuit are located inside the chamber (as shown in Figure 1), in order to extract the air with the greatest humidity content outside of the facility.

Figure 7b represents a set of streamlines starting from the supply discharge grille, also colored according to the relative humidity. Here, the fresh air is directed towards the drying intakes and reaches the least renovated zones of the technical room (left side in the picture), providing fresh air and helping homogenize the temperature and humidity of the whole facility. As the supply circuit is untreated air from the exterior, depending on the external psychrometric conditions, the introduced air will constitute a source of additional humidity to the system or will dilute its concentration. For this case, the air injected also has a low content of water vapor, increasing its concentration as it mixes with the air inside the room.

Table 3. Simulation case configuration

Variables	Values
Supply flow (m^3/h)	3000
Exhaust flow (m^3/h)	3000
Drying flow (m^3/h)	3000
Initial room temperature (K)	294
Swap temperature (K)	294
Dried air temperature (K)	299
Supply temperature (K)	292
Wall temperature (K)	292
Initial room relative humidity (%)	85
Swap relative humidity (%)	82
Dried air relative humidity (%)	60
Supply relative humidity (%)	55

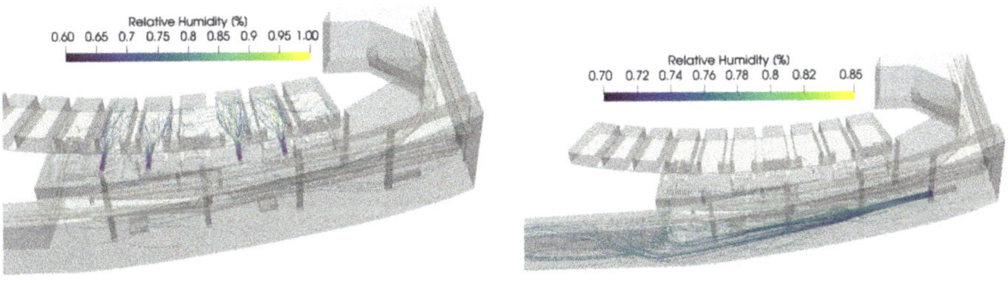

(a) (b)

Figure 7. Illustrative results from CFD simulation: (**a**) streamlines colored with relative humidity starting from the drying discharge grilles; (**b**) streamlines colored with relative humidity starting from the supply discharge grille.

Although it is not represented, from Table 3, it can be seen that the flow rate of pumped air for the supply and exhaust circuits is the same. This net balance between the exhaust and supply circuit generates a low net flux between the room and the gallery, reducing the exchange of humidity to its minimum. This configuration can be used to isolate the technical room as much as possible from the gallery.

Finally, to better visualize the different distributions across the room, 2D slices at different heights plotting the velocity, temperature, and relative humidity are shown and analyzed in Appendix A.1 for the presented case.

3.2. Architecture Optimization

As explained in Section 2.5.2, the model's architecture was optimized using Algorithm 1 with a set of pre-selected sensor locations. The final architecture obtained is depicted in Figure 8, comprising three dense layers followed by dropout layers.The configuration regarding the training hyperparameters is detailed in Table 4. The Huber loss [44] was used as loss function to train the model. The reason for this choice is that it is used for weather forecasting [45], and according to Borah et al. [46], it provides a better generalization with lesser sensitivity towards noise and outliers, which matches the requirements for the present work.

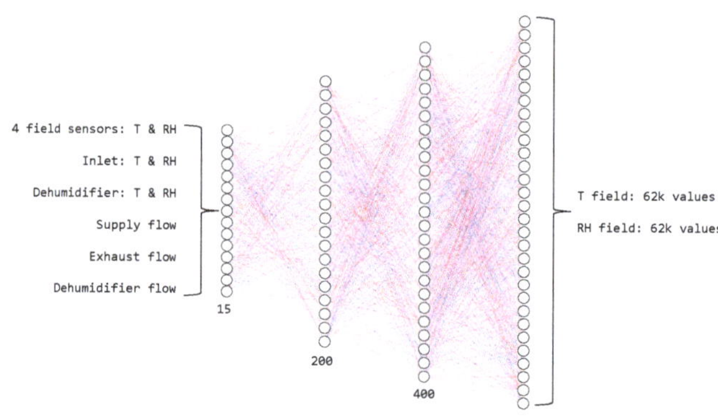

Figure 8. Final NN architecture used as the surrogate model, including summary of inputs to the model and number of outputs extracted.

Table 4. Hyperparameter configuration of the optimized surrogate model.

Hyperparameter	Configuration
Optimizer	Adam
Learning Rate	4.69×10^{-4}
Loss Function	Huber
Training Iterations	1000
Number of Batches	4
Dropout Rate	0.456

3.3. Sensor Location Optimization

Regarding the second stage of the optimizing process, sensor locations where selected to optimize the input information to the surrogate model regarding the state of the room.

This allowed us to fix the degrees of freedom regarding this aspect and maximize the predictive performance of the model. After applying the same algorithm to different sets of sensor locations, the optimal positions obtained are detailed in Table 5. In all the different combinations, three sensors were swapped between locations, while one sensor remained fixed near the swap surface to monitor the air exchange with the gallery. Note that two sensors share the same X and Y coordinates, but are located at different heights: one at the same height as the supply grille, and the other at the same height as the dehumidifier system, meaning that the neural network understands the relation between those circuits and the equilibrium state of the room.

Table 5. Optimal sensor positions.

Sensor	Coordinate X	Coordinate Y	Coordinate Z	Scheme Position
0	61.7	27.0	1.75	I
1	61.7	14.0	1.75	H
2	43.5	4.25	2.80	B
3	43.5	4.25	4.55	B

3.4. Model Accuracy and Validation

After determining the optimal sensor positions, the final version of the surrogate model was trained and its predictive capability was validated against CFD simulation results. To quantitatively evaluate the predictions of the CFD–AI surrogate model, two main metrics were used: Mean Absolute Error and Mean Relative Error, computed for both the relative humidity ([%]) and temperature ([K]) according to the following equations:

$$MAE = \frac{1}{N_C}\frac{1}{N_P}\sum_{i=1}^{C}\sum_{j=1}^{P}|(y_{i,j} - y'_{i,j})| \tag{6}$$

and

$$MRE\ [\%] = \frac{1}{N_C}\frac{1}{N_P}\sum_{i=1}^{C}\sum_{j=1}^{P}\frac{|(y_{i,j} - y'_{i,j})|}{y_{i,j}}100 \tag{7}$$

with C being the number of cases in the validation subset, P the number of predicted points at each case, $y_{i,j}$ the ground truth value of the ith case at the jth point, and $y'_{i,j}$ the predicted value. Using these metrics, the overall MAE obtained for the validating subset yielded an average error of 0.34 Kelvin for the temperature and 2.2 percentage points for the relative humidity. Translating these absolute errors to relative by computing the MRE, this yields a corresponding 0.11% for the temperature and a 2.3% for the relative humidity.

To illustrate the overall spatial accuracy of the CFD–AI surrogate model, the complete predicted data from a case with average MRE was compared with its ground truth. As the predictive model is trained using the extracted data from the point grid shown in Figure 5, the outputs of the model correspond also to the temperature and relative humidity to that sampling grid but unordered. Consequently, in Figure 9, the predictions from the selected case are presented in a scatter plot, without any spatial ordering, for temperature (**a**) and relative humidity (**b**). Different levels of relative error (red) have been defined to help the reader visualize the general error committed by the predictive model for this case. Also, a linear regression was adjusted between predicted and ground truth values (green), also indicating the fitted coefficients for each regression and the coefficient of determination (R^2). It can be observed that the majority of the points predicted present a low deviation from the ground truth, with errors typically around 2.5%, although zones with higher errors can also be found up to 5%. This indicates that the majority of the predicted domain displays an acceptable level of error with respect to the ground truth values. Although the slope coefficients of the linear regressions show room for improvement, they indicate that the model is able to identify which ventilation circuits are active and what is the result of the specific weather conditions of that day, providing a good tendency of the approximate evolution that the system is going to have.

In addition, by knowing the coordinates of each point comprising the point grid, the predictions can be re-arranged into their original 3D location. In Figure 10, the predicted relative humidity for the same case is shown at four different slices as an example of how the data generated by the CFD–AI surrogate model can be relocated spatially. Here, clips of the original domain can also be performed to isolate specific areas of interest. The capability of recovering the spatial information is key for the future implementation of this predictive model in a digital twin core. This will allow us to define complex control rules and evaluation metrics for the state of the room. As an example, in Figure 11, two specific filters have been applied to the predicted values, showing all the points that have a temperature higher than 20 °C (**a**) and those that also have a relative humidity higher than 88 [%] (**b**). The implementation of such kinds of conditional thresholds will allow the investigation of critical areas inside the domain and have a more specific control of key regions of the room.

To better illustrate the forecasting capabilities of the predictive model, inferences carried out by the predictive model for the example case have been re-arranged and displayed in Appendix A.2. The predictions are displayed in horizontal slices at 2.8 m and 4.55 m, selected as they contain the greatest variations in temperature and humidity. Predictions are accompanied by ground truth and relative error slices. A visual inspection of them reveals a strong agreement achieved by the predictive model.

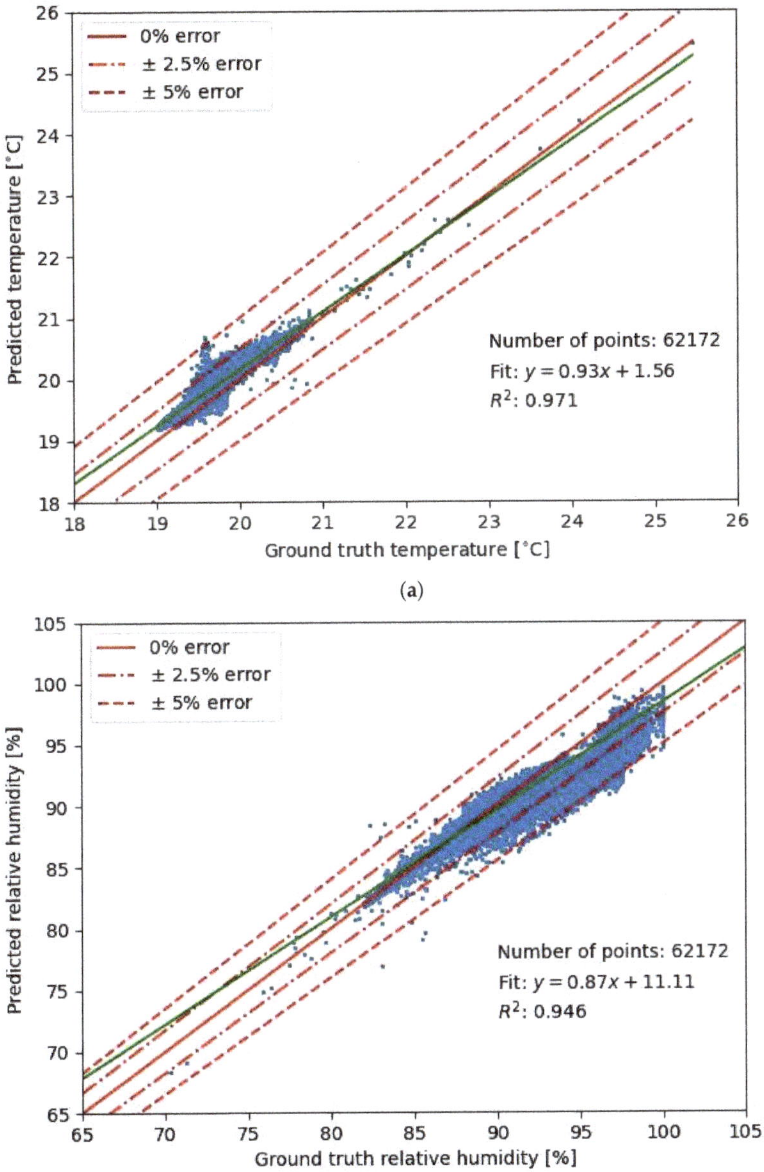

(a)

(b)

Figure 9. Ground truth vs predicted values for (**a**) temperature and (**b**) relative humidity at the selected case.

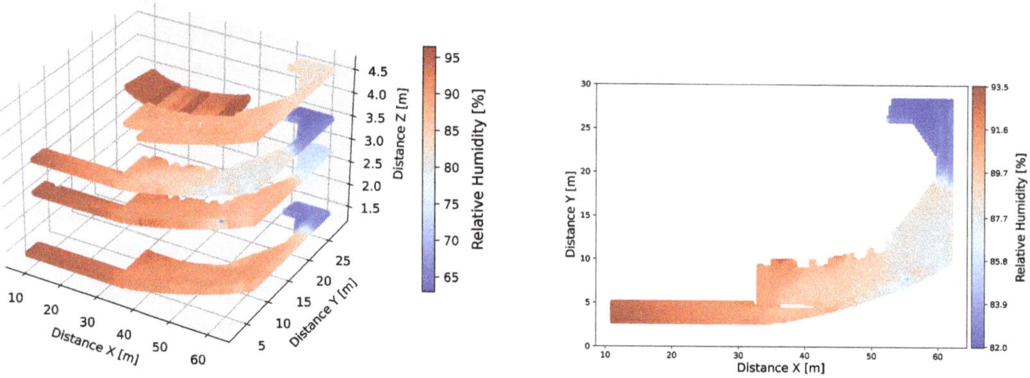

(**a**) (**b**)

Figure 10. Predicted relative humidity for 4 different slices located in the original point grid (**a**) and detail plane at Z = 2.8 m (**b**).

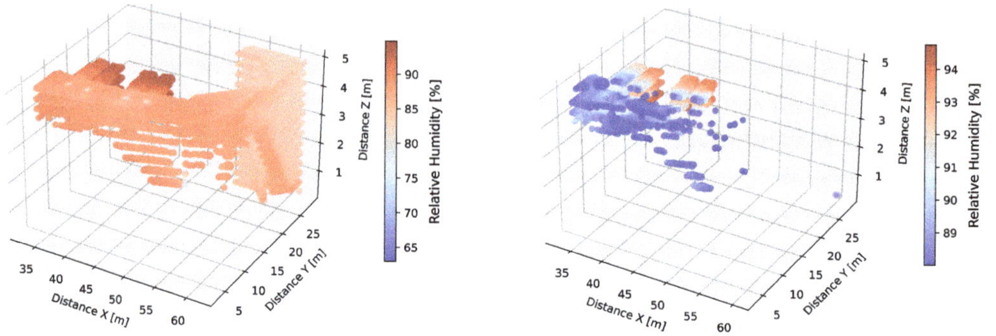

(a) (b)

Figure 11. Specific points in the predicted data that have a temperature higher than 20 °C (**a**) and those that also have a relative humidity higher than 88[%] (**b**).

4. Discussion

CFD simulations allowed us to gain insight on the dynamics of the system and how each ventilation circuit affected it. As observed, the net flow rate between supply and exhaust circuits allows us to control the exchanged flux between the technical room and the gallery: suctioning or pumping air from or to the gallery, or isolating the room. It is thus important to consider for the ventilation strategy the amount of air from the gallery that might enter the room and check if its temperature and relative humidity are beneficial or not.

Complementing these circuits, the drying circuit can effectively control and reduce high humidity in critical regions as the interior chamber. This circuit can act as an auxiliary tool to help control excessive humidity in a localized zone without the need of using high flow rates to have several renovations of air per hour.

The implemented evaporation/condensation model yielded coherent results, similar to [10–13], showing an accumulation of relative humidity inside the chamber due to the presence of the water surface. It also allows for the identification of the walls that present more condensation of water vapor, which in the future can be used by the control algorithm to evaluate the humidity content near these zones and act to prevent condensation.

It is important for the control algorithm to forecast the effect of the supply circuit in the system as a function of the amount of injected flow and its psychrometric conditions. When the external air has a low amount of water vapor, it will be beneficial to pump air in (e.g. in summer), whereas if the content of vapor is high (winter), the supply circuit will add extra humidity inside the room, increasing drying requirements and power consumption.

The explored methodology intends to use a surrogate model based on CFD simulations to provide a steady-state 3D distribution of temperature and relative humidity of the room. This ML tool can serve to implement a predictive control based on a digital twin.

Based on the insight gained with the CFD simulations, the set of variables that determine the configuration of the room climate were selected: initial conditions, supply and drying characteristics, and actuator flow rate. A total of 15 variables were chosen: 8 for the determination of initial conditions, 4 for the supply and drying, and 3 for the actuators.

To develop the surrogate model, an analysis of real psychrometric data was acquired over one year. This analysis served to develop a set of CFD simulations comprising conditions that are representative of the true behavior of the room during the year. The CFD dataset encompassed maximum, minimum, and medium conditions for each season so that the surrogate model could effectively handle realistic environmental variations. By incorporating the effects of water evaporation on the water surface and water condensation on walls and equipment, the CFD simulations were able to reproduce the expected behavior of the room and the impact of the different ventilation circuits on the psychrometric equilibrium state.

A total of 142 cases were generated from the defined combinations described, which comprised the dataset used to train the surrogate model. The chosen architecture—an MLP—was enhanced with dense layers and dropout layers to mitigate overfitting. The architecture was optimized in two stages. The first stage served to tune the training hyperparameters and model size using the Optuna framework, resulting in an efficient configuration that performs predictions in milliseconds, as opposed to the hours required by the CFD simulations.

The second stage aimed at optimizing the locations of the sensors determining the initial conditions. This secondary optimization improved the accuracy of the surrogate model, and also identified locations in close relation with the ventilation circuits. The selected points prove that the CFD–AI surrogate model is effectively learning the physical interaction between the ventilation circuits and the psychrometric state of the room.

The resulting surrogate model was capable of delivering predictions with average errors of 0.34 degrees Kelvin for the temperature and 2.2 percentage points for the relative humidity. The capability of recovering the spatial location of the inferred points helped in comparing the model's predictions with the CFD results. Upon these comparisons, the predicted model showed its ability to reproduce the general trends of the system, detecting the effect of the ventilation grilles and showing overall low relative errors. Thus, the spatial distribution of the temperature and relative humidity at general equilibrium state were satisfactorily predicted. Finally, two examples of possible filtering options for the predicted data were shown, demonstrating that the generated fields can be used for a more accurate spatial analysis of the system, allowing for more refined control strategies in the future. As the control strategy for a ventilation system will take into account different zones of the domain, the presence of regions with higher error will be dampened after averaging around several points. Thus, the obtained average errors can be considered sufficiently small to help on the control decisions by presenting the approximated evolution of the room.

The architecture used to create this surrogate model has demonstrated to be capable of learning to predict spatial distributions with high amounts of data from a reduced number of inputs. This demonstrates its versatility, in line with recent works conducted [39–41]. The presented results demonstrate the potential of surrogate models in flow prediction, allowing its future implementation in ventilation control to ensure optimal performance.

5. Conclusions

This paper demonstrated the feasibility of a new methodology to create a predictive model envisioned to be implemented on a digital twin for an unsupervised ventilation control. This predictive model is based on the developed surrogate model that combines data from CFD simulations and sensors. To demonstrate the methodology, CFD simulations of an underground technical facility have been performed, including the modeling of the relative humidity by implementing the evaporation and condensation of water vapor.

The training dataset was created to emulate the behavior of the actual system against different weather conditions and ventilation control points. An ML model was trained with this dataset to obtain a CFD–AI surrogate model capable of predicting the temperature and relative humidity distribution with low average errors—0.34 degrees Kelvin for the temperature and 2.2 percentage points for the relative humidity—by implicitly learning from the physical phenomena involved in the system. Each prediction was generated within approximately 131 ms, which is almost five orders of magnitude faster than the conventional CFD simulations used to train the model.

This work shows that it is possible to train such a predictive model to obtain inferences with high amounts of points from low sample data provided by sensors. This model takes the advantage of learning from smooth fields highly dependent on the ventilation circuits. Thus, the real dimensionality of the problem is reduced, as the predictive learns the general averaged spatial distribution and then applies specific changes accordingly with the inputs provided. More sensors can be included in the future to provide a better description of the system, generating more accurate simulations and supplying the predictive model with more information.

In the context of large ventilated rooms, this methodology opens the possibility for the implementation of such predictive models in control systems. Thus, a more detailed predictive control can be developed by having much more spatial information than with a reduced number of sensors.

Future work will focus on the real-time control of the room over short durations to validate and analyze the behavior of the digital twin. This will involve real-world implementation of the surrogate model to control the room's ventilation system and collecting empirical data to compare against the model's predictions. Based on these observations, the dataset will be expanded and refined to enhance the accuracy and reliability of the surrogate model and the implemented control.

This work will also be extended to more complex systems, moving beyond a single room to interconnected rooms or spaces with more complex geometries and larger domains. By scaling the approach, we aim to explore the digital twin's capabilities in managing ventilation in more sophisticated and dynamic environments.

Author Contributions: Conceptualization, R.M.-C. and J.L.-G.; sensor installation, F.M.; data management, J.M.; CFD simulations, J.L.-G. and A.G.-B.; software, J.L.-G., A.G.-B. and G.M.-A.; validation, R.M.-C., F.M. and J.L.-G.; investigation, J.L.-G. and A.G.-B.; resources, S.C. and E.B.; data curation, J.L.-G. and J.M.; writing—original draft preparation, J.L.-G. and A.G.-B.; writing—review and editing, J.L.-G. and G.M.-A.; visualization, J.L.-G. and G.M.-A.; supervision, R.M.-C., S.C. and E.B.; project administration, S.C. and E.B.; funding acquisition, S.C. and E.B. All authors have read and agreed to the published version of the manuscript.

Funding: This research was funded by the Spanish Agencia Estatal de Investigación, project PID2021-128405OB-I00. Jaume Luis-Gómez is supported by FPU21/03740 doctoral grant from the Spanish Ministry of Science, Innovation and Universities.

Data Availability Statement: The authors provide a repository via Zenodo (https://zenodo.org/records/15025264, 14 March 2025, DOI: 10.5281/zenodo.15024655) where readers can access the following: a list of cases along with the configuration of each boundary condition used to generate the dataset, the processed input and output data used to train the model, and a script to load both the model and data for prediction purposes. The data was extracted from the CFD simulations using the grid points described in Section 2.4, with all geometric details omitted. Due to the company's confidentiality policy, specific geometric information about the studied room has been removed. Therefore, the extracted dataset contains no geometric information.

Acknowledgments: Authors would like to thank the Spanish Agencia Estatal de Investigación for the project PID2021-128405OB-I00. J. Luis-Gómez would like to thank the Spanish Ministry of Science, Innovation and Universities for the supporting funds through the doctoral grant FPU21/03740.

Conflicts of Interest: Authors Francisco Martínez, Javier Mascarós and Elisa Borrás were employed by the company Avanqua Oceanogràfic, S.L. The remaining authors declare that the research was conducted in the absence of any commercial or financial relationships that could be construed as a potential conflict of interest.

Appendix A

Appendix A.1. Simulations Results

As explained in Section 3.1, a case from the dataset is presented to provide the reader with a general view of the behavior of the different circuits in the room. To illustrate the distribution of the studied fields across the room in three dimensions, we selected four horizontal slices at different heights corresponding to the case shown in Section 3.1: near the height of a person working at ground level (1.25 m), at the height where the supply and drying grills are (2.8 and 4.6 m), and an intermediate height (3.75 m). These slices offer a comprehensive view of how the various fields behave across the vertical space of the room.

In Figure A1, the effect of the supply circuit can be noticed at heights 1.25, 2.8, and 3.75, activating the flow in the main room, especially at the height where people may be working (1.25 m). Flow near the swap surface can be also appreciated, indicating an exchange with the gallery, but with a total net flow of zero. In Figure A1d), the effect of the drying grilles can be seen, directing the air into the chamber to lower the total moisture content inside and avoid water condensation.

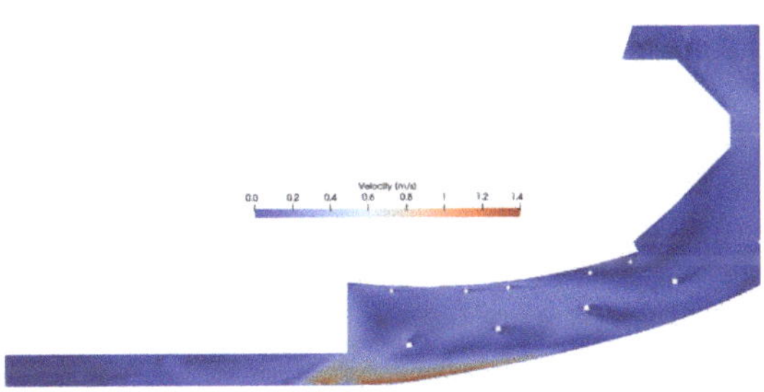

(a)

Figure A1. *Cont.*

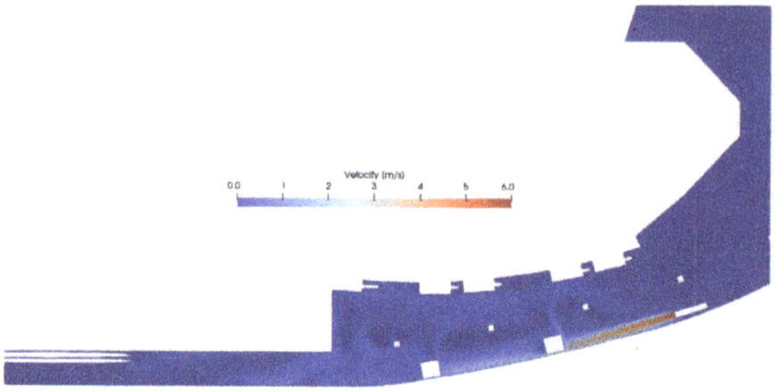

(b)

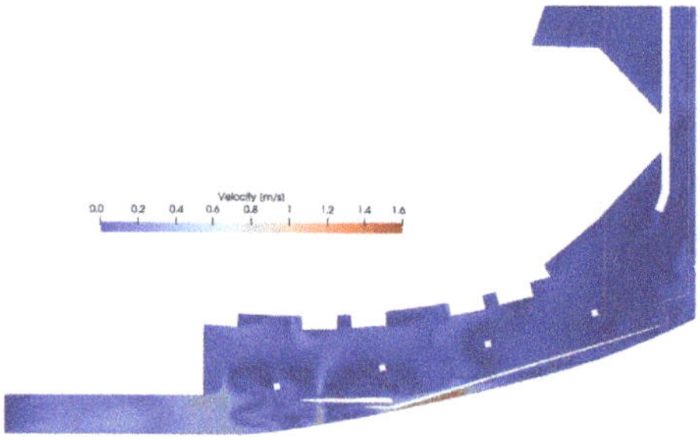

(c)

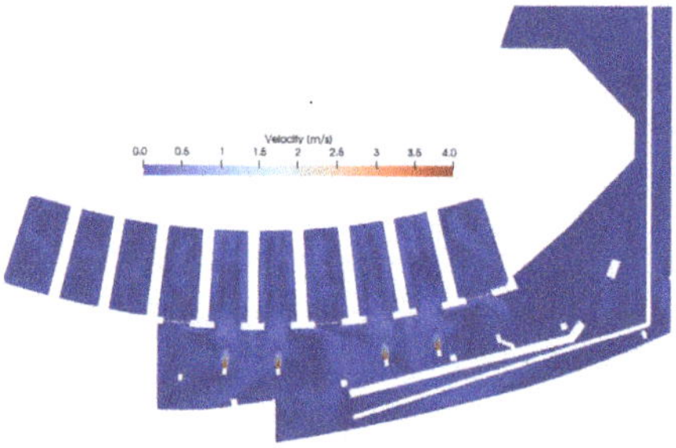

(d)

Figure A1. Velocity distribution at different heights: (**a**) at 1.25 m, (**b**) at 2.8 m, (**c**) at 3.75 m, and (**d**) at 4.6 m.

Similarly, the temperature distributions are presented in Figure A2. It is clear that there is an exchange of heat through the swap surface, either by diffusion and convection, as a gradient between the room temperature and the gallery can be observed at the first three slices. Acting on the opposite way, the supply circuit is injecting cold air, helping to maintain the room's temperature in a moderate value. Although the overall temperature difference is low, this effect can be extrapolated to more extreme conditions where the difference between the gallery and the exterior is greater, e.g., in summer or winter. At the highest slice, the dried air, which has also been warmed, is mixing and rising the temperature inside the chamber and retarding the appearance of water condensation, as the water vapor will need to lose more energy before condensating.

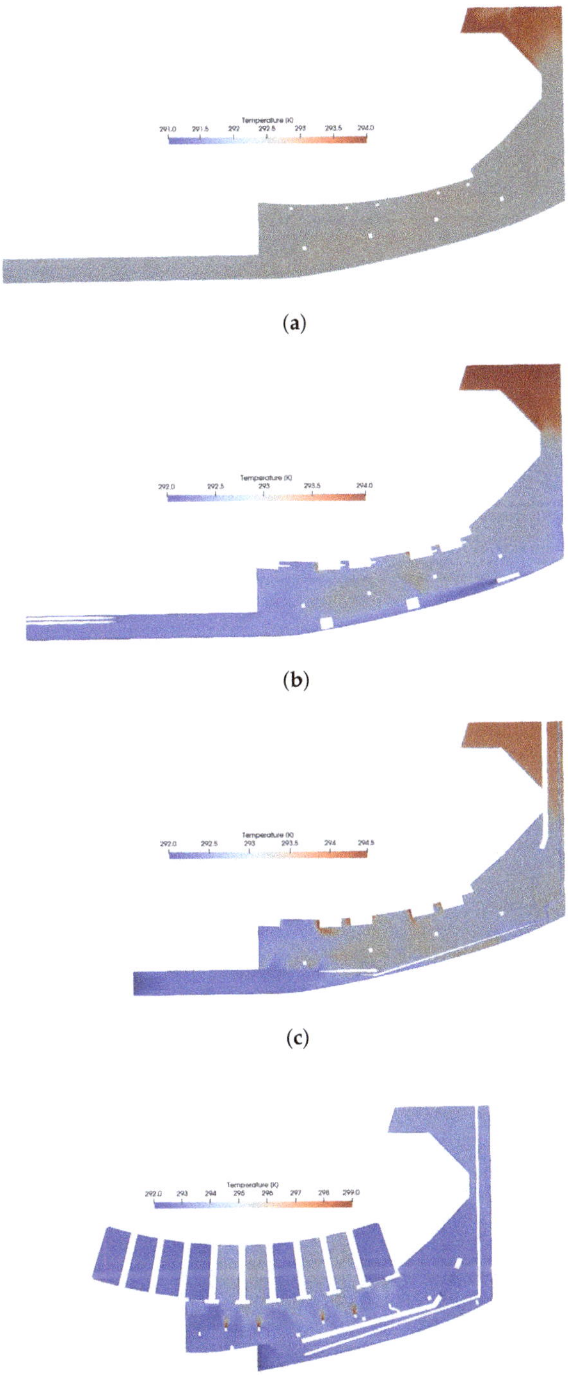

Figure A2. Temperature distribution at different heights: (**a**) at 1.25 m, (**b**) at 2.8 m, (**c**) at 3.75 m, and (**d**) at 4.6 m.

Finally, the humidity fields are shown in Figure A3. The previous analysis is also applicable to the relative humidity. In this case, the air entering through the swap surface has a higher vapor content than the air in the room, and a gradient can be observed, caused also by the injection of drier air through the supply circuit. The supplied air prevents the room's humidity from rising to extreme levels, although an appreciable part of the room has a relative humidity higher than 85%. Similarly to the effect on the temperature, the dehumidifier circuit is injecting dry air in to the chamber, lowering its moisture content and preventing condensation.

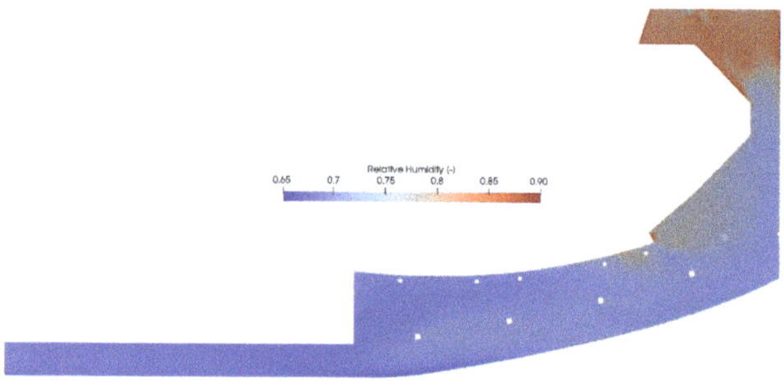

(a)

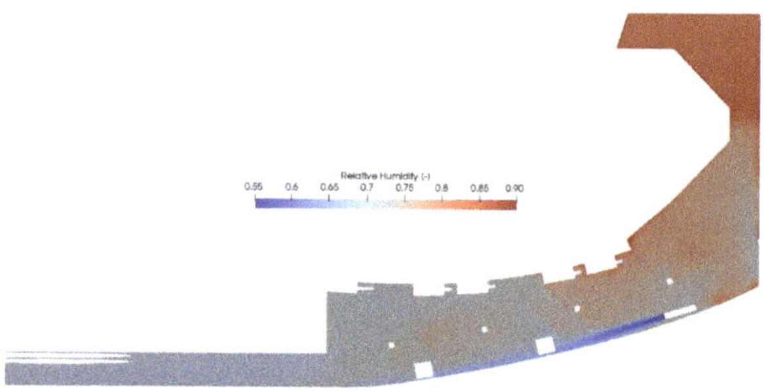

(b)

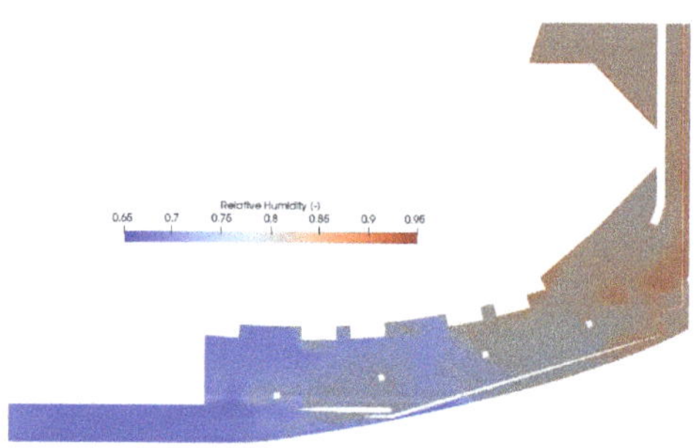

(c)

Figure A3. *Cont.*

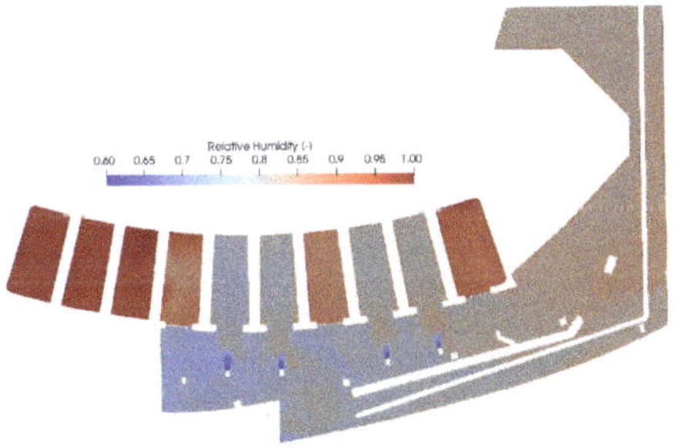

(**d**)

Figure A3. Humidity distribution at different heights: (**a**) at 1.25 m, (**b**) at 2.8 m, (**c**) at 3.75 m, and (**d**) at 4.6 m.

Appendix A.2. Surrogate Model Results

Similar to the simulation results section, this section presents the temperature and relative humidity fields predicted by the surrogate model alongside the ground truth (CFD results) at different heights. Additionally, the relative error between the predicted and CFD values is depicted to evaluate the accuracy of the surrogate model.

Figures A4 and A5 show the temperature fields at heights of 2.8 m and 4.6 m, respectively. Similarly, Figures A6 and A7 present the relative humidity fields for the same heights.

Regarding both temperature and humidity—at 2.8 m—higher discrepancies can be found in the vicinities of the supply jet, which may indicate the need for more cases in the training set involving the supply circuit. As per the second height—4.6 m—the larger relative errors appear inside the cavity, where there is more heterogeneity in the studied fields. Relative errors have been computed to provide a more visual comparison between ground truth and predicted variables. Overall, the majority of the room is correctly predicted, with low deviations and maximum relative errors inside an acceptable limit. The model presents a sufficient accuracy considering that the control strategy is going to be based on averaged zones of the room, so localized zones of error may yield a low impact on the control decision. Similar results can be found in the rest of the test cases.

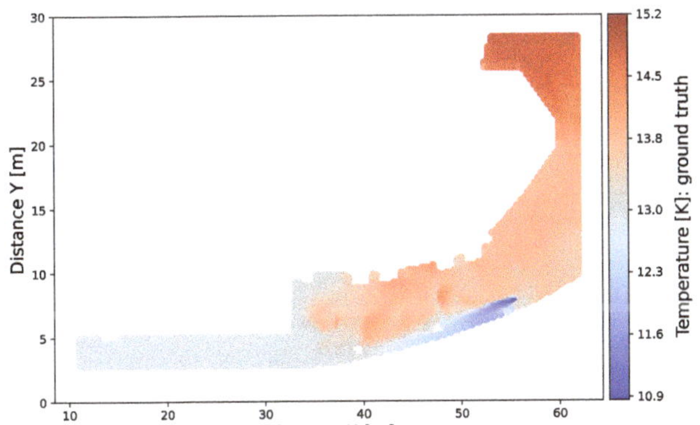

(**a**)

Figure A4. *Cont.*

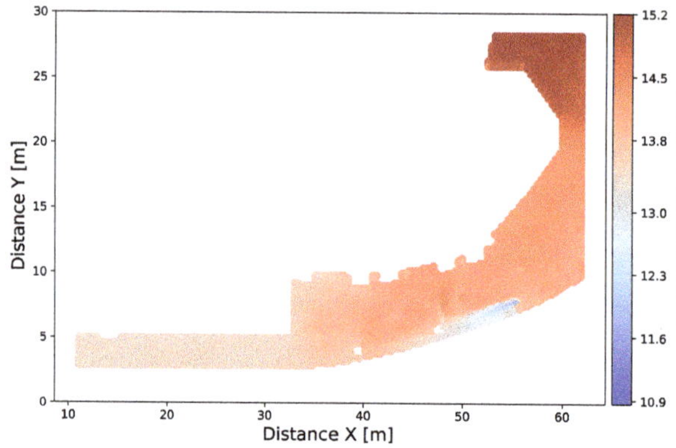

(**b**)

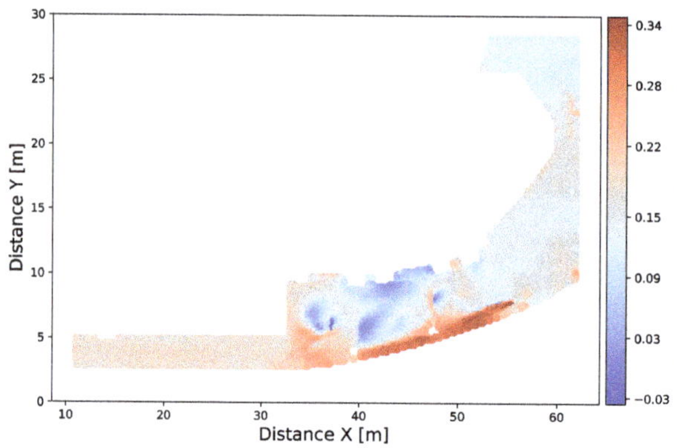

(**c**)

Figure A4. Temperature field at 2.8 m for (**a**) ground truth, (**b**) predicted values, and (**c**) relative error.

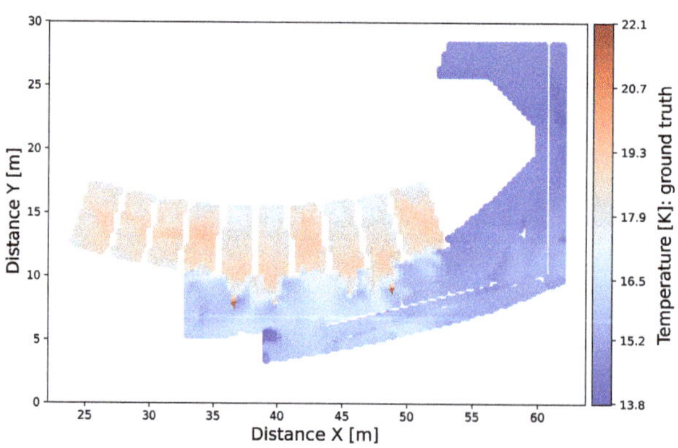

(**a**)

Figure A5. *Cont.*

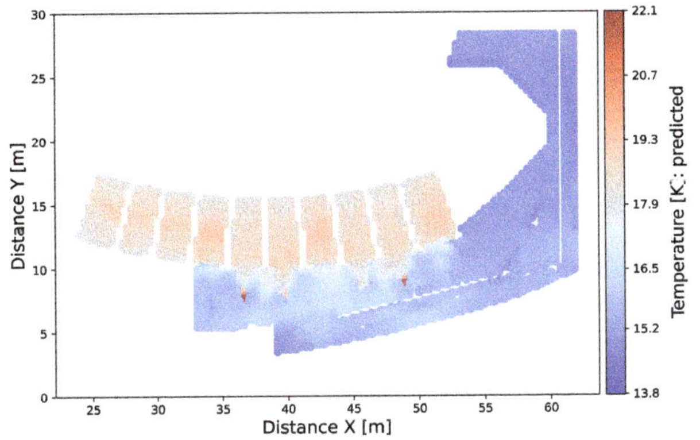

(b)

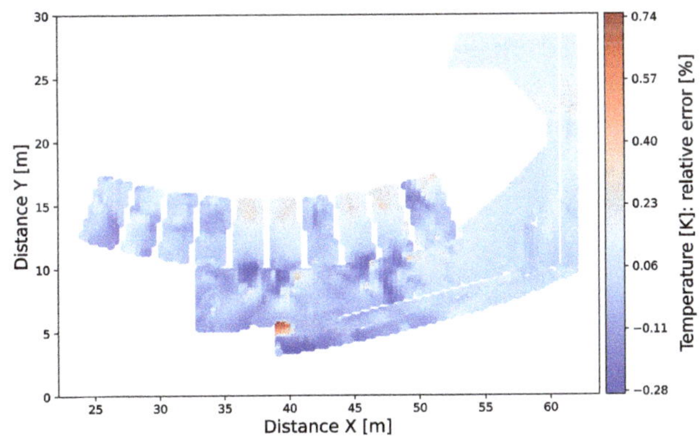

(c)

Figure A5. Temperature field at 4.6 m for (**a**) ground truth, (**b**) predicted values, and (**c**) relative error.

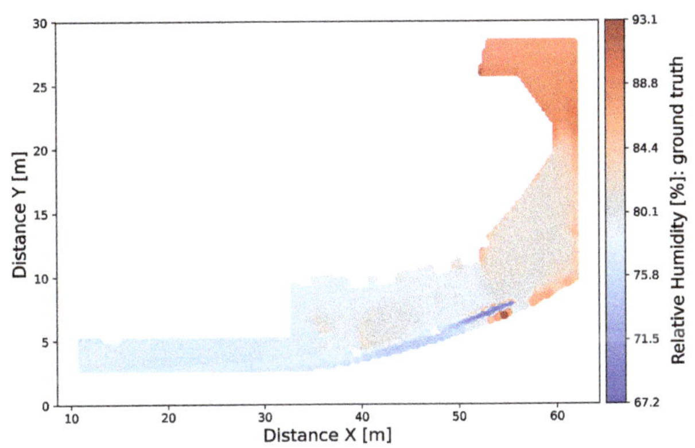

(a)

Figure A6. *Cont.*

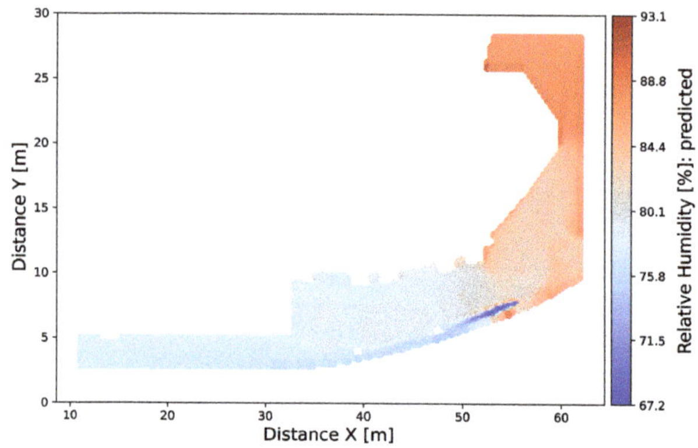

(**b**)

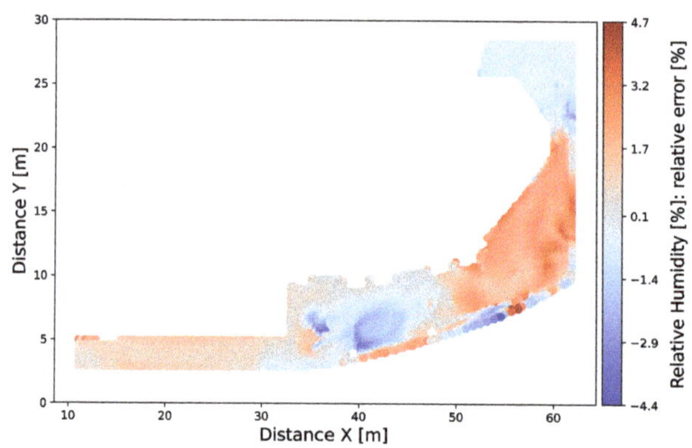

(**c**)

Figure A6. Relative humidity field at 2.8 m for (**a**) ground truth, (**b**) predicted values, and (**c**) relative error.

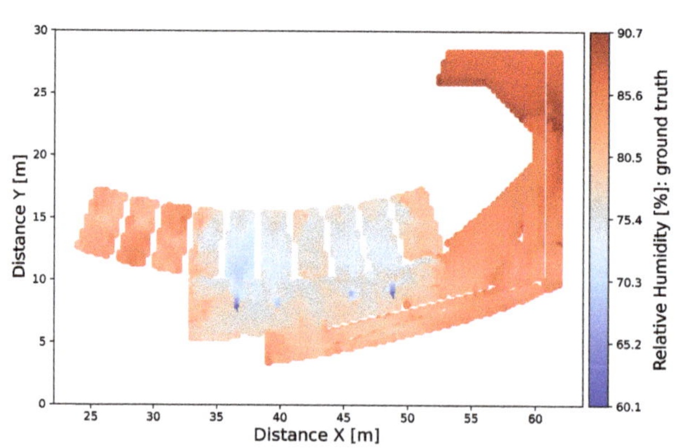

(**a**)

Figure A7. *Cont.*

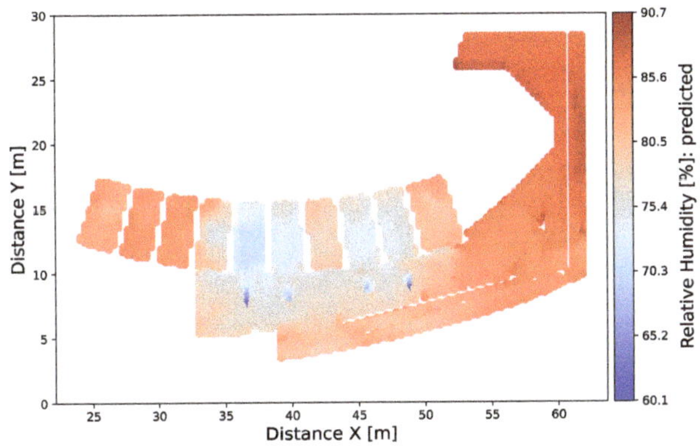

(**b**)

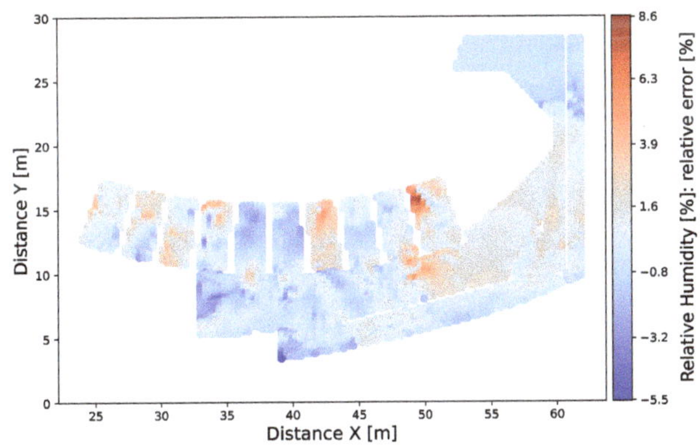

(**c**)

Figure A7. Relative humidity field at 4.6 m for (**a**) ground truth, (**b**) predicted values, and (**c**) relative error.

References

1. Sharma, A.; Kosasih, E.; Zhang, J.; Brintrup, A.; Calinescu, A. Digital Twins: State of the art theory and practice, challenges, and open research questions. *J. Ind. Inf. Integr.* **2022**, *30*, 100383. [CrossRef]
2. Pylianidis, C.; Osinga, S.; Athanasiadis, I.N. Introducing digital twins to agriculture. *Comput. Electron. Agric.* **2021**, *184*, 105942. [CrossRef]
3. Jiang, Y.; Yin, S.; Li, K.; Luo, H.; Kaynak, O. Industrial applications of digital twins. *Phil. Trans. R. Soc. A* **2021**, *379*, 20200360. [CrossRef] [PubMed]
4. Tao, F.; Qi, Q.; Wang, L.; Nee, A.Y. Digital Twins and Cyber–Physical Systems toward Smart Manufacturing and Industry 4.0: Correlation and Comparison. *Engineering* **2019**, *5*, 653–661. [CrossRef]
5. Charitonidou, M. Urban scale digital twins in data-driven society: Challenging digital universalism in urban planning decision-making. *Int. J. Archit. Comput.* **2022**, *20*, 238–253. [CrossRef]
6. Bárkányi, A.; Chován, T.; Németh, S.; Abonyi, J. Modelling for digital twins—Potential role of surrogate models. *Processes* **2021**, *9*, 476. [CrossRef]
7. Nielsen, P.V. Fifty years of CFD for room air distribution. *Build. Environ.* **2015**, *91*, 78–90. [CrossRef]
8. Chen, Q. Ventilation performance prediction for buildings: A method overview and recent applications. *Build. Environ.* **2009**, *44*, 848–858. [CrossRef]

9. Méndez, C.; San José, J.; Villafruela, J.; Castro, F. Optimization of a hospital room by means of CFD for more efficient ventilation. *Energy Build.* **2008**, *40*, 849–854. [CrossRef]
10. Yilmaz, D. Computational fluid dynamics modeling of surface condensation. *J. Braz. Soc. Mech. Sci. Eng.* **2020**, *42*, 351. [CrossRef]
11. Liu, J.; Aizawa, H.; Yoshino, H. CFD prediction of surface condensation on walls and its experimental validation. *Build. Environ.* **2004**, *39*, 905–911. [CrossRef]
12. Park, S.J.; Lee, I.B.; Lee, S.Y.; Kim, J.G.; Cho, J.H.; Decano-Valentin, C.; Choi, Y.B.; Lee, M.H.; Jeong, H.H.; Yeo, U.H. Air conditioning system design to reduce condensation in an underground utility tunnel using CFD. *IEEE Access* **2022**, *10*, 116384–116401. [CrossRef]
13. Rojas, G.; Grove-Smith, J. Improving ventilation efficiency for a highly energy efficient indoor swimming pool using CFD simulations. *Fluids* **2018**, *3*, 92. [CrossRef]
14. Limane, A.; Fellouah, H.; Galanis, N. Three-dimensional OpenFOAM simulation to evaluate the thermal comfort of occupants, indoor air quality and heat losses inside an indoor swimming pool. *Energy Build.* **2018**, *167*, 49–68. [CrossRef]
15. Ciuman, P.; Lipska, B. Experimental validation of the numerical model of air, heat and moisture flow in an indoor swimming pool. *Build. Environ.* **2018**, *145*, 1–13. [CrossRef]
16. Li, Z.; Heiselberg, P.K. *CFD Simulations for Water Evaporation and Airflow Movement in Swimming Baths*; Aalborg Universitet: Aalborg, Denmark, 2005.
17. Gallero González, F.J.; Maestre Rodríguez, I.; Foncubierta Blázquez, J.L.; Mena Baladés, J.D. Enhanced CFD-based approach to calculate the evaporation rate in swimming pools. *Sci. Technol. Built Environ.* **2020**, *27*, 524–532. [CrossRef]
18. Foncubierta Blázquez, J.L.; Maestre, I.R.; Gallero González, F.J.; Gómez Álvarez, P. Experimental test for the estimation of the evaporation rate in indoor swimming pools: Validation of a new CFD-based simulation methodology. *Build. Environ.* **2018**, *138*, 293–299. [CrossRef]
19. Elazm, M.A.; Shahata, A. Numerical and field study of the effect of air velocity and evaporation rate on indoor air quality in enclosed swimming pools. *Int. Rev. Mech. Eng* **2015**, *9*, 97–103. [CrossRef]
20. ASHRAE. *Applications Handbook*; American Society of Heating, Refrigerating, and Air-Conditioning Engineers: Peachtree Corners, GA, USA, 1978.
21. Sun, Z.; Wang, S. A CFD-based test method for control of indoor environment and space ventilation. *Build. Environ.* **2010**, *45*, 1441–1447. [CrossRef]
22. Quang, T.V.; Doan, D.T.; Phuong, N.L.; Yun, G.Y. Data-driven prediction of indoor airflow distribution in naturally ventilated residential buildings using combined CFD simulation and machine learning (ML) approach. *J. Build. Phys.* **2024**, *47*, 439–471. [CrossRef]
23. Vinuesa, R.; Brunton, S.L. Enhancing computational fluid dynamics with machine learning. *Nat. Comput. Sci.* **2022**, *2*, 358–366. [CrossRef] [PubMed]
24. Raissi, M.; Perdikaris, P.; Karniadakis, G.E. Physics-informed neural networks: A deep learning framework for solving forward and inverse problems involving nonlinear partial differential equations. *J. Comput. Phys.* **2019**, *378*, 686–707. [CrossRef]
25. Cuomo, S.; Cola, V.S.D.; Giampaolo, F.; Rozza, G.; Raissi, M.; Piccialli, F. Scientific Machine Learning Through Physics–Informed Neural Networks: Where we are and What's Next. *J. Sci. Comput.* **2022**, *92*, 88. [CrossRef]
26. Eivazi, H.; Tahani, M.; Schlatter, P.; Vinuesa, R. Physics-informed neural networks for solving Reynolds-averaged Navier-Stokes equations. *Phys. Fluids* **2022**, *34*, 075117. [CrossRef]
27. Huang, Z.; Shen, Y.; Li, J.; Fey, M.; Brecher, C. A Survey on AI-Driven Digital Twins in Industry 4.0: Smart Manufacturing and Advanced Robotics. *Sensors* **2021**, *21*, 6340. [CrossRef]
28. Kreuzer, T.; Papapetrou, P.; Zdravkovic, J. Artificial intelligence in digital twins—A systematic literature review. *Data Knowl. Eng.* **2024**, *151*, 102304. [CrossRef]
29. Molinaro, R.; Singh, J.S.; Catsoulis, S.; Narayanan, C.; Lakehal, D. Embedding data analytics and CFD into the digital twin concept. *Comput. Fluids* **2021**, *214*, 104759. [CrossRef]
30. Cuelles, A.; Güemes, A.; Ianiro, A.; Flores, O.; Vinuesa, R.; Discetti, S. Three-dimensional generative adversarial networks for turbulent flow estimation from wall measurements. *J. Fluid Mech.* **2024**, *991*, A1. [CrossRef]
31. Iserte, S.; González-Barberá, A.; Barreda, P.; Rojek, K. A study on the performance of distributed training of data-driven CFD simulations. *Int. J. High Perform. Comput. Appl.* **2023**, *37*, 503–515. [CrossRef]
32. Cremades, A.; Freibergs, R.; Hoyas, S.; Ianiro, A.; Discetti, S.; Vinuesa, R. Assessment of non-intrusive sensing in wall-bounded turbulence through explainable deep learning. *arXiv* **2025**, arXiv:2502.07610.
33. Iserte, S.; Macías, A.; Martínez-Cuenca, R.; Chiva, S.; Paredes, R.; Quintana-Ortí, E.S. Accelerating urban scale simulations leveraging local spatial 3D structure. *J. Comput. Sci.* **2022**, *62*, 101741. [CrossRef]

34. Rojek, K.; Wyrzykowski, R.; Gepner, P. AI-Accelerated CFD Simulation Based on OpenFOAM and CPU/GPU Computing. In *Computational Science–ICCS 2021: 21st International Conference*; Paszynski, M., Kranzlmüller, D., Krzhizhanovskaya, V.V., Dongarra, J.J., Sloot, P.M.A., Eds.; Springer: Berlin/Heidelberg, Germany, 2021; pp. 373–385.
35. Ferziger, J.H.; Perić, M.; Street, R.L. *Computational Methods for Fluid Dynamics*; Springer: Berlin/Heidelberg, Germany, 2019.
36. Carrier, W.H. The temperature of evaporation. *ASHVE Trans.* **1918**, *24*, 25–50.
37. Smith, C.; Jones, R.; Löf, G. Energy requirements and potential savings for heated indoor swimming pools. *ASHRAE Trans.* **1994**, *100*, 864–874.
38. Murtagh, F. Multilayer perceptrons for classification and regression. *Neurocomputing* **1991**, *2*, 183–197. [CrossRef]
39. Mansour, Y.; Lin, K.; Heckel, R. Image-to-image MLP-mixer for image reconstruction. *arXiv* **2022**, arXiv:2202.02018.
40. AlBasiouny, E.R.; Heliel, A.F.A.; Abdelmunim, H.E.; Abbas, H.M. Multilayer Perceptron Generative Model via Adversarial Learning for Robust Visual Tracking. *IEEE Access* **2022**, *10*, 121230–121248. [CrossRef]
41. Liu, B.; Wei, Y.; Zhang, Y.; Yang, Q. Deep Neural Networks for High Dimension, Low Sample Size Data. *Proc. IJCAI* **2017**, *2017*, 2287–2293.
42. Radhakrishnan, S.; Calafell, J.; Miró, A.; Font, B.; Lehmkuhl, O. Data-driven wall modeling for LES involving non-equilibrium boundary layer effects. *Int. J. Numer. Methods Heat Fluid Flow* **2024**, *34*, 3166–3202. [CrossRef]
43. Akiba, T.; Sano, S.; Yanase, T.; Ohta, T.; Koyama, M. Optuna: A Next-generation Hyperparameter Optimization Framework. In Proceedings of the 25th ACM SIGKDD International Conference on Knowledge Discovery and Data Mining, Anchorage, AK, USA, 4–8 August 2019.
44. Huber, P.J. Robust estimation of a location parameter. In *Breakthroughs in Statistics: Methodology and Distribution*; Springer: Berlin/Heidelberg, Germany, 1992; pp. 492–518.
45. Taggart, R.J. Point forecasting and forecast evaluation with generalized Huber loss. *Electron. J. Stat.* **2022**, *16*, 201–231. [CrossRef]
46. Borah, P.; Gupta, D. Functional iterative approaches for solving support vector classification problems based on generalized Huber loss. *Neural Comput. Appl.* **2020**, *32*, 9245–9265. [CrossRef]

fluids MDPI

Article

Machine Learning for Dynamic Pressure Coefficient Prediction in Vertical Water Jets

Amin Salemnia [1], Seyedehmaryam Hosseini Boldaji [2], Vida Atashi [3,*] and Manoochehr Fathi-Moghadam [4]

[1] Department of Water Engineering, Shahid Chamran University, Ahvaz 6135743136, Iran; amin.salemnia@gmail.com
[2] Department of Electrical and Computer Engineering, K. N. Toosi University of Technology, Tehran 163171419, Iran; smaryamhoseini@email.kntu.ac.ir
[3] Faculty of Civil Engineering Department, University of North Dakota, Grand Forks, ND 58202, USA
[4] Faculty of Water and Environmental Engineering Department, Shahid Chamran University, Ahvaz 6135743136, Iran; mfathi@scu.ac.ir
* Correspondence: vida.atashi@und.edu; Tel.: +1-701-885-2449

Abstract: Vertical water jets present significant challenges for hydraulic structures due to their potential to cause erosion and structural damage. This study aimed to predict the dimensionless pressure coefficient (C_p) of vertical water jets by examining the relationships between experimental parameters, such as Froude number, slope, and the ratio of waterfall height over the product of the Froude number and diameter, referred to as α, using machine learning models. Two hundred forty controlled experiments were conducted, with pressure data collected. To address the problem's non-linearity, six machine learning models were tested: linear regression, K-nearest neighbors, decision tree, support vector regression, random forest, and XGBoost. The XGBoost model outperformed others, achieving an R-squared of 0.953 and a Root Mean Squared Error (RMSE) of 0.191. Residual analysis validated its better performance, demonstrating that it delivered the most accurate predictions with minimal bias. Feature importance analysis revealed the Froude number was the most significant predictor, followed by slope and diameter. This study emphasizes the importance of the Froude number in predicting jet behavior and shows the efficacy of advanced machine learning models in capturing complex fluid dynamics, providing valuable insights for optimizing engineering applications such as water jet cutting and cooling systems.

Keywords: vertical water jets; machine learning models; pressure coefficient; Froude number

1. Introduction

In hydraulic engineering, overflow spillways, particularly in high-head dams, generate vertical water jets that are critical to the system's function. These water jets, as exemplified in Figure 1, possess substantial energy that, if not properly managed, can cause significant damage to the dam structure and downstream environments. Traditional methods of understanding and mitigating the impact of these jets have primarily relied on physical models, which can be expensive, time-consuming, and limited in their scope [1–4].

In recent years, traditional approaches to understanding and mitigating the impact of vertical water jets, such as physical modeling, numerical methods, and hydraulics software, have been supplemented by emerging technologies in artificial intelligence and machine learning. However, despite advancements in numerical modeling, accurately predicting the properties of vertical water jets and dynamic pressure coefficients remains a challenge due to the complex nature of turbulent flow [5–7]. The application of machine learning methods in this context represents a novel and promising approach to address these challenges.

Figure 1. Photo of free-falling jet in Kariba Dam—Zambia [8].

Estimating dynamic pressure coefficients is vital for understanding jet behavior and optimizing hydraulic structures like plunge pools. Traditional models often fall short of accurately representing complex turbulent flows, especially at high velocities. Data-driven algorithms and machine learning methods offer a promising solution to these limitations.

By applying large datasets, machine learning approaches can improve prediction accuracy and efficiency while uncovering new insights into hydraulic dynamics. The benefits include more precise estimations, rapid analysis, adaptability to new data, and the potential revelation of novel design concepts. Integrating machine learning with traditional engineering can lead to more resilient and efficient hydraulic structures that are better equipped to handle high-velocity flows. Ultimately, this approach expands our capabilities in hydraulic structure management, potentially revolutionizing the field by combining data-driven methods with domain expertise.

Turbulent jets have been extensively studied through both laboratory experiments and numerical simulations. Numerous researchers have investigated the jets under various conditions [9–18], focusing on properties such as dynamic pressure coefficients and break-up length in laboratory settings. In contrast to extensive laboratory experimental studies, relatively few numerical investigations have explored under-pressure flows [19,20]. Taebi (2022) conducted a review of Recent Advances in Deep Learning for Computational Hemodynamics [21]. The main finding of this review is that deep learning (DL) approaches, particularly physics-informed neural networks (PINNs), show significant potential in accelerating and enhancing computational hemodynamics simulations. These models can reduce computational time, improve the prediction accuracy of complex flow patterns, and assist in quantitative flow pattern estimation based on sparse data. The review highlights several applications, such as predicting hemodynamics in aortic aneurysms, coronary artery bypass surgery, and cerebral aneurysms, demonstrating the feasibility and effectiveness of DL in these contexts.

Between numerical studies, relatively few investigations have explored the behavior of free-falling jets under various conditions. Jia et al. (2001) examined the scouring process caused by the dynamic pressure of jets in plunge pools using the CCHE 3D model and highlighted that pressure fluctuations, which are closely related to the velocity field, play a significant role in plunge pools, while Salehi Neyshabouri et al. (2003) conducted a similar study with a two-dimensional (2D) numerical model [22,23]. Neither study investigated the effect of jet air content on scour. Despite these efforts, the impact of jet air content and other variables remains underexplored.

Cui et al. (2012, 2014) performed numerical simulations to predict the mechanisms of sand-bed erosion [24,25]. Within a Lagrangian discrete framework, each particle can theoretically be tracked by considering force equilibrium. However, the computational demands for realistic sand-bed erosion simulations are substantial. Previous simulations used much larger particles to study granular behavior in the erosion process with reasonable accuracy. Due to computational limitations, the Lagrangian discrete approach is typically limited to a small number of particles [26]. Conversely, employing various machine learning methods based on artificial intelligence presents a more feasible alternative for simulating jet behavior. While various studies have explored jet behavior using numerical methods [27–30], there have been no detailed investigations utilizing machine learning techniques to predict dynamic pressure based on plunge pool pressure fluctuations and distributions.

This study marks a significant advancement in hydraulic engineering by applying various machine learning models to predict the dynamic pressure coefficients of vertical water jets. Our innovative approach leverages data from controlled experiments that were conducted at the hydraulic laboratory of Shahid Chamran University of Ahwaz, Iran. By comparing the performance of multiple machine learning algorithms, including linear regression, K-nearest neighbors, decision trees, support vector regression, random forest, and XGBoost, we aim to identify the most effective method for predicting jet behavior.

The primary goal of this research is to overcome the limitations of traditional numerical and physical models in accurately representing complex turbulent flows, especially in high-velocity scenarios. By harnessing the power of machine learning, we seek to improve both the accuracy and efficiency of pressure coefficient predictions. This novel approach not only enhances our predictive capabilities but also has the potential to uncover new insights into hydraulic dynamics. Ultimately, our research aims to revolutionize hydraulic structure management by combining data-driven methods with domain expertise, leading to more resilient and efficient designs for better withstanding the challenges posed by high-velocity water jets.

2. Materials and Methods

In this section, we will outline the methodologies employed in our study, covering the theoretical framework, the experimental facility, data collection procedures, preprocessing techniques, and feature selection criteria. Additionally, we will describe the selection, training, evaluation, and hyperparameter tuning of various machine learning models used to predict dynamic pressures.

2.1. Theory

Water jets, characterized by turbulent flow resulting from a continuous momentum source, exhibit varied states upon impact with a surface or entry into a plunge pool, depending on their Froude numbers and falling heights. These jets may arrive fully broken up, termed as developed jets, presenting a fully turbulent form without a core, or they may retain a solid core, leading to higher impact pressure.

Currently, there are no widely accepted methods for predicting dynamic pressure coefficients or mean and fluctuating pressures on a plunge pool floor. Consequently, designers often have to develop solutions on a case-by-case basis. One major complication in developing a general solution is that plunging jets tend to partially or completely break up upon entry into the pool. The degree of jet break-up significantly influences both the extent and magnitude of pressure. Additionally, the jet tends to entrain substantial amounts of air, which mitigates pressure amplitudes.

Modeling falling jets is challenging due to the effects of surface tension and turbulence on jet break-up and air entrainment characteristics. Furthermore, the pressure fluctuation spectrum within the pool is influenced by the scale of turbulence, making it difficult to simulate these effects accurately in physical models. However, careful selection of model size and interpretation of results can help minimize or account for scale effects. The stability and internal turbulence of a free-falling jet are crucial determinants of its characteristics. Rayleigh (1892) [31] conducted pioneering investigations into jet stability, followed by more recent studies by researchers such as Baron (1949) [32], Chen and Davis (1964) [33], Hoyt and Taylor (1977) [34], McKeogh (1978) [35].

The experimental equation of the break-up length for circular jet, established by Ervine et al., 1997 [36], is as follows:

$$\frac{L_b}{D_0 F_0} = \frac{1.05}{C^{0.82}} \tag{1}$$

where L_b is the break-up length in meters, D_0 is nozzle diameter in meters, F_0 is Froude number at issuance conditions, and C is the turbulence parameter defined as follows:

$$C = 1.14 T_u F_0^2 \tag{2}$$

$$F_0 = \frac{U_0}{\sqrt{g D_0}} \tag{3}$$

where T_u is turbulent intensity. The mean dynamic pressure on the pool floor is described by a pressure coefficient, C_p, defined as follows:

$$Cp = \frac{H_m - y}{\sqrt{V^2{}_j/2g}} \tag{4}$$

$$V_j = \sqrt{U_0^2 + (2gH)} \tag{5}$$

where H_m and y are the mean head and depth at the plunge pool in meters, respectively, and V_j is the velocity of the impingement jet in m/s. It is worth mentioning that all the experiments were conducted in the absence of water depth in the plunge pool, so "y" is equal to 0 in Equation (4). The dynamic pressures were measured at the surface where the water jet directly impacted it. The parameters are shown in Table 1.

Table 1. An outline of the range of jet test conditions.

Fall Height (cm)	Diameter (cm)	Flow (lit/s)	Angle (Degree)	Mean Head Depth (cm)
8, 15, 25, 35, 45	7.9	11, 14, 17, 21	30, 45, 90	0
8, 15, 25, 35, 45	5.1	11, 14, 17, 21	30, 45, 90	0
8, 15, 25, 35, 45	4.3	11, 14, 17, 21	30, 45, 90	0
8, 15, 25, 35, 45	4.1	11, 14, 17, 21	30, 45, 90	0

2.2. *Experimental Facility*

Figure 2 illustrates the experimental configuration employed for circular jet experiments. The jets originated from elongated, smooth nozzles equipped with an upstream tube bundle designed to minimize turbulence. These nozzles were available in diameters of 4.1, 4.3, 5.1, and 7.9 cm. A recirculating pump facilitated flow rates of up to 21 L per second, yielding Reynolds numbers ranging from tens to 10^6. The vertical fall distance was adjustable, with a maximum height of 45 cm. Table 1 outlines the range of jet test conditions. There was no plunge pool depth because of the calculation of the maximum dynamic pressure.

In our study, the dynamic pressure of the flow was measured using a precision pressure sensor strategically placed at the center point of the jet. This sensor was instrumental in capturing accurate and detailed dynamic pressure data, which is critical for the training and validation of our machine learning models. By positioning the sensor at this central location, we ensured the collection of representative pressure readings that reflect the core behavior of the vertical water jet, thereby enhancing the reliability of our predictive analysis.

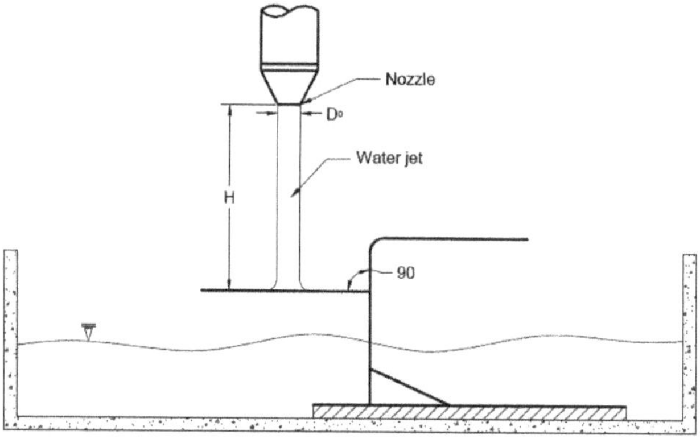

Figure 2. Schematic representation of the data capture model and influencing parameters.

2.3. Data Collection, Preprocessing, and Feature Selection

Data for this study were collected from 240 controlled laboratory experiments, each representing unique conditions of waterfall height (h), nozzle diameter (d), slope of the measuring surface (s), and flow rate from the nozzle (q). Each experiment was conducted over a 20-s period, with a pressure sensor recording the pressure (p(m)) every 0.02 s, resulting in 1000 samples per experiment.

In preprocessing, outliers were first identified and removed from each experiment using the 1.5xIQR criterion. Subsequently, the maximum pressure value from the filtered dataset of each experiment was selected to represent the maximum observed pressure for that setup. This selection was intended to focus analysis on peak pressure impacts, which are critical in the study of fluid dynamics under varying conditions.

To prepare the data for modeling, dimensionless inputs were calculated to standardize the data across different experimental setups. The Froude number and the ratio of waterfall height over the product of the Froude number and diameter as shown in Equation (6), referred to as alpha (α), were computed. Along with the slope, these parameters were selected based on expert opinions and their relevance to the physics of water jets as features for the machine learning models. All input features were then normalized using the StandardScaler from scikit-learn, which removes the mean and scales each feature to unit variance, ensuring a uniform scale across the dataset and enhancing model performance and comparability.

Initial visual analysis using scatter plots of the calculated C_p values revealed non-linear relationships between C_p and its controlling factors (Froude number and α) under different slope conditions. The data variability suggested that a simple linear modeling approach might be inadequate for capturing the underlying dynamics of jet behavior. To rectify this and improve model interpretability, a logarithmic transformation of the C_p values was applied. This transformation was aimed at linearizing the relationships, simplifying the statistical modeling,

$$\alpha = \frac{h}{F_0 \times d} \tag{6}$$

2.4. Model Selection, Training, Evaluation, and Hyperparameter Tuning

Six different machine learning models were evaluated: Ridge Regression, K nearest neighbors, Support Vector Machine (SVM), decision tree, random forest, and Extreme Gradient Boosting (XGBoost). The dataset was split into training (60% of the data), validation (20%), and test sets (20%) using a randomized approach to ensure the statistical independence of the test set. This was facilitated by employing scikit-learn's train_test_split function with a specified random_state to maintain reproducibility and prevent data leakage. The models were initially tuned and validated using the training and validation sets, optimizing for both R-squared and Root Mean Squared Error (RMSE). Extensive hyperparameter tuning was performed using random search, leading to the following optimized models:

- Ridge Regression: Ridge (alpha = 0.5)
- K nearest neighbors: KNeighborsRegressor (n_neighbors = 7)
- Decision tree: DecisionTreeRegressor (min_impurity_decrease = 0.0002)
- Support Vector Machine: SVR (kernel = 'rbf', gamma = 'auto', C = 5, epsilon = 0.06)
- Random forest: RandomForestRegressor (n_estimators = 2500, min_impurity_decrease = 0.0001)
- XGBoost: XGBRegressor (n_estimators = 2500, learning_rate = 0.01, max_leaves = 5, subsample = 0.7, colsample_bytree = 0.8)

Machine learning models have shown significant promise in predicting dynamic pressure coefficients and other fluid dynamics parameters, particularly in the context of vertical water jets. These models offer powerful tools for understanding and predicting complex flow behaviors, each with its unique strengths:

1. Ridge Regression has proven effective in fluid dynamics applications, especially for predicting pressure coefficients. Its ability to handle multicollinearity, a common issue in fluid dynamics data, makes it particularly valuable.
2. K-Nearest Neighbors (KNN), with its non-parametric nature, excels at capturing complex, non-linear relationships in fluid dynamics problems. It has been successfully used to predict flow characteristics and pressure distributions.
3. Support Vector Machine (SVM) has demonstrated its efficacy in predicting dynamic pressure across various fluid flow scenarios. Its capacity to manage high-dimensional data and model non-linear relationships has made it a popular choice among researchers.
4. Decision trees offer interpretability and can handle both numerical and categorical data, making them useful for understanding the factors influencing pressure coefficients. They have been effectively employed in flow prediction and classification tasks.
5. Random forest, an ensemble method based on decision trees, has shown excellent performance in predicting flow characteristics and pressure distributions. Its robustness against overfitting and ability to handle high-dimensional data have led to its increased adoption in fluid dynamics applications.
6. Extreme Gradient Boosting (XGBoost) has emerged as a powerful tool for predicting various fluid dynamics parameters, including pressure coefficients. Its high performance and ability to model complex relationships in data have made it increasingly popular in recent years.

The application of these machine learning models to vertical water jets represents an innovative approach to addressing the challenges in predicting dynamic pressure coefficients. By leveraging the strengths of each model, researchers can improve the accuracy and efficiency of predictions, potentially overcoming the limitations of traditional numerical and physical models. As these methods evolve, they promise to revolutionize hydraulic structure management by integrating advanced computation with domain expertise. This could lead to more resilient and efficient designs that are more suited for handling high-velocity water jets.

Each model was retrained on the combined training and validation dataset and evaluated on the test set using R-squared and RMSE. Additionally, the residuals of the predictions were analyzed through histogram plots to assess the distribution of prediction errors. To ensure the reproducibility of our results, the test dataset were kept separate and only used for final evaluation. The consistent performance of the models on both the validation and test datasets suggests the robustness and reliability of our findings. Data and model codes are available upon request to facilitate further research and verification by other researchers.

3. Results and Discussion

This section presents the results obtained from evaluating various machine learning models for predicting the dimensionless pressure coefficient (C_p) of vertical water jets. The performance of each model is assessed, and the impact of different features on model accuracy is analyzed. Additionally, insights into the behavior of vertical water jets based on the model predictions are discussed.

3.1. *Performance of Machine Learning Models*

Several machine learning models, including linear regression, K-Nearest Neighbors (KNN), decision tree, support vector regression (SVR), random forest, and XGBoost, were evaluated to predict the pressure coefficient (C_p) based on experimental conditions characterized by the Froude number, slope, and alpha. The performance of each model was assessed using root mean squared error (RMSE) and R-squared (R^2) as evaluation metrics.

Root Mean Squared Error (RMSE) measures a model's error in predicting quantitative data by representing the square root of the average squared differences between predicted and observed values. A smaller RMSE indicates that predicted values are closer to observed data points, reflecting better model performance. R-squared (R^2), or the coefficient of determination, indicates the proportion of variance in the dependent variable explained by the independent variables in a regression model. An R^2 of 1 signifies perfect prediction accuracy. Generally, a higher R-squared value indicates a better model fit.

These metrics are crucial for evaluating and comparing the efficacy of different machine learning models. Each model's performance was systematically compared to determine which most effectively predicts water jet pressure dynamics under varied experimental conditions.

The quantitative results of the model evaluations are presented in Table 2 and Figure 3. XGBoost outperformed all other models, achieving the highest R-squared value of 0.953 and the lowest RMSE of 0.191, indicating its superior ability to capture complex relationships and variations within the data accurately.

Table 2. Model performance metrics for predicting C_p.

Model	R-Squared	RMSE
Linear Regression (LR)	0.774	0.421
K-Nearest Neighbors (KNN)	0.900	0.280
Decision Tree	0.902	0.277
Support Vector Regression (SVR)	0.914	0.260
Random Forest	0.941	0.215
XGBoost	0.953	0.191

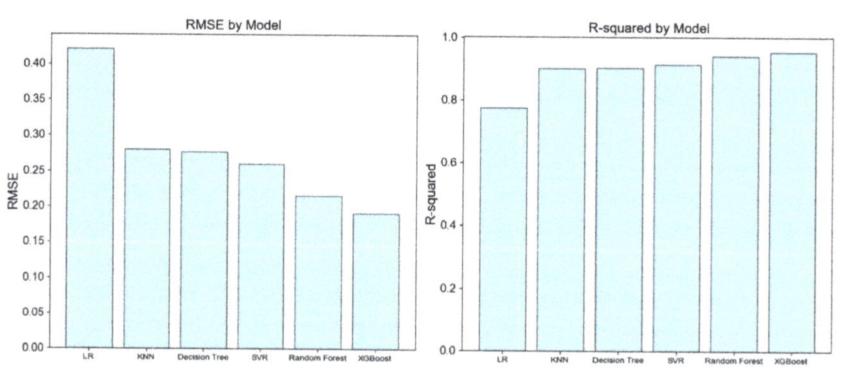

Figure 3. Comparison of model performance for predicting C_p: RMSE and R-squared values across different machine learning models.

Linear regression exhibited the least favorable performance with an R-squared value of 0.774 and an RMSE of 0.421, suggesting that the model's assumptions of linearity did not hold true for the complex dynamics of fluid pressure in this setting. Conversely, models like SVR and random forest showed strong performance with R-squared values above 0.9, underscoring their ability to handle non-linear data more effectively.

The scatter plot of predicted versus actual values (Figure 4) visually confirms the quantitative assessments. XGBoost's predictions are notably closer to the diagonal line, demonstrating its higher precision and consistency across various test cases.

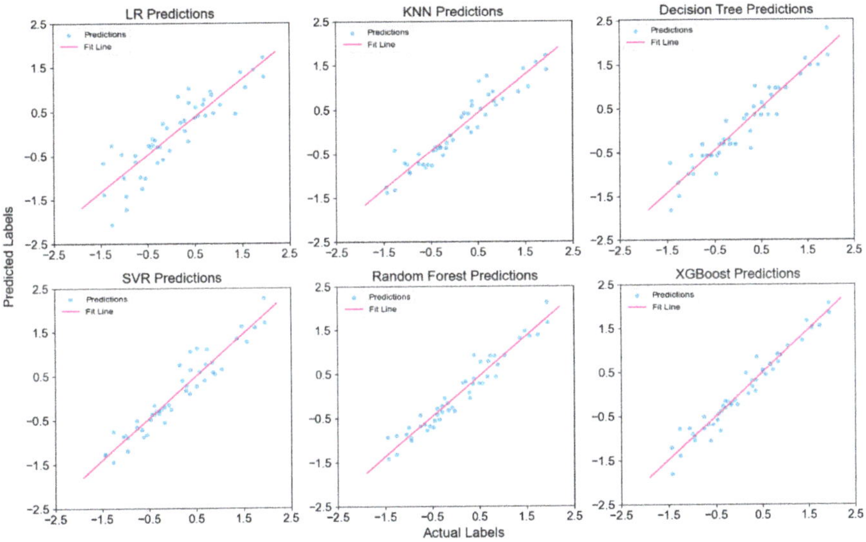

Figure 4. Scatter plots of predicted vs. actual values for C_p with fitted lines across different machine learning models.

3.2. Analysis of Feature Impact and Importance

Understanding the relative importance of each feature used in the predictive models is crucial for interpreting model behavior and guiding further model improvements. This analysis focuses on evaluating how each feature contributes to the predictive accuracy of the models, particularly the random forest and XGBoost models, both of which provide an explicit measure of feature importance.

The random forest model's ability to quantify the importance of each feature offers valuable insights into which variables most significantly impact the model's predictions. Similarly, the XGBoost model also provides built-in functionality for assessing feature importance, allowing for comparative analysis across models. Table 3 demonstrates the relative importance of different features in predicting the pressure coefficient (C_p) as estimated by the random forest model. Figure 5 presents the estimated importance of the Froude number, slope, and alpha (α), which were identified as key predictors in our study.

Table 3. Relative Feature Importance for predicting C_p estimated by the random forest model.

Feature	Importance (%)
Froude number	57
Slope	23
alpha	20

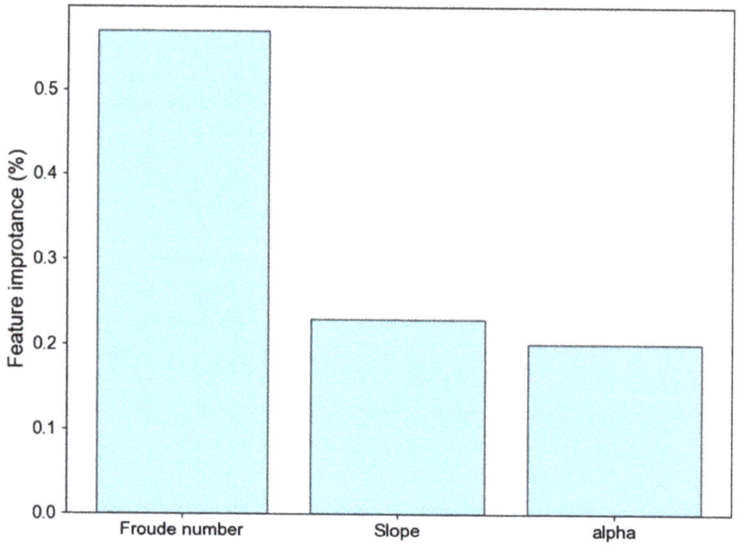

Figure 5. Relative Feature Importance estimated by the random forest model for predicting C_p of vertical water jets, expressed as percentages.

Additionally, the feature importance analysis using the XGBoost model has shown similar trends, with the Froude number, slope, and alpha contributing 43%, 35%, and 22%, respectively. These results are presented in Table 4, providing a consistent view across both high-performing models.

Table 4. Relative Feature Importance for Predicting C_p Estimated by the XGBoost Model.

Feature	Importance (%)
Froude number	43
Slope	35
alpha	22

The analysis shown in Tables 3 and 4 revealed that the Froude number is the most significant predictor, accounting for approximately 57% and 43% of the importance in the random forest and XGBoost models, respectively. This underscores its critical role in the dynamics of fluid motion and pressure in vertical water jets. The slope of the surface impacts the pressure measurements as well, contributing about 23% and 35% to the model's predictions in random forest and XGBoost, respectively. Alpha, while less influential than the other two features, still plays a notable role by accounting for 20% and 22% of the importance in these models.

These findings have implications for both model optimization and engineering design decisions related to fluid jets and hydraulic systems. Knowledge of feature importance can be used to refine model architectures, potentially simplifying models by eliminating less important variables or focusing on interaction effects between the most critical features. Additionally, understanding which factors most significantly affect pressure can inform design decisions in applications involving fluid jets, such as in hydraulic systems or water-based cooling technologies.

3.3. Residual Analysis

Residual analysis is a crucial step in assessing the fit of a model and understanding any underlying patterns that might suggest bias or inefficiency in the predictions. The residuals from the SVM, random forest, and XGBoost models were analyzed to evaluate how well these models capture the underlying dynamics of the data and to identify any systematic errors.

SVM model residuals: The residuals for the SVM model (Figure 6) were examined to assess model fit. The plot shows a distribution that appeared relatively normal. To quantitatively assess this observation, we conducted a Shapiro–Wilk test. The Shapiro–Wilk test resulted in a statistic of 0.948, with a p-value of 0.034, indicating that the normality assumption is not fully supported at the 5% significance level. This suggests that while the SVM model generally performs well, there might be non-normally distributed errors, which implies the presence of minor systematic errors under specific conditions that could be further explored.

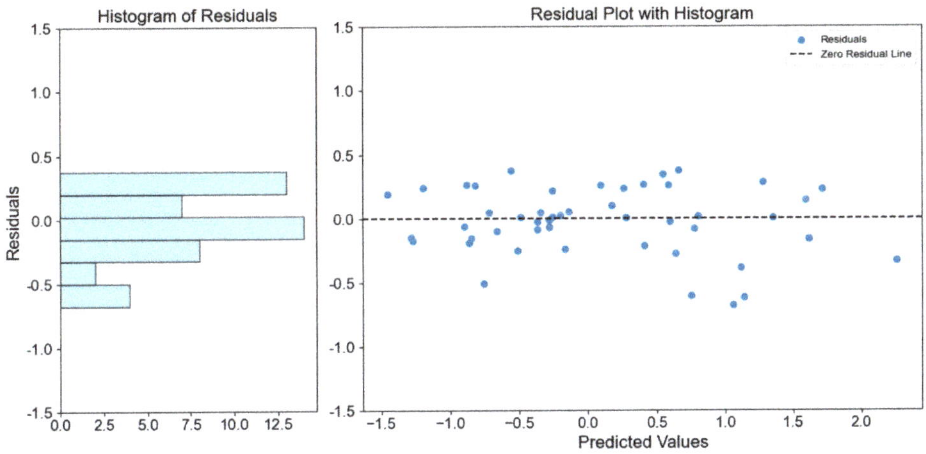

Figure 6. Residual Analysis for the SVM model: Histogram of residuals and residual plot against predicted values.

Random forest model residuals: The residuals plot for the random forest model (Figure 7) illustrates the distribution of residuals associated with predictions of the pressure coefficient. The histogram alongside the scatter plot indicates a normal distribution of residuals, which suggests that the model adequately captures the variability in the data without systematic errors or biases. To further validate this observation, a Shapiro–Wilk test was performed, yielding a test statistic of 0.976 and a p-value of 0.425, confirming the residuals' normal distribution at the conventional significance levels.

XGBoost Model Residuals: The XGBoost model's residuals (Figure 8) demonstrate an excellent fit, with the histogram closely resembling a normal distribution. This indicates a highly effective model in terms of accuracy and consistency across different scenarios. The minimal deviation from normality suggests that the XGBoost model efficiently handles the non-linear complexities of the dataset. Further quantitative analysis using the Shapiro–Wilk test confirms this observation, yielding a test statistic of 0.968 and a p-value of 0.208, which do not reject the hypothesis of normality at the conventional alpha levels (e.g., 0.05).

The analysis of residuals plays a critical role in validating the assumptions of the modeling process and ensuring the robustness of the predictions. The normal distribution of residuals across these models, particularly in the random forest and XGBoost, confirms that they are well-suited for this type of predictive task, capturing the essential patterns without being affected by non-systematic fluctuations. However, the slight deviations observed in the SVM model's residuals suggest areas for model tuning, possibly through parameter adjustments or by incorporating additional data preprocessing steps. Such refinements could further improve the model's performance and reduce any minor biases that could affect its predictions in real-world applications.

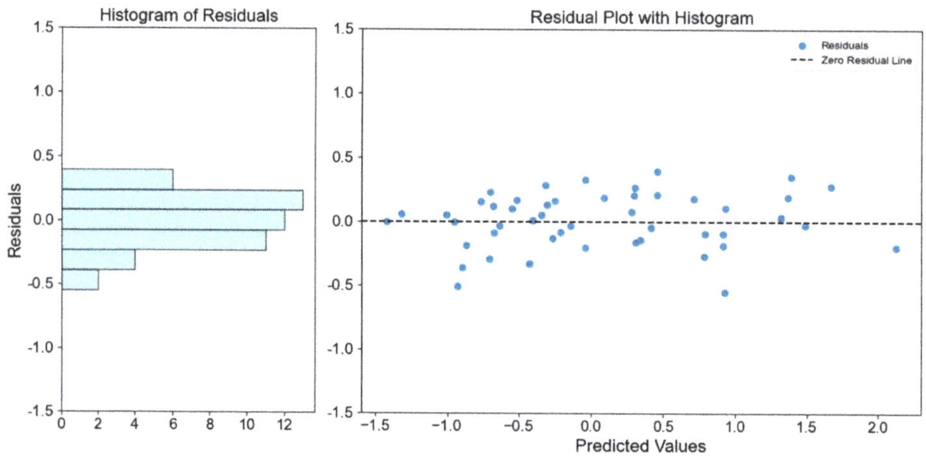

Figure 7. Residual Analysis for the random forest model: Histogram of residuals and residual plot against predicted values.

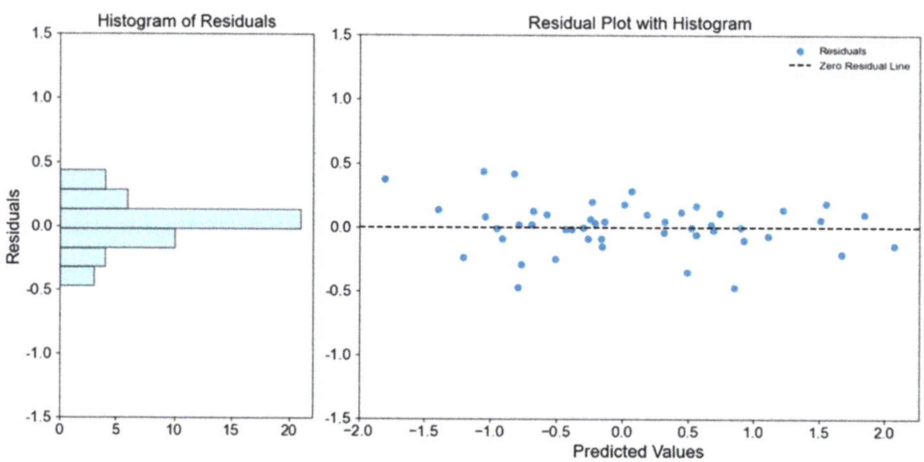

Figure 8. Residual analysis for the XGBoost Model: Histogram of residuals and residual plot against predicted values.

3.4. Insights into Jet Behavior

The behavior of vertical water jets, as characterized by the pressure coefficient (C_p), exhibits complex dynamics that are influenced by several factors, including the Froude number, the slope of the impact surface, and the ratio of height over the Froude number times the diameter (alpha), which serves as a dimensionless parameter relevant to the physics of water jets and is used as one of the features for machine learning models in predicting the behavior of vertical water jets. Our analysis has provided profound insights into how these factors interact and influence the resulting pressure dynamics.

Non-linear Relationships and Transformations: Initial observations from the scatter plot of calculated C_p from p(m) (Figure 9) demonstrate that the relationship between C_p and the controlling factors (Froude number, alpha) is inherently non-linear. This plot shows varied behaviors under different slope conditions, which are marked by triangle, square, and circle symbols representing slopes of 90, 60, and 30 degrees, respectively. This variability suggests that a simple linear approach may not adequately capture the underlying physics of jet behavior.

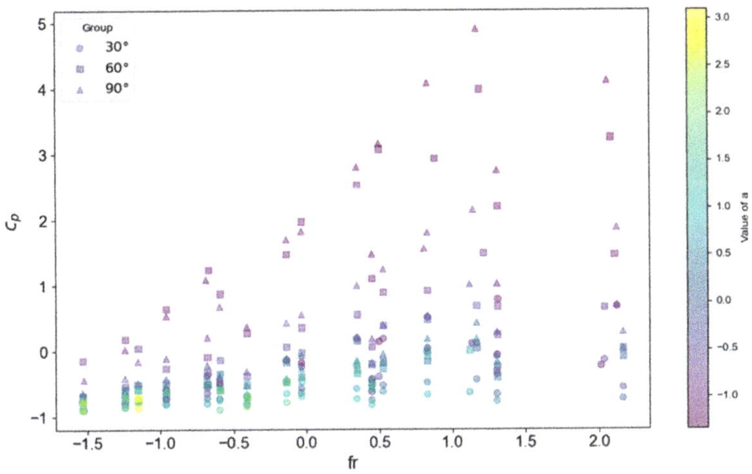

Figure 9. Scatter plot of Normalized Dimensionless Pressure Coefficient (Cp) against Normalized Froude Number, color-coded by Alpha Values and marked by Slope Angles.

To address the non-linearity observed, a logarithmic transformation of C_p values was applied. Figure 10 displays the log-transformed values of C_p, plotted against the Froude number and alpha, with the slope still indicated by distinct markers. This transformation reveals a more linear relationship, simplifying the interpretation and subsequent modeling. The transformed data not only align better with linear assumptions but also enhance the model's ability to predict with greater accuracy.

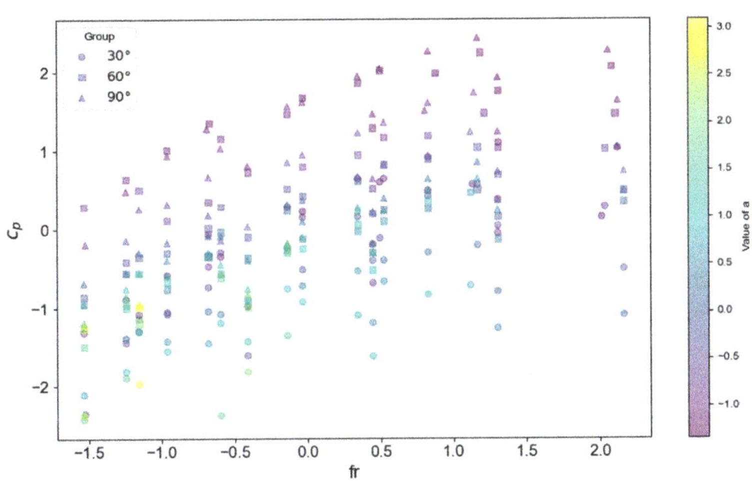

Figure 10. Scatter plot of Normalized Log-Transformed Dimensionless Pressure Coefficient (log (C_p)) against Normalized Froude Number, color-coded by Alpha Values and marked by Slope Angles.

Rationale and Implications of Logarithmic Transformation: The decision to employ a logarithmic transformation of the C_p values stems from the exponential nature of the changes observed in pressure dynamics as influenced by the Froude number and alpha. In fluid dynamics, especially in scenarios involving jet behavior, pressures can increase exponentially with slight changes in velocity or height parameters, making it challenging to model these effects linearly. By transforming C_p to a logarithmic scale, we linearize these exponential relationships, thus fitting a linear model more effectively and reducing the impact of extreme values. This not only improves the interpretability of the data by reducing skewness and stabilizing variance but also aligns the statistical modeling process more closely with the physical phenomena under study. Moreover, achieving a more linear relationship enhances the robustness of the models, improving predictive accuracy across a range of conditions and facilitating the use of linear regression techniques, which are generally less complex and computationally intensive than their non-linear counterparts. This methodological approach ensures that our models reflect both the theoretical understanding of jet behavior and provide practical tools for engineers and practitioners to manage fluid dynamics effectively.

Model Interpretation and Jet Behavior Insights: The transformed data provide a clearer view of how specific changes in the experimental setup affect the pressure dynamics of vertical water jets. For instance, increasing the Froude number generally results in higher C_p values, indicating higher impact pressures. Similarly, changes in the slope significantly alter the pressure dynamics, which is critical for applications involving spray dynamics or fluid impact studies.

Implications for Fluid Dynamics and Engineering: The insights gained from these analyses are valuable for engineering applications where precise control of fluid dynamics is necessary. Understanding the impact of non-linear relationships and the conditions under which they manifest allows engineers to better design systems for optimal performance, such as in water jet cutting, cooling systems, or hydraulic machinery. Additionally, these findings can guide further experimental designs for exploring other aspects of jet behavior that may not have been fully captured in the initial studies. Such follow-up studies could investigate the effects of varying environmental conditions or material properties on the behavior of jets.

4. Conclusions

This study explored the complex dynamics of vertical water jets by investigating the relationships between various experimental parameters such as the Froude number, slope, and alpha, and the resultant pressure coefficient (C_p). Through a rigorous evaluation of multiple machine learning models, the research identified the critical factors influencing jet behavior and demonstrated the efficacy of advanced predictive models in capturing these complexities.

Key Findings:

- Non-linear Dynamics and Transformations: The study highlighted the non-linear nature of the relationship between the input features and the target variable. The logarithmic transformation of the C_p values helped linearize this relationship, enhancing model performance and providing clearer insights into the underlying physical processes.
- Feature Importance: The analysis revealed that the Froude number is the most critical predictor, which influences jet dynamics more significantly than other variables. Understanding the impact of the Froude number, along with the slope of impact and the dimensionless parameter alpha, is crucial for predicting and manipulating jet behavior in practical applications.
- Model Performance: Among the evaluated models, the XGBoost model exhibited outstanding performance with an R-squared value of 0.953 and an RMSE of 0.191. This high level of accuracy attests to the model's capability to handle the non-linear interactions between the input features effectively.

The findings from this research have direct applications in fields such as hydraulic engineering, environmental science, and industries where precise control of fluid dynamics is critical. For instance, in water jet cutting applications, the ability to predict and manipulate jet behavior based on the Froude number and slope can lead to improved cutting efficiency and precision. Similarly, in cooling systems or erosion control mechanisms, understanding the impact of these parameters can guide the design and optimization of fluid impact processes.

Further investigations could explore additional variables that may affect jet behavior, such as temperature, viscosity, and material properties of the impacted surface. Additionally, extending the study to include the effects of external environmental factors like wind and ambient pressure could provide more comprehensive models of jet dynamics. Incorporating computational fluid dynamics (CFD) simulations or conducting validation experiments with real-world scenarios could further strengthen the predictive capabilities of the models developed in this study.

Overall, this research demonstrates the synergy between advanced machine learning techniques and traditional engineering principles in solving complex problems. By successfully applying these methods to the study of vertical water jets, we have advanced our understanding of fluid dynamics and showcased the practical utility of predictive modeling in various engineering domains. The methodologies developed and insights gained from this study offer a valuable framework for future research and application in related fields, ultimately contributing to more efficient and sustainable solutions in fluid systems engineering.

Author Contributions: Conceptualization, A.S.; methodology, A.S. and S.H.B.; software, A.S. and S.H.B.; validation, A.S. and S.H.B.; formal analysis, A.S. and S.H.B.; investigation, A.S. and S.H.B.; data curation, A.S. and S.H.B.; writing—original draft preparation, A.S. and S.H.B.; writing—review and editing, V.A.; visualization, A.S. and S.H.B.; supervision, V.A. and M.F.-M.; project administration, V.A. and M.F.-M. All authors have read and agreed to the published version of the manuscript.

Funding: This research received no external funding.

Data Availability Statement: The data presented in this study are available on request from the corresponding author.

Conflicts of Interest: The authors declare no conflicts of interest.

References

1. Lu, S.; Wang, C.; Wang, W.; Li, M.; Zhang, D. Analysis on the Shape and Impact Pressure of the High-Pressure Water Jet during the Hydraulic Flushing Cavity Technique. *Geofluids* **2021**, *2021*, 7496540. [CrossRef]
2. Dong, C.; Yu, G.; Zhang, H.; Zhang, M. Scouring by submerged steady water jet vertically impinging on a cohesive bed. *Ocean Eng.* **2020**, *196*, 106781. [CrossRef]
3. Godasiaei, S.H.; Kamali, H. Water jet angle prediction in supersonic crossflows: Euler–Lagrange and machine learning approaches. *Eur. Phys. J. Plus* **2024**, *139*, 251. [CrossRef]
4. Wei, W.; Xu, W.; Deng, J.; Liu, B. Experimental study of impact pressures on deep plunge pool floors generated by submerged inclined jets with controlled aeration. *J. Hydraul. Eng.* **2020**, *146*, 04020021. [CrossRef]
5. Wang, H.; Qian, Z.; Zhang, D.; Wang, T.; Wang, C. Numerical study of the normal impinging water jet at different impinging height, based on Wray–Agarwal turbulence model. *Energies* **2020**, *13*, 1744. [CrossRef]
6. Luo, X.; Ye, W.; Huang, R.; Wang, Y.; Du, T.; Huang, C. Numerical investigations of the energy performance and pressure fluctuations for a waterjet pump in a non-uniform inflow. *Renew. Energy* **2020**, *153*, 1042–1052. [CrossRef]
7. Carvalho, R.F.; Lopes, P.M.; Beg, M.N.A. Digital Flow in a Pool Induced by a Vertical Jet. *Water* **2024**, *16*, 1386. [CrossRef]
8. Siamilandu, O. Excavation Work on Kariba Dam Complete. 2023. Available online: https://www.newsday.co.zw/thestandard/local-news/article/200021415/excavation-work-on-kariba-dam-complete (accessed on 27 August 2024).
9. Xu, G.; Antonia, R. Effect of different initial conditions on a turbulent round free jet. *Exp. Fluids* **2002**, *33*, 677–683. [CrossRef]
10. Mistry, D.; Dawson, J.R. Experimental investigation of entrainment processes of a turbulent jet. In *APS Division of Fluid Dynamics Meeting Abstracts*; APS Division of Fluid Dynamics: College Park, MD, USA, 2014.
11. Lemanov, V.; Terekhov, V.; Sharov, K.; Shumeiko, A. An experimental study of submerged jets at low Reynolds numbers. *Tech. Phys. Lett.* **2013**, *39*, 421–423. [CrossRef]
12. Rajaratnam, N.; Khan, A. Intersecting circular turbulent jets. *J. Hydraul. Res.* **1992**, *30*, 373–387. [CrossRef]
13. Raiford, J.P.; Khan, A.A. Investigation of circular jets in shallow water. *J. Hydraul. Res.* **2009**, *47*, 611–618. [CrossRef]
14. Ead, S.; Rajaratnam, N. Plane turbulent wall jets in shallow tailwater. *J. Eng. Mech.* **2002**, *128*, 143–155. [CrossRef]
15. Bhuiyan, F.; Habibzadeh, A.; Rajaratnam, N.; Zhu, D.Z. Reattached turbulent submerged offset jets on rough beds with shallow tailwater. *J. Hydraul. Eng.* **2011**, *137*, 1636–1648. [CrossRef]
16. Hussein, H.J.; Capp, S.P.; George, W.K. Velocity measurements in a high-Reynolds-number, momentum-conserving, axisymmetric, turbulent jet. *J. Fluid Mech.* **1994**, *258*, 31–75. [CrossRef]
17. Salemnia, A.; Fathi Moghodam, M. Effect of nozzle diameter and falling height on the dynamic pressure coefficient of vertical free water jets. *Water Soil Sci.* **2015**, *24*, 185–195.

18. Salemnia, A.; Fathi, M.M. Investigation of break up length of free water jet. *Iran. J. Irrig. Drain.* **2019**, *13*, 309–318.
19. Carraro, M.; De Vanna, F.; Zweiri, F.; Benini, E.; Heidari, A.; Hadavinia, H. CFD modeling of wind turbine blades with eroded leading edge. *Fluids* **2022**, *7*, 302. [CrossRef]
20. Cravero, C.; De Domenico, D.; Marsano, D. The use of uncertainty quantification and numerical optimization to support the design and operation management of air-staging gas recirculation strategies in glass furnaces. *Fluids* **2023**, *8*, 76. [CrossRef]
21. Taebi, A. Deep learning for computational hemodynamics: A brief review of recent advances. *Fluids* **2022**, *7*, 197. [CrossRef]
22. Neyshabouri, A.S.; Da Suva, A.F.; Barron, R. Numerical simulation of scour by a free falling jet. *J. Hydraul. Res.* **2003**, *41*, 533–539. [CrossRef]
23. Jia, Y.; Kitamura, T.; Wang, S.S. Simulation of scour process in plunging pool of loose bed-material. *J. Hydraul. Eng.* **2001**, *127*, 219–229. [CrossRef]
24. Cui, X.; Li, J.; Chan, A.; Chapman, D. A 2D DEM–LBM study on soil behaviour due to locally injected fluid. *Particuology* **2012**, *10*, 242–252. [CrossRef]
25. Cui, X.; Li, J.; Chan, A.; Chapman, D. Coupled DEM–LBM simulation of internal fluidisation induced by a leaking pipe. *Powder Technol.* **2014**, *254*, 299–306. [CrossRef]
26. Taghipour, F.; Ellis, N.; Wong, C. Experimental and computational study of gas–solid fluidized bed hydrodynamics. *Chem. Eng. Sci.* **2005**, *60*, 6857–6867. [CrossRef]
27. Aziz, T.; Raiford, J.; Khan, A. Numerical simulation of turbulent jets. *Eng. Appl. Comput. Fluid Mech.* **2008**, *2*, 234–243. [CrossRef]
28. Islam, M.R.; Zhu, D.Z. A numerical study of confined wall jets in a shallow basin. *J. Hydraul. Res.* **2011**, *49*, 595–600. [CrossRef]
29. Ghahremanian, S.; Moshfegh, B. Evaluation of RANS models in predicting low Reynolds, free, turbulent round jet. *J. Fluids Eng.* **2014**, *136*, 011201. [CrossRef]
30. Faghani, E.; Saemi, S.D.; Maddahian, R.; Farhanieh, B. On the effect of inflow conditions in simulation of a turbulent round jet. *Arch. Appl. Mech.* **2011**, *81*, 1439–1453. [CrossRef]
31. Rayleigh. XXXVI. Investigations in Capillarity:—The size of drops.—The liberation of gas from supersaturated solutions.—Colliding jets.—The tension of contaminated water-surfaces. *Lond. Edinb. Dublin Philos. Mag. J. Sci.* **1899**, *48*, 321–337. [CrossRef]
32. Baron, T.; Comings, E.W.; Johnstone, H. *Atomization of Liquid Jets and Droplets*; Engineering Experiment Station, University of Illinois: Champaign, IL, USA, 1949.
33. Chen, T.-F.; Davis, J.R. Disintegration of a turbulent water jet. *J. Hydraul. Div.* **1964**, *90*, 175–206. [CrossRef]
34. McKeogh, E.J.; Elsawy, E.M. Air retained in pool by plunging water jet. *J. Hydraul. Div.* **1980**, *106*, 1577–1593. [CrossRef]
35. Hoyt, J.W.; Taylor, J. Waves on water jets. *J. Fluid Mech.* **1977**, *83*, 119–127. [CrossRef]
36. Ervine, D.; Falvey, H.; Withers, W. Pressure fluctuations on plunge pool floors. *J. Hydraul. Res.* **1997**, *35*, 257–279. [CrossRef]

fluids

MDPI

Article

Engine Mass Flow Estimation through Neural Network Modeling in Semi-Transient Conditions: A New Calibration Approach

T. Savioli [1], M. Pampanini [1], G. Visani [2], L. Esposito [3] and C. A. Rinaldini [4,*]

1 Atris Engineering srl, 41122 Modena, Italy; tommaso.savioli@atrisengineering.it (T.S.); marco.pampanini@atrisengineering.it (M.P.)
2 Paul G. Allen School of Computer Science and Engineering, University of Washington, Seattle, WA 98195, USA; gvisan01@cs.washington.edu
3 Motori Minarelli spa, 40012 Calderara di Reno, Italy; l.esposito@minarelli.com
4 Department of Engineering "Enzo Ferrari", University of Modena and Reggio Emilia, 42121 Reggio Emilia, Italy
* Correspondence: carloalberto.rinaldini@unimore.it; Tel.: +39-059-205-6152

Abstract: Nowadays, engine experimental research represents a very expensive field within the automotive industry, but it remains fundamental for engine and vehicle development. The present work aims to investigate a novel approach for engine control system calibration, by adopting machine learning techniques to model physical parameters of the engine starting from experimental data measured at the test bench. The main goal is to create a methodology which accelerates the calibration process without losing accuracy. A model that estimates air mass flow is created by adopting either a tree ensemble model or an artificial neural network trained on a small dataset, which was previously acquired at the test bench using a random calibration of the volumetric efficiency map. The model's performance is first validated on a larger, random dataset. Then, the volumetric efficiency calculated from the air mass flow model estimation is used to calibrate the transfer function of the Engine Control Unit. Finally, the sensitivity of the model error correlated with the number of data points acquired is used in order to determine the best practice for a Design Of Experiment, which minimizes data acquisition. The methodology proposed can lead to reduced time and costs of the whole calibration process of the engine, without losing accuracy. The analysis was conducted on the entire vehicle, which is crucial for drivability, especially in motorcycles since they are highly sensitive to air-to-fuel ratio adjustments. This work demonstrates that machine learning models can be adopted for the fine-tuning of the calibration process, which is normally performed manually.

Keywords: engines; neural networks; random forest; gradient boosted decision; vehicle; calibration

1. Introduction

1.1. Engine Calibration

The calibration of internal combustion engines is a complex and critical process to ensure that vehicles meet emissions regulations, deliver optimal performance, and ensure fuel efficiency. Traditionally, engine calibration requires numerous experimental tests and simulations to find the right balance between these parameters [1]. The main constraints are related to pollutant emissions, which have guided the evolution of engine control systems for several years [2]. Engine calibration for passenger cars and motorcycles has to work within emission limits, but without sacrificing drivability. The increased stringency of these constraints has prompted the development of several algorithms and control strategies aimed at predicting physical phenomena occurring in the engines, as well as calibration strategies [3]. Manual calibration of the control system requires many experimental measurements at the test bench, which make development costs high. It is thus fundamental to find alternative strategies to reduce development costs. Starting from test automation and measurements, several strategies have begun to emerge which can predict engine phenomena, thus reducing the number of measurements required [4]. In response to these challenges, the evolution of digital technologies has introduced machine learning algorithms as powerful tools to enhance precision, reduce time, and lower the costs associated with engine calibration pre-processing. Machine learning algorithms possess several properties that make them particularly attractive for applications in modelling and control of complex non-linear systems [5].

1.2. Artificial Neural Networks (ANNs)

Neural networks are particularly effective in modeling complex nonlinear systems such as internal combustion engines, where the relationships between input variables (e.g., pressure, temperature, fuel mixture composition) and output variables (e.g., engine power, emissions, fuel efficiency) are not easily determinable through traditional methods. ANNs can learn these complex behaviors by training on experimental data, making them ideal tools for real-time prediction and optimization. Several studies have demonstrated the effectiveness of neural networks in engine calibration. For example, in [6] a neural network model is developed for pre-calibration of diesel engine emissions and performance, using a multi-objective genetic algorithm to optimize the outcomes. The authors of [7] provided an overview of neural network applications in the calibration of spark-ignition engines, highlighting their use in virtual sensing and as substitutes for lookup tables. Additionally, ref. [8] explored the use of ANNs to accelerate the experimental calibration of propulsion systems, showing a significant reduction in calibration time. In [9], the development of a virtual methodology based on physical models and ANNs to optimize engine calibration is reported. The role of neural networks is not limited to engine calibration, but also extends to the modeling and optimization of engines: for example, ANNs are used in [10] to predict efficiency and emission of a gasoline engine, while in [11] they are used to analyze the potential of alternative fuels for a diesel engine; Ref. [12] report the use of artificial neural networks to create a predictive model on the basis of which fuel consumption of motor vehicles can be determined. Brahma et al. (2008) [13] compared neural networks with other modeling techniques, showing that ANN offers greater accuracy in engine calibration than global regression models. Another interesting aspect of ANNs is their ability to replace black-box controls in engine calibration systems. Meli et al. (2024) [14] demonstrated how neural networks can be used to model complex Engine Control Unit (ECU) functions, improving calibration accuracy. Neural networks have also been used to estimate volumetric efficiency in engines, significantly reducing the experimental effort required for base calibration, as highlighted by De Nola et al. (2019) [15]. Zhou et al. (2008) [16] introduced a neural network-based modeling approach for vehicle emission testing and engine calibration, showing significant improvements in accuracy and efficiency. Another important contribution was made by Pan et al. (2021) [17], who developed an engine calibration model using Gaussian process regression, showing how ANNs can overcome the inaccuracies associated with traditional combustion models. Jacob et al. (2020) [18] conducted an interdisciplinary review of engine management system calibration strategies, highlighting how ANN can enhance the accuracy of calibration techniques for various alternative fuels. Rahimi-Gorji et al. (2017) [19] used ANNs to model the effects of atmospheric conditions on engine performance, demonstrating how these networks can be applied to optimize engine calibration under different operating conditions. The application of neural networks to engine management systems was also explored by Garg et al. (2012) [20], who described various use cases of ANNs to improve engine calibration and control. Shayler et al. (2000) [21] examined the use of ANNs in engine management systems, highlighting how these networks can significantly reduce calibration work and improve overall efficiency. Another significant example is provided by Schoeggl and Ramschak (2000) [22], who used ANNs to assess vehicle drivability during development, calibration, and quality testing, showing how these networks can improve the reliability of results. Atkinson and Mott (2005) [23] applied an ANN to optimize the dynamic calibration of diesel engines, demonstrating significant improvements in performance and emissions. Lowe and Zapart (1999) [24] conducted a comparative study on the use of ANNs for confidence interval estimation in engine management systems, showing how these networks can improve calibration accuracy. Friedrich et al. (2016) [25] investigated the use of ANNs for medium-speed engine optimization, demonstrating how these networks can enhance the modeling and prediction of engine performance. Vinoth et al. (2022) [26] developed a recurrent neural network-based soft sensor for flow estimation in rocket engine injector calibration, demonstrating the applicability of ANNs in contexts beyond internal combustion engine calibration.

These studies clearly demonstrate the potential of neural networks in engine calibration, offering advanced solutions to address modern challenges in engine management. Integrating ANNs into calibration processes not only promises to improve operational efficiency but also opens new possibilities for continuous optimization of propulsion systems. The application of ANNs in internal combustion engine development is still ongoing and can lead to not only improved performance of engines, but also decreased experimental effort for the development of engine control systems, which is a field still in continuous evolution.

1.3. Tree Ensemble Models

While ANNs are a popular and effective choice, other "traditional" machine learning algorithms can also be used to model non-linear relationships in data. Particularly fitting to the present work are tree ensemble models, such as Random Forests [27] and Gradient-Boosted Decision Trees [28]. These models have been shown to consistently deliver better performance than ANNs in certain scenarios, such as with tabular data and in low-to-medium data regimes [29]. The favorable performance in low-data scenarios makes the uses of these models appealing, as one of the main goals of the present study is minimizing data acquisition.

In the engine calibration field, these models have been used to predict combustion feedback information to control spark timing [30]. They have also been applied to extrapolate information from knock sensor signals for combustion knock control [31].

1.4. The Present Study

This work deeply investigates the calibration of volumetric efficiency of a motorcycle engine, by adopting a predictive model based on an ANN or a tree ensemble trained on data collected at the test bench. The main innovation of this work lies in the application of machine learning models for the fine-tuning of the entire pre-production vehicle, representing a significant advancement over traditional methodologies. In contrast to the majority of studies in the literature, which focus on calibrating a standalone engine in a test cell, the use of machine learning models at the final stage of vehicle calibration on a rolling bench, particularly for a motorcycle, represents a novelty for the field. This innovative approach bridges the gap between conventional engine testing and full-vehicle performance evaluation, providing a more integrated and efficient calibration process that reduces both time and costs while maintaining high precision.

The data used for the modelling are measured on a rolling bench with the complete pre-production vehicle. The main constraint regarding the calibration of the air mass flow estimation is related to the sensitivity of the whole vehicle at small variations of torque, and in catalyst efficiency if the estimation does not have the accuracy needed. Due to these factors, normally the fine-tuning calibration of production or pre-production vehicles is performed by hand, at the test bench and in road tests. The present work demonstrates that it is possible to use machine learning models to save many days of work on motorcycle calibration without losing accuracy. Another interesting point to underline is that the estimation model was created starting from a random calibration of the vehicle, which is the worst condition that a calibrator operator could have in their hands.

In the following paragraphs, the procedure adopted is described, starting with how data was collected at the engine bench, and how the machine learning models were created; particular attention was given to the validation process and choice of model type. The model trained with the data acquired from a randomly calibrated vehicle is used to estimate the air mass flow rate through the motorcycle intake manifold across the entire engine operating range. From air mass flow, it is then easy to calculate injection time. Finally, this information defines the calibration map which is written into the ECU and used for final validation in real driving conditions of the vehicle—with a physical driver at the rolling bench—demonstrating the accuracy of this approach.

2. Materials and Methods

The main aim of this project is to use machine learning models to calibrate the injection volumetric efficiency of the engine targeting the lambda (air-to-fuel ratio), which is required by the ECU system. The air-to-fuel ratio primarily affects emissions, particularly when the engine is operating during the emission cycle loads. It also influences vehicle drivability and, crucially, performance—factors that are especially important for motorcycle engines. Regarding emissions, it is very important to maintain lambda within the stoichiometric zone (normally 1/1.5% in rich zone) in order to maintain catalyst efficiency. Drivability, especially for motorbikes, is very affected by torque variations due to air-to-fuel ratio oscillations; these variations are transmitted to the rear wheel, resulting in a harsh feeling when driving the motorcycle.

All the data used for this work were collected in the Motori Minarelli factory using an engine test cell equipped with a roller bench for testing the complete vehicle. The test bench—provided by Ono Sokki (Yokohama, Japan) [32]—is normally used for vehicle development; the motorcycle's controls are fully automated except for gear shifting. For accurate estimation of air flow and replication of the motorcycle's final behavior, it is essential that the vehicle is in the final production configuration, with all the components installed, such as the air filter, intake manifold, fairings, etc. As mentioned in the introduction, the tests were carried out with the whole vehicle because the focus of the work is on the final calibration process, with the intent of mass production.

The vehicle was braked by an eddy current dynamometer suitable for motorcycles, which develops up to 70 kW of peak power. The braking torques were set as a function of vehicle speed, following the coastdown curve, which was measured experimentally on a specific test road [33]. The test cell is equipped with traditional ventilation systems and additional systems to ensure a safe environment and compliance with test specifications. A large fan is positioned in front of the vehicle to simulate the variable air supply the motorcycle would experience when moving at a certain speed on the road; this feature is very important because air flows around the vehicle can affect air mass flow in the engine. Figure 1 shows the full experimental setup with the motorcycle at the test bench.

The setup that allows effective control over the vehicle during testing consists of a development ECU, which enables real-time monitoring and modification of engine parameters and calibration through analog/digital acquisition systems. Communication with the ECU is ensured by a Controller Area Network (CAN) connection to a PC, which uses the ETAS INCA (Stuttgart, Germany) [34] software for calibration, monitoring, diagnostics, and data acquisition. Other measurements, fuel consumption, engine torque, power, etc., were also acquired at the same time with a digital acquisition module by the test bench acquisition software.

In addition, an automatic system for motorcycle control manages the throttle, brake, and gearbox to achieve prescribed operational points and follow riding cycles efficiently. A test campaign was carried out driving the motorcycle with an automated cycle which can cover engine load from 1% to 100% at every engine speed (from 3000 RPM to 10,000 RPM) chosen for the acquisition. The test carried out for data acquisition consists of fixing the engine/vehicle speed (the gear does not change during the test) and varying the engine load from 1% to 100% in 60 s; this method allows the engine to reach a semi-transient condition. The data acquisition is performed continuously with a sampling frequency of 100 ms. The total duration of the test is 20 min.

Figure 1. Whole test bench setup.

The data were collected with a random calibration of the ECU in terms of volumetric efficiency of the engine. Random calibration was created starting from the production calibration with only one constraint: the value inserted is saturated at +−20%. The choice was made because of the measurement range of the UEGO sensor adopted (Bosch LSU 4.9).

In this work, different machine learning models were compared in terms of accuracy to find the best one for this particular application:

1. Gradient-Boosted Decision Tree Regressor [28]: This is a popular machine learning model consisting of an ensemble of decision trees trained sequentially. A decision tree is a non-parametric model that learns simple binary decision rules over the inputs—structured hierarchically in a tree—to approximate the target. The tree is constructed using the inputs and targets of the training data. To make a prediction for a test datapoint, the datapoint is "carried" along a branch of the tree according to how the inputs behave under the learned binary rules. Then, once a leaf is reached, the prediction is made using the average target values of the training datapoints that belong to that leaf. Decision trees are known to be prone to overfitting, and to be unable to learn very flexible decision boundaries: however, this can be significantly improved by "ensembling" multiple trees together. In particular, gradient-boosting is a technique whereby decision trees are sequentially fitted to the residual error (aka the "gradient") of the previous tree.
2. Artificial Neural Network [35]: This class of models provides the highest level of flexibility in terms of the input-target relationships that can be learned. In particular, we focus on Feed-Forward Fully Connected Neural Networks. The fundamental building block of these is the perceptron, which takes as input a set of features, linearly combines them into a single value using learned weights and biases, and finally transforms said value with a simple nonlinear "activation" function. Multiple perceptrons can be combined in parallel, forming a layer, and layers are stacked, with the outputs of perceptrons in one layer being fed as inputs to the perceptrons in the next layer. Neural networks are trained using the backpropagation algorithm, whereby a loss function between the network's output and the target is determined—in our case the Mean Squared Error loss—and the gradient of the loss value is backpropagated through the network using automatic differentiation; finally, the network parameters—weights

and biases of the linear layers—are updated by taking a step in the direction of the gradient, using a predetermined learning rate.

3. Results

3.1. ECU Strategies and Calibrations

The main signals involved in estimating the engine combustion air flow are engine speed, usually expressed in Rotation Per Minute (RPM) and the Throttle Position Sensor (TPS) percentage. In some cases, the tests required either actively riding the motorcycle positioned on the roller, thus simulating normal road use, known as Auto Load Regulation (ALR), or using an electric actuator mechanically connected to the throttle control, adjusted via a potentiometer. This test setup, which can rotate the throttle grip from fully closed (0%) to fully open (100%) from outside the cell, is called Auto Speed Regulation (ASR). In this work, the data acquisition was performed by keeping vehicle speed constant, and by performing a sweeping of the load from 0 to 100% using ASP regulation.

Another parameter involved in estimating the engine combustion air flow is the Manifold Absolute Pressure (MAP), usually expressed in kPa and measured by a pressure sensor located in the motorcycle's intake duct. This value mainly depends on the throttle opening and external atmospheric pressure. The ECU has a custom strategy for the estimation of the engine load. The load is considered in terms of TPS percentage if the TPS is greater than a threshold, which is a function of engine speed and TPS itself. If the threshold line is not crossed, the load is estimated by a MAP sensor. This strategy is mainly due to the sensitivity of TPS at low load; at this point, a small percentage of TPS opening corresponds to high gradients of air mass flow, thus leading to a worsening of the load estimation. Instead, the MAP sensor has a better resolution in small throttle opening, and thus is a more reliable parameter for estimating load.

Additionally, the motorcycle's exhaust line is equipped with a linear lambda sensor (UEGO—Universal Exhaust Gas Oxygen sensor) capable of measuring the amount of oxygen in the exhaust gases, thereby providing the ECU with an estimate of the burnt mixture composition. This information allows the ECU to adjust the air-to-fuel ratio accordingly through automatic calibration with a Closed Loop (CL) algorithm. The CL strategy is active when the engine is running in stable condition, because in the case of high load gradient or high engine speed gradient the UEGO measurements lose accuracy. The adoption of a CL in this case could create instability in the air-to-fuel ratio. In these conditions, the ECU estimation is made only by calibrated models, and the amount of fuel injected is corrected only with environmental conditions (because these parameters are stable, and they do not enter into CL control).

UEGO signal and injection time are essential for calculating an indirect measure of the air flow into the engine. As mentioned above, the air mass flow is not directly measured with a specific sensor, due to calibration concerns. Indeed, introducing such a sensor would alter the intake line, affecting the mass flow it is trying to measure, thus not providing an accurate representation of what the mass flow would be without the sensor. The calculation of air mass flow is reported below in Equation (1):

$$\dot{m}_{air} = \lambda \, \alpha_{st} \, t_{inj} \, \rho_F \, GAIN_{inj} \, \frac{1}{10^6 \, 60} \tag{1}$$

where:

- $\dot{m}_{air}$ is the air flow rate expressed in $\frac{mg}{cycle}$
- λ is the lambda of the mixture
- α_{st} is the stoichiometric air-fuel ratio, 14.65 in this case since the fuel is gasoline
- t_{inj} is the injection time expressed in μs
- ρ_F is the fuel density expressed in $\frac{\mathrm{g}}{\mathrm{dm}^3}$, in this case, it is 750 $\frac{\mathrm{g}}{\mathrm{dm}^3}$
- $GAIN_{inj}$ is the injector gain, which is the quantity corresponding to the fuel flow rate deliverable by the model being used, expressed in $\frac{\mathrm{cm}^3}{\mathrm{min}}$, in this case, it is 280 $\frac{\mathrm{cm}^3}{\mathrm{min}}$

The response of the lambda sensor contains an intrinsic reading delay, which can be adjusted in various ways [36]. In this work, we take into consideration the time delay measurement presented by the universal exhaust gas oxygen sensor along with uncertainties in the volumetric efficiency. For that purpose, observers are designed by means of a super-twisting sliding mode estimation scheme. Also, two control schemes based on a general nonlinear model and a similar nonlinear affine representation for the dynamics of the normalized air to fuel ratio were designed in this work by using the super-twisting sliding mode methodology. Such dynamics depend on the control input, that is, the injected fuel mass flow, its time derivative, and its reciprocal. The two latter terms are estimated by means of a robust sliding mode differentiator. The observers and controllers are designed based on an isothermal mean value engine model. Numeric and hardware in-the-loop simulations were carried out with this model, the parameters being taken from a real engine. The results obtained show good output tracking and rejection of disturbances when the engine is in closed-loop with the proposed control. There are three main factors that influence this delay. The first is associated with the time interval between the opening of the intake valve and the exhaust valve, and is considered a combustion delay. The second factor represents the time it takes for the exhaust gases to reach the lambda sensor, depending on the air speed in the ducts, and is known as transmission delay. The third and final factor is related to the sensor itself, whose actual output is an aperiodic exponential curve, indicating a slow dynamic associated with a first-order system. The time constant can be approximately considered as 20 ms for a conventional sensor. The first two factors are clearly proportional to the engine's rotational speed n, and can be expressed as follows [37]:

- Combustion Delay:

$$T_c = \frac{3}{4}\frac{120}{n} = \frac{90}{n} \tag{2}$$

- Transmission delay:

$$T_{exh} = \frac{1}{v_{exh}} \approx T_c = \frac{90}{n} \tag{3}$$

3.2. Data Training Acquisition and Creation of Models

The calibration, as already described, of injection timing in this engine is accomplished using two different strategies. The first is called Speed Density (SD) and is based on engine speed (RPM) and MAP. This strategy is primarily used at low loads where the estimation derived only from the throttle position is insufficient. The second strategy is called Alpha Speed (AS), it is expressed in terms of engine speed (RPM) and TPS, and it covers the remaining operating range of the engine. The two strategies are separated by an RPM-TPS threshold vector, which determines which strategy has to be used for determining the injection timing [38,39]. For training the machine learning models, two different data acquisition processes were performed, one for each strategy. Figures 2 and 3 show the investigated operating points for the SD and AS strategies, respectively. The data acquisition was designed to cover all boundaries of the engine in terms of load and speed. The data acquisition system sampling frequency is set to 100 ms. For each strategy (AS, SD), the engine speed is held fixed at different values, and the load is swept from 0% to 100% within each strategy's boundaries, thus resulting in a "striped" acquisition pattern. The total time of data acquisition is approximately 20 min, and each parameter is sampled at 100 ms [8]. The maneuvers performed can be considered almost in steady-state condition. The approach adopted is well-suited to the experimental setup (roller bench) and offers significant time savings in data acquisition without compromising accuracy, thanks to the slow transition between measurement points. This method contrasts with traditional data acquisition for machine learning models, which is usually performed under steady-state conditions on the engine test bench. The Design of Experiment (DOE) has been chosen for time vs accuracy in case a whole vehicle is used for the data collecting. A full factorial approach is not useful since every point would need acceleration and braking of the vehicle, and the time to reach the steady state would increase for every point reached. It is crucial to clearly define the boundaries of the problem being studied to avoid extrapolations that could reduce the accuracy of the regression model estimates. The data acquired were filtered before being used by the algorithms to isolate only the points contained in each strategy. The two datasets were used to discriminate the adoption of the TPS sensor or MAP sensor signal as input for the ML model.

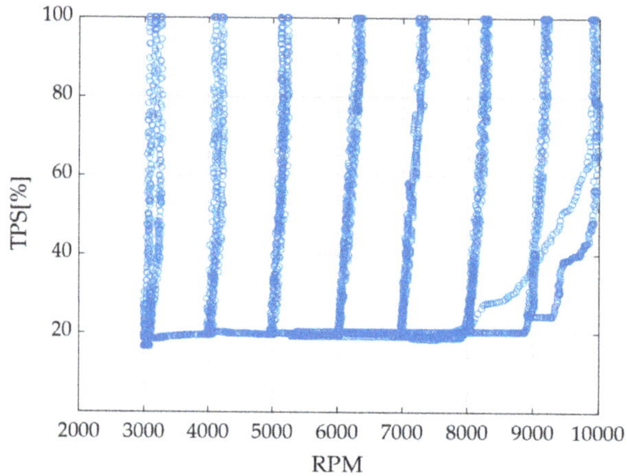

Figure 2. Measured operation points (high load).

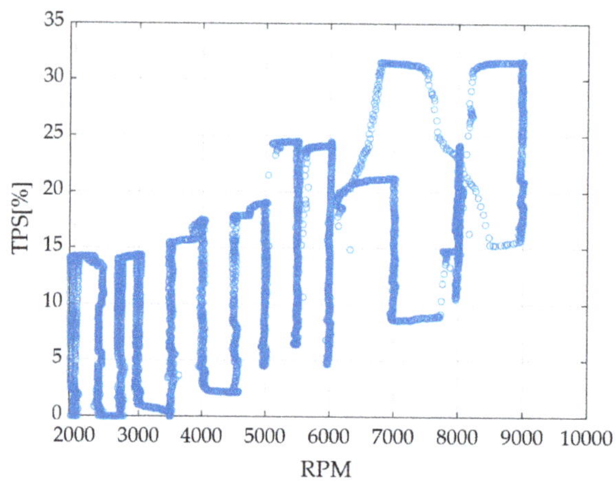

Figure 3. Measured operation points (low load).

The training datasets are derived from data acquisition, applying some post-processing constraints and calculations. First, the measurement errors were filtered out, and secondly, the UEGO signal was delayed as described in the previous section. The data was divided between AS and SD, and the model was created with different load input signals (TPS or MAP). Figure 4 shows the range within which machine learning models can be used successfully. On the left, a brief acquisition related to the Speed Density strategy is shown, where only a few load sweeps at constant RPM can be observed. This dataset is used to train the predictive model. On the right, a dataset produced in MATLAB R2023b is shown which, with a certain level of discretization, queries the machine learning algorithm at all the engine operating points that fall within the injection strategy (in this case SD).

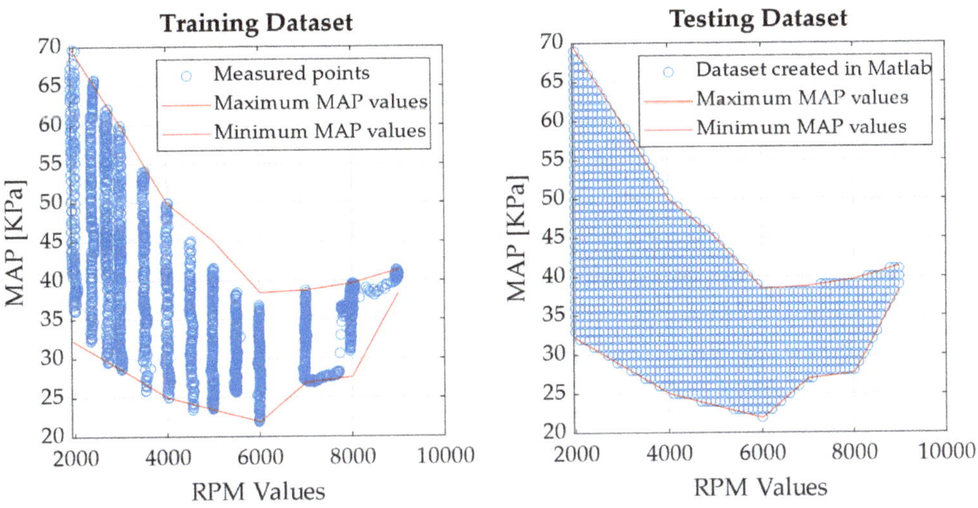

Figure 4. Training dataset vs. testing dataset.

The machine learning model was created using the throttle position sensor, the MAP sensor, and the engine rotational speed as inputs. The generated output is the airflow rate. Models with additional inputs (ambient pressure, ambient temperature, engaged gear, throttle opening derivative) were tested and developed, but they resulted in poorer performance. This outcome is due to the fact that the combustion engine under analysis does not have variable valve timing systems, and the airflow rate is solely linked to those parameters. Furthermore, the pressure and temperature conditions in the test cell are constant, as there is a conditioning system in the test room. Models were created using the functions "fitrnet" and "fitrensemble" included in MATLAB Toolbox "Statistics and Machine Learning". The first was used for feed-forward fully-connected neural networks, and the second for tree ensembles. We used Bayesian Optimization—natively integrated in both functions—to find the best configurations of hyperparameters. This algorithm is iterative and able to search through various possibilities to minimize the objective function, which in this case is the expression log(1+loss), where the loss is the mean square error (MSE) that is calculated with the current iteration hyperparameter configuration. In this work, multiple iterations of optimization are used to ensure that the results were comparable and repeatable across different runs of the code, and this was indeed the case. In fact, the Bayesian algorithm implemented in MATLAB almost always provides a different combination of hyperparameters compared to the previous result, but the performance that the models achieve is very similar. For this reason, the described models can be considered representative of their respective categories.

For the Alpha Speed injection strategy, several experimental trials have identified the neural network as the best model. The one in consideration is composed by two hidden layers (the first with 155 neurons and the second with 239), ReLU activation function, Glorot initialization of weights, and LBFGS as training solver algorithm. For Speed Density, a tree ensemble with Least Squares Boost algorithm (gradient boosted trees), 49 sequential decision trees, a learning rate of approximately 0.21, one as the minimum number of observations per leaf node, 69 as the maximum number of decision splits, and two as the number of features to randomly select for each split, seems to be the best choice in terms of model performance (Table 1).

Table 1. Model setup.

Injection Strategy	Alpha Speed	Speed Density
Model type	Neural network	Gradient boosted trees
Model features	• Bayesian optimization • Feed-forward fully connected • [115,239] neurons • ReLU activation function • Glorot weights initialization • LBFGS training solver algorithm	• Bayesian optimization • LS Boost algorithm • 49 sequential trees • Learning rate of 0.21 • 1 minimum observation per leaf • 69 maximum decision splits • 2 features per split

Figure 5 shows on the x axis the measured air mass flow, and on the y axis the estimated air mass flow; for the sake of brevity, only the predictions of the gradient boosted tree model are shown here. The model was created by excluding outlier points. It can be observed that there is good agreement between the model predictions and real air mass flow.

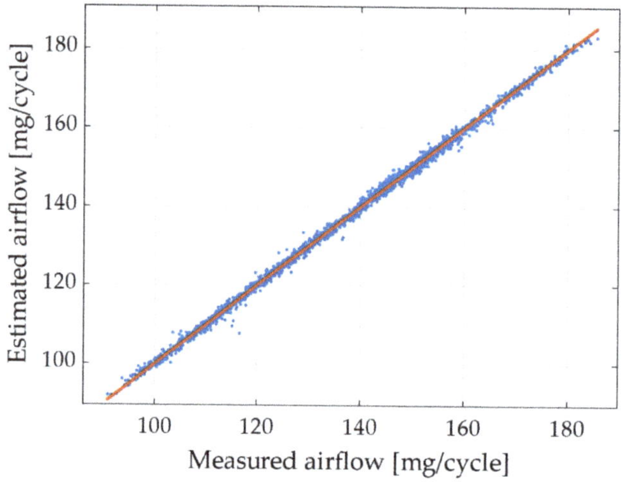

Figure 5. Measured vs. predicted airflow with gradient boosted trees.

4. Discussion

4.1. Model Perfomance Evalutation

Once the machine learning models produce acceptable performance in predicting the air flow rate, the estimated injection times are derived using Equation (1) and the target lambda, and are then inputted into the ECU. With these, the complete map can be compiled and uploaded to the ECU. If, as in the case at hand, injection times obtained from the bench using the standard experimental method are available and thus serve as the target values for the models to achieve, comparison errors can be calculated. These errors can guide potential improvements in the structure and the choice of hyperparameters for the machine learning models.

$$err_{perc} = \frac{t_{inj_{target}}}{t_{inj_{model}}} * 100 - 100 \quad (4)$$

In this way, the following results were obtained, clearly highlighting where the models used produce injection times that deviate most from the targets. Regarding the Alpha Speed strategy, a neural network was used with results shown in Figure 6.

It can be seen that only certain engine speeds were measured, but even for the unstudied speeds, as long as they are interpolated between speeds in the training data, the estimated injection times produce very low errors. Speeds at the borders of the training range instead show higher errors; this result was expected because machine learning models struggle at extrapolation. The highest error values occur just below the maximum engine RPM, at throttle openings of 90% and above. This is because, in that range, being close to the limit, very few data were acquired on the test bench. At 10,000 RPM, no samples were measured. The average absolute percentage error is 2.5%, while the standard deviation is 2.8.

TPS\RPM	**3000**	**3500**	**4000**	**4500**	**5000**	**5500**	**6000**	**6500**	**7000**	**8000**	**8500**	**9000**	**10,000**
20	−0.1	−0.5	−2.8	0.1	−0.8	−0.6	−0.6	−0.7	0.1	−0.1	−0.2	−3.9	0
25	4.4	3.6	1.9	3.7	3.4	3.3	3.9	4.7	2	1.8	−1	−2.4	0
27.5	3.9	3.3	2.3	3.4	1.5	1.1	2.1	1.2	0.7	−0.2	−3.1	−6.4	0
30	3.1	4.1	3.1	4.2	2.6	0.7	2.7	2	0.2	1	1	0.4	0
35	2.5	2.3	1.9	2.5	0.9	1.4	0.8	2	1.1	0.3	1.9	−1.7	0
40	3	2.2	3.5	3	1.4	1.9	2.8	2.8	2.2	0.5	2.2	3.5	−2.3
45	4.3	1.7	3.5	3.5	2.1	2.2	2.5	1.3	1.5	2	2	2.7	1.7
50	5.2	1.4	1.4	3.4	2.5	1.1	1.2	2.7	0.2	−0.8	1.6	2.2	−1.5
60	6.1	1.8	3.8	4.3	3.3	2.5	2	4.3	5.5	1.2	1.4	3.6	−0.9
70	4.2	1.8	5	3.8	3	4.2	2.4	2.9	4.8	−0.3	2.4	2.9	−7.9
80	3.1	3.1	4.1	5.2	1.1	3	2.7	2.8	3.7	0.4	0.9	1.2	−9.4
90	2.6	0.1	2.3	2	0.4	1.6	2	3.7	3.1	3.1	2.7	−0.8	−14.3
100	6.1	3.3	1.8	2.2	1.1	2.3	1.7	2.9	3.9	1.2	5.1	0	−13.2

Figure 6. Percentage error in comparison to original calibration map (Alpha Speed). The cells highlighted in red represent model-estimated injection times that are lower than the target values, while the green cells indicate injection times that are higher than the target.

In Figure 7, we try to understand whether a relationship exists between percentage error and injection time. The goal is to exclude the possibility that low errors are due to a link with high injection times and vice versa. This could distract the user from the actual evaluation of the model. As can be seen, the two variables appear to be uncorrelated, as the scatter plot, despite its dispersion, does not exhibit any particular trend with increasing injection time.

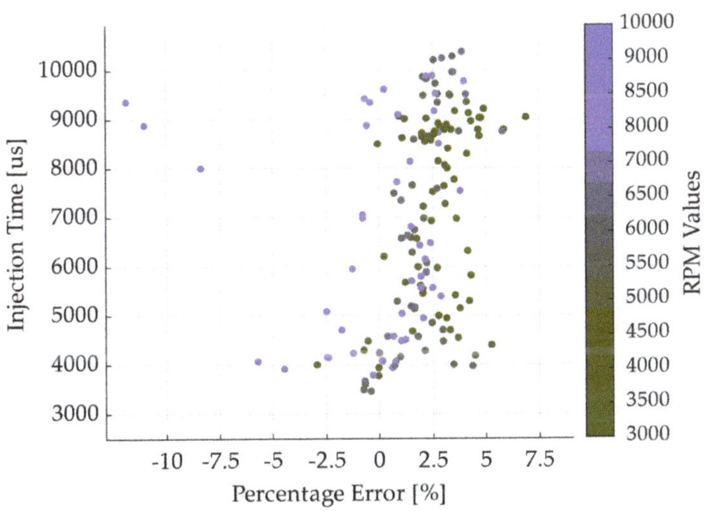

Figure 7. Percentage error vs. injection timing (Alpha Speed).

Figures 8 and 9 show analogous results for the Speed Density strategy, using a gradient boosted decision tree model.

MAP\RPM	2000	2400	3000	3500	4000	4500	5000	5500	6000	6500	7000	8000	8500	9000
25	0	0	0	0	0	−5.3	−5.7	−0.5	2.4	−2.2	−8.6	0	0	0
28	0	−1.1	2.9	−0.3	1	−0.6	−4.2	−3.6	−1.2	−3.2	−6	−8.8	0	0
32	−3.7	−2.9	1.7	−1.9	0.5	−0.6	−0.5	−0.5	1.6	1.1	6.2	4.1	0.3	0
34	−1.7	−1.8	−2.7	−4.5	−3.4	−3.4	−1.7	−1.7	−0.5	−1.7	3.8	3.7	1.7	0
37	−0.1	−2.6	−2	−0.7	−1.8	−2.7	0.4	0.1	−0.7	−1.5	3	4.9	3.6	−5.7
38	1.6	−3.6	−3.4	−4.2	−3.7	−2.3	−1	−0.4	−1.9	−3.9	−0.4	−1.7	3.2	−2
40	−0.6	−0.8	−1.2	−1	−1.1	−1.5	0.9	1.7	2	0	−3.2	8.7	12.4	1.6
43	−2.3	−2.7	−3.3	−2.6	−1.7	−1.5	1	2.7	0	0	0	0	0	0
45	−1.1	−2.3	−2.5	−1.9	−2.5	1.7	−1.4	0	0	0	0	0	0	0
47	−1.6	−2.2	−2.3	−1.7	1	0.6	1.2	0	0	0	0	0	0	0
50	−1.3	−1.2	−3.6	−0.2	1.3	0	0	0	0	0	0	0	0	0
55	1.4	0	0.6	2.1	0	0	0	0	0	0	0	0	0	0
60	2.1	1.2	1	0	0	0	0	0	0	0	0	0	0	0
65	2.7	0.8	0	0	0	0	0	0	0	0	0	0	0	0

Figure 8. Percentage error in comparison to original calibration map (Speed Density). The cells highlighted in red represent model-estimated injection times that are lower than the target values, while the green cells indicate injection times that are higher than the target.

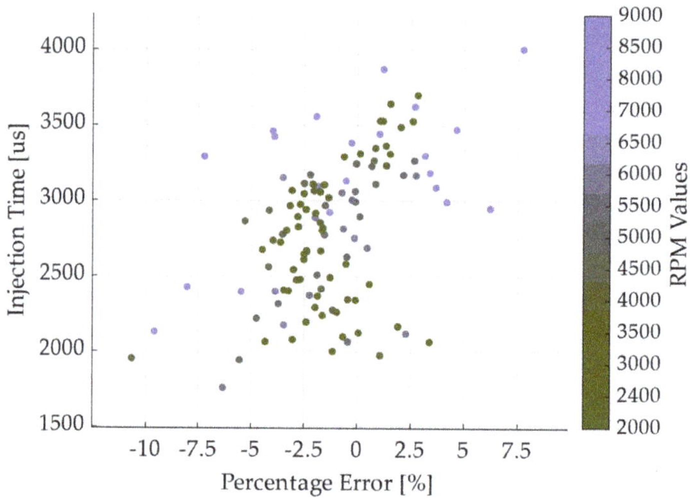

Figure 9. Percentage error vs. injection timing (Speed Density).

4.2. Motorcycle Calibration Validation

Finally, the calibrated map, calculated from the models starting from the air-to-fuel ratio required by the ECU calibration, was flashed into the vehicle. The vehicle was tested at the test bench in Road Load Control (RLC) with a real driver, to evaluate the performance of the calibration created by ML models. During these acquisitions, the usual engine parameters were measured, including MAP, various temperatures and pressures, and the lambda value measured by the sensor, as well as the target lambda channel, which is automatically calculated by the control unit through interpolation based on RPM and TPS relative to the point of interest. At Motori Minarelli, as in many other companies in the sector, calibrators use two thresholds around the target lambda value to describe an acceptable range for the measured lambda signal produced by combustion. In this case, the acceptable numbers must fall within +/−3.5% of the target during engine operation [16].

The results at high engine loads are reported in Figure 10 (AN strategy), whereas results at low loads are shown in Figure 11 (SD strategy). In terms of the latter, it was necessary to redo the training data acquisition for the machine learning models, repeating the maneuvers more slowly and achieving a finer discretization of points than the first attempt. In fact, at low loads and low engine speeds, small variations in TPS percentage correspond to large changes in the MAP signal (tens of KPa in pressure).

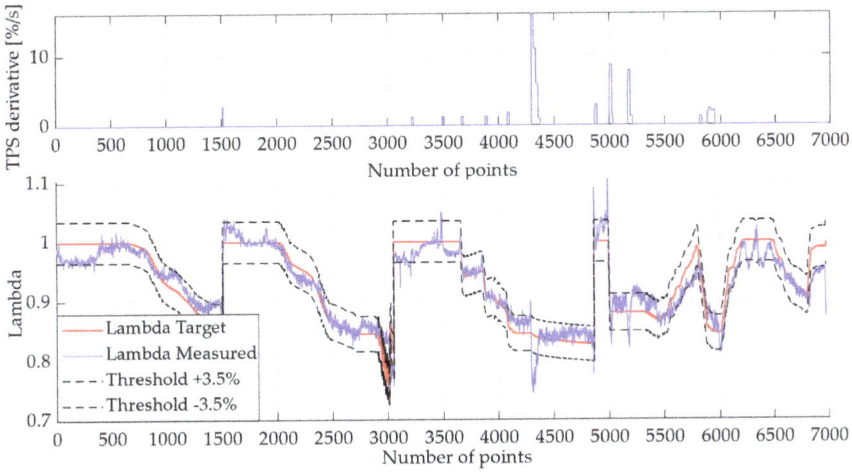

Figure 10. Vehicle data acquisition for Alpha Speed with ECU map created by IA.

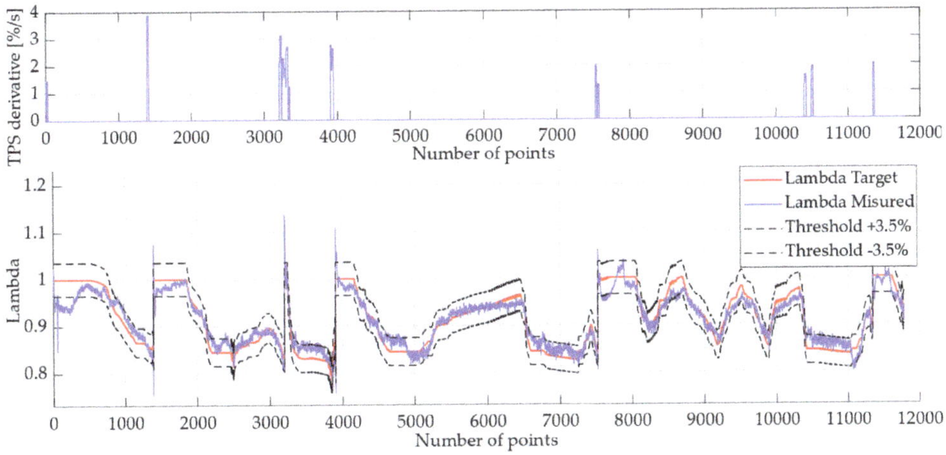

Figure 11. Vehicle data acquisition for Speed Density with ECU map created by IA.

The maneuvers were performed without high load gradient, because the fluid compensation film, which corrects the fuel mass flow, was not modeled and thus not evaluated. Each figure consists of two graphs. The upper graph shows the derivate of TPS [%/s] to evaluate if the maneuver was performed with slow gradient; the lower graph shows the lambda target required by the system (in red), the lambda signal measured (in blue), and an offset of the target of +−3.5% (dotted black) which represents the limit for an acceptable calibration. It can be seen that the lambda measurement coincides with the dotted lines; this evidence shows the good accuracy of the estimation. Only part of the dataset shows a lambda signal outside of the acceptable limits, but these points are not of interest because they are related to gear shifting or engine speed limiter interventions, which affect lambda signal because of cutoff or misfire, which can occur in these conditions.

Using the proposed approach, the reduction in calibration time, when compared to a conventional calibration process which requires approximately four days to complete, has been estimated at around 80%. It is important also to note that, in vehicle development, this calibration process is repeated multiple times, one for each modification made during the design phase (e.g., update of intake manifold, frame, fairings, driver seat, fuel tank, wiring, exhaust system). As a result, the benefits of employing machine learning models are compounded, ultimately leading to significant savings in both time and costs.

4.3. DOE Evaluation

A brief Design of Experiment (DOE) was then conducted to study how the performance of the models changed as a function of the number of points acquired for the training datasets (Figure 12). The objective was to obtain a grid that can be used in any acquisition context, which determines a minimum standard number of samples. This goal was achieved by attempting to discretize as uniformly as possible, especially at the boundaries, to avoid model extrapolation. A substantial portion of points was therefore eliminated, up to approximately 95% of the samples, i.e., from 4800 observations to about 70 operating points. The results obtained show the potential margin for improvement in the selection of key samples for model training [40].

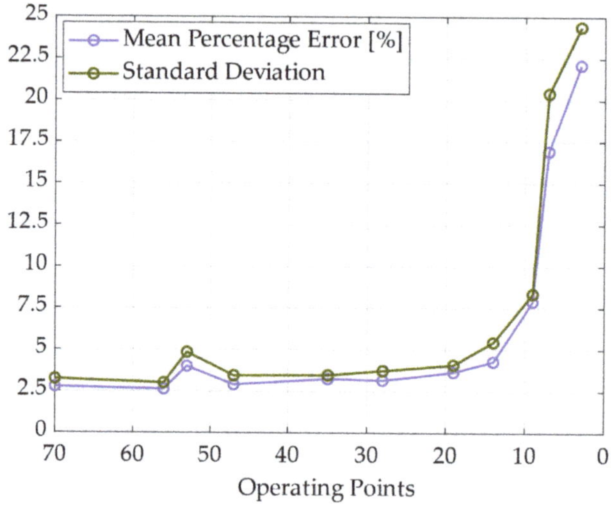

Figure 12. Mean percentage error vs. operating points.

The main reason for this result is related to the type of experimental setup. When using an entire vehicle on the roller bench, many acquisition points are essentially overlapping. However, it is interesting to note that acquisition can be performed more quickly and with the help of acquisition systems that also operate at low frequencies. The decision to use the entire vehicle was made because the goal of calibrating the complete system requires using the vehicle with all its parts. Conversely, for a less advanced engine development phase, the engine test bench would be more appropriate to optimize the DOE structure and measurement times.

This work has enabled a reduction in the engine air flow calibration procedures by about one-fifth compared to the traditional procedure, without losing points of accuracy.

The errors recorded during the modeling campaign are, in fact, comparable to the uncertainties of the measurement chain.

5. Conclusions

The main aim of this work is to demonstrate the potential of machine learning models for calibrating a complete motorcycle setup, chosen for its high accuracy requirements in production vehicles. Focusing on motorcycles, air-to-fuel ratio calibration is crucial not only for emissions and consumption but also for the instantaneous torque of the engine; therefore, small oscillations in lambda can significantly affect vehicle drivability.

The presented results show a completely new methodological approach, from the experimental setup to the model validation, and its direct application to the production vehicle. It has been demonstrated that this methodology can drastically reduce the calibration effort needed to fine-tune the vehicle at the production stage, without reducing accuracy. The machine learning models employed, particularly regression and neural network algorithms, have demonstrated their capability to accurately predict the optimal calibration parameters based on limited experimental data.

These models can prove advantageous not only in test bench conditions but also in a whole vehicle, demonstrating significant potential for other calibration.

The reduction in calibration time compared to a conventional calibration process has been estimated at around 80%, as traditional calibration takes approximately four days to complete.

Author Contributions: Conceptualization T.S., C.A.R., M.P., L.E. and G.V.; Data curation T.S., M.P., L.E. and G.V.; Writing T.S., C.A.R., M.P., L.E. and G.V.; Validation C.A.R., M.P., L.E. and G.V. All authors have read and agreed to the published version of the manuscript.

Funding: This research received no external funding.

Data Availability Statement: The data presented in this study are available on request from the corresponding author.

Acknowledgments: All authors would like to acknowledge Motori Minarelli for the support offered during the experimental process of this project.

Conflicts of Interest: Authors T. Savioli and M. Pampanini were employed by the company Atris Engineering srl. Author L. Esposito was employed by the company Motori Minarelli spa. The remaining authors declare that the research was conducted in the absence of any commercial or financial relationships that could be construed as a potential conflict of interest.

Abbreviations

ANN	Artificial Neural Network
ECU	Engine Control Unit
CAN	Controlled Area Network
TPS	Throttle Position Sensor
ALR	Auto Load Regulation
ASR	Auto Speed Regulation
UEGO	Universal Exhaust Gas Oxygen sensor
MAP	Manifold Absolute Pressure
CL	Closed Loop
SD	Speed Density
AS	Alfa Speed
DOE	Design Of Experiment
RLC	Road Load Control

References

1. Martyr, A.J.; Plint, M.A. *Engine Testing: Theory and Practice*; Elsevier: Amsterdam, The Netherlands, 2011; ISBN 978-0-08-052472-6.
2. Johnson, T.; Joshi, A. Review of Vehicle Engine Efficiency and Emissions. *SAE Int. J. Engines* **2018**, *11*, 1307–1330. [CrossRef]

3. Ashok, B.; Ashok, S.D.; Kumar, C.R. A review on control system architecture of a SI engine management system. *Annu. Rev. Control.* **2016**, *41*, 94–118. [CrossRef]
4. Castagné, M.; Bentolila, Y.; Chaudoye, F.; Hallé, A.; Nicolas, F.; Sinoquet, D. Comparison of Engine Calibration Methods Based on Design of Experiments (DoE). *Oil Gas Sci. Technol. Rev. d'IFP Energies Nouv.* **2008**, *63*, 563–582. [CrossRef]
5. Suykens, J.A.; Vandewalle, J.P.; De Moor, B.L. *Artificial Neural Networks for Modelling and Control of Non-Linear Systems*; Springer International Publishing: Cham, Switzerland, 2012; ISBN 978-1-4757-2493-6.
6. Samadani, E.; Shamekhi, A.H.; Behrouzi, M.; Chini, R. A method for pre-calibration of DI diesel engine emissions and performance using neural network and multi-objective genetic algorithm. *SID* **2009**, *28*, 61–70.
7. Turkson, R.F.; Yan, F.; Ali, M.K.A.; Hu, J. Artificial neural network applications in the calibration of spark-ignition engines: An overview. *Eng. Sci. Technol. Int. J.* **2016**, *19*, 1346–1359. [CrossRef]
8. De Simio, L.; Iannaccone, S.; Iazzetta, A.; Auriemma, M. Artificial neural networks for speeding-up the experimental calibration of propulsion systems. *Fuel* **2023**, *345*, 128194. [CrossRef]
9. Boccardo, G.; Piano, A.; Zanelli, A.; Babbi, M.; Cambriglia, L.; Mosca, S.; Millo, F. Development of a virtual methodology based on physical and data-driven models to optimize engine calibration. *Transp. Eng.* **2022**, *10*, 100143. [CrossRef]
10. Yang, R.; Yan, Y.; Sun, X.; Wang, Q.; Zhang, Y.; Fu, J.; Liu, Z. An Artificial Neural Network Model to Predict Efficiency and Emissions of a Gasoline Engine. *Processes* **2022**, *10*, 204. [CrossRef]
11. Jaliliantabar, F.; Ghobadian, B.; Najafi, G.; Yusaf, T. Artificial Neural Network Modeling and Sensitivity Analysis of Performance and Emissions in a Compression Ignition Engine Using Biodiesel Fuel. *Energies* **2018**, *11*, 2410. [CrossRef]
12. Ziółkowski, J.; Oszczypała, M.; Małachowski, J.; Szkutnik-Rogoż, J. Use of Artificial Neural Networks to Predict Fuel Consumption on the Basis of Technical Parameters of Vehicles. *Energies* **2021**, *14*, 2639. [CrossRef]
13. Brahma, I.; Sharp, M.C.; Richter, I.B.; Frazier, T.R. Development of the nearest neighbour multivariate localized regression modelling technique for steady state engine calibration and comparison with neural networks and global regression. *Int. J. Engine Res.* **2008**, *9*, 297–323. [CrossRef]
14. Meli, M.; Wang, Z.; Bailly, P.; Pischinger, S. *Neural Network Modeling of Black Box Controls for Internal Combustion Engine Calibration*; SAE Technical Paper 2024-01–2995; SAE International: Warrendale, PA, USA, 2024. [CrossRef]
15. de Nola, F.; Giardiello, G.; Gimelli, A.; Molteni, A.; Muccillo, M.; Picariello, R. Volumetric efficiency estimation based on neural networks to reduce the experimental effort in engine base calibration. *Fuel* **2019**, *244*, 31–39. [CrossRef]
16. Zhou, H.; Li, X.; Lee, C.-F.F. Investigation on soot emissions from diesel-CNG dual-fuel. *Int. J. Hydrogen Energy* **2019**, *44*, 9438–9449. [CrossRef]
17. Pan, T.; Cai, Y.; Chen, S. Development of an Engine Calibration Model Using Gaussian Process Regression. *Int. J. Automot. Technol.* **2021**, *22*, 327–334. [CrossRef]
18. Jacob, A.; Ashok, B. An interdisciplinary review on calibration strategies of engine management system for diverse alternative fuels in IC engine applications. *Fuel* **2020**, *278*, 118236. [CrossRef]
19. Rahimi-Gorji, M.; Ghajar, M.; Kakaee, A.-H.; Ganji, D.D. Modeling of the air conditions effects on the power and fuel consumption of the SI engine using neural networks and regression. *J. Braz. Soc. Mech. Sci. Eng.* **2017**, *39*, 375–384. [CrossRef]
20. Garg, A.B.; Diwan, P.; Saxena, M. Artificial neural networks based methodologies for optimization of engine operations. *Int. J. Sci. Eng. Res.* **2012**, *3*, 1–5.
21. Shayler, P.; Goodman, M.; Ma, T. The exploitation of neural networks in automotive engine management systems. *Eng. Appl. Artif. Intell.* **2000**, *13*, 147–157. [CrossRef]
22. Schoeggl, P.; Ramschak, E. *Vehicle Driveability Assessment Using Neural Networks for Development, Calibration and Quality Tests*; SAE Technical Paper 2000-01–0702; SAE International: Warrendale, PA, USA, 2000. [CrossRef]
23. Atkinson, C.; Mott, G. *Dynamic Model-Based Calibration Optimization: An Introduction and Application to Diesel Engines*; SAE Technical Paper 2005-01–0026; SAE International: Warrendale, PA, USA, 2005. [CrossRef]
24. Lowe, D.; Zapart, C. Point-Wise Confidence Interval Estimation by Neural Networks: A Comparative Study based on Automotive Engine Calibration. *Neural Comput. Appl.* **1999**, *8*, 77–85. [CrossRef]
25. Friedrich, C.; Auer, M.; Stiesch, G. Model Based Calibration Techniques for Medium Speed Engine Optimization: In-vestigations on Common Modeling Approaches for Modeling of Selected Steady State Engine Outputs. *SAE Int. J. Engines* **2016**, *9*, 1989–1998. [CrossRef]
26. Vinoth, B.; Uma, G.; Umapathy, M. Recurrent Neural Network based Soft Sensor for flow estimation in Liquid Rocket Engine Injector calibration. *Flow Meas. Instrum.* **2022**, *83*, 102105. [CrossRef]
27. Breiman, L. Random forests. *Mach. Learn.* **2001**, *45*, 5–32. [CrossRef]
28. Friedman, J.H. Greedy function approximation: A gradient boosting machine. *Ann. Stat.* **2001**, *29*, 1189–1232. [CrossRef]
29. Grinsztajn, L.; Oyallon, E.; Varoquaux, G. Why do tree-based models still outperform deep learning on typical tabular data? *Adv. Neural Inf. Process. Syst.* **2022**, *35*, 507–520.
30. Liu, J.; Ulishney, C.; Dumitrescu, C.E. Random Forest Machine Learning Model for Predicting Combustion Feedback Information of a Natural Gas Spark Ignition Engine. *J. Energy Resour. Technol.* **2020**, *143*, 012301. [CrossRef]
31. Kefalas, A.; Ofner, A.B.; Pirker, G.; Posch, S.; Geiger, B.C.; Wimmer, A. Estimation of Combustion Parameters from Engine Vibrations Based on Discrete Wavelet Transform and Gradient Boosting. *Sensors* **2022**, *22*, 4235. [CrossRef]
32. Ono Sokki Official Website. Available online: https://www.onosokki.net/ (accessed on 7 October 2024).

33. Preda, I.; Covaciu, D.; Ciolan, G. *Coast Down Test—Theoretical and Experimental Approach*; Transilvania University Press: Brașov, Romania, 2010.
34. INCA Base Product. ETAS Group. Available online: https://www.etas.com (accessed on 7 October 2024).
35. Jain, A.; Mao, J.; Mohiuddin, K. Artificial neural networks: A tutorial. *Computer* **1996**, *29*, 31–44. [CrossRef]
36. Espinoza-Jurado, J.; Dávila, E.; Rivera, J.; Raygoza-Panduro, J.J.; Ortega, S. Robust Control of the Air to Fuel Ratio in Spark Ignition Engines with Delayed Measurements from a UEGO Sensor. *Math. Probl. Eng.* **2015**, *2015*, 989674. [CrossRef]
37. Meng, L.; Wang, X.; Zeng, C.; Luo, J. Adaptive Air-Fuel Ratio Regulation for Port-Injected Spark-Ignited Engines Based on a Generalized Predictive Control Method. *Energies* **2019**, *12*, 173. [CrossRef]
38. Zhai, Y.-J.; Yu, D.-L. A Neural Network Model Based MPC of Engine AFR with Single-Dimensional Optimization. In *Advances in Neural Networks—ISNN 2007*; Liu, D., Fei, S., Hou, Z.-G., Zhang, H., Sun, C., Eds.; Springer: Berlin/Heidelberg, Germany, 2007; pp. 339–348. ISBN 978-3-540-72383-7. [CrossRef]
39. Uzun, A. Air mass flow estimation of diesel engines using neural network. *Fuel* **2014**, *117*, 833–838. [CrossRef]
40. Pontes, F.J.; Amorim, G.F.; Balestrassi, P.P.; Paiva, A.P.; Ferreira, J.R. Design of experiments and focused grid search for neural network parameter optimization. *Neurocomputing* **2016**, *186*, 22–34. [CrossRef]

fluids

MDPI

Article

Acceleration of Modeling Capability for GDI Spray by Machine-Learning Algorithms

Yassine El Marnissi [1], Kyungwon Lee [1] and Joonsik Hwang [1,2,3,*]

[1] Department of Mechanical Engineering, Mississippi State University, Starkville, MS 39762, USA; ye44@msstate.edu (Y.E.M.); kl1598@msstate.edu (K.L.)
[2] Center for Advanced Vehicular Systems (CAVS), Mississippi State University, Starkville, MS 39762, USA
[3] Institute for Clean Energy Technology (ICET), Mississippi State University, Starkville, MS 39762, USA
* Correspondence: hwang@me.msstate.edu

Abstract: Cold start causes a high amount of unburned hydrocarbon and particulate matter emissions in gasoline direct injection (GDI) engines. Therefore, it is necessary to understand the dynamics of spray during a cold start and develop a predictive model to form a better air-fuel mixture in the combustion chamber. In this study, an Artificial Neural Network (ANN) was designed to predict quantitative 3D liquid volume fraction, liquid penetration, and liquid width under different operating conditions. The model was trained with data derived from high-speed and Schlieren imaging experiments with a gasoline surrogate fuel, conducted in a constant volume spray vessel. A coolant circulator was used to simulate the low-temperature conditions (−7 °C) typical of cold starts. The results showed good agreement between machine learning predictions and experimental data, with an overall accuracy R^2 of 0.99 for predicting liquid penetration and liquid width. In addition, the developed ANN model was able to predict detailed dynamics of spray plumes. This confirms the robustness of the ANN in predicting spray characteristics and offers a promising tool to enhance GDI engine technologies.

Keywords: machine learning; Gasoline Direct Injection (GDI); projected liquid volume; tomographic reconstruction

1. Introduction

In recent decades, Gasoline engines has been recognized as one of the efficient and widely used internal combustion (IC) engines. Approximately 13 million gasoline cars were sold in the United States in 2023 [1]. However, it is important to note that Greenhouse gas emissions increased from 2022 to 2023 by 1.1 percent (36.8 billion Metric Tons) [2]. Gasoline direct injection (GDI) plays a significant role in contributing to these emissions and typically offers better fuel efficiency compared to Port Fuel Injection (PFI) engines [3]. Fuel is injected directly into the combustion chamber in GDI engines. This technique prevents wall wetting within the port, reduces the time required for fuel transport, provides precise regulation of fuel entering the combustion chamber, and has the potential for cleaner combustion. Multiple injections, along with high fuel injection pressure, improve control over the injection process, leading to a higher vaporization rate and thus overcoming cold start problems. Most of the technical challenges associated with conventional Port Fuel Injection (PFI) are overcome in GDI, particularly in achieving higher engine volumetric efficiencies. In GDI, the absence of vapor in the intake charge allows more air induction into the combustion chamber. Furthermore, the direct injection prevents engine knocking by using advanced evaporative cooling which lowers the in-cylinder temperature. However, GDI engines are facing challenges during start, which can lead to changes in the spray pattern and, consequently, delayed ignition. This delay can have an impact on the production of hydrocarbon (HC), carbon monoxide, and nitrogen oxide (NO_x). The spray break-up and mixing properties are affected by cold fuel temperature, which results in the deterioration of the combustion process. One of the significant challenges to understanding the deterioration of the combustion process is the inappropriate injection parameter or undesired fuel properties. The analysis of spray evolution is notably complex due to the involvement of turbulent dispersed multiphase flows, as well as the injector design factors such as nozzle orifice, and the counterbore hole [4]. The spray formation and characteristics depend on the fuel properties and breakup, which are governed by the atmospheric, cold-start conditions. To address these challenges of spray behavior in the GDI engines, the adoption of machine learning could offer substantial improvement in predictive accuracy.

Literature on cold start spray pays attention to the examination of spray characteristics using non-invasive optical diagnostics [5,6]. The reliability of these optical diagnostic methods is crucial for acquiring experimental data, which forms the foundation for machine learning, particularly in training, and developing accurate and reliable predictive models of spray behavior. Tong et al. [7] focused on evaluating the impact of fuel volatility on the cold start performance of GDI engines using Planar Laser-Induced Fluorescence (PLIF) and Mie scattering. They found that during the initial phase of the cold start, low-volatility fuels result in unstable combustion and higher unburned hydrocarbon (UHC) emissions. Bruno et al. investigated the effects of injection pressures ranging from 2 MPa to 10 MPa on spray penetration and fuel distribution during cold starts using optical diagnostics. The experimental results indicated that degraded cold start performance was observed at pressures below 5 MPa, while the best performance was achieved at higher injection pressures combined with a late start of injection (SOI). Various optical diagnostics such as the Schlieren shadowgraph and high-speed imaging were utilized to examine the dispersion of liquid vapor and droplet atomization, providing detailed insights into spray dynamics and accurate data for training machine learning models.

Computational fluid dynamics (CFD) simulation also offers a detailed and cost-effective approach to understanding spray dynamics, designing new sprays, and optimizing device operating conditions. The predominant method to simulate sprays is the Lagrangian simulations [8]. Daniele [9] investigated the air-fuel mixing under a Spray G injector using a 3D CFD model. The simulation results agreed with the experimental measurements in terms of wall film formation and spray characteristics. Filippo et al. [10] studied the impact of the initial droplet distribution on the predicted spray physics using Lagrangian and Eulerian methods in liquid fuel atomization. They compared the results with the spray conditions of the baseline spray G from the ECN. The study found that accurate predictions of liquid and vapor penetration can be achieved through a careful selection of the initial mean droplet size, in conjunction with standard evaporation and breakup models. Despite the development of CFD, achieving precise details of spray evolution under cold start remains challenging due to complex spray dynamics associated with changes in fuel properties according to fuel and ambient temperatures.

Given this complexity, as simulations incorporate more complex phenomena, computational expenses rise. Selective use of machine learning offers a practical method to reduce computational costs [11,12]. To obtain more information on spray behavior and atomization, such as flow patterns, velocities, turbulence, and the breakup of liquid gasoline into smaller droplets, large datasets are necessary for CFD simulations. In contrast, machine learning can effectively predict both macroscopic and microscopic spray characteristics with only limited training data [13]. For a more in-depth insight into spray behavior, researchers commonly rely on Reynolds-Averaged Navier–Stokes (RANS). RANS averages flow properties over time to solve the governing equations for fluid flow and also provide a deeper understanding of the cold start behavior [14]. This detailed data from RANS simulations can serve as a useful foundation for supervised training in machine learning, which enhances its predictive capabilities in fluid dynamics [15]. Meanwhile the large eddy simulation (LES) approach seems to be better at predicting flow separation and is well-suited for unsteady or transient flows as it captures the large-scale turbulent structures [16]. Moreover, LES can be effectively combined with machine learning to enhance simulations of spray phenomena and reduce memory occupation in overall combustion simulations [17,18].

Understanding the quantitative effects of various conditions on fuel spray behavior requires reliable experimental data that can serve as input to mathematical models capable of predicting fuel spray characteristics and spray dynamics. Recently, machine learning has become a viable approach to predict spray dynamics [19,20]. The continuous advancement of precise measurements and simulations is rapidly increasing the generation of vast amounts of data to cover a wide array of fields associated with classification, regression, and computer vision. The application of machine-learning algorithms to analyze extensive datasets is driving innovative advancements across a diverse range of fields. Nowadays, artificial neural network (ANN) models are used to investigate spray behavior in relation to various parameters, including nozzle design and engine-specific parameters. ANNs are employed due to their robust capacity for nonlinear fitting, with this fitting ability contingent upon the number of hidden layers, neurons, and hidden functions. Junjian [21] studied the spray tip penetration of octanol diesel by the ANN model. The study established 15 ANN models, and the results showed that models had a high prediction accuracy with an R^2 value of 0.99901, which was in good agreement with the experimental results of the 20% octanol-blended fuel. The use of ANN models allowed for better prediction results compared to existing mathematical models, avoiding prediction errors. The work of Hwang et al. [22] investigated the capability of a machine-learning algorithm to predict the spray morphology under ECN spray G conditions. They developed a predictive model based on a variant of linear regression and trained it using data from spray experiments carried out in a constant-flow spray vessel. The computed tomographic algorithm was used to construct a 3D spray from the projected liquid volume data at three viewing angles 0°, 11.25°, and 22.5°. The results from the predictive model showed excellent agreement with the experimental data, demonstrating the efficacy of linear regression with a set of nine input features to predict spray performance. Thangaraja [23] applied ANN to generate a wide range of thermophysical properties such as density, viscosity, specific heat, thermal conductivity, and more, for biodiesel fuels based on their composition and temperature. Following ANN applications, Seo et al. [24] estimated cold start emissions using an ANN. They developed various models that accurately predict the surge in carbon monoxide levels. Giannelli et al. [25] conducted an analysis of cold start emissions by using a series of Gaussian functions and a semi-empirical equation involving tractive power. The study considered the effect of catalyst warm-up and fast idle on emissions. Sabatini et al. [26] developed a semi-empirical temperature model to simulate thermal transients in a catalytic converter. They found that during the cold start, the model can estimate the high-off temperature.

In this study, the objective is to evaluate the capability of an ANN to accurately predict cold start spray dynamics and provide detailed quantitative information based on operational conditions such as fuel temperature and injection pressure. While previous studies have employed ANNs to model spray characteristics, a significant gap persists in developing models capable of accurately predicting the intricate 3D spray dynamics specific to cold-start conditions in GDI engines. By integrating data obtained from high-speed and Schlieren imaging into the ANN model, this research aims to develop a reliable tool that can predict 3D spray dynamics, liquid penetration, and liquid width. The predicted spray dynamics from the ANN model are compared with experimental data to ensure the model's validity. This tool is crucial for analyzing how the air-fuel mixture distribution behaves during the cold start of GDI engines, a key factor in reducing emissions and enhancing engine performance.

2. Experimental Setup

2.1. Injector Configuration and Test Fuel

The spray test was conducted under engine conditions in a constant volume vessel, controlled in pressure and temperature. A high-pressure GDI fuel injector is used to inject fuel at seven different pressure levels, ranging from 50 bar to 350 bar. Its geometric parameters are listed in Table 1.

Table 1. Injector Nozzle Geometry.

Parameter	Specification
Nozzle type	Multi-hole nozzle
Number of Nozzles	8
Nozzle Shape	Axisymmetric
Nozzle Angle	26.6°
Inner Orifice Diameter	150.6 μm
Counter-bore Diameter	360.4 μm

The injector was preconditioned by circulating coolant to guarantee uniformity in the experiment. The fuel and injector temperature were measured using a 1/16 thermocouple. The lowest temperature reached by the injector tip was −7 °C. Despite efforts to lower it further, this was limited by the heat exchange with the coolant lines. A Teledyne ISCO 65X syringe pump (Teledyne labs, Mason, OH, USA) was used to generate the target fuel injection pressure.

A surrogate fuel designated as PACE-20 was used as the testing fuel due to its excellent performance in terms of RON, distillation curve evaluations, and its ability to accurately capture the realistic evaporative spray characteristics. Table 2 lists the properties of PACE-20 fuel.

Table 2. PACE-20 Fuel Properties.

	PACE-20
RON	92.1
MON	84.2
H:C	1.97
PMI	1.5
Ethanol [wt%]	10.2
Total Aromatics [wt%]	28.6
Total Olefins [wt%]	4.9
HOV [KJ/kg]	407.5
Density@ 15 °C [kg/L]	0.74

2.2. Optical Techniques

The experiments were carried out in a constant-volume vessel. Figure 1 represents the high-speed imaging setup. Schlieren imaging and high-speed extinction techniques are employed to capture the vapor and liquid phases, respectively. Two optical windows are specifically designated for these tasks. The high-speed visualization is achieved using five quartz windows, each having a diameter of 90 mm. The liquid phase spray is identified through diffused backlight illumination extinction imaging. A green LED with 60 ns TTL modulation is used to provide high visibility, paired with a Fresnel lens with a diameter and focal length of 150 mm each. The light is then directed through an engineered diffuser to evenly distribute it. To further enhance the accuracy and clarity of the image, we employ a band-pass filter with a bandwidth of 20 nm.

1. Top of Form

A Phantom (V611) camera (Vision Research Inc., Wayne, NJ, USA), renowned for its high-speed imaging capabilities, was utilized in this study. It was set with a higher shutter speed (14,000 fps) and a resolution of 608 × 608 pixels. For detailed imaging of the spray evolution, the camera was equipped with two lenses: an autofocus Nikkor lens with a 50 mm focal length and a wide maximum aperture of f/1.8.

The DBIE optical method was implemented to selectively capture the liquid phase (indicated by the green LED) and isolate it from the vapor phase. An injector with a rotating adapter was installed in the vessel chamber, enabling the adjustment of viewing angles in a clockwise direction at 0°, 11.25°, and 22.5°. These specific angles facilitate 3D tomographic reconstruction.

In this study, a Schlieren z-type configuration is employed for the visualization of vapor phases due to its high sensitivity to changes in the refractive index of air that occur when fuel vaporizes [27]. A red LED light is focused through a Nikkor lens and aperture. The light is then passed through a concave mirror that collimates the light and directs it through the test region. The light was then directed to a second concave mirror, which collects the parallel light rays. To enhance the sensitivity, a knife edge with a 50% cut-off at the focal point was used. The Schlieren and extinction imaging setups are identical and the components of the optical arrangement for the Schlieren setup are summarized in Table 3.

Figure 1. Experimental Setup for Extinction and Schlieren Imaging.

Table 3. Schlieren Optical Configuration Imaging Setup.

Component	Description
Camera and Lens	High-Speed Camera; Prime Lens: AF-S Nikkor 85 mm f/1.8G (Nikon)
Light Source and Lens	Red LED, 500 ns; Collection Lens: Nikkor, 50 mm f/1.8
Aperture and Collimation	0.5 mm diameter aperture; Collimating Mirror: GSO, 152 mm f/6.0 (Concave)
Light Path and Enhancement	Through spray region; Knife-Edge at focal point, ~50% cut-off

2.3. Schlieren Photography

In a Schlieren imaging setup, light travels uniformly through the spray test region. However, when it enters a medium with a different refractive index, the light rays are refracted. The corresponding linear relationship (Gladstone–Dale relation) relates the refractive index of the medium with the gas density [28]:

$$n - 1 = k\rho \tag{1}$$

where n represents the refractive index, ρ denotes the density, and k is the molar refractivity. Moreover, the change in the direction of a ray propagating in the z direction can be expressed as follows:

$$\varepsilon_x = \frac{L}{n_0}\frac{\partial n}{\partial x}\ \varepsilon_y = \frac{L}{n_0}\frac{\partial n}{\partial y}\ \text{Plane}\ (x - y) \tag{2}$$

where L and n_0 represent the optical distance and the index of refraction of the surroundings, respectively. Equations (1) and (2) demonstrate a clear correlation between the variations in the deflection angle and the gas density, which can be effectively captured through Schlieren imaging. Consequently, the deflection of light rays that pass through the spray region can be visualized and is dependent on the sensitivity of the Schlieren setup, resulting in an image with a distinctive background.

In order to isolate the spray region and remove background Schlieren fluctuation, an appropriate background intensity I_0 was subtracted from the spray image I. The detection of the vapor envelope was achieved using the "imgradient" function, which is available in the image-processing tool in MATLAB. Subsequently, an area detection method based on "bwperim" function was implemented. This function allows the identification of spray macroscopic characteristic parameters such as vapor penetration length which is defined as the distance between the nozzle tip and the furthest point on the vapor boundary at each injection. It is noteworthy that prior to using machine learning, the resultant images obtained by high-speed and Schlieren imaging might undergo preprocessing, including intensity normalization, noise reduction, and z-score normalization [29]. Such preprocessing is important for achieving accurate spray predictions [30].

3. Image Processing Technique

3.1. Extinction Imaging Method

Extinction imaging is a quantitative method for measuring liquid fuel concentration that provides more accurate measurements of spray characteristics compared to the standard Mie scattering method. The Mie scattering imaging technique is known to face challenges due to the inherent nature of its scattering process, which generates undesired signals in areas with dense particle concentrations. Consequently, extinction imaging is now recognized as a fundamental diagnostic technique in the field of spray research. The Projected Liquid Volume (PLV) along a defined line of sight is determined based on factors such as droplet size, the extinction coefficient, and liquid particle size. The formula to determine the optical thickness uses the Beer-Lambert law that relates the reduction in light intensity to the characteristics of the medium it passes through. The formula is defined as follows:

$$\tau = -\mathrm{In}\left(\frac{I}{I_0}\right) \tag{3}$$

where I and I_0 represent the transmitted and incident light intensities, respectively. This level of transmission is suitable for detecting the spray "edge" above the camera noise floor. However, it is important to take into account factors like vapor-phase beam steering. The optical depth, referred to as τ, is calculated and used to determine the Projected Liquid Volume (PLV). This suggests a spatial integration of the Liquid Volume Fraction (LVF) across the measurement domain, particularly in the cross-stream direction (y), as expressed in the following formula:

$$\mathrm{PLV} = \tau\frac{\pi d^3/6}{C_{ext}} = \int_{-y_\infty}^{y_\infty} (\mathrm{LVF}).dy \tag{4}$$

where d and C_{ext} are droplet diameter and extinction coefficient, respectively. The extinction coefficient (C_{ext}) represents the extent to which droplets in the medium attenuate light through absorption. Its value depends on the size of the droplets, the wavelength of light, the refractive index of the medium, and the collection angle of the optical setup. It was assumed, based on measurements taken during the injection across the plume, that the Sauter Mean Diameter (SMD) consistently approximated 12 µm. In addition, C_{ext} was determined to be 272×10^{-6} mm^2, based on a droplet diameter of 12 µm. This value was obtained for individual experimental setups using MiePlot v4.6 available at [31].

3.2. 3D Tomographic Reconstruction

The CT algorithm was used to construct a 3D spray from the liquid volume fraction (LVF) data at three different viewing angles 0°, 11.25°, and 22.5°. The reconstruction technique was carried out with MATLAB R2023b using a function like "iradon" for performing the inverse Radon transform and to support GPU computing. The reconstruction procedure was executed on axial sections extending from the nozzle tip to a position z = 60 mm downstream, with a hamming filter parameter value of 0.3. This approach offers detailed information about the distribution and movement of the plumes and better spatial and temporal resolution.

4. Machine Learning Method and Computational Framework

4.1. Machine Learning Networks

A multilayer feed-forward ANN has emerged as a modeling technique in predicting spray dynamics due to its high efficiency in identifying the nonlinear relationships between sets of input and output data. ANN accepts an input, propagates it through numerous layers of hidden neurons, and generates a prediction that links the input of all the neurons.

A neural network is essential because it involves adjusting the weights and biases of the neurons based on the input data. More specifically, Feed-forward ANN architecture comprises three types of layers: the input layer, the output layer, and hidden layers in between, as shown in Figure 2.

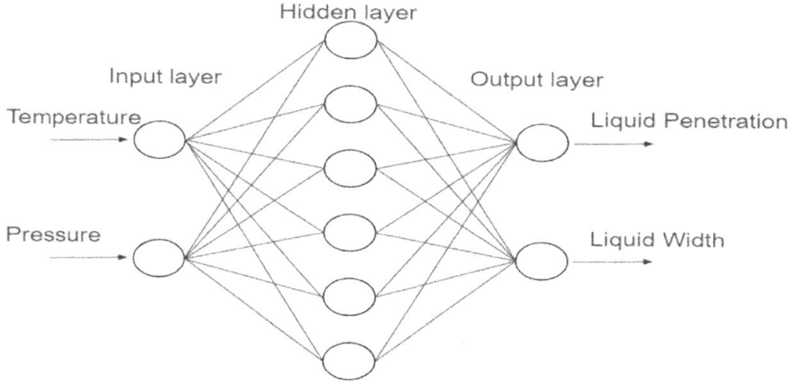

Figure 2. ANN structure with a single hidden layer.

An artificial neuron consists of numerous analog signal-processing elements known as "neurons". A neuron within the hidden layers can receive one or more input data $[X_1, X_2, \ldots, X_m]$. These inputs are transferred to the artificial neuron and subsequently multiplied by their respective synaptic weights W_i. An additional input signal, known as bias input w_0, is introduced to the artificial neuron alongside the input data. The neuron subsequently employs an activation function f(z), converts the weighted sum of input signals into output signals, and forwards the outcome to all the neurons in the subsequent layer.

$$y = f(z) = f\left(w_0 + \sum_{i=1}^{N} w_i x_i\right) \quad (5)$$

This process enables the network to learn from the data and enhances its ability to make accurate predictions and informed decisions.

There are different types of activation functions including linear, tangent, sigmoid, and hyperbolic. Among the mentioned functions, the sigmoid takes any real value as input and output values in the range 0 to 1.

In this paper, the sigmoid function is employed for all hidden layers and linear for the output layer, as it exhibits a satisfactory performance in terms of prediction. The coefficient of determination and mean square error are taken as a valuable indicator of how well the input variables (injection pressure and fuel temperature) in our model explain the variations in the dependent variables, which is spray tip penetration.

The coefficient of determination and mean square error can be calculated as follows:

$$R^2 = 1 - \frac{\sum_{i=1}^{m}(Y_i - Y_{i,model})^2}{\sum_{i=1}^{m}(Y_i - \overline{Y}_i)^2} \quad (6)$$

$$MSE = \frac{1}{n}\left\{\left(Y_i - Y_{i,model}\right)^2\right\} \quad (7)$$

Y_i: The value of the variable to be predicted.
$Y_{i,model}$: Predicted value of Y_i.
$\overline{Y}_i$: The mean of the sample.
n: The number of values in the sample.

4.2. Computational Setup

Computational analyses were performed on a high-performance workstation equipped with an AMD Ryzen Threadripper 7980X (3.2 GHz, 64 cores), GeForce RTX 4090 24 GB GDDR6X, and 64 GB DDR5/6800MHz Kingston FURY Renegade memory, on a GIGABYTE TRX50 AERO D motherboard.

In this study, the projected liquid volume was measured through experiments and used as the target for the ANN. This ANN model is trained using input and target tables, which facilitates the management of extensive PLV data under various operational conditions. This approach enables the development of flexible networks that are optimized for efficient file size management, reduce storage demands, and improve computational efficiency while maintaining high accuracy. For training, we used two features—temperature and pressure—which served as inputs for the trained neural network. Each training step incorporates projected liquid volume (PLV) values obtained from experiments. It should also be noted that since our model is linear and does not contain complex terms such as products or power terms, we did not standardize these features.

For PACE-20 fuel, we predict LVF cold spray at −7 °C at pressures of 100 bar and 350 bar. Each condition, such as (−7 °C, 100 bar), is represented by a 1 × 300 cell array that captures the evolution of the spray. Within each cell of the array, there is a 608 × 608-pixel image of the spray, taken at distinct intervals within the 0 to 2 ms spray evolution.

4.3. Image Regression Modeling

In this study, the regression model for spray images of size (M × N) was trained using temperature and pressure as input vectors, and the associated PLV data vector as the output. This study focuses on predicting the behavior of cold spray across a range of pressures and at various times. For each temperature, the ANN is trained using input tables paired with a range of injection pressures (50 bar to 350 bar). For the target, we organize the PLV data into 608 tables, each with dimensions of 608 × 7. We construct each table by extracting the first column from each set of PLV data images and placing it into the first of these 608 tables. We then proceed with the second column for the second table, and so forth. After assembling these tables, each one is trained individually. Once the ANN training is complete, we employ the reverse process, recompiling the trained columns back into 608 × 608 images. Figure 3 illustrates the schematic of the training algorithm for a spray image.

The mean processing time for the ANN to process seven images of specified size 608 × 608, corresponding to the seven pressures at a fixed temperature is 1 h. The training process involved 600 iterations and took nearly 3 h to predict three images from three different angles for PLV data (spray images). An early stopping mechanism was integrated to monitor validation performance. If the coefficient of determination (R^2) fell below 0.991 for any of the 608 tables, the training was stopped, and adjustments were made to the number of neurons in the hidden layers until the model converged with an R^2 value greater than 0.991 before resuming the training process. The use of enhanced GPUs can accelerate this procedure. Model accuracy can be improved by optimizing the hidden layers or considering different activation functions. As shown in Figure 4, the ANN prediction accurately captures the overall shape and distribution of the spray core region. However, at the edges, there is a presence of minor noise, which may be attributed to the variability in pixel intensities resulting from the optical techniques used during the experimental image acquisition.

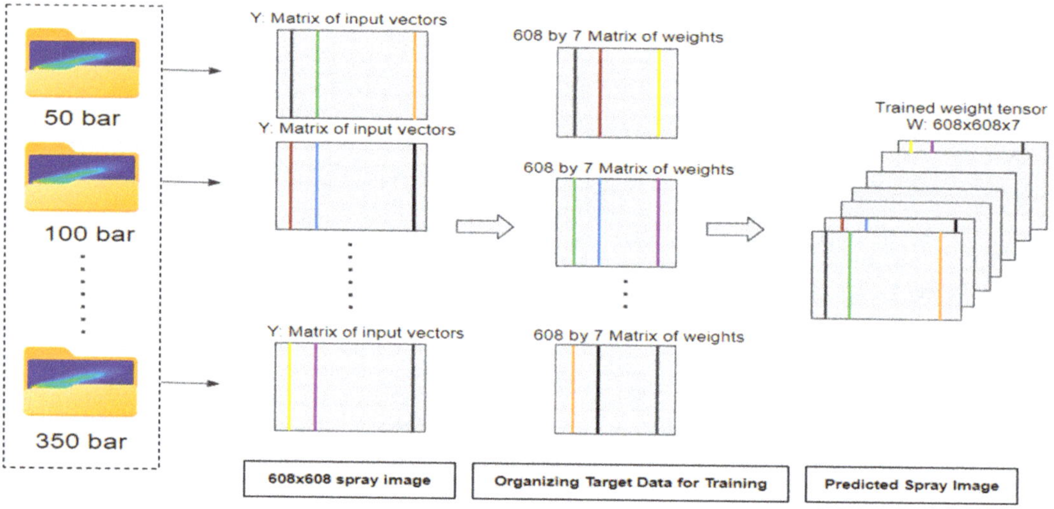

Figure 3. ANN Training Framework for Spray Topology Reconstruction.

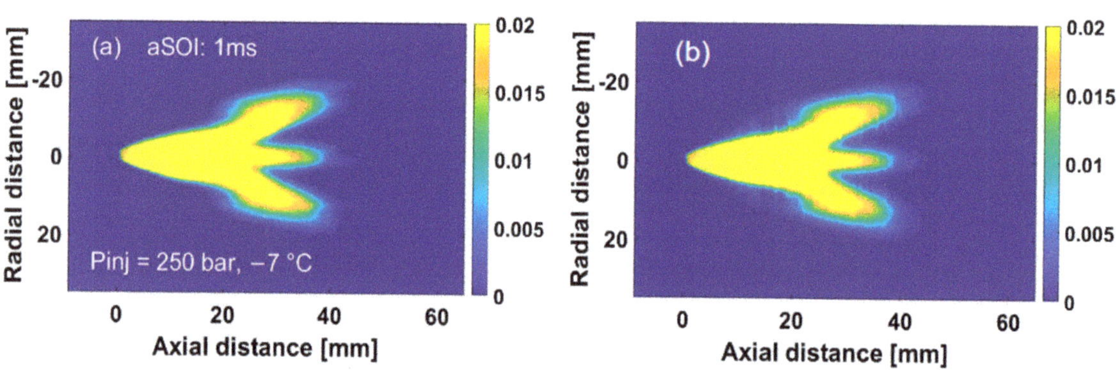

Figure 4. (**a**) Experimental PLV Map. (**b**) PLV Map Predicted by ANN (Machine Learning).

The trained regression model predicts the spray dynamics at any time step, thanks to the strategy of training a time series regression model and methodically repeating this process. This offers a significant advantage compared to CFD models that require time-stepping simulations. The ANN enables rapid optimization of complex processes, which allows for quicker adjustments with fewer simulations for more efficient outcomes [32]. It should also be noted that recent studies have developed a hybrid method combining machine learning (ML) and numerical methods. This approach aims to address errors in under-resolved simulations and reduce computational time. Dmitrii et al. [33] introduced a new approach for solving nonlinear partial differential equations. In this study, ML is employed to enhance the approximations within CFD simulations and provide accurate solutions at each point on a grid that would be considered too coarse to accurately resolve the dynamics.

5. Results and Discussion

5.1. Experimental and Machine Learning Prediction of Liquid and Vapor Spray

The liquid penetration and liquid width significantly influence the spray morphology. Accordingly, the line-of-sight data analysis was performed using a PLV threshold of 0.2×10^3 mm^3 (liquid)/mm^2. The PLV data represents the average of 20 experimental runs under each set of working conditions. To ensure reliability, we used the mean standard errors of both liquid penetration and width, denoted as SEM = $\sigma/\sqrt{n}$ values. These mean standard errors were found too small to be discernible in the figures. Measurements of liquid penetration length and liquid width were obtained under different injection pressure conditions P_{inj} = [50, 150, 350] bar and T_{fuel} = [−7, 60] °C. Figure 5 illustrates the temporal evolution of PLV and Schlieren patterns captured at the initial stages of spray formation under 50 bar and −7 °C. Both liquid penetration and liquid width were synchronized at the time just after the actual start of injection. The experimental images that were acquired indicated the presence of an injection delay, characterized as the time between the start of electronic command and the start of injection (SOI). As a result, it was observed that the early spray formation for the hot fuel was faster in comparison to the cold fuel scenarios due to the reduced viscosity of the hot fuel, which allows fast needle opening. In addition, hot fuel spray promotes better atomization and faster spray formation from multi-hole injectors [34]. Meanwhile, the cold fuel spray affects stability due to the straighter edges produced, which indicate weaker vortex formation and reduced mixing with air [35]. The results showed that as injection pressure increases, both liquid and vapor penetration lengthen due to enhanced air entrainment and increased flow momentum. It is also observed that the liquid penetration initially increases linearly during the first phase, which includes acceleration and the first transition stages, followed by the quasi-steady stage and second transition, where the penetration reaches its maximum and subsequently becomes stagnant.

By comparing the liquid and vapor penetration at high and low temperatures, it can be observed that with an increase in injection pressure, the liquid/vapor at high temperatures extends further in comparison to low temperatures [36]. At high injection pressure, a significant difference in liquid width was observed between the distances of 15 mm and 30 mm. Furthermore, in the PLV and Schlieren intermittency images at high temperatures, the spray encountered collapsing, resulting in an accelerated axial penetration of liquid/vapor due to spray momentum. Based on these, detailed experimental data and insights serve as the foundation of the ANN model to predict the dynamics of spray under various operational conditions.

It is widely agreed that factors such as injection pressure and fuel temperature affect spray penetration and width. However, it is evidently impractical to account for every factor in the development of an ANN model. The objective of our model is not to create a general model encompassing liquid penetration and width under various conditions, but rather to specifically provide quantitative information on the effects of injection and environmental conditions on the behavior of the spray. With a focus on liquid penetration, width, and 3D LVF predictions under cold-start conditions using the ML algorithm, successful predictions in this domain are expected to be applicable to vapor spray behaviors as well. The PLV dataset resulting from these experiments was used to train an ANN model that predicts the spray under various operational conditions P_{inj} = [50, 150, 350] bar and T_{fuel} = [−7, 60] °C. It is worth noting that the modeling approach is designed to achieve high accuracy. A series of tests conducted on various ANN architectures revealed that a configuration with two layers was particularly effective in accurately predicting the spray image. Table 4 illustrates the ANN architecture, comprising an input layer with two neurons, a hidden layer with two neurons, and an output layer with a variable neuron. The training of the

ANN is performed using the Levenberg-Marquardt training algorithms available within the MATLAB Neural Fitting Tool (nftool). The choice of the best ANN network is based on the coefficient of determination R^2-values for training, validation, and testing. The model is developed using experimental data from PACE-20. In this process, 70% of the data is randomly selected for the training set, while 15% is randomly chosen for both testing and validation. The ANN training process for predicting spray characteristics (liquid penetration and width) involved 200 iterations, and it took less than 1 min to complete.

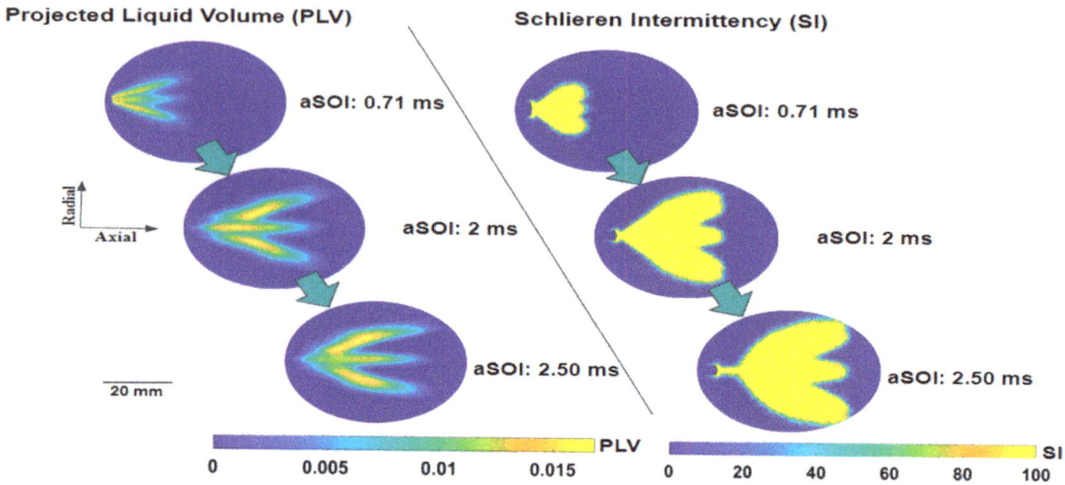

Figure 5. Temporal Evolution of PLV and Schlieren Patterns Under 50 bar and −7 °C.

Table 4. ANN configuration.

Criteria	Details
Image dimensions	608 × 608
Data input	7 PLV values (images of spray)
Feature set	2 characteristics: Fuel temperature (°C) and injection pressure (bar)
ANN Architecture	Input layer 2 × hidden layer 2
Training Algorithm function	Levenberg Marquardt
Training Data Points (70%)	PLV (−7 °C): 209,093 Liquid Penetration and Width: 2940
Validation Data Points (15%)	PLV (−7 °C): 44,806 Liquid Penetration and Width: 630

Figure 6 illustrates the temporal evolution of liquid penetration and width over time after the start of injection (aSOI) at these conditions. Solid lines are true values from experiments and symbols are predicted by the machine learning model. In the analysis of liquid penetration and width, the results indicate that the largest deviations between the predicted values by the ANN and the experimental data for both the liquid length and liquid width throughout the injection phase are 3.1 mm with an error of 5.16%, and 0.9 mm with an error of 3.10%, respectively. Similarly, as observed, the developed model accurately predicts the vapor penetration length and vapor width, with small deviations between the experimental and predicted results, showing an error of just 0.7%. Moreover, predicting spray behavior using CFD under various conditions, such as cold starts and hot fuel scenarios, can be challenging due to complex physical processes that include spray dynamics, the effects of temperature and pressure variations, and droplet behavior affected by air entrainment, which increases the turbulence of spray motion [37]. In contrast, the developed machine learning model has shown good agreement with the experimental data in accurately predicting the penetration and width of both liquid and vapor. It is noteworthy that the results indicate a reasonably strong fit for all datasets that include both liquid and vapor spray, with an overall R^2 of more than 0.99. Figure 7 illustrates the training accuracy of the ANN model in predicting liquid penetration.

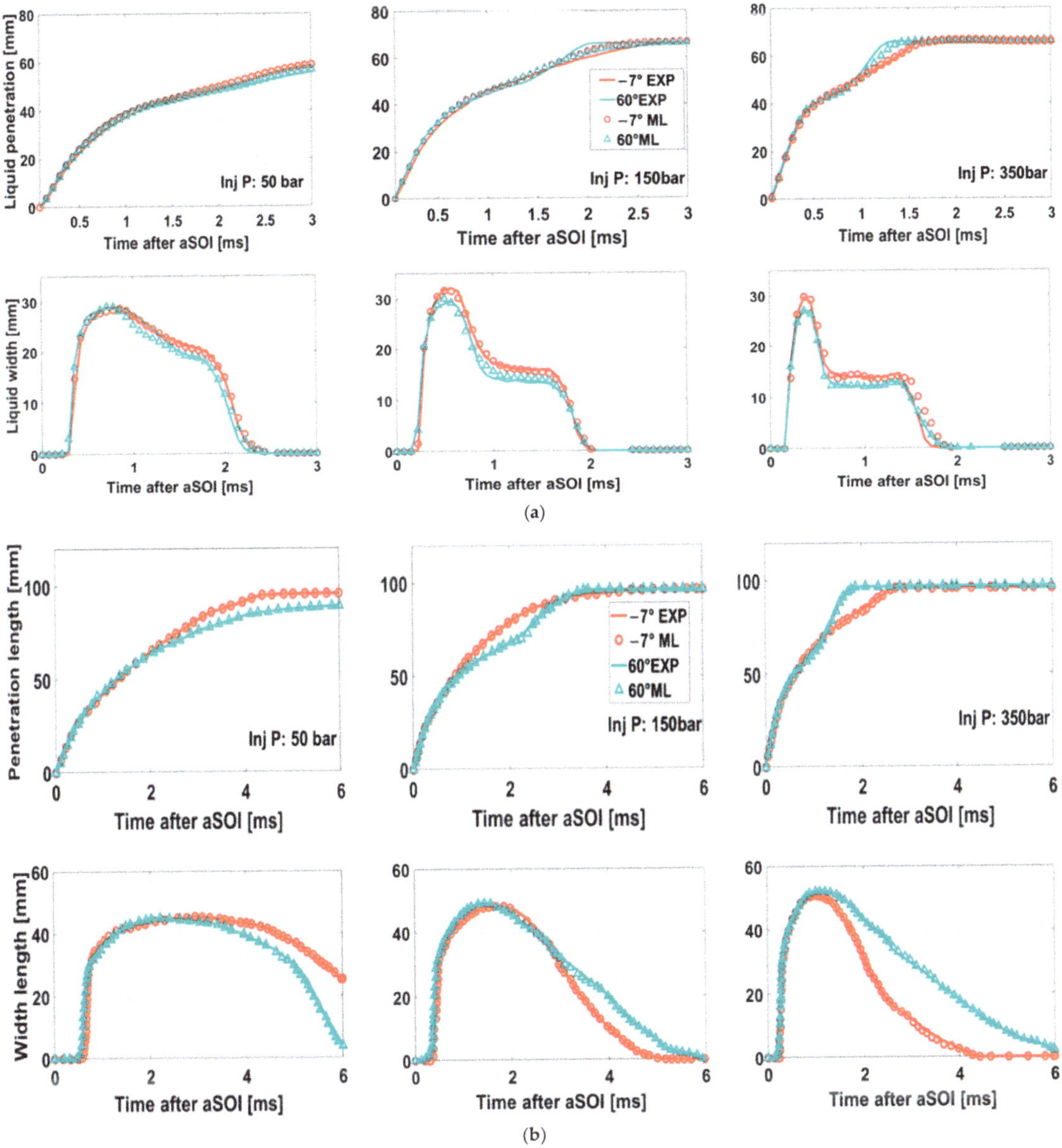

Figure 6. Analysis of Penetration and Width: Experimental and ANN-Predicted Values. (**a**) liquid penetration; liquid width @ 15 mm (**b**) vapor penetration; vapor width @ 30 mm.

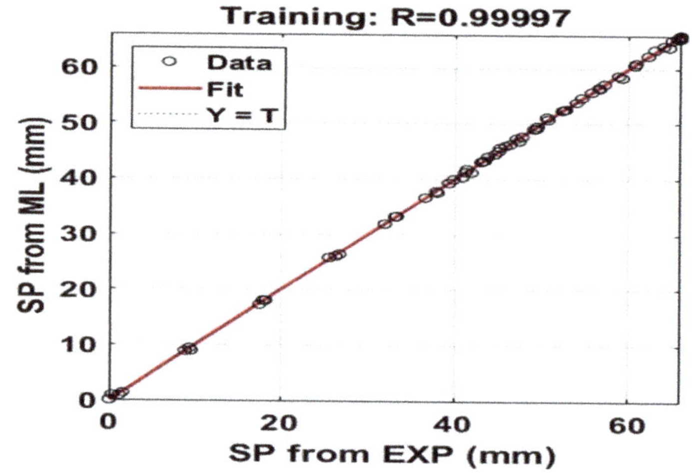

Figure 7. ANN Training Regression Plot for Spray Penetration (SP) Prediction.

5.2. Liquid Volume Fraction Prediction and Validation

The spatio-temporal evolution of the spray is displayed in Figure 8. Several interesting findings arise from the 3D liquid volume fraction (LVF). Notably, it is observed that at higher fuel temperatures and injection pressure, the spray morphology exhibits a strong plume collapsing toward the core area of the spray. The variation in fuel plume shape is influenced by the number of injector holes and their length-to-diameter (L/D) ratios. As the number of holes in the injector increases, the fuel plume broadens, leading to a larger collapse length. It should be noted that at higher injection pressures, the fuel plumes ejected from the orifices may interact with each other. This trend is clearly noticeable in 3D LVF at 350 bar. The collapsed spray then merges and forms a continuous closed region that blocks ambient gas from entering the central region. This leads to a decrease in pressure within the central area as the spray moves downward. The collapse of the plume can be attributed to an increase in the plume cone angle or a deviation in the plume direction from the injector axis. The edges of the plume cones overlap, leading to an increase in velocity at the intersection of the plumes. Then, the plumes start to merge into a single formation, which is characterized by a single velocity peak. It is important to emphasize that with an increase in ambient density, the central recirculation zone disappears, leading to a collapse of the plumes. More specifically, when the plumes are injected into a denser environment, the spray penetration tends to decrease. The ANN model is expected to replicate these observations, showing its potential to accurately predict complex fluid dynamics.

In this study, we used a MATLAB code enhanced for GPU processing to process and reconstruct three-dimensional spray dynamics from ANN-predicted images at different angles of 0°, 11.25°, and 22.5°, as illustrated in Figure 9. It should be noted that the predicted images from these three angles contained slight noise, which affected the clarity and intensity consistency. The source of this issue is the light source in the imaging experimental setup, which does not emit light uniformly but rather radiates over a range of angles due to diffuse radiation. This impacts the accuracy of quantifying the liquid phase in the spray during the experiment and consequently affects the predicted images, which may not accurately reflect the edges of the liquid phase [38]. To mitigate this issue, we applied a noise reduction technique using the 'imfilter' function. This algorithm efficiently handled large 608 × 608-pixel image datasets. Through advanced filtering, weighted sums, and symmetry operations, the code assembled cross-sectional planes from these predicted images, which enabled the creation of detailed 3D visualizations from 2D views. The resulting figure shows a 3D spray prediction at 100 bar, −7 °C, and 350 bar, −7 °C, captured at aSOI 1.14 ms and 1.87 ms, respectively. This innovative approach, which combines ANN-based image prediction with advanced 3D spray reconstruction, represents a valuable step in the field of fluid dynamics research. It is important to note that the 3D spray distribution offers quantitative insights into the temporal evolution of both cold and hot fuel spray behaviors, and such LVF data serves to validate and refine CFD models. It was observed that at higher injection pressure and fuel temperature, plume collapsing becomes more prominent. This phenomenon is characterized by a strong plume movement towards the core area. The results, as shown in Figure 10, show a strong agreement between the experimental data and the spray behavior predicted by the model. The model successfully captures the complete plume dynamics, with distinct plumes at low pressure and plumes collapsing at high pressure.

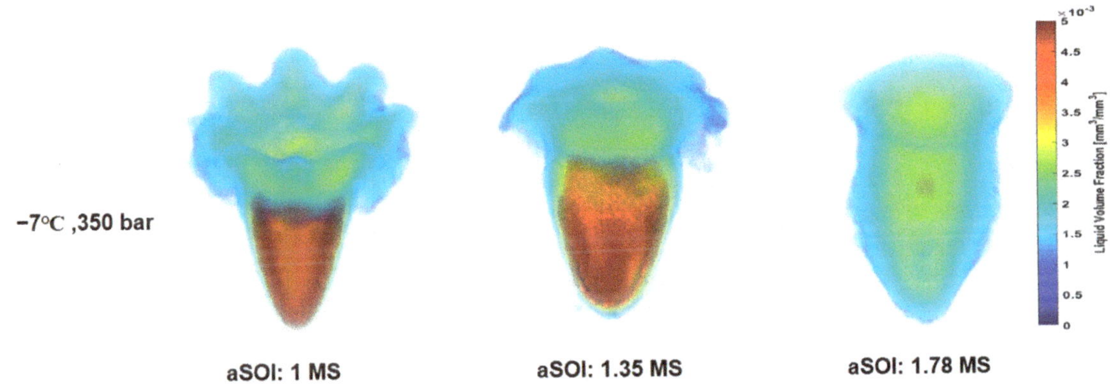

Figure 8. Temporal Evolution of 3D LVF during Cold Start.

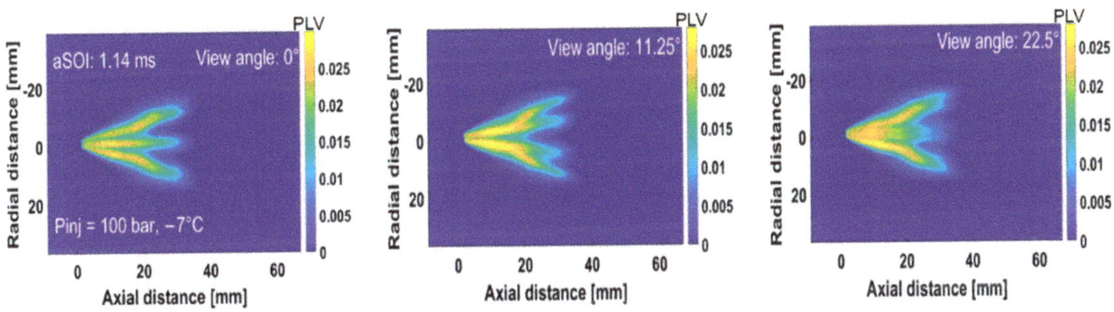

Figure 9. ML predicted spray images at (−7 °C, 100 bar), 1.14ms: Angles 0°, 11.25°, and 22.5°.

To summarize, ANN spray analysis is a powerful algorithm used to predict spray morphology and serves as an effective intermediate tool. However, it does not completely substitute for CFD simulations because its performance depends on the training data used to modify the neuron weights. For data that falls outside the target dataset, its accuracy may not be reliable. A trained ANN can serve as a useful bridge between experiments and CFD simulations. It can enhance the understanding of plume dynamics and offer validation for CFD simulation results in cases where experimental data is unavailable. Additionally, it can improve CFD simulations by providing valuable initial data.

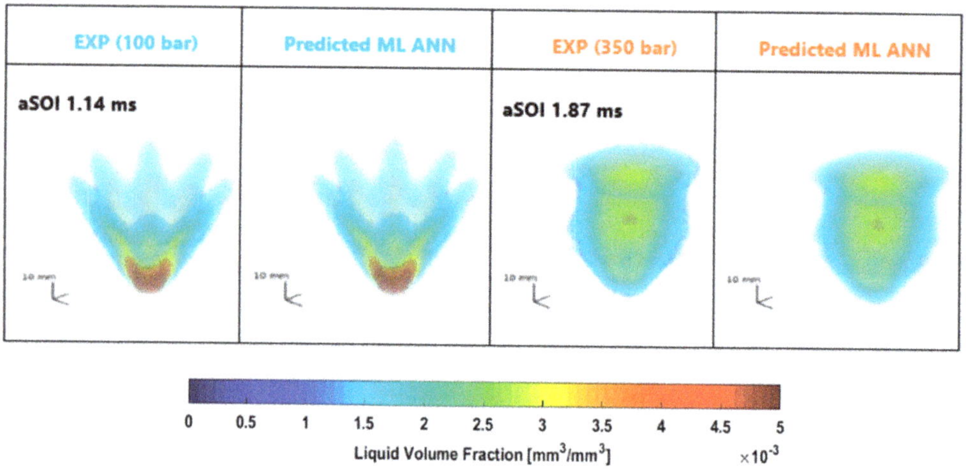

Figure 10. Experimental and ANN Prediction of 3D LVF Distribution.

6. Conclusions

In this study, a machine learning technique was implemented to predict the gasoline spray morphology under various injection pressures and fuel temperatures. A set of experimental datasets obtained from various optical diagnostic techniques, such as high-speed and Schlieren imaging, was utilized to train and validate the ANN model. The developed machine-learning algorithm was adept at predicting line-of-sight measurements and 3D liquid volume, providing quantitative information about 3D spray through the computed tomography (CT) algorithm. This approach might provide additional insight into future studies and can be adapted for models that predict both cold-start spray dynamics and hot fuel spray.

The main conclusions of this study are as follows:

1. The developed ANN predicts the liquid/vapor penetration length and width with high accuracy ($R^2 > 0.99$), showing strong agreement with experimental results.
2. The ANN model accurately predicts plume behavior that captures distinct plumes at low injection pressure (100 bar) and their collapse at high injection pressure (350 bar) which is still challenging for CFD simulations.
3. The developed machine learning model enables real-time predictions of 3D LVF distributions under various injection pressures and fuel temperatures. The proposed ANN model reduces the need for extensive computational optimization time and minimizes costs compared to CFD.
4. The machine learning model presented in this study can be adapted to various fuel types and operating conditions and can serve as a versatile spray model. It can also help overcome challenges related to cold start emissions.

Author Contributions: Conceptualization, Methodology, Investigation, Writing—original draft, Y.E.M.; data curation, K.L.; Supervision, Validation, Writing—review & editing, J.H. All authors have read and agreed to the published version of the manuscript.

Funding: This research received no external funding.

Data Availability Statement: Data will be made available on request.

Conflicts of Interest: The authors declare no conflict of interest.

References

1. Patrick, M. December 2023 Market Beat. Available online: https://www.nada.org/media/9244/download (accessed on 28 September 2024).
2. NOAA Research. Record Carbon Dioxide Emissions Impeding Progress on Meeting Climate Goals. Available online: https://research.noaa.gov/2023/12/05/record-fossil-carbon-dioxide-emissions-impeding-progress-on-meeting-climate-goals-report/#:~:text=Greenhouse%20gas%20emissions%20from%20fossil,by%20the%20Global%20Carbon%20Project (accessed on 20 February 2024).
3. Wen, M.; Zhang, C.; Yue, Z.; Liu, X.; Yang, Y.; Dong, F.; Liu, H.; Yao, M. Effects of Gasoline Octane Number on Fuel Consumption and Emissions in Two Vehicles Equipped with GDI and PFI Spark-Ignition Engine. *J. Energy Eng.* **2020**, *146*, 04020069. [CrossRef]
4. El Marnissi, Y.; Hwang, J. Microscopic Imaging on Diesel Spray and Atomization Process. *Processes* **2024**, *12*, 359. [CrossRef]
5. He, X.; Xu, K.; Xu, Y.; Zhang, Z.; Wei, W. Effects of nozzle diameter on the characteristic time scales of diesel spray two-stage ignition under cold-start conditions. *Fuel* **2023**, *335*, 126700. [CrossRef]
6. Shi, Z.; Lee, C.F.; Wu, H.; Wu, Y.; Zhang, L.; Liu, F. Optical diagnostics of low-temperature ignition and combustion characteristics of diesel/kerosene blends under cold-start conditions. *Appl. Energy* **2019**, *251*, 113307. [CrossRef]
7. Tong, K.; Quay, B.; Zello, J.; Santavicca, D. *Fuel Volatility Effects on Mixture Preparation and Performance in a GDI Engine During Cold Start*; SAE Technical Paper 2001-01-3650; SAE International: Warrendale, PA, USA, 2001. [CrossRef]
8. Lien, H.P.; Li, Y.; Pati, A.; Sadiki, A.; Hasse, C. Numerical studies of gasoline direct-injection sprays (ECN Spray G) under early- and late-injection conditions using Large Eddy Simulation and droplets-statistics-based Eulerian–Lagrangian framework. *Fuel* **2024**, *357*, 129708. [CrossRef]
9. Piazzullo, D.; Costa, M.; Allocca, L.; Montanaro, A.; Rocco, V. A 3D CFD Simulation of GDI Sprays Accounting for Heat Transfer Effects on Wallfilm Formation. *Source SAE Int. J. Engines* **2017**, *10*, 2166–2175. [CrossRef]
10. Gerbino, F.; Tretola, G.; Morgan, R.; Atkins, P.; Vogiatzaki, K. Influence of the initial droplet distribution on the prediction of spray dynamics in Eulerian-Lagrangian simulations. *Int. J. Multiph. Flow* **2021**, *141*, 103642. [CrossRef]
11. Schmidt, D.; Maulik, R.; Lyras, K. Machine learning accelerated turbulence modeling of transient flashing jets. *Phys. Fluids* **2021**, *33*, 127104. [CrossRef]
12. Ravindran, A.C.; Kokjohn, S.L. Combining machine learning with 3D-CFD modeling for optimizing a DISI engine performance during cold-start. *Energy AI* **2021**, *5*, 100072. [CrossRef]
13. Richards, B.; Emekwuru, N. Using Machine Learning to Predict Synthetic Fuel Spray Penetration from Limited Experimental Data Without Computational Fluid Dynamics. In Proceedings of the International Conference on Energy and Sustainable Futures, Dubai, United Arab Emirates, 5–6 December 2023; pp. 51–59. [CrossRef]
14. Laget, O.; Pacaud, P.; Perrin, H. Démarrage à froid d'un moteur Diesel à bas taux de compression: Investigations expérimentales et numériques RANS tridimensionnelles. *Oil Gas Sci. Technol.* **2009**, *64*, 407–429. [CrossRef]
15. Thuerey, N.; Weißenow, K.; Prantl, L.; Hu, X. Deep learning methods for reynolds-averaged navier–stokes simulations of airfoil flows. *AIAA J.* **2020**, *58*, 25–36. [CrossRef]
16. Paredi, D.; Lucchini, T.; D'Errico, G.; Onorati, A.; Pickett, L.; Lacey, J. Validation of a comprehensive computational fluid dynamics methodology to predict the direct injection process of gasoline sprays using Spray G experimental data. *Int. J. Engine Res.* **2020**, *21*, 199–216. [CrossRef]
17. Zhang, Y.; Xu, S.; Zhong, S.; Bai, X.S.; Wang, H.; Yao, M. Large eddy simulation of spray combustion using flamelet generated manifolds combined with artificial neural networks. *Energy AI* **2020**, *2*, 100021. [CrossRef]
18. Es-Sahli, O.; Sescu, A.; Afsar, M.Z.; Buxton, O.R.H. Investigation of wakes generated by fractal plates in the compressible flow regime using large-eddy simulations. *Phys. Fluids* **2020**, *32*, 105106. [CrossRef]
19. Hwang, J.; Lee, P.; Mun, S.; Karathanassis, I.K.; Koukouvinis, F.; Tagliante, F.; Nguyen, T.; Pickett, L. A New Pathway for Prediction of Gasoline Sprays using Machine-Learning Algorithms. *SAE Int. J. Adv. Curr. Pract. Mobil.* **2022**, *5*, 343–356. [CrossRef]
20. Chang, M.; Park, S. Predictions and analysis of flash boiling spray characteristics of gasoline direct injection injectors based on optimized machine learning algorithm. *Energy* **2023**, *262*, 125304. [CrossRef]
21. Tian, J.; Liu, Y.; Bi, H.; Li, F.; Bao, L.; Han, K.; Zhou, W.; Ni, Z.; Lin, Q. Experimental study on the spray characteristics of octanol diesel and prediction of spray tip penetration by ANN model. *Energy* **2022**, *239*, 121920. [CrossRef]
22. Hwang, J.; Lee, P.; Mun, S.; Karathanassis, I.K.; Koukouvinis, P.; Pickett, L.M.; Gavaises, M. Machine-learning enabled prediction of 3D spray under engine combustion network spray G conditions. *Fuel* **2021**, *293*, 120444. [CrossRef]
23. Thangaraja, J.; Zigan, L.; Rajkumar, S. A machine learning framework for evaluating the biodiesel properties for accurate modeling of spray and combustion processes. *Fuel* **2023**, *334*, 126573. [CrossRef]
24. Seo, J.; Yun, B.; Kim, J.; Shin, M.; Park, S. Development of a cold-start emission model for diesel vehicles using an artificial neural network trained with real-world driving data. *Sci. Total Environ.* **2022**, *806*, 151347. [CrossRef]
25. Giannelli, R.A.; Stubleski, R.; Saunders, A. Semi-empirical Analysis of Cold Start Emissions. *Int. J. Fuels Lubr.* **2014**, *7*, 591–599. [CrossRef]
26. Sabatini, S.; Kil, I.; Dekar, J.; Hamilton, T.; Wuttke, J.; Smith, M.A.; Hoffman, M.A.; Onori, S. A new semi-empirical temperature model for the three way catalytic converter. *IFAC-PapersOnLine* **2015**, *48*, 434–440. [CrossRef]
27. Settles, G.S. *Schlieren and Shadowgraph Techniques*; Springer: Berlin/Heidelberg, Germany, 2001.

28. Payri, R.; Salvador, F.J.; Bracho, G.; Viera, A. Differences between single and double-pass schlieren imaging on diesel vapor spray characteristics. *Appl. Therm. Eng.* **2017**, *125*, 220–231. [CrossRef]
29. Sada, A.; Kinoshita, Y.; Shiota, S.; Kiya, H. Histogram-Based Image Pre-processing for Machine Learning. In Proceedings of the 2018 IEEE 7th Global Conference on Consumer Electronics, Nara, Japan, 9–12 October 2018.
30. Jourdain, C.; Weiss, J.; Seers, P. Comparison of image preprocessing methods for fuel droplet characterization. In Proceedings of the 47th AIAA Thermophysics Conference, Denver, CO, USA, 5–9 June 2017; American Institute of Aeronautics and Astronautics Inc. (AIAA): Reston, VA, USA, 2017. [CrossRef]
31. Available online: http://www.philiplaven.com/mieplot.htm (accessed on 10 October 2024).
32. Badra, J.; Sim, J.; Pei, Y.; Viollet, Y.; Pal, P.; Futterer, C.; Brenner, M.; Som, S.; Farooq, A.; Chang, J. Combustion System Optimization of a Light-Duty GCI Engine Using CFD and Machine Learning. *SAE Tech. Pap.* **2020**. [CrossRef]
33. Kochkov, D.; Smith, J.A.; Alieva, A.; Wang, Q.; Brenner, M.P.; Hoyer, S. Machine learning-accelerated computational fluid dynamics. *Proc. Natl. Acad. Sci. USA* **2021**, *118*, e2101784118. [CrossRef] [PubMed]
34. Aleiferis, P.G.; Shukla, J.; Brewer, M.; Cracknell, R.F. Spray development of iso-octane, ethanol, hydrous ethanol and water from a multi-hole injector under ultra cold fuel temperature conditions. *Fuel* **2021**, *303*, 120983. [CrossRef]
35. Hemdal, S.; Denbratt, I.; Wärnberg, J. Stratified Cold Start Sprays of Gasoline-Ethanol Blends. *Int. J. Fuels Lubr.* **2009**, *2*, 683–696. [CrossRef]
36. Lee, K.; Pintor, D.L.; Assanis, D.; Cho, S.; Hwang, J. Fuel temperature and injection pressure influence on the cold start GDI sprays. *Appl. Energy Combust. Sci.* **2023**, *16*, 100206. [CrossRef]
37. Kook, S.; Pickett, L.M. Liquid length and vapor penetration of conventional, Fischer-Tropsch, coal-derived, and surrogate fuel sprays at high-temperature and high-pressure ambient conditions. *Fuel* **2012**, *93*, 539–548. [CrossRef]
38. Lehnert, B.; Weiss, L.; Berrocal, E.; Wensing, M. Quantifying extinction imaging of fuel sprays considering scattering errors. *Int. J. Engine Res.* **2023**, *24*, 4413–4420. [CrossRef]

MDPI AG
Grosspeteranlage 5
4052 Basel
Switzerland
Tel.: +41 61 683 77 34

Fluids Editorial Office
E-mail: fluids@mdpi.com
www.mdpi.com/journal/fluids

www.ingramcontent.com/pod-product-compliance
Lightning Source LLC
Chambersburg PA
CBHW061607220926
51850CB00005B/448